PHYSICS RESEARCH AND TECHNOLOGY

PROCEEDINGS OF THE 2015 INTERNATIONAL CONFERENCE ON "PHYSICS, MECHANICS OF NEW MATERIALS AND THEIR APPLICATIONS," DEVOTED TO THE 100TH ANNIVERSARY OF THE SOUTHERN FEDERAL UNIVERSITY

PHYSICS RESEARCH AND TECHNOLOGY

Additional books in this series can be found on Nova's website
under the Series tab.

Additional e-books in this series can be found on Nova's website
under the eBooks tab.

PHYSICS RESEARCH AND TECHNOLOGY

PROCEEDINGS OF THE 2015 INTERNATIONAL CONFERENCE ON "PHYSICS, MECHANICS OF NEW MATERIALS AND THEIR APPLICATIONS," DEVOTED TO THE 100TH ANNIVERSARY OF THE SOUTHERN FEDERAL UNIVERSITY

IVAN A. PARINOV
SHUN-HSYUNG CHANG
AND
VITALY YU. TOPOLOV
EDITORS

Copyright © 2016 by Nova Science Publishers, Inc.

All rights reserved. No part of this book may be reproduced, stored in a retrieval system or transmitted in any form or by any means: electronic, electrostatic, magnetic, tape, mechanical photocopying, recording or otherwise without the written permission of the Publisher.

We have partnered with Copyright Clearance Center to make it easy for you to obtain permissions to reuse content from this publication. Simply navigate to this publication's page on Nova's website and locate the "Get Permission" button below the title description. This button is linked directly to the title's permission page on copyright.com. Alternatively, you can visit copyright.com and search by title, ISBN, or ISSN.

For further questions about using the service on copyright.com, please contact:
Copyright Clearance Center
Phone: +1-(978) 750-8400   Fax: +1-(978) 750-4470   E-mail: info@copyright.com.

NOTICE TO THE READER

The Publisher has taken reasonable care in the preparation of this book, but makes no expressed or implied warranty of any kind and assumes no responsibility for any errors or omissions. No liability is assumed for incidental or consequential damages in connection with or arising out of information contained in this book. The Publisher shall not be liable for any special, consequential, or exemplary damages resulting, in whole or in part, from the readers' use of, or reliance upon, this material. Any parts of this book based on government reports are so indicated and copyright is claimed for those parts to the extent applicable to compilations of such works.

Independent verification should be sought for any data, advice or recommendations contained in this book. In addition, no responsibility is assumed by the publisher for any injury and/or damage to persons or property arising from any methods, products, instructions, ideas or otherwise contained in this publication.

This publication is designed to provide accurate and authoritative information with regard to the subject matter covered herein. It is sold with the clear understanding that the Publisher is not engaged in rendering legal or any other professional services. If legal or any other expert assistance is required, the services of a competent person should be sought. FROM A DECLARATION OF PARTICIPANTS JOINTLY ADOPTED BY A COMMITTEE OF THE AMERICAN BAR ASSOCIATION AND A COMMITTEE OF PUBLISHERS.

Additional color graphics may be available in the e-book version of this book.

Library of Congress Cataloging-in-Publication Data

Names: International Conference on "Physics, Mechanics of New Materials and
    Their Applications" (2015 : Azov, Russia) | Parinov, Ivan A., 1956-
    editor. | Chang, Shun-Hsyung, editor. | Topolov, Vitaly Yu. (Vitaly
    Yuryevich), 1961- editor. | kIlUzhnyfi federal'nyfi universitet (Russia),
    honouree.
Title: Proceedings of the 2015 International Conference on "Physics,
    Mechanics of New Materials and Their Applications", devoted to the 100th
    anniversary of the Southern Federal University / editors: Ivan A. Parinov,
    Shun-Hsyung Chang, and Vitaly Yu. Topolov (Vorovich Mathematics, Mechanics
    and Computer Science Institute, Southern Federal University, Russia, and
    others).
Other titles: 2015 International Conference on "Physics, Mechanics of New
    Materials and Their Applications | Physics research and technology.
Description: Hauppauge, New York : Nova Science Publishers, Inc., [2016] |
    Series: Physics research and technology | 2015 International Conference on
    "Physics and Mechanics of New Materials and Their Applications" (PHENMA
    2015) held May 19-22, 2015, in Azov, Russia. | Includes bibliographical
    references.
Identifiers: LCCN 2015051059 (print) | LCCN 2016001095 (ebook) | ISBN
    9781634845779 (hardcover) | ISBN 1634845773 (hardcover) | ISBN
    9781634845908 ()
Subjects: LCSH: Materials--Congresses. | Physics--Materials--Congresses. |
    Mechanics--Materials--Congresses.
Classification: LCC TA401.3 .I54956 2015 (print) | LCC TA401.3 (ebook) | DDC
    660--dc23
LC record available at http://lccn.loc.gov/2015051059

Published by Nova Science Publishers, Inc. † New York

CONTENTS

Preface		**xiii**
I. Processing Techniques		**1**
Chapter 1	Local Atomic and Electronic Structure of the Pt/C Nanoparticles Grown in He and H2 Gas Phase Reducing Agent *N. M. Nevzorova, I. N. Leontyev and G. E. Yalovega*	**3**
Chapter 2	Effects of Combined Modifying of Niobate Materials Non-Containing Toxic Elements *Kh. A. Sadykov, I. A. Verbenko, L. A. Reznichenko, G. M. Konstantinov, A. A. Pavelko, L. A. Shilkina and S. I. Dudkina*	**9**
Chapter 3	Synthesis, Structure, and Optical Characteristics of Barium Strontium Niobate Thin Films *G. N. Tolmachev, A. P. Kovtun, I. N. Zaharchenko, I. M. Aliev, S. P. Zinchenko, A. V. Pavlenko, I. A. Verbenko and L. A. Reznichenko*	**17**
Chapter 4	Optimization of Conditions for Obtaining $PbMn_{1/2}Nb_{1/2}O_3$ by Direct Synthesis of a Mixture of Oxides and Sintering by Conventional Ceramic Technology *L. A. Shilkina, I. A. Verbenko, G. M. Konstantinov, S. I. Dudkina and L. A. Reznichenko*	**23**
Chapter 5	The Effect of Mechanical Activation on the Properties of Multiferroic $PbFe_{0.5}Ta_{0.5}O_3$ Ceramics *A. A. Gusev, I. P. Raevski, E. G. Avvakumov, V. P. Isupov, S. I. Raevskaya, S. P. Kubrin, D. A. Sarychev, V. V. Titov, H. Chen, C.-C. Chou and V. Yu. Shonov*	**31**
Chapter 6	Fabrication of Piezoelectric Ceramics with Lowered Sintering Temperature for MEMS *V. V. Eremkin, V. G. Smotrakov, M. A. Marakhovskiy and A. E. Panich*	**37**

Chapter 7	New Ceramic Matrix Piezocomposites for Underwater Ultrasonic Applications *E. I. Petrova, A. A. Naumenko, M. A. Lugovaya* *and A. N. Rybyanets*	**47**
Chapter 8	Lead-Free Piezoceramic Materials for Ultrasound Applications *M. V. Talanov, L. A. Shilkina and L. A. Reznichenko*	**55**
Chapter 9	Creation and Physical Properties of Highly Effective Lead-Free Solid Solutions *Kh. A. Sadykov, I. A. Verbenko, G. M. Konstantinov,* *L. A. Reznichenko and A. A. Pavelko*	**61**
Chapter 10	Digital Piezoelectric Material Based on Extracoarse-Grained Piezocomposite "Porous Ceramic – Polymer" *D. I. Makariev, A. N. Reznitchenko, A. A. Naumenko, E. I. Petrova* *and A. N. Rybyanets*	**69**
Chapter 11	Numerical Simulation of Films Laser Annealing on the Sapphire Surface *S. P. Malyukov, A. V. Sayenko and Yu. V. Klunnikova*	**75**
Chapter 12	Numerical Analysis of Inorganic Glassy Dielectric Laser Processing *S. P. Malyukov, Y. V. Klunnikova and T. H. Bui*	**81**
Chapter 13	Features of Sapphire and Glassy Dielectric Junction Formation for Elements of Microelectronics *S. P. Malyukov, Yu. V. Klunnikova, A. V. Sayenko* *and D. A. Bondarchuk*	**89**
Chapter 14	Features of Technology of SAW Devices *V. G. Dneprovski and G. Ya. Karapetyan*	**95**
Chapter 15	Utilization of Leonardite for Producing Slow-Release Fertilizer *Petchporn Chawakitchareon, Jittrera Buates* *and Rewadee Anuwattana*	**103**
II. Physics		**113**
Chapter 16	Piezoelectric Performance and Features of Microgeometry of Novel Composites Based on Ferroelectric ZTS-19 Ceramics *Petr A. Borzov, Alexander A. Vorontsov, Vitaly Yu. Topolov* *and Olga E. Brill*	**115**
Chapter 17	Figures of Merit and Related Parameters of Modern Piezo-Active 1–3-Type Composites for Energy-Harvesting Applications *Vitaly Yu. Topolov, Christopher R. Bowen and Ashura N. Isaeva*	**123**
Chapter 18	Piezoelectric Sensitivity and Anisotropy of a Novel Lead-Free 2–2 Composite *Vitaly Yu. Topolov, Christopher R. Bowen* *and Franck Levassort*	**131**

Contents

Chapter 19 Characterization Techniques for Advanced Materials and Devices **141**
M. A. Lugovaya, E. I. Petrova, T. V. Rybyanets,
G. M. Konstantinov and A. N. Rybyanets

Chapter 20 New Functional Materials **151**
K. P. Andryushin, I. N. Andryushina, L. A. Reznichenko,
L. A. Shilkina and O. N. Razumovskaya

Chapter 21 Piezoelectric Ceramic Materials with High Ferroelectric Hardness
and Efficiency **157**
H. A. Sadykov, I. A. Verbenko, S. I. Dudkina and L. A. Reznichenko

Chapter 22 Experimental Study of High Temperature Porous Piezoceramics **161**
I. A. Shvetsov, M. A. Lugovaya, N. A. Shvetsova,
G. M. Konstantinov and A. N. Rybyanets

Chapter 23 Structural Peculiarities of Porous Ferroelectric Ceramics **169**
G. M. Konstantinov, Y. B. Konstantinova, N. A. Shvetsova,
N. O. Svetlichnaya and A. N. Rybyanets

Chapter 24 Complex Parameters of Porous PZT Piezoceramics Measured for
Different Vibrational Modes **175**
M. A. Lugovaya, A. A. Naumenko, E. I. Petrova and A. N. Rybyanets

Chapter 25 Experimental Study of Ultrasonic Attenuation and Dispersion
Relationships for Ceramic Matrix Composites **183**
E. I. Petrova, A. A. Naumenko, S. A. Shcherbinin
and A. N. Rybyanets

Chapter 26 Correlation Bonds: Structure – Electrophysical Properties of
the Binary PZT System **191**
I. N. Andryushina, L. A. Reznichenko, L. A. Shilkina, S. I. Dudkina,
K. P. Andryushin and O. N. Razumovskaya

Chapter 27 Refinement of the $R3c \rightarrow R3m$ Transition Line in the Phase
Diagram of PZT Solid Solutions **199**
A. A. Pavelko, L. A. Shilkina, K. P. Andryushin, I. N. Andryushina,
S. I. Dudkina, L. A. Reznichenko and O. N. Razumovskaya

Chapter 28 Absorption Spectra of Microwave Energy of Piezoelectric Ceramics
Synthesized under Different Sintering Temperature **207**
E. N. Sidorenko, V. G. Smotrakov, V. V. Eremkin, I. I. Natkhin,
M. E. Agarkova, A. A. Naumenko and E. I. Petrova

Chapter 29 Pyroelectric Activity and Dielectric Response of
the $PbZr_{1-x}Ti_xO_3$ System in the Range of $0.37 \leq x \leq 0.57$ **215**
Yu. N. Zakharov, A. A. Pavelko, A. G. Lutokhin,
V. G. Kuznetsov, L. A. Shilkina and L. A. Reznichenko

Chapter 30	Polycrystalline Solid Solutions Based on Alkali Metal Niobates Modified with Glass-Forming Additives *I. A. Verbenko, L. A. Reznichenko, A. G. Abubakarov, L. A. Shilkina, S. I. Dudkina and A. A. Pavelko*	**221**
Chapter 31	Effect of Zr and (Ti, Zr) Doping on Ferroelectric and Magnetic Phase Transitions in $Pb(Fe_{1/2}Nb_{1/2})O_3$ *I. P. Raevski, V. V. Titov, S. I. Raevskaya, V. V. Laguta, M. Marysko, S. P. Kubrin, H. Chen, C.-C. Chou, M. A. Malitskaya, A. V. Blazhevich, D. A. Sarychev, L. E. Pustovaya, I. N. Zakharchenko, E. I. Sitalo and V. Yu. Shonov*	**225**
Chapter 32	Influence of MnO_2 on Dielectric Characteristics, Grain and Crystal Structure of $PbFe_{0.5}Nb_{0.5}O_3$ Ceramics *N. A. Boldyrev, A. V. Pavlenko, L. A. Reznichenko and L. A. Shilkina*	**233**
Chapter 33	Effect of Dynamic Fatigue on Dielectrtic and Pyroelectric Properties of $(1 - x) PbFe_{1/2}Nb_{1/2}O_3 - xPbTiO_3$ Ceramics *A. F. Semenchev, I. P. Raevski, S. I. Raevskaya, T. A. Minasyan, S. P. Kubrin, H. Chen, C-C Chou, M. A. Malitskaya and V. V. Titov*	**239**
Chapter 34	Influence of Li_2CO_3 on Dielectric and Ceramic Characteristics of $PbFe_{0.5}Nb_{0.5}O_3$ *N. A. Boldyrev, A. V. Pavlenko and L. A. Shilkina*	**245**
Chapter 35	Dielectric Characteristics of $(Ba_{0.5}Sr_{0.5})Nb_2O_6$ Ceramics at Temperatures from $-250°C$ to $180°C$ *A. G. Abubakarov, A. V. Pavlenko, L. A. Reznichenko and A. S. Nazarenko*	**253**
Chapter 36	Structure, Microstructure and Composition of $Bi_{1-x}Pr_xFeO_3$ Ceramics *S. V. Titov, I. A. Verbenko, L. A. Reznichenko, L. A. Shilkina, V. A. Aleshin, V. V. Titov, S. I. Shevtsova and A. P. Kovtun*	**259**
Chapter 37	Dielectric Characteristics of Ceramic Solid Solutions of $(1 - x)BiFeO_3 - xPb(Fe_{0.5}Nb_{0.5})O_3$ ($x = 0.5–1.0$ and $\Delta x = 0.05$) *A. V. Pavlenko, L. A. Reznichenko, V. S. Stashenko, A. V. Markov and N. A. Boldyrev*	**267**
Chapter 38	The Structure, Microstructure, Dielectric Spectroscopy and Thermal Properties of $BiFeO_3$ Modified with Dy *S. V. Khasbulatov, A. A. Pavelko, L. A. Shilkina, L. A. Reznichenko, G. G. Hajiyev, A. G. Bakmaev, M.-R. M. Magomedov, Z. M. Omarov and V. A. Aleshin*	**275**

Chapter 39	Preparation, Structure, Microstructure, Dielectric and Thermal Properties of Bismuth Ferrite with Holmium, Erbium and Ytterbium	**283**
	S. V. Khasbulatov, A. A. Pavelko, L. A. Shilkina, V. A. Aleshin, I. A. Verbenko, G. G. Hajiyev, Z. M. Omarov, A. G. Bakmaev, O. N. Razumovskaya and L. A. Reznichenko	
Chapter 40	Magnetodielectric Interactions in $Bi_{0.5}La_{0.5}MnO_3$ Ceramics in Temperature Range of $10 - 120$ K	**289**
	A. V. Pavlenko, A. V. Turik, L. A. Reznitchenko and Yu. S. Koshkid'ko	
Chapter 41	Mathematical Simulation of Processes of the Electromass Transfer in Controlled Electrochemical Resistance	**293**
	T. P. Skakunova, Yu. Ya. Gerasimenko, D. D. Fugarov and I. V. Tarasov	

III. Mechanics

		303
Chapter 42	Mechanical Properties of the Bulk Composite Nanomaterial Consisting of Albumin and Carbon Nanotubes	**305**
	Alexander Gerasimenko, Levan Ichkitidze, Vitaly Podgaetsky and Sergei Selishchev	
Chapter 43	Dynamic Contact Problem for a Heterogeneous Layer with a Liquid Sheet on a Non-Deformable Foundation	**315**
	M. A. Sumbatyan, A. Scalia and E. A. Usoshina	
Chapter 44	The Development of Methods for the Determination of Thermal and Tribological Characteristics of the Friction Surfaces	**323**
	P. G. Ivanochkin, S. I. Builo, I. V. Kolesnikov and N. A. Myasnikova	
Chapter 45	Investigation of the Indenter Temperature and Speed Effect during Instrumented Indentation on the Mechanical Properties of Carbon Steels	**331**
	E. V. Sadyrin, L. I. Krenev, B. I. Mitrin, I. Yu. Zabiyaka, S. M. Aizikovich and S. O. Abetkovskaya	
Chapter 46	Numerical-Analytical Solution for Deflections and Force Factors due to Concentrated Load on Cantilever Plate	**337**
	Yu. E. Drobotov and G. A. Zhuravlev	
Chapter 47	Theory and Experiment in the Ultrasonic Nondestructive Testing for Arrays of Spatial Defects of the Elastic Materials	**353**
	N. V. Boyev	
Chapter 48	Boundary Element Method in Solving Problems of Poroviscoelasticity	**359**
	L. A. Igumnov, A. A. Belov and A. A. Ipatov	

Chapter 49	Description of Non-linear Viscoelastic Deformations by the 3D Mechanical Model *A. D. Azarov and D. A. Azarov*	367
Chapter 50	Determination of Strain Characteristics *A. I. Kozinkina and Y. A. Kozinkina*	377
Chapter 51	Frequency Dependences of the Complex Material Constants for Porous PZT Piezoceramics *A. A. Naumenko, M. A. Lugovaya, E. I. Petrova, I. A. Shvetsov and A. N. Rybyanets*	383
Chapter 52	Finite Element Modeling of Lossy Piezoelectric Elements *A. A. Naumenko, S. A. Shcherbinin, A. V. Nasedkin and A. N. Rybyanets*	391
Chapter 53	Electromechanical Response Characterization of Piezoelectric Materials *S. A. Shcherbinin, I. A. Shvetsov, M. A. Lugovaya, A. A. Naumenko and A. N. Rybyanets*	399
Chapter 54	Dielectric, Piezoelectric and Elastic Properties of Pzt/Pzt Ceramic Piezocomposites *N. A. Shvetsova, M. A. Lugovaya, I. A. Shvetsov, D. I. Makariev and A. N. Rybyanets*	407
Chapter 55	Surface Acoustic Waves Method for Piezoelectric Material Characterization *N. A. Shvetsova, A. N. Reznitchenko, I. A. Shvetsov, E. I. Petrova and A. N. Rybyanets*	415
Chapter 56	Dispersion Relation and Resonance Frequencies of Surface Acoustic Waves Excited in Barium Titanate Films *P. E. Timoshenko, V. V. Kalinchuk, V. B. Shirokov, M. O. Levi and A. V. Pan'kin*	423
Chapter 57	Investigation of Features of Surface Wave Fields in a Media with Inhomogeneities *O. V. Bocharova, V. V. Kalinchuk, A. V. Sedov and I. E. Andjikovich*	433
Chapter 58	Vibroacoustics of Composite Polymeric Shells of Rotation Reinforced by Discrete Circular Ribs *V. G. Safronenko and O. I. Safronenko*	439
Chapter 59	Forced Oscillations of Shells with Auxetic Properties *A. S. Yudin and S. A. Yudin*	447
Chapter 60	Oscillations of Compressible Liquid Free Surface Generated by Plate *G. N. Trepacheva*	453

Contents

V. Applications **459**

Chapter 61   Development of the Model and Structure of the GaAs P-I-N Photodetector for Integrated Optical Commutation Systems   **461**
I. Pisarenko and E. Ryndin

Chapter 62   Knock Sensors Based on Lead-Free Piezoceramics   **469**
Yu. I. Yurasov, A. V. Pavlenko, I. A. Verbenko,
L. A. Reznichenko and H. A. Sadykov

Chapter 63   Biological Solders for Laser Welding of Biological Tissues   **475**
Alexander Gerasimenko, Levan Ichkitidze, Vitaly Podgaetsky,
Evgenie Pyankov, Dmitrie Ryabkin and Sergei Selishchev

Chapter 64   HIFU Transducers Designs and Treatment Methods
for Hemostasis of Deep Arterial Bleeding   **485**
A. N. Rybyanets, A. E. Berkovich, T. V. Rybyanets,
D. I. Makariev and A. N. Reznitchenko

Chapter 65   Theoretical Modeling and Experimental Study of High Intensity
Focused Ultrasound Transducers: An Update   **493**
S. A. Shcherbinin, A. A. Naumenko, N. A. Shvetsova,
A. E. Berkovich and A. N. Rybyanets

Chapter 66   Modeling of Stress State of Welded Joint in Pipeline Taking into
Account Uneven Distribution of Mechanical Properties   **501**
V. V. Deryushev, E. E. Kosenko, A. V. Cherpakov, V. V. Kosenko,
M. M. Zaitseva, L. E. Kondratova and S. V. Teplyakova

Chapter 67   Method and Computer Software for Definition of Stress-Strain State
in Layered Anisotropic Constructions at Pulse Loading   **509**
I. P. Miroshnichenko

Chapter 68   Based on Analytical Modeling Identifying the Location of Multiple
Cracks in the Rod Construction   **515**
A. N. Soloviev, A. V. Cherpakov and I. A. Parinov

Chapter 69   Vibrodiagnostics of Truss Model with Damages: Experiment   **521**
A. N. Soloviev, A. V. Cherpakov, E. V. Rozkov and I. A. Parinov

Chapter 70   Analysis of Functional and Test Deformation Responses of
Frame-Rod Models for Early Diagnostics of Integral
Metallic Constructions   **531**
E. V. Saulina, Yu. V. Esipov and A. I. Cheremisin

Chapter 71   Mathematical Modeling of Plane-Parallel Electrochemical
Electrolytic Cell with Perforated Cathode   **537**
A. N. Gerasimenko, Yu. Ya. Gerasimenko, E. Yu. Gerasimenko
and A. I. Emelyanov

Chapter 72	Algorithm of Redistribution of Connections between Outputs on the Base of Swarm Intellect *O. A. Purchina, A. Y. Poluyan, D. D. Fugarov and Y. N. Bugaeva*	**545**
Chapter 73	Sustainable Livelihood Framework Analysis for Improving Local Sector Industries, Case Study: Local Sector Industries at Ledok Kulon, East Java-Indonesia *R. A. Retno Hastijanti*	**557**
Chapter 74	Cluster Development of Small and Medium Manufacturing Industry in Surabaya City, East Java, Indonesia *Erni Puspanantasari Putri*	**565**
Index		**573**

PREFACE

Advanced materials and composites, including piezoelectrics, nanomaterials, nanostructures, functional materials, polymeric composites and so on, are very important for modern sciences, technologies and techniques. Their properties improve difficultly without intense chemical, physical, mechanical researches and development of modern numerical approaches and methods of mathematical modeling. Tremendous interest to similar investigations grows constantly, caused numerous applications and due to fast development of theoretical, experimental and numerical methods, requiring improvement of experimental equipment, theoretical and numerical approaches, computer hard- and software. These achievements create a new scientific knowledge. They allow one to understand and estimate very fine processes and transformations, occurring during processing, loading and operation of modern materials and devices under intense internal and external influences that lead to arising critical conditions and states. The modern devices and goods with characteristic sizes, changing from nano- and micro- up to macroscale ranges, possess very high accuracy, longevity and extended possibilities to operate in wide temperature and pressure ranges.

This collection presents selected reports of the 2015 International Conference on "Physics, Mechanics of New Materials and Their Applications" (PHENMA-2015), taken place in Azov, Russia, 19-22 May, 2015 (http://phenma2015.math.sfedu.ru), devoted to 100-year Anniversary of the Southern Federal University, and sponsored by the Russian Department of Education and Science, Russian Foundation for Basic Research, Ministry of Science and Technology of Taiwan, South Scientific Center of Russian Academy of Sciences, New Century Education Foundation (Taiwan), Ocean and Underwater Technology Association, Unity Opto Technology Co., EPOCH Energy Technology Corp., Fair Well Fishery Co., Formosa Plastics Co., Woen Jinn Harbor Engineering Co., Lorom Group, Longwell Co., Taiwan International Ports Co., Ltd., University of 17 Agustus 1945 Surabaya (Indonesia), Khon-Kaen University (Thailand), Don State Technical University (Russia) and South Russian Regional Centre for Preparation and Implementation of International Projects.

The thematic of the PHENMA-2015 continues ideas of previous international symposia PMNM-2012 (http://pmnm.math.rsu.ru), PHENMA-2013 (http://phenma.math.sfedu.ru), and PHENMA-2014 (http://phenma2014.math.sfedu.ru) whose results have been published in the edited books "Physics and Mechanics of New Materials and Their Applications," Ivan A. Parinov, Shun Hsyung-Chang (Eds.), Nova Science Publishers, New York, 2013, 444 p. ISBN: 978-1-62618-535-7, "Advanced Materials – Physics, Mechanics and Applications,"

Springer Proceedings in Physics. Vol. 152. Shun-Hsyung Chang, Ivan A. Parinov, Vitaly Yu. Topolov (Eds.), Springer, Heidelberg, New York, Dordrecht, London, 2014, 380 p. ISBN: 978-3319037486, and "Advanced Materials – Studies and Applications," Ivan A. Parinov, Shun-Hsyung Chang, Somnuk Theerakulpisut (Eds.), Nova Science Publishers, New York, 2015, 527 p. ISBN: 978-1-63463-749-7, respectively.

The presented papers are divided into four scientific directions: (i) processing techniques, (ii) physics, (iii) mechanics, and (iv) applications.

Into framework of the first theme are considered, in particular synthesis, structure, and characteristics of different niobate materials, including no containing toxic elements; influence of mechanical activation on properties of multiferroic ceramics; processing technologies for advanced materials applied in MEMS, underwater ultrasonic and ultrasound applications; features of manufacture of the lead-free solid solution, "porous ceramic – polymer" composites; laser annealing of the sapphire crystals and also features of processing the devices on surface acoustic waves.

The second direction covers influence of microgeometry on characteristics of PZT ceramics; theoretical studies of piezo-active 1-3 and 2-2 composites; wide spectrum of electric physical and temperature characteristics of new functional materials, in particular demonstrating a high ferroelectric hardness and efficiency; authors discuss high-temperature porous piezoceramics and ceramic matrix composites, PZT systems for which phase diagrams, absorption spectra of microwave energy, pyroelectric activity and dielectric response are studied; moreover, there are present investigations of alkali metal niobates and bismuth ferrites, modified by various additives.

From viewpoint of mechanics in the third section are studied bulk composite nanomaterial consisting of albumin and carbon nanotubes; dynamic contact problem for heterogeneous layer with liquid sheet; thermal and tribological characteristics of friction surfaces; effects of indentation on mechanical properties of carbon steels; deflections and force factors arising due to concentrated loading of cantilever plate; ultrasonic nondestructive testing the spatial defects of elastic materials; moreover, authors study the problems of poroviscoelasticity, non-linear viscoelastic strains, mechanical characteristics of porous piezoceramics with finite element modeling of lossy piezoelectric elements, surface acoustics waves excited in ferroelectric films and media with inhomogeneities, and also vibroacoustics of composite polymeric shells and forced oscillations of shells with auxetic properties.

The fourth direction demonstrates novel results in modeling and experimental studies of GaAs p-i-n photodetector for optical communication; knock sensors based on lead-free piezoceramics; biological solders for laser welding of biological tissues; HIFU transducers for hemostasis of deep arterial bleeding; welded joint in pipeline; layered anisotropic constructions at pulse loading; rod constructions and truss models with damages; early diagnostics of integral metallic constructions; moreover, authors investigate electrochemical electrolytic cells and other applications.

The book is addressed to students, post-graduate students, scientists and engineers, taking part in R and D of nanomaterials, piezoelectrics, magnetic and other advanced materials, and also different devices, which are based on them, demonstrating broad applications in different areas of science, technique and technology. The book includes new studies and results in the fields of Condensed Matter Physics, Materials Science, Physical and Mechanical Experiment,

Processing Techniques and Engineering of Nanomaterials, Piezoelectrics, other Advanced Materials and Composites, Numerical Methods and results, and also different applications, developed devices and goods.

Ivan A. Parinov (Ed.)

Shun-Hsyung Chang (Ed.)

Vitaly Yu. Topolov (Ed.)

August, 2015

I. PROCESSING TECHNIQUES

In: Proceedings of the 2015 International Conference ... ISBN: 978-1-63484-577-9
Editors: Ivan A. Parinov, Shun-Hsyung Chang et al. © 2016 Nova Science Publishers, Inc.

Chapter 1

LOCAL ATOMIC AND ELECTRONIC STRUCTURE OF THE PT/C NANOPARTICLES GROWN IN HE AND H2 GAS PHASE REDUCING AGENT

N. M. Nevzorova, I. N. Leontyev and G. E. Yalovega[*]
Department of Physics, Southern Federal University,
Rostov-on-Don, Russia

ABSTRACT

In situ investigation of nucleation and growth of Pt/C nanoparticles in the process of thermal decomposition of the precursor $Pt(acac)_2$ were carried out in both helium and hydrogen atmospheres. The influence of the carbon support and a reducing atmosphere on the temperature of nucleation and growth of Pt NPs during synthesis process were shown. Inactive carbon Vulcan XC 72 was used as a support. To control directly the process of Pt/C NPs formation, the XAS (EXAFS and XANES) technique was applied. It was found the formation of Pt nanoparticles on carbon support surface. The temperature of nucleation and growth of Pt/C nanoparticles in helium and hydrogen atmosphere was determined.

Keywords: Pt/C nanoparticles, XAFS, synthesis, platinum-based catalysts

1. INTRODUCTION

Nanocomposite materials are currently the object of scientific research all over the world [1]. They are used in medicine, the production of batteries, automotive industry, including the production of electronics and new materials [2]. In particular, platinum-based nanocomposites are promising catalysts for low temperature fuel cell in a hydrogen industry. Nanocatalysts have a small size providing high reactive. The required size of the nanoparticles (order of

---

[*] E-mail: yalovega1968@mail.ru.

several nanometers) depends on the preparation conditions. There are several types of platinum based catalysts: pure platinum on different support, ad-atoms [3], core–shell [4, 5], random alloy [6], and surface Pt-skin. High-surface-area carbon black can be impregnated with catalyst precursors by mixing both of them in an aqueous solution [7]. Following the impregnation step, a reduction step is required to reduce the catalyst precursor to its metallic state. The most common liquid phase reducing agents are $NaBH_4$, hydrazine etc.

To control the microstructure characteristics of catalysts the nature of carbon support, temperature, pH and water-organic solvent content [8] are usually changed. These changes could affect the crystal growth, orientation as well as grain size.

To study process of formation catalysts with optimized composition and microstructure it is necessary to understand the mechanism of precursor reduction and particle growth. We have used XAFS technique to control the process of nanoparticle formation at different synthesis conditions (T, heating time, atmosphere). He and H_2 were used as reducing gases.

2. METHODS

We have investigated nucleation and growth of Pt/C nanoparticles in the process of thermal decomposition of the precursor in both helium and hydrogen atmospheres. As a precursor of platinum we have chosen the platinum acetylacetonate $Pt(acac)_2$. The carbon support (Vulcan XC-72) was impregnated with the $Pt(acac)_2$ precursor at room temperature and stirred ultrasonically until complete evaporation of the tetrahydrofuran solvent.

The platinum L_3-edge X-ray absorption spectra were recorded in transmission mode at the DUBBLE beam-line (Dutch-Belgian beam-line, BM26A) [9] of the 6 GeV ESRF synchrotron (Grenoble, France) during a uniform filling mode, giving a typical storage ring current of 200 down to 160 mA within one synchrotron run. The synchrotron radiation emitted by the bending magnet (magnetic field strength: 0.4 T) was monochromatized with a double crystal Si(111) monochromator. The *in-situ* X-ray absorption spectroscopy (XAS) measurements performed in a quartz capillaries 1 mm diameter. The reducing gas passed through capillaries while heating. Nucleation and growth Pt nanoparticals performed with helium and hydrogen gas flows at atmospheric pressure. The process of nanoparticles nucleation and growth was followed *in-situ* by recording XANES scans in every 6 min. Specific temperatures and gas flow rates is present below. The XAS spectra of Pt foil and $Pt(acac)_2$ pellet were recorded in transmission mode at ambient temperature and pressure as references. The energy scale for the Pt L_3-edge (11564 eV) absorption spectra calibrated to the first maximum of the first derivative spectrum of a metallic Pt foil. The EXAFS data were collected in equidistant k steps of 0.05 $\mathring{A}^{-1}$ up to $k = 13$ $\mathring{A}^{-1}$. Analyses of the EXAFS experimental data performed by comparison of theoretical and experimental spectra using FEFF-8.2 code [10] at the Debye temperature 300 K. Collected spectra were corrected, energy calibrated, pre-edge subtracted and post-edge normalized using Athena package [11] based on the IFEFFIT program [12]. For modeling of corresponding Pt/C based nanoparticles, data from atoms.inp database [13] used.

3. EXPERIMENTAL RESULTS AND DISCUSSION

At a first step, the kinetics of platinum nanoparticles formation has been studied in a helium atmosphere. A helium-gas flow rate was 5 ml/min. To determine local atomic structure in initial point of the heating we have recorded the EXAFS spectra at room temperature. Then, sample was heated and during the heating process the measurement of the XANES spectra were carried out. The heating was finished when formation of Pt/C nanoparticles was observed and the temperature of the formation was determined. These phenomena could be identified when changes in the shape of XANES spectra were finalized.

After heating, the sample was cooled down to room temperature (25°C). To determine the local atomic structure of nanoparticles upon reaching room temperature were recorded EXAFS spectra. To study nucleation and growth of Pt/C nanoparticles in hydrogen atmosphere we have used the similar procedure.

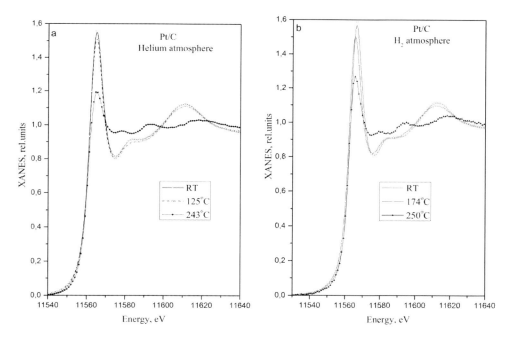

Figure 1. Normalized XANES spectra of Pt L_3-edge in the process of platinum nanoparticles nucleation in a helium atmosphere (a) and hydrogen atmosphere (b).

Figure 1 shows XANES spectra of the Pt L_3-edge in the process of platinum nanoparticles nucleation in a helium (a) and hydrogen atmosphere (b). As one can see, while temperature increases the shape of the absorption spectra changes. At room temperature, the shape of the spectra of the both samples corresponding to the spectrum of Pt(acac)$_2$. As seen in Figure 1 (a), in helium atmosphere precursor decomposition begins at about 120°C that reflected in slight XANES spectrum changes. The shape of spectrum at about 240°C is close to the spectrum of Pt nanoparticles. A further increase in temperature did not lead to a change in shape of the spectra. As seen in Figure 1(b) in hydrogen atmosphere precursor decomposition begins at higher temperatures of about 170°C and the final formation of nanoparticles completes at temperatures of about 250°C. Moreover, for both atmospheres the

XANES spectra exhibits a significant decrease in the intensity of the main peak with increasing temperature. The spectrum of the sample before heating shows high white line intensity, indicating a high valence state of Pt. The white line intensity slightly decreases after start of heating process. Changes in the white line intensity during heating are related to the structural changes of the platinum.

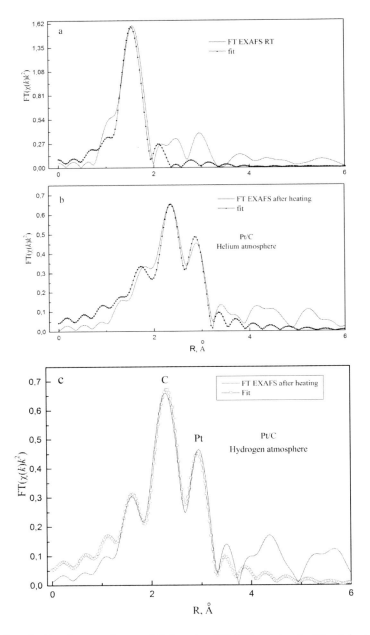

Figure 2. Modeling the Fourier transforms of Pt L_3-edge EXAFS spectra for Pt catalysts supported on Vulcan XC-72 before heating (a) and after heating in helium (b) and hydrogen (c) atmospheres.

We simulated EXAFS spectra using code IFEFFIT to determine the neighborhood of the absorbing atom and to clarify Pt-C interaction. Figure 2 shows the change of Fourier transform of EXAFS spectra observed for catalysts before heating (a) and after heating in helium (b) and hydrogen (c) atmospheres. In accordance to our data before reduction there are two peaks which correspond to Pt-O and Pt-C bond for $Pt(acac)_2$ (at $R = 1.95$ Å and $R = 2.0$ Å, respectively). We obtained distances Pt-O, that are in a good agreement with the distances obtained in the works [14] and [15]. Coordination number for Pt-O is $N = 4$. Figure 2 (b) represents clear peaks, which correspond to Pt-Pt at $R = 2.73$ Å and Pt-C at $R = 2.07$ Å with coordination numbers 6 and 0.8, respectively. The presence of these two peaks confirms formation of Pt nanoparticles on carbon support surface. Figure 2 (c) shows the simulation results for $Pt(acac)_2$ on the carbon support in hydrogen atmosphere after process of heating. There are two obvious peaks corresponding to Pt-C at $R = 2.03$ Å and Pt-Pt at $R = 2.75$ Å with coordination numbers 0.9 and 6, respectively. Obviously, after heating, the platinum particles formed on the carbon support. We obtained good simulation results; in all cases, R-factor was lesser than 4% (0.036, 0.024 and 0.008).

CONCLUSION

1. In a helium atmosphere, the decomposition of the precursor $Pt(acac)_2$ begins at about 120°C. Formation of nanoparticles completes at temperatures of about 240°C. 2. In a hydrogen atmosphere, the decomposition of the precursor $Pt(acac)_2$ begins at about 170°C. Formation of nanoparticles completes at temperatures of about 250°C. 3. Distances Pt-Pt for the nanoparticles formed in helium atmosphere is shorter in comparison with the nanoparticles formed in hydrogen atmosphere (2.73 Å and 2.75 Å, respectively).

ACKNOWLEDGMENTS

The research, leading to these results, received funding from the Ministry of Education and Science of the Russian Federation under grant agreement No. 11.2432.2014/K. We thank ESRF (Grenoble, France) for the allocation of synchrotron radiation beam-time (Dutch-Belgian Beam Lines, DUBBLE 26A). The DUBBLE staffs are acknowledged for technical support.

REFERENCES

[1] Rodriguez, J.A. *Surf. Sci. Rep.* 1996, *vol. 24.* 223 – 287.
[2] Suib, Steven L. New and Future Developments in Catalysis: Batteries, Hydrogen Storage and Fuel Cells. 2013, 217 – 231.
[3] Watanabe, M.; Motoo, J.S. *Electroanal. Chem.* 1975, *vol. 60*, 267 – 273.
[4] Mani, P.; Srivastava, R.; Strasser, P. *J. Phys. Chem. C.* 2008, *vol. 112*, 2770–2778.

[5] Strasser, P.; Koh, S.; Anniyev, T.; Greeley, J.; More, K.; Yu, C.; Ziu, L.; Kaya, S.; Nord-lund, D.; Ogasawara, H.; Toney, M.F.; Nilsson, A. *Nat. Chem.* Published Online 25 April, 2010.

[6] Duong, H.T.; Rigsby, M.A.; Zhou, W.-P.; Wieckowski, A. J. *Phys. Chem. C.* 2007, *vol. 111*, 13460–13465.

[7] Zhang, J. *PEM Fuel Cell Electrocatalysts and Catalyst Layers.* Springer-Verlag, London Limited, 2008.

[8] Leontyev, I. Journal of Alloys and Compounds. 2010, vol. 500, 241-246.

[9] Nikitenko, S.; Beale, A.M.; van der Eerden, A.M.J.; Jacques, S.D.M.; Leynaud, O.; O'Brien, M.G.; Detollenaere, D.; Kaptein, R.; Weckhuysen, B.M.; Bras, W. *J. Synchrotron Rad.* 2008, *vol. 15*, 632–640.

[10] Ankudinov, A.L.; Bouldin, C.; Rehr, J.; Sims, J.; Hung, H. *Phys. Rev. B.* 2002, *vol. 65*, 104107–104118.

[11] Ravel, B.; Newville, M. *J. Synchrotron Radiat.* 2005, *vol. 12*, 537–541.

[12] Newville, M. *J. Synchrotron Radiat.* 2001, *vol. 8*, 322–324.

[13] http://cars9.uchicago.edu/~newville/adb/search.html.

[14] Onuma, S.; Horioka, K.; Inoue H.; Shibata, S. *Bull. Chem. Soc. Jpn.* 1980, *vol. 53*, 2679.

[15] Beck, I.E.; Kriventsov, V.V.; Ivanov, D.P.; Zaikovsky, V.I.; Bukhtiyarov V.I. *Nuclear Instruments and Methods in Physics Research A.* 2009, *vol. 603*, 108–110.

In: Proceedings of the 2015 International Conference … ISBN: 978-1-63484-577-9
Editors: Ivan A. Parinov, Shun-Hsyung Chang et al. © 2016 Nova Science Publishers, Inc.

Chapter 2

EFFECTS OF COMBINED MODIFYING OF NIOBATE MATERIALS NON-CONTAINING TOXIC ELEMENTS

Kh. A. Sadykov[*], *I. A. Verbenko, L. A. Reznichenko, G. M. Konstantinov, A. A. Pavelko, L. A. Shilkina and S. I. Dudkina*

Research Institute of Physics, Southern Federal University,
Rostov-on-Don, Russia

ABSTRACT

Regularities of changes in the grain structure, piezoelectric and dielectric properties of the $(Na_{1-x}Li_x)NbO_3$-based solid solutions modified with $(Bi_2O_3 + Fe_2O_3)$ are established. It is shown that the modification causes a formation of microstructures with a "cementing" feature of intercrystalline layers. It is found that ceramics based on the $(Na_{1-x}Li_x)NbO_3$ system are a promising base for the creation of materials with high Q_M, g_{33}, the anisotropy of the piezoelectric coefficients and low electrical conductivity while maintaining sufficient piezoelectric activity and $K_p \sim 0.2$.

Keywords: lead-free ferroelectrics, niobates, grain structure, piezoelectric properties, dielectric properties

1. INTRODUCTION

Lead-containing solid solution systems like $PbTiO_3$-$PbZrO_3$ (PZT) form the basis of almost all applicable ferroelectric ceramic materials (FECMs). Due to toughening of environment requirements to electrical products, operating conditions of devices and materials used in them [1], there is a necessity for development of fundamentally new

---

[*] Corresponding author: Kh. A. Sadykov. Research Institute of Physics, Southern Federal University, 344090 Rostov-on-Don, Russia. E-mail: hizir-2010@mail.ru.

FECMs, which do not contain toxic elements (e.g., lead). The aim of the present study is to develop and investigate lead-free FECMs based on $(Na_{1-x}Li_x)NbO_3$ solid-solution system.

2. OBJECTS AND METHODS

Our research targets were solid solutions based on $(Na_{1-x}Li_x)NbO_3$ system modified by $(Bi_2O_3 + Fe_2O_3)$ oxides added in the amount of 0.5-3 wt.%. Solid phase synthesis was carried out in two stages: $T_{synt.1,2}$ = 850-70°C (depending on composition) during $\tau_1 = \tau_2 = 6$ h. Sintering was performed by means of conventional ceramic technology at temperature T_{sp} = 1160-1200°C (depending on the composition) during 1 h.

3. RESULTS AND DISCUSSION

One of the most prospective contenders for the role of matrix of highly effective lead-free materials are $(Na_{1-x}Li_x)NbO_3$ solid solutions that have strong ferroelectric (FE) properties and sufficiently high Curie temperature [2].

Figure 1 shows micrographs of cleavage and pore surfaces of ceramic samples modified by 1.0 and 1.5 wt. percentage of combined $(Bi_2O_3 + Fe_2O_3)$ modifier, respectively. The data was obtained from secondary electrons using a scanning electron microscope. One can see that the ceramic samples have extremely inhomogeneous microstructure with a large number of fine pores. The cleaved facet passes through grains. Grain boundaries are indistinct. Grain size is less than 1 μm, so there is no way to reveal the internal structure of the grains, even at maximum magnification. As the content of modifier increases up to 1.5 wt.% the microstructure remains heterogeneous, but herewith the grain size increases up to (15-20) μm and grain boundaries become clearer. The grains have a non-idiomorphic shape and vary greatly in size. Further modification (2.0 wt.%, Figure 2a) leads to increase of microstructure heterogeneity and grain boundaries diffusion [2].

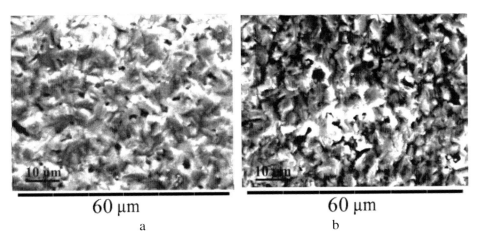

Figure 1. Micrographs of cleaved surface of the solid solutions modified by a-1.0, b, c-1.5, c-2.0 wt.% of $(Bi_2O_3 + Fe_2O_3)$ resulting in secondary electrons in a scanning electron microscope.

Our analysis of polished and etched surface images of the samples (Figure 2 b and c) reveals the compact formations close to Fe$_2$O$_3$ in composition. Formation of a separate phase based on Fe$_2$O$_3$ indicates its inertness with respect to newly formed solid solution. On the contrary, bismuth oxide equally distributed over the sample and disposed to the formation of liquid phase was probably responsible for the observed effects: abrupt decrease of the grain size, clear heterogeneity of microstructure, and increased strength of grain boundaries.

Figure 3 shows the basic electrical characteristics of materials based on the (Na$_{1-x}$Li$_x$)NbO$_3$ system modified with the combined modifier (Bi$_2$O$_3$ + Fe$_2$O$_3$). It is clear that addition of the modifier increases anisotropy of the piezoelectric effect, i.e., piezoelectric coefficients d_{3j} follows the condition $d_{33}/|d_{31}| = 5.1$.

On addition of the modifier (Bi$_2$O$_3$ + Fe$_2$O$_3$) almost all characteristics of the studied ceramic materials vary linearly with an extremum at $x = (1.5-2.0)$ wt.%. Such dinamics can be result of both morphotropic phase transition localized in the specified concentration range and changing nature of the modifier embedded in polycrystalline structure. It should be noted that at $x = 1.5$ wt.%, we derived a $Q_M > 700$ at a sufficiently high K_p (~ 0.2). In combination with other electrophysical characteristics (low $\varepsilon_{33}^T/\varepsilon_0$ ~ 160 and high V_1^E ~ 5.6 m/s) this offers the prospect of application of the studied objects in high-frequency electromechanical force transducers used, for example, in ultrasonic oil well logging [3].

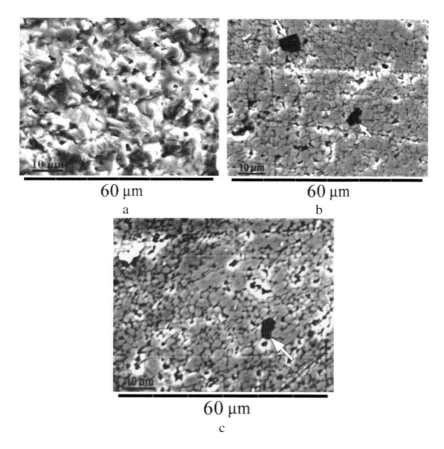

Figure 2. Micrographs of cleaved (a) and polish (b,c) surfaces of the solid solutions modified by 2.0 wt.% of (Bi$_2$O$_3$ + Fe$_2$O$_3$) resulting in secondary electrons in a scanning electron microscope.

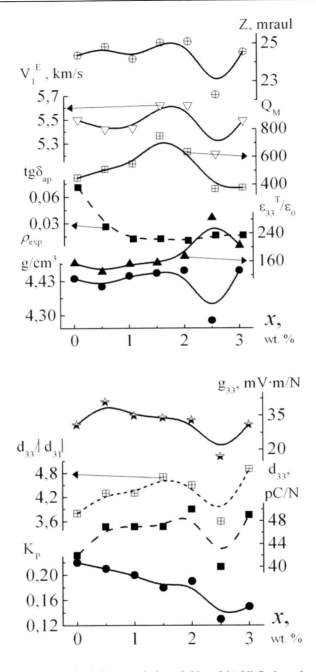

Figure 3. Dependences of the electrical characteristics of $(Na_{1-x}Li_x)$ NbO$_3$-based solid solutions on molar concentration of the combined modifier (Bi$_2$O$_3$ + Fe$_2$O$_3$).

Figure 4 shows dependencies of $\varepsilon/\varepsilon_0$ of (Na, Li) NbO$_3$-based ceramic samples on frequency and temperature. There are non-modified objects and objects modified by the combined modifier (Bi$_2$O$_3$ + Fe$_2$O$_3$) [4].

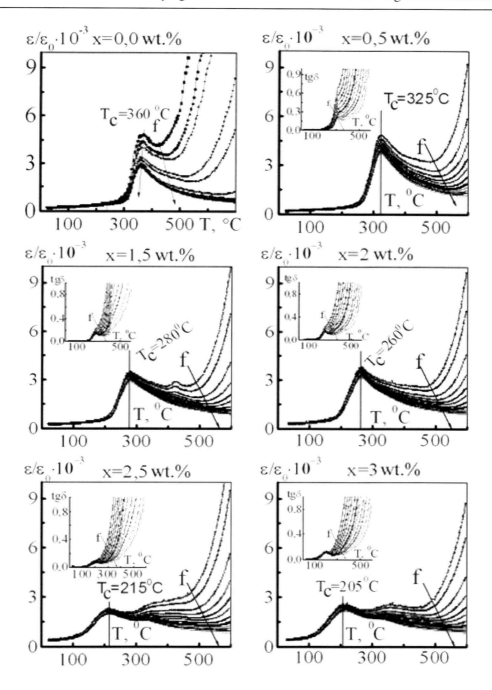

Figure 4. Dependencies of ε/ε0 and tgδ on temperature (*T*, °C) at (20-1000) Hz frequency range of (Na, Li) NbO₃-based ceramic samples modified by (Bi₂O₃ + Fe₂O₃) (Tsynt = 1205°C).

Figure 4 data analisys shows that the studied objects are characterised by dielectric spectra with expressed dispersed maximum of $\varepsilon/\varepsilon_0$ without relaxor-like behavior in the ferroelectric-paraelectric (FE-PE) phase transition (PT) area. At temperatures above the PT, there is widening of $\varepsilon/\varepsilon_0$ dispersion and sharp increase in $\varepsilon/\varepsilon_0$, which is particularly noticeable at low frequencies of the measuring electric field. Such behavior of $\varepsilon/\varepsilon_0$ in niobate ceramics refers to electrical conductivity increase due to oxygen losses and partial niobium recovery [2]. Within temperature range of (20-200)°C in ceramics containing 0-1.5 wt.% of the modifier, we observe a stroke of $\varepsilon/\varepsilon_0$ which is almost parallel to the x-axis. This indicates high stability of the objects within the aformentioned temperature range. Since the content of the modifier increases up to 2 wt.%, a second maximum of $\varepsilon/\varepsilon_0$ occurs at high temperatures. Upon the further modification, the maximum intensifies and shifts to the high-temperature region. It is accompanied by temperature decrease of the FE-SE PT and strengthening of its dispersion, as well as non-monotonic change in the $\varepsilon/\varepsilon_0$ dispersion depth (with minimum at $x = 1.5$ wt.%).

Appearance of the additional maximum of $\varepsilon/\varepsilon_0$ can be result of either PT between two non-polar phases or reconstruction of the defect structure in polycrystalline material, so it needs an additional clarification by means of high-temperature X-ray studies.

CONCLUSION

The regularities of changes in the grain structure of the $(Na_{1-x}Li_x)$ NbO_3-based solid solutions modified by $(Bi_2O_3 + Fe_2O_3)$ were established. Modification was determind as a reason of formation of microstructures with "cementing" feature of intercrystalline layers based on liquid phases enriched by Bi.

Ceramics based on the $(Na_{1-x}Li_x)NbO_3$ system were found as a promising base for creation of materials with high Q_M, g_{33}, anisotropy of the piezoelectric coefficients and low electrical conductivity while maintaining sufficient piezoelectric activity and $K_p \sim 0.2$. In particular, developed materials with 1.5 wt.% of $(Bi_2O_3 + Fe_2O_3)$ are charcterized by $Q_M > 700$ and $K_p \sim 0.2$ and regarded as prospective contenders for use in power ultrasonic transducers. Materials with 2.0 wt.% of $(MnO_2 + Fe_2O_3)$ are characterized by large values of $g_{33} \sim 40$ mV·m/N and $d_{33}/|d_{31}| \sim 5$ and applicable in non-destructive testing devices.

ACKNOWLEDGMENTS

The Ministry of Education and Science of the Russian Federation financially supported this work: Grant of the President of the Russian Federation No. MK-3232.2015.2; themes Nos. 1927, 213.01-2014/012-VG and 3.1246.2014/K (the basic and project parts of the State task).

REFERENCES

[1]   Directive 2002/95/EC of the European Parliament and of the Council of 27 January 2003 on the Restriction of the Use of Certain Hazardous Substances in Electronic Equipment, *Official Journal of the European Union*, 2003, vol. 37, 19-23.

[2]   Sadykov, Kh. A. et al., *Composite Material Structures*, 2013, vol. 3, 45-55 (in Russian).

[3] Sadykov, Kh. A. et al., *Bulletin of the Russian Academy of Sciences: Physics*, 2013, vol. 77, 1253-1255 (in Russian).

[4] Fesenko, E. G. *Perovskite Family and Ferroelectricity*, Atomizdat, Moscow, 1972, pp. 1-248 (in Russian).

In: Proceedings of the 2015 International Conference ...   ISBN: 978-1-63484-577-9
Editors: Ivan A. Parinov, Shun-Hsyung Chang et al.   © 2016 Nova Science Publishers, Inc.

Chapter 3

SYNTHESIS, STRUCTURE, AND OPTICAL CHARACTERISTICS OF BARIUM STRONTIUM NIOBATE THIN FILMS

G. N. Tolmachev[1,2], A. P. Kovtun[1,2], I. N. Zaharchenko[2], I. M. Aliev[2], S. P. Zinchenko[1], A. V. Pavlenko[1,2,], I. A. Verbenko[2] and L. A. Reznichenko[2]*

[1]Southern Scientific Center of the Russian Academy of Sciences,
Rostov-on-Don, Russia
[2]Research Institute of Physics, Southern Federal University,
Rostov-on-Don, Russia

ABSTRACT

Using ceramic $Ba_{0.5}Sr_{0.5}Nb_2O_6$ (BSN) in an oxygen atmosphere at elevated pressure (0.5 Torr), we obtain films by means of RF sputtering on a MgO (001) substrate. Optical emission spectra of a ceramic target sprayed at a discharge are measured. The structure and optical properties of the films BSN/MgO (001) at room temperature are studied. A specular intensity of the H-polarized radiation in the ranges of 55-75 degrees and 500-800 nm is measured. It's suggested that the BSN films grow on a substrate from a dispersed phase. It is stated that the obtained films are highly textured, and crystallographic planes (001) of BSN and MgO are parallel. We found an angle providing an independence of the intensity of the reflected signal on the film thickness. The tangent of the angle is equal to the refractive index of the film material. If the film material is transparent, then the optical characteristics of the substrate do not change during deposition of the film, and if the film surface is smooth then the intersection of the specular reflection exists.

Keywords: thin films, ferroelectric strontium barium niobate, discharge with runaway electrons.

---

* Corresponding author: E-mail: tolik_260686@mail.ru.

1. INTRODUCTION

Thin films of strontium barium niobate ($Ba_{1-x}Sr_xNb_2O_6$, 0.25 $<x<$ 0.75) are studied as a potential material for many microdevice applications such as holographic storage, pyroelectric infrared detectors, electrooptic modulators, and beam steering because of its large pyroelectric coefficient, excellent piezoelectric and electrooptic properties, and photorefractive sensitivity [1-4]. Magnetron [5], pulsed laser deposition [6] and other methods are used to form such structures, but high-frequency RF-sputtering under high oxygen pressure [7] technique shows itself better in the preparation of the nanoscale $Ba_{1-x}Sr_xTiO_3$ films [7-9]. This technique can also possibly develop technical and mathematical tools in-situ monitoring of the film growth [10]. The present chapter reports our research results on obtaining, structure and optical properties of the $(Ba_{0.5}Sr_{0.5})Nb_2O_6$ (BSN) thin films, obtained on the (001) MgO crystal face.

2. METHODS OF PRODUCTION AND RESEARCH

Thin films are deposited on the BSN-chipped MgO (001) substrate within three (BSN3), five (BSN5), ten (BSN10), and fifteen (BSN15) minutes of spraying. BSN ceramic targets were obtained in the Department of Smart Materials and Nanotechnology of the Institute of Physics, Southern Federal University. The gas discharge RF-film deposition was carried out using BSN Plasma-50-SE according to the procedure described in detail in work [7]. Angular intensity curves of the mirroring polarized (H-polarization) radiation with wavelengths in a range of 500–800 from the test films at room temperature were obtained using the measuring complex described in work [10]. The emission spectra of the discharge plasma Y (λ) in the range of λ = 300-900 nm were measured directly during the film deposition method described in work [7]. X-ray studies were carried out by means of the room temperature diffractometerUltimaIV (Rigaku).

3. EXPERIMENTAL RESULTS AND DISCUSSION

Figure 1 shows the typical optical glow discharge with the sputtering BSN spectrum providing information about the process of transferring the material from the SBN-target to the MgO substrate. The virtual absence in the spectrum of atomic and ionic lines of the target elements (Ba, Sr, and Nb) is shown in Figure 1. Against the background of a large number of lines of oxygen (both atomic and ionized) during the deposition of the BSN-films in the category, only the weak line of strontium (λ = 421.5 nm and λ = 460.7 nm), barium (λ = 430.9 nm and λ = 455.4 nm) are observed, while the line of niobium (for example, at λ = 416.3 nm and λ = 405.89 nm) is not detected. These results are indicative of the fact that, as in a case of the synthesis of thin films of TR barium strontium titanate [9], there is BSN-film "growing" of the dispersed phase, i.e., the phase in which the short-range order is stored such as a target.

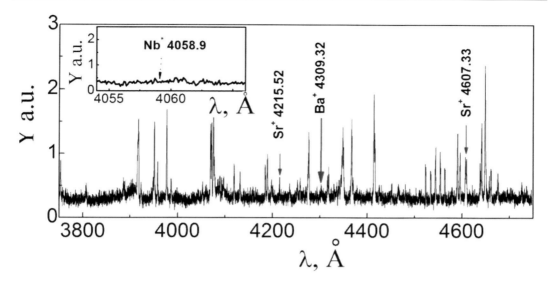

Figure 1. Spectrum of optical glow of discharge with sputtering BSN at wavelengths λ = 375.0 – 475.0 nm, oxygen pressure 0.5 Torr, and RF power 160 W. An arrow in the inset marks the spot line Nb. Non-marked lines belong to oxygen (atoms or ions).

Figure 2 shows diffraction patterns of the ceramic target and films of BSN10 and BSN15. It is seen that only 00l film reflexes are present at the diffraction patterns recorded by 2θ/ω method, and therefore, the (001) plane is the plane of the conjugation with the substrate. The results of the φ-scan of 320 reflexes captured at the in-plane geometry testify that the films are monocrystalline (Figure 3 presents the φ-scan for BSN15 films as example). Furthermore, the block's azimuthal disorientation does not exceed about 3.5°.

The angular position symmetry of reflexes corresponds to the presence of the 4-fold symmetry axis being perpendicular to the (001) surface. The analysis of angular positions of the φ-scan peaks showed the presence of twins in the films. The parameter c determined by 00l reflexes and the parameter a determined by 320 and 410 reflexes appeared to be close to the target parameters. The structural perfection of the films on the [001] direction is characterized by an average size of coherent scattering regions and average microdeformations values, found out by the approximation method. The obtained results are shown in Table 1.

Table 1. Structural Characteristics of Materials

Unit-cell parameter	BSN target	BSN3	BSN5	BSN10	BSN15
c, Å	3.943	3.955	3.968	3.960	3.962
a, Å	12.455	12.276	12.419	12.580	12.511
D, Å		200	230	400	280
ε		$3 \cdot 10^{-3}$	$2 \cdot 10^{-3}$	$4 \cdot 10^{-3}$	$2 \cdot 10^{-3}$

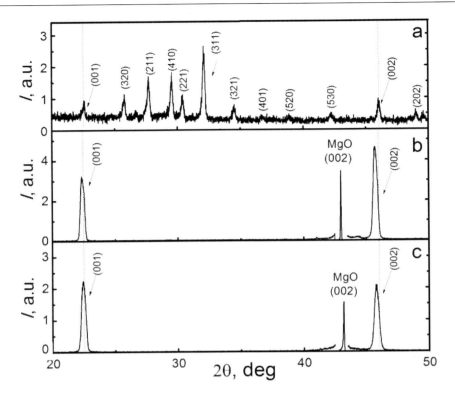

Figure 2. XRD patterns of the BSN-target (a), BSN10-film (b), and BSN15-film (c).

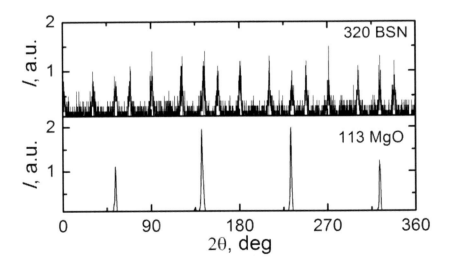

Figure 3. φ-scan of 320 reflexes of the BSN15 thin film (top part, geometry in plane) and of 113 reflexes of MgO (bottom part, geometry out of plane).

The curves of the intensity of the specular reflection from the film were obtained at an angle that provides an intersection of the all curves of the films with different thicknesses for the wavelength of 800, 700 or 600 nm. The nature of this intersection is contained in a simple model of an optically homogeneous transparent film on an optically homogeneous substrate (details of this model are represented in work [11]). With the specular reflection, the angle whose tangent is the refractive index of the film material (Brewster angle), a film reflectivity is not depending on the thickness. Indeed, the dependence on the thickness is determined at the processes of interference between two beams interfaces as follows: "film material – substrate" and "film material – air." In the case, where the angle of the reflection with the angle of the refraction is 90°, the amplitude of the reflection from the "air – film material" interface vanishes. It leads to the "destruction" of interference because of the lack of the dependence of the reflectance on the film thickness [12].

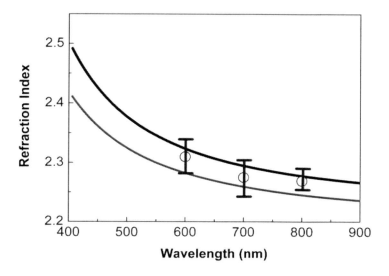

Figure 4. Dispersion of the refraction index of $Sr_{0.61}Ba_{0.39}Nb_2O_6$ [13] (solid curve) and o-tangents of angles of the intersection of the reflection.

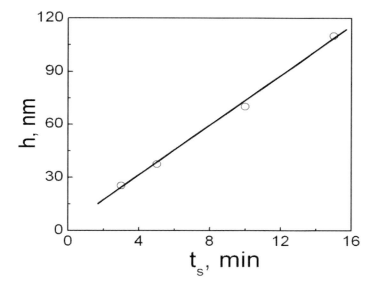

Figure 5. Dependence of the film thickness h on the deposition time t_s.

In Figure 4, calculated tangents of the angles of the reflection intersection are shown by markers. For comparison, according to work [13], curves show the dispersion of refractive indices of $Sr_{0.61}Ba_{0.39}Nb_2O_6$ for ordinary (upper curve) and extraordinary (lower curve) polarized light. The thickness of the film at the initial stage can be easily calculated at the certain refractive. Figure 5 shows the dependence of the thickness of the film on the sputtering time: the films with the deposition time of 3, 5, 10 and 15 minutes correspond to thicknesses of 25, 38, 70 and 110 nm, respectively. The rate of the film growth in this case, at the initial stage of spraying, is about $8 - 9$ nm/min (Figure 5).

The results obtained should be used at the preparation of thin films of single-crystal BSN substrates on MgO. The results of optical studies may form a basis for a development of technical and mathematical tools in-situ monitoring of film growth in the future.

ACKNOWLEDGMENT

This work was supported by the Ministry of Education and Science of the Russian Federation (Federal program: State contract No. 14.575.21.0007; Projects No. 3.1246.2014/K, No. 2132) and RFBR (research project No. 15-08-05711-a).

REFERENCES

[1]   Glass, AM. *J. Appl. Phys.,* 1969, vol. 40, 4699–713.
[2]   Neurgaonkar, RR; Hall, WF; Oliver, JR; Ho, WW; Cory, WK. *Ferroelectrics.,* 1988, vol. 87, 167.
[3]   Xu, YC; Chen, J; Xu, R; Mackenzie, JD. *Phys. Rev. B,* 1991, vol. 44, 35 – 41.
[4]   Lenzo, PV; Spencer, EG; Ballman, AA. *Appl. Phys. Lett.,* 1967, vol. 11, 23–24.
[5]   Willmott, PR; Herger, JR; Patterson, BD; Windiks, PR. *Phys. Rev. B.,* 2005, vol. 71, 144114.
[6]   Willmott, PR; Herger, JR. *Rev. of Modern Phys.,* 2000, vol. 72, 315.
[7]   Mukhortov, VM; Golovko, YuI; Mamotov, AA; Tolmachev, GN; Biryukov, SV; Masychev SI. *Proceedings of the South Scientific Center of the Russian Academy of Sciences.,*2007, vol. 2, 224-265 (in Russian).
[8]   Mukhortov, VM; Golovko, YI; Tolmachev, GN; Klevtzov, AN. *Ferroelectrics,* 2000, vol. 247, 75-83.
[9]   Mukhortov, VM; Golovko, YI; Tolmachev, GN; Mashenko, AI. *Technical Physics.,* 1998, vol. 43(9), 1097-1101.
[10] Zinchenko, SP; Kovtun, AP; Tolmachev, GN. *Technical Physics Letters.,* 2014, vol. 40(1), 21-24.
[11] Landau, LD; Lifshitz, EM. *Electrodynamics of Continuous Media.,*Nauka, Moscow, 1982 (in Russian).
[12] Aliev, IM; Kovtun, AP; Pavlenko, AV; Tolmachev, GN. *IJAER.,* 2014, vol. 9, 26219-26224.
[13] Kip, D; Aijlkemeyer, S; Buse, K; Mersch, F; Pankrath, R; Kratz. E. *J. Phys. Stat. Sol.,* 1996, vol. 154, 5-7.

In: Proceedings of the 2015 International Conference …   ISBN: 978-1-63484-577-9
Editors: Ivan A. Parinov, Shun-Hsyung Chang et al.   © 2016 Nova Science Publishers, Inc.

Chapter 4

OPTIMIZATION OF CONDITIONS FOR OBTAINING $PbMn_{1/2}Nb_{1/2}O_3$ BY DIRECT SYNTHESIS OF A MIXTURE OF OXIDES AND SINTERING BY CONVENTIONAL CERAMIC TECHNOLOGY

L. A. Shilkina, I. A. Verbenko, G. M. Konstantinov, S. I. Dudkina and L. A. Reznichenko*
Research Institute of Physics, Southern Federal University,
Rostov-on-Don, Russia

ABSTRACT

The chapter shows the possibility of making ceramics $PbMn_{1/2}Nb_{1/2}O_3$ directly from oxides while bypassing Columbine technology. The study of crystal structures and microstructures in the temperature range of sintering $(970–1220)°C$ allowed us to select the optimum process conditions to ensure maximum density of the ceramic samples and the stability of the structure. It was found that the least defected structure formed at $T_s = 1100°C$, the most defective at $T_s = (1140–1150)°C$. The highest density of ceramics is achieved in the temperature range of $(1160–1180)°C$.

Keywords: $PbMn_{1/2}Nb_{1/2}O_3$, ceramics, technology, structure, microstructure

1. INTRODUCTION

$PbMn_{1/2}Nb_{1/2}O_3$ (PMnN) is a triple-oxide of oxygen-octahedral type. Many of them $(PbMg_{1/3}Nb_{2/3}O_3, PbFe_{1/2}Nb_{1/2}O_3$ and other) have been extensively studied in recent years because they are the base of many functional (e.g., piezoelectric) materials. However, PMnN,

---

*E-mail: ilich001@yandex.ru.

in comparison with other similar compounds is studied as insufficient. Weak interest to it is caused by the presence of elements with varying degrees of oxidation— especially manganese—in its composition. They predetermine the high electrical conductivity of PMnN. It makes PMnN unsuitable for use as a piezoelectric material. Multiferroic materials have attracted renewed interest in recent years. High-temperature multiferroics are of particular interest. "Ferromagnetic ions" such as Mn^{2+}, Mn^{3+}, (Mn^{4+}), Fe^{2+}, Fe^{3+}, Co^{2+}, Co^{3+}, Co^{4+}, Ni^{2+} are a condition of the multiferroics existence.

It is known that the synthesis of ternary oxides having the perovskite structure and containing niobium passes to form an intermediate phase with a pyrochlore structure. To eliminate the pyrochlore phase, the intermediate synthesis of the compounds with the columbite structure is performed. However, the use of highly charged B-cations, such as Fe^{3+}, Mn^{3+}, makes simple synthesis from respective oxides possible.

Thus, the aim of this study is to establish conditions obtaining pure PMnN ceramics by direct synthesis from a mixture of simple oxides and sintering by conventional ceramic technology.

2. METHODS

Ceramics PMnN synthesized using a conventional solid-state reaction technique. The starting chemicals were PbO (99.9%), MnO_2 (99%), Nb_2O_5 (99%). The stoichiometric mixture was thoroughly mixed in a 96% aqueous solution of ethanol, dried and calcined at $T_1 = (900–970)°C$ for 5h, then reblended calcined in $T_2 = (950–1100)°C$ for 5 h. The ceramics were sintered at $T_s = (950–1230)°C$ for 2 h. All operations conducted in a free access of air atmosphere. Selection temperature synthesis performed on the base of X-ray analysis.

The crystalline structure of the sintered samples at room temperature was determined using powder X-ray diffraction (XRD) with $Co_{K\alpha}$ radiation. The microstructure observed by scanning electron microscopy TM-1000 Hitachi.

The choice of the sintering temperature of ceramics $PbMn_{0.5}Nb_{0.5}O_3$ conducted on the base of analysis of the following characteristics:

I. content of impurity phases (the relative intensity of the strongest line was estimated, I/I_1, where I is the line intensity of the impurity phase, I_1 is the intensity of the strongest line of the perovskite phase);

II. parameter, a, and volume, V, of the perovskite cell;

III. the full width at half maximum (FWHM) of the X-ray line 200, B_{200}, which characterizes the degree of homogeneity of the compounds (variation of cell parameters);

IV. size of coherent scattering regions, D, in the <100>, which characterizes the degree of perfection of the crystal structure (calculation was based on the Scherrer formula – $D = \lambda/(\beta\cos\theta)$ [1], where λ is the X-ray wavelength, β is the physical broadening of X-ray line, θ is the diffraction angle corresponding to the measured line);

V. dislocation density, $\rho_{disl} = 3/D^2$ [2];

VI. experimental, ρ_{exp}, X-ray, $\rho_{X\text{-}ray}$ and relative, ρ_{rel}, density of ceramics, first determined by hydrostatic weighing in octane, the second formula $\rho_{x\text{-}ray} = 1.67 M/V$, where M is the molecular weight, V-cell volume, $\rho_{rel} = \rho_{exp}/\rho_{X\text{-}ray} \times 100$.

3. EXPERIMENTAL RESULTS AND DISCUSSION

Using X-ray diffraction impurity pyrochlore phase was found at $T_2 < 950°C$. It did not disappear even with subsequent sintering at $T_s < 1050°C$. With further optimization of the synthesis temperature regimes ($T_1 = T_2 = 970°C$) the amount of impurity was reduced to $< 2\%$ and during sintering ($T_s \geq 1050°C$) it was completely eliminated.

Figure 1 shows the diffraction patterns for all sintering temperatures. It well seen that the ceramics with sintering temperatures $(970–1050)°C$ have low intensity and larger width of X-ray lines. Diffraction patterns for ceramics with sintering temperature higher than $1050°C$ are characterized lower width of X-ray lines and their intensities are higher. Their changes are not monotonically.

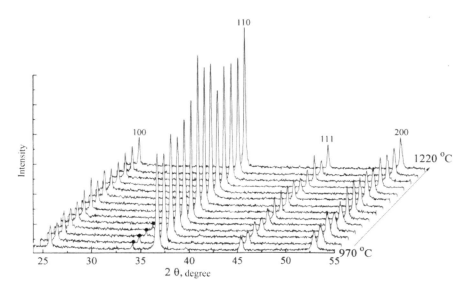

Figure 1. X-ray diffraction patterns of ceramics PMnN at different sintering temperatures, dark circles denote line of pyrochlore phase.

Figure 2 (top) shows the perovskite cell volume, V, of the relative intensity of the line pyrochlore phase, I/I_1, experimental, ρ_{exp}, $\rho_{X\text{-}ray}$, and relative, ρ_{rel}, densities as a function of T_s ceramics. The insets show ρ_{exp} and ρ_{rel} in the range temperature of the sintering corresponding to the maximum density ceramics. In the range I-III, cell volume changes very little, this is similar to the Invar effect (IE). At $T_s = (1020–1100)°C$, V increases monotonically, and in the interval $1130 < T_s < 1160°C$ it behaves unstable, namely a sharp rise is replaced by a decrease $\Delta V = 0.2$ Å^3, significantly exceeding the measurement error. Maxima in the dependence $\rho_{exp}(T_s)$ and $\rho_{rel}(T_s)$ are in the range $T_s = (1160–1180)°C$, immediately after decreasing the V cell. Figure 2 (bottom) shows the dependence of FWHM, B_{200}, X-ray line 200, the size of the coherent scattering region (CSR), D, and the dislocation density, ρ_{disl}, as a function of the sintering temperature. It is seen that the ceramics sintered at $T_s \leq 1050°C$ have small dimensions of CSR, high dislocation density and a large FWHM B_{200} that is indicating a lack of the uniformity of ceramic composition. Above $T_s = 1050°C$ B_{200}, ρ_{disl} sharply reduces, D increases, the least defect structure (minimum ρ_{disl} and maximum D) forms at $T_s = 1100°C$. In the range $T_s = (1100–1130)°C$ where an increase in V cells occurs with the highest rate, the defectiveness of the structure again increases (maximum ρ_{disl} and minimum D).

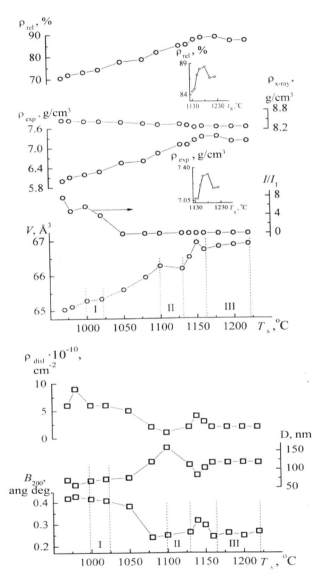

Figure 2. Dependencies of the volume of the perovskite cell V, the relative intensity line of the pyrochlore phase I/I_1, experimental values of ρ_{exp}, ρ_{X-ray}, and relative ρ_{rel} densities on sintering temperatures of ceramics. Insets present ρ_{exp} and ρ_{rel} in the range of sintering temperatures corresponding to the maximum density ceramics (top) and Dependencies of the FWHM, B_{200}, X-ray line 200, the size of the coherent scattering region (CSR), D, and the dislocation density, ρ_{disl}, on the sintering temperature (bottom).

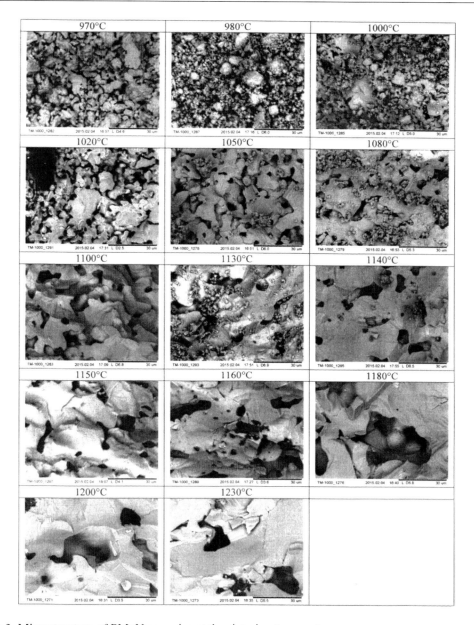

Figure 3. Microstructure of PMnN ceramics at the sintering temperatures.

Comparing Figure 2, we can conclude that the maximum density of the ceramic achieves when its structure saturates by the defects. The dependence of the ceramic density on its real (defect) structure has been established earlier in solid solutions of $PbTi_{1-x}Zr_xO_3$ [3]. Thus, at the sintering temperatures, the corresponding region IE-I leads to an increase of the ceramic uniformity. The region IE-II, probably associates with a change of the real (defect) structure of PMnN. In the region IE-III, at $T_s > 1180°C$, the formation of the structure completed, so that the change of the cell volume and density of ceramics practically finished.

The ceramic microstructure (Figure 3) showed a good correlation with changes in the structural parameters. At low sintering temperatures in the cleavage there is visible fine fraction (pyrochlore), which disappears when $T_s = 1050°C$. The microstructure is very heterogeneous, the grain size of the main phase varies from 3 to 15 μm, cleavage of the sample occurs mainly along grain boundaries of the main phase. When $T_s = 1020°C$, microstructure becomes more uniform, but there are nuclei of the second phase, which at $T_s = 1050°C$ in the form of spherulite growths reach a diameters of (35–40) μm. A similar pattern was observed in [4] during the annealing of PZT films and explained a migration of lead or its oxide on the film surface. At $T_s = 1080°C$, the sizes of these growths reduced, their number increased, and the distribution on the cleavage became more uniform.

At $T_s = 1100°C$, it was formed homogeneous structure, without inclusions, the average grain size was approximately (5–10) μm, and X-ray data of ceramics had the least defective structure.

In the SEM images of this sample, we can see a great number of pores a circular shape, which indicates that the ceramic sintering took place in the presence of a sufficiently large quantity of liquid. It is known [5], that lead oxide has the melting point ~ 1150°C, which apparently provides the conditions of liquid-phase sintering the ceramics. At $T_s = 1130°C$, spherulites reappear, but smaller, fairly evenly distributed over the cleaved surface, that completely disappear at $T_s = 1150°C$. Starting with $T_s = 1160°C$ and above, dark areas appear in the SEM images of ceramics.

At $T_s = 1200°C$ within these areas, a domain structure with clear boundaries is clearly visible which, as shown in work [6], are crystallographic shear planes. Analysis of the data suggests that increase of V with increasing T_s occurs due to the formation of additional oxygen vacancies.

CONCLUSION

1. By using direct solid-phase synthesis of a mixture of simple oxides followed by sintering by conventional ceramic technology, it is possible to produce ceramic $PbMn_{1/2}Nb_{1/2}O_3$ composition.

2. A variable degree of oxidation of Mn, and high stability of the intermediate pyrochlore phase require the careful selection of temperature and time of synthesis, as well as sintering of ceramics $PbMn_{1/2}Nb_{1/2}O_3$.

3. The results obtained allowed us to select optimal modes of synthesis and sintering PMnN, ensuring lack of impurity phases and high density of the ceramics.

ACKNOWLEDGMENT

This work was financially supported by the Ministry of Education and Science of the Russian Federation (Basic and Design State Tasks Nos. 1927 (213.01-11/2014-21), 213.01-2014/012-SH, and 3.1246.2014/K; Grant of the President of the Russian Federation № MK-3232.2015.2.

REFERENCES

[1] Guinier, A. *Radiocristallographie*. Paris. Dunon. 1956, pp. 1 – 604.
[2] Mirkin, L.A., *Guide to X-ray Analysis of Polycrystalline*. Publishing House of the Physical and Mathematical Literature, Moscow. 1961, pp. 1 – 864 (in Russian).
[3] Andryushina, I.N.; Reznichenko, L.A.; Shilkina, L.A.; Andryushin, K.P.; Dudkina, S.I. *Ceramics International*. 2013, *vol. 39*, 1285-1292.
[4] Senkevich, S.V.; Kanareykin, A.G.; Kaptelov, E.Yu.; Pronin, I.P. Proceedings of the Russian State Pedagogical University named after A.I. Herzen. 2013, No. 157, 101-106.
[5] Kulikov, I.S., *Thermodynamics Oxides*. Moscow, 1986, p. 115-125 (in Russian).
[6] Meitzler, A.H. *Ferroelectrics*. 1975, *vol. 11*, 503-510.

In: Proceedings of the 2015 International Conference …   ISBN: 978-1-63484-577-9
Editors: Ivan A. Parinov, Shun-Hsyung Chang et al.   © 2016 Nova Science Publishers, Inc.

Chapter 5

THE EFFECT OF MECHANICAL ACTIVATION ON THE PROPERTIES OF MULTIFERROIC PBFE$_{0.5}$TA$_{0.5}$O$_3$ CERAMICS

A. A. Gusev[1], I. P. Raevski[2,], E. G. Avvakumov[1],*
V. P. Isupov[1], S. I. Raevskaya[2], S. P. Kubrin[2],
D. A. Sarychev[2], V. V. Titov[2], H. Chen[3], C.-C. Chou[4]
and V. Yu. Shonov[2]

[1]Institute of Solid State Chemistry and Mechanochemistry SB RAS,
Novosibirsk, Russia
[2]Research Institute of Physics and Physical Faculty,
Southern Federal University, Rostov-on-Don, Russia
[3]Institute of Applied Physics and Materials Engineering,
Faculty of Science and Technology, University of Macau,
Macau, China
[4]National Taiwan University of Science and Technology,
Taipei, Taiwan

ABSTRACT

Dielectric properties and ^{57}Fe Mössbauer spectra of PbFe$_0$Ta$_{0.5}$O$_3$ (PFT) multiferroic ceramics obtained from mechanically activated powders have been studied. We found out that such ceramics exhibit a non-relaxor dielectric behavior in contrast to PFT ceramics obtained by usual solid state sintering.

Studies of the temperature dependencies of doublet intensity in the ^{57}Fe Mössbauer spectrum enabled us to determine the temperature of antiferromagnetic phase transition. We found out that the temperature of this phase transition could be changed in a wide range by mechanical activation and subsequent annealing.

---

[*] E-mail: igorraevsky@gmail.com.

Keywords: $PbFe_0Ta_{0.5}O_3$ (PFT) multiferroic ceramics, ^{57}Fe Mössbauer spectra, mechanically activated powders, solid state sintering, antiferromagnetic phase transition

1. INTRODUCTION

Ternary perovskite oxide $PbFe_{0.5}Ta_{0.5}O_3$ (PFT) is a promising multiferroic material [1-3]. However, both ferroelectric and magnetic properties of PFT are not well understood yet. During cooling, PFT undergoes the sequence of phase transitions: from the cubic paraelectric to tetragonal ferroelectric phase, then to the monoclinic ferroelectric phase and, finally, to the G-type antiferromagnetic phase [1-7]. In contrast to other $PbB'^{3+}_{0.5}B''^{5+}_{0.5}O_3$ perovskites with B'^{3+} = Sc, Yb; B''^{5+} = Nb, Ta [8, 9], Fe^{3+} and Ta^{5+} cations in PFT as well as Fe^{3+} and Nb^{5+} in $PbFe_{0.5}Nb_{0.5}O_3$ (PFN), randomly occupy B-sites of the ABO_3 perovskite lattice. However, while in PFN ferroelectric phase transition is non-diffused [10], both single crystals and ceramics of PFT exhibit relaxor-like dielectric behavior [3, 5, 7, 11, 12]. At the same time, in contrast to classical relaxors such as disordered $PbSc_{0.5}Ta_{0.5}O_3$, the $\varepsilon(T)$ maximum in PFT is located not in the cubic phase but rather within the temperature range of tetragonal phase [4, 6]. Recent studies of the effect of the bias field on the properties of PFT ceramics revealed that PFT combines the properties of both ferroelectric with a sharp phase transition and relaxor [13].

Similar to PFN, antiferromagnetic phase transition temperature (T_N) in PFT is more than 100 K higher as compared to lead-free $AFe^{3+}_{0.5}B^{5+}_{0.5}O_3$ (A = Ca, Sr, Ba; B^{5+} = Nb, Ta) perovskites [14, 15].

This dramatic difference is attributed to the possibility of magnetic superexchange via an empty 6p state of Pb^{2+} ions [14, 15] or to the clustering of Fe ions [16]. Both these mechanisms seem to be the origin of the large scattering of T_N values (130 − 180 K) obtained for PFT in different works [2, 3, 7, 17].

One of the possible ways to change the degree of a local compositional ordering and/or clustering of B-site ions in ternary perovskites is mechanical activation of the starting oxides [18]. We have shown recently that T_N value of PFN can be changed substantially by mechanical activation and subsequent annealing [19]. However, there is no data on the effect of mechanical activation on the properties of PFT. The scope of the present work is a study of the dielectric properties and magnetic phase transition temperature of PFT ceramics sintered at different temperatures from the mechanically activated mixture of oxides.

2. METHODS

Stoichiometric mixture of high-purity PbO, Fe_2O_3 and Ta_2O_5 oxide powders was used for mechanical activation. In order to compensate PbO losses during sintering 3wt.% excess of PbO was added to this mixture. Mechanical activation was carried out using the high-energy planetary-centrifugal ball mill AGO-2 under a ball acceleration of 40 g. A mixture of powdered reagents (10 grams) was placed into a steel cylinder together with 200 grams of steel balls and activated for 20 minutes. Annealing of the samples, pressed at 1000 kg/cm^2, was carried out in a closed alumina crucible at different temperatures for 2 hours. X-ray phase analysis was performed using DRON-3 diffractometer and Cu-K$_\alpha$ radiation. Dielectric studies of the samples with fired on Ag electrodes were carried out with the aid of a Wayne Kerr 6500B impedance analyzer in the $10^2 - 10^6$ Hz range in the course of both heating and cooling at a rate of 2 – 3°C/min. Mössbauer ^{57}Fe spectra were measured with the aid of MS-1104EM rapid spectrometer and analyzed using the original computer program UNIVEM. Both dielectric and Mössbauer studies were carried out in the 12K – 320K range using the closed-cycle helium cryostat-refrigerator Janis Ccs-850.

3. EXPERIMENTAL RESULTS AND DISCUSSION

X-ray diffraction studies of the samples pressed from a mechanically activated mixture of PbO, Fe$_2$O$_3$ and Ta$_2$O$_5$ have shown that after annealing for 2 hours at temperature $T_a = 500°C$, practically pure perovskite phase is formed.

Samples annealed at $T_a = 900°C$ and at higher temperatures appeared to form rather dense ceramics with a density of about 92-95% of the theoretical one. We carried out comparative studies of the dielectric properties of i) PFT ceramic sample, sintered from the mechanically activated mixture of oxides at 1100°C and having the mean grain size 3 – 4 µm, and ii) Li-doped PFT ceramic sample, obtained by one-stage sintering at 1080°C [2], which had the the mean grain size 4 – 5 µm. The results obtained are present in Figure 1. As one can see from Figure 1a, Li-doped PFT ceramic sample exhibits a very large dielectric response and a pronounced frequency dispersion of ε. Temperature T_m of the $\varepsilon(T)$ maximum is similar to that reported for PFT ceramics and single crystals in the majority of the published data [3, 5, 7, 11, 12, 13]. The T_m frequency shift, $\Delta T_m = T_m(10^6 \text{ Hz}) - T_m(10^3 \text{ Hz})$ is also close to ΔT_m values reported for PFT ceramics [3, 5, 11-13]. The increase of T_m with the frequency is well fitted with the Vogel-Fulcher relation, typical of relaxors. The parameters of the Vogel-Fulcher relation – an attempt frequency $f_0 = 5 \cdot 10^{11}$ Hz and activation energy $W = 0.018$ eV for Li-doped PFT ceramics are very close to the ones reported for ternary perovskite PbSc$_{0.5}$Ta$_{0.5}$O$_3$ [13]. Thus, dielectric behavior of the Li-doped PFT ceramic sample is similar to the majority of published data on the properties of PFT.

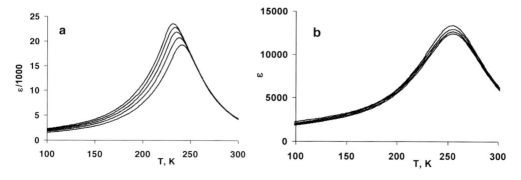

Figure 1. Temperature dependences of dielectric permittivity measured at $10^2, 10^3, 10^4, 10^5$, and 10^6 Hz for (a) one-step-sintered Li-doped PFT ceramics and (b) ceramic sample pressed from the mixture of mechanically activated oxides. Frequency increases from the uppermost curve to the lowermost one.

In contrast to this, PFT ceramic sample sintered from the mechanically activated mixture of oxides exhibits frequency dispersion only in the vicinity of the $\varepsilon(T)$ maximum (Figure 1b). Moreover, the frequency shift of T_m for this sample does not exceed 0.5 K in the frequency range $10^3 - 10^6$ Hz. Such behavior is very similar to the one observed for PFN [10] and is sharply different from all the data reported up to now for PFT ceramics and single crystals. At room temperature Mössbauer ^{57}Fe spectra of all the PFT samples studied were doublets with quadrupole splitting of 0.44 mm/s and isomer shift of 0.4 mm/s (relative to metallic iron), corresponding to the Fe^{3+} ions occupying the octahedral sites of perovskite lattice. This result corroborates the data of the X-ray photoelectron studies showing for PFN single crystals and ceramics the presence of only trivalent iron within the sensitivity of the method [20]. Below the Neel temperature, T_N, the Mössbauer spectrum transforms from doublet to sextet. This transformation is accompanied by a dramatic decrease of the magnitude η of Mössbauer spectra intensity normalized to its value at 300 K (Figure 2a). The position of the abrupt drop in the temperature dependence of η allows one to obtain the value of T_N from the Mössbauer experiment. This method was successfully used to determine the values of T_N for several multiferroics and their solid solutions [2, 5, 6, 14, 15, 17, 19]. The results obtained were very similar to the data obtained by traditional methods such as measuring the temperature dependences of magnetization [2, 6, 22], as well as the line-width of either the electron spin resonance spectrum [21, 22] or Raman spectrum [23]. Temperature dependences of η for the mixture of PbO, Fe_2O_3 and Ta_2O_5 mechanically activated for 20 minutes and subsequently annealed at different temperatures are shown in Figure 2a. The steps on these dependences correspond to the temperature of magnetic phase transition. For comparison, results for the PFT crystal, chosen as a reference, are shown. One can see that both diffusion and position of steps on the $\eta(T)$ curves depend on annealing temperature T_a. The highest T_N value possesses the sample annealed at 650°C. With an increase in T_a above 650°C, T_N starts to decrease, that is, the effect of mechanical activation is annealed. For $T_a = 1100$°C, the value of T_N becomes even lower than the one for the PFT crystal.

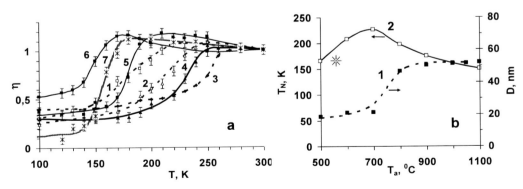

Figure 2. (a) Temperature dependences of η – maximal intensity of the doublet in Mössbauer spectrum related to its value at 300 K for the samples pressed from the mixture of PbO, Fe_2O_3, and Ta_2O_5 mechanically activated for 20 minutes, after annealing for 2 hours at different annealing temperatures T_a: (1) 500°C; (2) 600°C; (3) 650°C; (4) 700°C; (5) 900°C, (6) 1100°C; curve 7 shows the $\eta(T)$ dependence for PFT single crystal; (b) dependences of the mean size D of the X-ray coherent scattering blocks (1) and the Neel temperature T_N (2) on T_a for the same samples; asterisk marks T_N value for PFT single crystal.

Annealing is accompanied by an increase in the size of the X-ray coherent scattering blocks (Figure 2b). Thus for $T_a < 650°C$ the major factor determining the T_N value seems to be the increase of the size of the regions of the perovskite phase (coherent length), while for $T_a \geq 700°C$ it is the decrease in the degree of iron clustering with an increase in T_a.

CONCLUSION

The use of mechanical activation enabled us to modify substantially both ferroelectric and magnetic properties of $PbFe_{0.5}Ta_{0.5}O_3$. Thus, obtained $PbFe_{0.5}Ta_{0.5}O_3$ ceramics do not show the relaxor-like behavior. One of the possible explanations of unusual properties of $PbFe_{0.5}Ta_{0.5}O_3$ may be a coexistence of tetragonal and cubic phases in a wide temperature range. For some samples, an increase of magnetic phase transition temperature by more than 70 K was achieved. This result is in line with the model assuming that the T_N value of $PbFe_{0.5}Ta_{0.5}O_3$ depends substantially on the degree of Fe^{3+} and Ta^{5+} ions clustering [16].

ACKNOWLEDGMENT

The work was partially supported by Russian Foundation for Basic Research (grant 13-03-00869a), research projects 2132 and 3.1137.2014K of the Ministry of education and science of Russian Federation, and Research Committee of the University of Macau under Research and Development Grant for Chair Professor No RDG007/FST-CHD/2012.

REFERENCES

[1] Sanchez, D. A.; Ortega, N.; Kumar, A.; Roque-Malherbe, R.; Polanco, R.; Scott, J. F.; Katiyar, R. S. *AIP Advances,* 2011, vol. 1, 042169-1 – 042169-14.

[2] Raevski, I. P.; Titov, V. V.; Malitskaya, M. A.; Eremin, E. V.; Kubrin, S. P.; Blazhevich, A. V.; Chen, H.; Chou, C.-C.; Raevskaya, S. I.; Zakharchenko, I. N.; Sarychev, D. A.; Shevtsova, S. I. *J. Mater. Sci.,* 2014, vol. 49, 6459 - 6466.

[3] Martinez, R.; Palai, R.; Huhtinen, H.; Liu, J.; Scott, J. F.; Katiyar, R. S. *Phys. Rev. B.,* 2010, vol. 82, 134104-1 - 134104-6.

[4] Geddo-Lehmann, A.; Sciau, P. *J. Phys.: Condens. Matter,* 1999, vol. 11, 1235 - 1245.

[5] Kubrin, S. P.; Raevskaya, S. I.; Kuropatkina, S. A.; Raevski, I. P.; Sarychev, D. A. *Ferroelectrics,* 2006, vol. 340, 155 - 159.

[6] Raevski, I. P.; Molokeev, M. S.; Misyul, S. V.; Eremin, E. V.; Blazhevich, A. V.; Kubrin, S. P.; Chen, H.; Chou, C.-C.; Raevskaya, S. I.; Titov, V. V.; Sarychev, D. A.; Malitskaya, M. A. *Ferroelectrics,* 2015, vol. 475, 52 - 60.

[7] Nomura, S.; Takabayashi, H.; Nakagawa, T. *Jpn. J. Appl. Phys.,* 1968, vol. 7, 600 - 604.

[8] Bokov, A. A.; Shonov, V. Yu.; Rayevsky, I. P.; Gagarina, E. S.; Kupriyanov, M. F. *J. Phys.: Condens. Matter,* 1993, vol. 5, 5491 - 5504.

[9] Raevski, I. P.; Prosandeev, S. A.; Emelyanov, S. M.; Savenko, F. I.; Zakharchenko, I. N.; Bunina, O. A.; Bogatin, A. S.; Raevskaya, S. I.; Gagarina, E. S.; Sahkar, E. V.; Jastrabik, L. *Integrated Ferroelectrics,* 2003, vol. 53, 475 - 487.

[10] Raevski, I. P.; Kubrin, S. P.; Raevskaya, S. I.; Prosandeev, S. A.; Malitskaya, M. A.; Titov, V. V.; Sarychev, D. A.; Blazhevich, A. V.; Zakharchenko, I. N. *IEEE Trans. Ultrason. Ferroelect. Freq. Contr.*, 2012, vol. 59, 1872 - 1878.

[11] Raevskii, I. P.; Eremkin, V. V.; Smotrakov, V. G.; Malitskaya, M. A.; Bogatina, S. A.; Shilkina, L. A. *Crystallogr. Rep.*, 2002, vol. 47, 1007- 1011.

[12] Zhu, W. Z.; Kholkin, A.; Mantas, P. Q.; Baptista, J. L. *J. Eur. Ceram. Soc.*, 2000, vol. 20, 2029 - 2034.

[13] Raevskaya, S. I.; Titov, V. V.;Raevski, I. P.; Lutokhin, A. G.; Zakharov, Yu. N.; Shonov, V. Yu.; Blazhevich, A. V.; Sitalo, E. I.; Chen, H.; Chou, C.-C.; Kovrigina, S. A.; Malitskaya, M. A. *Ferroelectrics,* 2015, vol. 475, 31 - 40.

[14] Raevski, I. P.; Kubrin, S. P.; Raevskaya, S. I.; Titov, V. V.; Sarychev, D. A.; Malitskaya, M. A.; Zakharchenko, I. N.; Prosandeev, S. A. *Phys. Rev. B.*, 2009, vol. 80, 024108-1-024108-6.

[15] Raevski, I. P.; Kubrin, S. P.; Raevskaya, S. I.; Prosandeev, S. A.; Sarychev, D. A.; Malitskaya, M. A.; Stashenko, V. V.; Zakharchenko, I. N. *Ferroelectrics,* 2010, vol. 398,16 - 25.

[16] Raevski, I. P.; Kubrin, S. P.; Raevskaya, S. I.; Sarychev, D. A.; Prosandeev, S. A.; Malitskaya, M. A. *Phys. Rev. B.,* 2012, vol. 85, 224412-1-224412 - 5.

[17] Raevski, I. P.; Kubrin, S. P.; Raevskaya, S. I.; Stashenko, V. V.; Sarychev, D. A.; Malitskaya, M. A.; Seredkina, M. A.; Smotrakov, V. G.; Zakharchenko, I. N.; Eremkin, V. V. *Ferroelectrics*, 2008, vol. 373, 121 - 126.

[18] Gao, X.; Xue, J.; Wang, J. *Mater. Sci. Eng. B.,* 2003, vol. 99, 63 - 69.

[19] Gusev, A. A.; Raevski, I. P.; Avvakumov, E. G.; Isupov, V. P.; Kubrin, S. P.; Chen, H.; Chou, C.-C.; Sarychev, D. A.; Titov, V. V.; Pugachev, A. M.; Raevskaya, S. I.; Stashenko, V. V. The effect of mechanical activation on the synthesis and properties of multiferroic lead iron niobate. In: *Advanced Materials - Physics, Mechanics and Applications. Springer Proceedings in Physics*, Shun-Hsyung Chang, Ivan A. Parinov, Vitaly Yu. Topolov (Eds.). Heidelberg, New York, Dordrecht, London: Springer Cham. 2014, vol. 152, 15 - 26.

[20] Kozakov, A. T.; Kochur, A. G.; Googlev, K. A.; Nikolsky, A. V.; Raevski, I. P.; Smotrakov, V. G.; Yeremkin, V. V. *J. Electron Spectrosc. Related Phenom.*, 2011, vol. 184, 16 - 23.

[21] Laguta, V. V.; Rosa, J.; Jastrabik, L.; Blinc, R.; Cevc, P.; Zalar, B.; Remskar, M.; Raevskaya, S. I.; Raevski, I. P. *Mater. Res. Bull.*, 2010, vol. 45, 1720 - 1727.

[22] Laguta, V. V.; Glinchuk, M. D.; Maryško, M.; Kuzian, R. O.; Prosandeev, S. A.; Raevskaya, S. I.; Smotrakov, V. G.; Eremkin, V. V.; Raevski, I. P. *Phys. Rev. B.*, 2013, vol. 87, 064403-1- 064403 - 8.

[23] Druzhinina, N. S.; Yuzyuk, Yu. I.; Raevski, I. P.; El Marssi, M.; Laguta, V. V.; Raevskaya, S. I. *Ferroelectrics*, 2012, vol. 438, 107 – 114.

In: Proceedings of the 2015 International Conference …   ISBN: 978-1-63484-577-9
Editors: Ivan A. Parinov, Shun-Hsyung Chang et al.   © 2016 Nova Science Publishers, Inc.

Chapter 6

FABRICATION OF PIEZOELECTRIC CERAMICS WITH LOWERED SINTERING TEMPERATURE FOR MEMS

V. V. Eremkin[*]*, V. G. Smotrakov,*
M. A. Marakhovskiy and A. E. Panich
Southern Federal University, Rostov-on-Don, Russia

ABSTRACT

A multilayer design of a piezoelectric transducer, based on reverse piezoelectric effect, is suitable for mass production. The technology based on tape casting enables us to combine sintering of active ceramic layers and forming a system of inner electrodes and to obtain monolithic miniature electro-mechanical systems (MEMS) with a lowered operating voltage. However, the necessity of correlating a ceramics sintering temperature with an electrode melting point appears in this case. The sintering temperatures of the most well-known piezoelectric materials should be essentially decreased. In the present chapter, we analyze the possibility to lower the sintering temperature for two soft piezoelectric materials; one of them corresponds to the Type II, and another belongs to the Type VI according to the official classification of US Department of Defense [1]. We study the effect of a mechanical activation of initial oxide mixtures in a planetary mill on conditions of the synthesis and a dispersity of powders, modes of sintering and a microstructure of the ceramic samples. A homogeneous distribution of elements in the charge, which can be achieved by mechanical or "wet" chemical methods, does not guarantee obtaining the uniphase, chemically uniform product after the thermal treatment. It is concerned with peculiarities of the chemical reaction passage at the solid-state synthesis of PZT-based solid solutions. To perform the full ceramic material synthesis at relatively low temperatures and decreasing the particle size, we use preliminary obtained compounds or solid solutions as the precursors for the subsequent synthesis. Such a mode of the synthesis enables us to change the typical consecution of chemical transformations. The sintering of both materials can be performed at temperatures from 1000°C to 1200°C. However, if in the case of the Type II material acceptable for applications, the functional parameters are reached at 1000°C, then for the Type VI material it occurs only

---

[*] Corresponding Author address: E-mail: smotr@ip.rsu.ru.

at 1100°C. Both the aforementioned materials are suitable for the formation of monolithic MEMS with internal Ag-Pd electrodes.

Keywords: actuator, piezoelectric, multilayered ceramics, lead zirconate titanate, microstructure, high-energy milling, ZTP-19, ZTPSt-2

1. INTRODUCTION

At the creation of monolithic multilayer piezoelectric actuators, the temperature of lead-containing ceramics sintering has not to exceed 1130°C for account of an interaction with internal Ag-Pd electrodes [2], while it is more than 1200°C for the majority of materials. Further lowering of a sintering temperature permits one to decrease the content of Pd in electrodes and to minimize a PbO evaporation into surroundings.

The simplest way consists in using melting at low temperature additions to a synthesized material [3, 4]. However, the presence of a liquid phase complicates the control over the ceramics microstructure, the chemical and phase composition. The other one is concerned with the higher quality of the powders obtained. According to [5], the powders, intended for preparation of an advanced ceramics, must have: a fine particle size (≤ 1 µm), a narrow size distribution, an absence of agglomerates (at least hard ones), the spherical or equiaxial particle shape, the high purity chemical composition and the single phase. Such powders can be obtained by chemical or mechanical methods. The first ones [6, 7] suppose a utilization of soluble inorganic or organometal compounds, containing the elements, which come in the chemical composition of materials. The molecular level of a homogenization for elements is realized in a solution. "Wet" chemical methods differ by the mode used for the transformation of a solution into a homogeneous solid state for a following synthesis. Generally, this methods are more expensive then mechanical ones and demand special chemical reagents. A chemical technology markedly complicates with the rise of the number of the material components; moreover, it is difficult to obtain soluble compounds of some metals. The progress in mechanical methods of the powder preparation is concerned with the use of the small (~ 1mm) balls, manufactured from steady to wear zirconia ceramics [8]. Long (120 – 240 h) milling with the small balls enables us to lead the PZT particle size as far as 0.2 µm. Using of high-speed mills decreases the time of the treatment for several hours.

In this chapter we study the possibility of the low temperature sintering for soft piezoelectric ceramics with the compositions $Pb_{0.95}Sr_{0.05}Zr_{0.53}Ti_{0.47}O_3 + 1$ wt. % Nb_2O_5 (ZTP-19) [9] and $0.98Pb_{0.86}Sr_{0.10}Ba_{0.04}Zr_{0.555}Ti_{0.445}O_3 + 0.02BiNi_{1/4}W_{1/3}O_3$ (ZTPSt-2) [10, 11], which markedly differ in values of a dielectric permittivity and piezoelectric constants d_{ij}.

2. SAMPLE PREPARATION

Starting materials used by us were $BaCO_3$, Bi_2O_3, WO_3, Nb_2O_5 (extra-pure grade), TiO_2 (capacitor grade), PbO, $SrCO_3$ (analytical grade), and ZrO_2 and NiO (pure grade). At different stages of the process, the materials were ground in the Fritsch Pulverisette 5 planetary mill with the yttria-stabilized zirconia grinding bawl and balls at 400 rpm. The phase analysis was performed on the X-ray diffract meter DRON-2.0 (Co K_α radiation). A particle size distribution in the powders was studied by means of the Fritsch laser particle sizer Analysette 22 Compact. The grain size in the ceramic samples (on fracture surfaces) was found using a scanning electron microscopy on JEOL JCM-6390.

$Zr_{0.518}Ti_{0.459}Nb_{0.023}O_{2.012}$ solid solution containing all the elements that come in the position B of the perovskite lattice was used as a precursor for the synthesis of ZTP-19. After a preliminary mixing of ZrO_2, TiO_2 and Nb_2O_5, the charge was treated in the planetary mill during 5 h and calcined at 1300°C during 4 h. After the addition of PbO (including 1 wt. % of the surplus to compensate an evaporation) and $SrCO_3$, as well as mixing in the low-speed ball mill, the powder was grained in the planetary mill for 2 h. The phase composition of the powder corresponds to the morphotropic phase boundary after the synthesis at 750°C (4 h). A part of the powder was additionally treated in the planetary mill for 2 h to decrease the agglomerate size from 2.83 (sample I) down to 1.81 μm (sample II).

ZTPSt-2 contains the appreciable quantity of alkaline-earth ions in the A position of the perovskite unit cell, and these ions are responsible for decreasing the Curie temperature and increasing the dielectric permittivity at the room temperature. Due to the high barium and strontium carbonates decomposition temperature, it is difficult to guarantee the complete material synthesis when the traditional route based on simple oxides (carbonates) is chosen. Earlier we have encountered this problem [12] in an attempt to prepare anisotropic piezoelectric ceramics, belonging to the $PbTiO_3 - CaTiO_3$ system. Similar to the work [12], where $CaTiO_3$ was used as a precursor, in the present work $Ba_{0.286}Sr_{0.714}TiO_3$ was preliminarily synthesized. The synthesis temperature was 1150°C, exposure occurred for 4 h. $Ba_{0.286}Sr_{0.714}TiO_3$ and other oxides, including 1 wt. % of the surplus of PbO, were mixed in a ball mill. The charge was treated in the planetary mill for 2 h. The phase composition corresponds to the morphotropic phase boundary with the equal quantity of tetragonal and rhombohedral phases after firing at 850°C for 4 h. A part of the powder was additionally treated in the planetary mill for 2 h to decrease the agglomerate size from 2.41 (sample III) down to 0.36 μm (sample IV). The primary particle size for all powders (I – IV) does not exceed 0.3 μm.

The ceramics sintering was performed at the atmosphere of the PbO vapor. The heating rate was 200°C/h, and the samples were at the highest temperature for 3 h (ZTP-19) or 2 h (ZTPSt-2). Dependencies of the density ρ and the middle grain size D_g of ceramics from the sintering temperature are shown in Figures 1 and 2.

It is clear that the methods proposed enable us to obtain dense ceramics of both materials in the temperature range from 1000°C up to 1250°C. The grain size increases from 0.8 μm to 2.3 μm for ZTP-19 and from 1 μm to 5 μm for ZTPSt-2 at the increasing sintering temperature. Figure 2 also suggests that the more disperse powder causes the lower grain size. It is probably concerned with the increase of the PbO evaporation from the samples and the lack of a liquid phase at sintering. The ceramics, prepared from these powders, are also characterized by lowered properties.

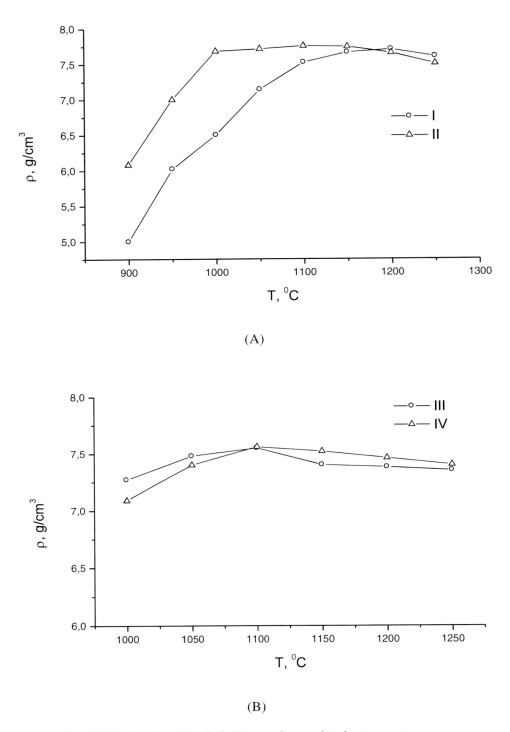

Figure 1. Density of ZTP-19 (A) and ZTPSt-2 (B) ceramics vs. sintering temperature.

Fabrication of Piezoelectric Ceramics with Lowered Sintering Temperature for MEMS 41

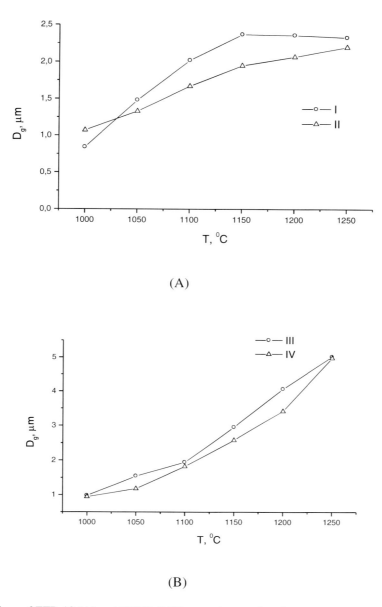

Figure 2. Grain sizes of ZTP-19 (A) and ZTPSt-2 (B) ceramics *vs.* sintering temperature.

3. RESULTS AND DISCUSSION

The samples intended for measurements had the form of disks with the diameter 10 mm and the thickness 1 mm. The Ag electrodes were plotted on the opposite disk surfaces. They were poled in the electric field $E = 10$ kV/cm at cooling through the Curie temperature. The dielectric permittivity $\varepsilon_{33}^T/\varepsilon_0$ was determined at the frequency 1 kHz. The piezoelectric coefficient d_{31}, planar electromechanical coupling factor k_p and mechanical quality factor Q_m were measured by the resonance method. The experimental results obtained are shown in Figures 3-6.

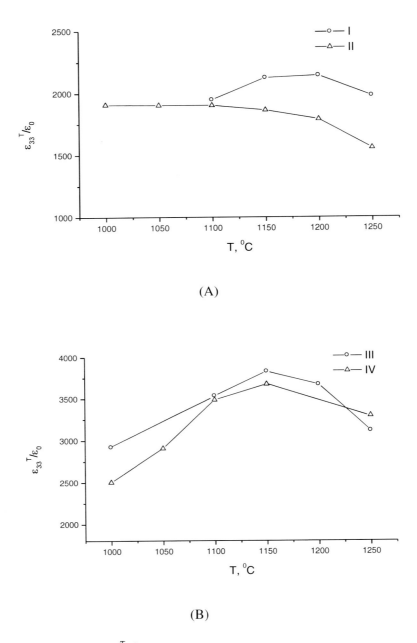

Figure 3. Dielectric permittivities $\varepsilon_{33}^T/\varepsilon_0$ of ZTP-19 (A) and ZTPSt-2 (B) ceramics vs. sintering temperature.

Fabrication of Piezoelectric Ceramics with Lowered Sintering Temperature for MEMS

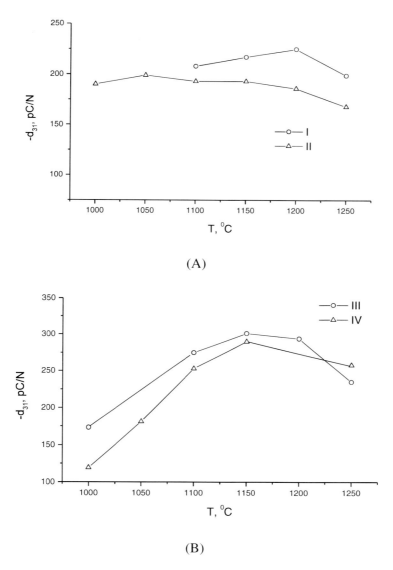

Figure 4. Piezoelectric constant d_{31} of ZTP-19 (A) and ZTPSt-2 (B) ceramics vs. sintering temperature.

The piezoelectric ceramics properties on the one hand are directly proportional to grain size, and on the other hand deteriorate at a high temperature on the reason of ceramics decomposition caused mainly by the PbO loss. These factors operate in opposite directions with the increasing sintering temperature. It is clear, that for the ZTP-19 ceramics the PbO loss is the main factor that defines the character of the above-shown dependencies. The difference in values of $\varepsilon_{33}^T/\varepsilon_0$, $|-d_{31}|$ for the samples II and I is more expressive than that for the samples III and IV. Moreover, maxima of the $\varepsilon_{33}^T/\varepsilon_0$, k_p and $|-d_{31}|$ dependencies on sintering temperature for samples II and I are related to different temperatures. The more fine powder II leads to an appearance of maximal properties at 1000 – 1050°C, and the more coarse powder I causes the maximal properties only at 1150 – 1200°C. For the ZTPSt-2 ceramics the PbO loss also exists, however up to 1150°C relations between the sintering temperature and the functional parameters mainly depend on the grain size increase, which is more than two time stronger for this material in comparison with the ZTP-19 ceramics. Since further softening of piezoelectric materials is generally realized at the expense of substitution of a part of $PbZrO_3$ on a relaxor ferroelectric in solid solution, one can expect the decrease of the PbO loss in this case and the increase of differences in properties of the ceramics prepared at low and high temperatures.

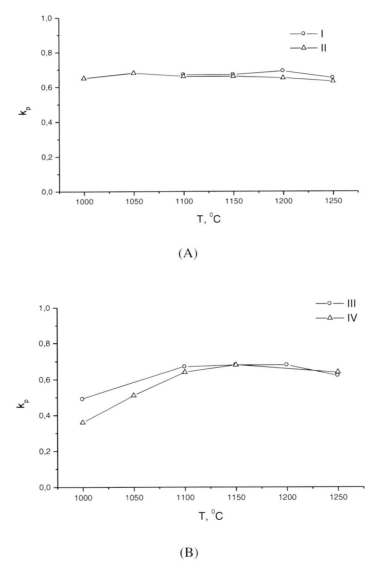

Figure 5. Planar electromechanical coupling factor k_p of ZTP-19 (A) and ZTPSt-2 (B) ceramics vs. sintering temperature.

Fabrication of Piezoelectric Ceramics with Lowered Sintering Temperature for MEMS   45

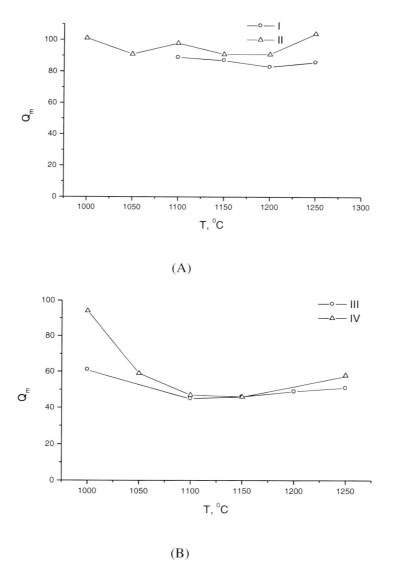

Figure 6. Mechanical quality factor Q_m of ZTP-19 (A) and ZTPSt-2 (B) ceramics *vs.* sintering temperature.

Dependencies of Q_m on the sintering temperature (Figure 6) are also indicative of the more "hard" properties, i. e., the lowered mechanical losses, for piezoelectric ceramics obtained from the highly disperse powder. This effect is probably concerned not only with the lower grain size but also with the formation of additional vacancies in the oxygen sub-lattice owing to the more active PbO evaporation. In contrast to the ZTP-19 ceramics, ZTPSt-2 has the clear minimum on the dependence of Q_m on the sintering temperature.

CONCLUSION

Thus, for both piezoelectric ceramic materials, the temperature of sintering can decrease to 1000°C due to high-energy milling at the synthesis. However, if in the case of ZTP-19 the necessary functional parameters, which are acceptable for applications, are obtained at this temperature, then for ZTPSt-2 it occurs only at 1100°C. This is probably determined with the fact that ZTPSt-2 grain size changes are more than two times stronger than ZTP-19 ones in the sintering range examined. Both materials are suitable for monolithic multilayer structures with internal Ag-Pd electrodes. At the same time, the minimal sintering temperature, which secures necessary for practice properties, has to reveal tendency of increasing for materials with larger values of dielectric permittivity and coarse grains, and these circumstances would limit the use of these materials in monolithic multilayer transducers.

REFERENCES

[1] *Piezoelectric Ceramic Material and Measurements Guidelines for Sonar Transducers.* Mil. Std. 1376B (SH), 1995.

[2] Wersing, W.; Wahl, H.; Schnoller, M. *Ferroelectrics.* 1988, *vol. 87*, 271 – 294.

[3] Takahashi, S. *Japan. J. Appl. Phys.* 1980, *vol. 19*, 771-772.

[4] Wang, C.-H.; Wu, L. *Japan. J. Appl. Phys.* 1993, *vol. 32*, 3209-3213.

[5] Rahaman, M. N. *Ceramic Processing and Sintering.* Marcel Dekker, New York, Basel, Hong Kong, 1995.

[6] Bernier, J. C. *Mat. Sci. Engineering A.* 1989, *vol. 109*, 233-241.

[7] Rhine, W. E.; Bowen, H. K. *Ceram. Intern.* 1991, *vol. 17*, 143-152.

[8] Tashiro, S.; Sasaki, N.; Tsuji, Y.; et al. *Japan. J. Appl. Phys.* 1987, *vol. 26(26-2)*, 142 – 144.

[9] Glozman, I. A. *Piezoceramics.* Moscow: Energy, 1972 (In Russian).

[10] Savenkova, G. E.; Didkovskaya, O. S.; Klimov, V. V.; Venevtsev, Yu. N. Piezoceramic Material. USSR Patent SU567706 (In Russian).

[11] *Solid Solution on the Base of Lead, Strontium Zirconate-Titanate ZTPSt-2.* Technical Conditions 6-09-27-145-86 (In Russian).

[12] Smotrakov, V. G.; Eremkin, V. V.; Doroshenko, V. A.; et al. *Inorganic Materials.* 1994, *vol. 30*, 230 – 231.

In: Proceedings of the 2015 International Conference ... ISBN: 978-1-63484-577-9
Editors: Ivan A. Parinov, Shun-Hsyung Chang et al. © 2016 Nova Science Publishers, Inc.

Chapter 7

NEW CERAMIC MATRIX PIEZOCOMPOSITES FOR UNDERWATER ULTRASONIC APPLICATIONS

E. I. Petrova[*]*, A. A. Naumenko, M. A. Lugovaya and A. N. Rybyanets*
Institute of Physics, Southern Federal University,
Rostov-on-Don, Russia

ABSTRACT

A new family of polymer-free ceramic matrix piezocomposites with properties combining better parameters of PZT, PN type ceramics, and 1-3 composites for wide-band underwater ultrasonic transducers is presented. New "damped by scattering" ceramic matrix piezocomposites are characterized by previously unachievable low mechanical quality factor (Q^I_M = 2-5) combined with high piezoelectric (d_{33} = 250-350) and electromechanical (k_t = 0.45-0.5) parameters, high Curie temperature (T_C = 340°C), and low acoustic impedance (Z_A = 15-20 MRayl) in a wide working frequency range (0.2-12 MHz). Complex sets of elastic, dielectric, and piezoelectric parameters of new piezocomposites were systematically studied using impedance spectroscopy approach and ultrasonic method. A line of wide-band underwater ultrasonic transducers with high sensitivity and resolution were manufactured and tested. High acoustic efficiency, low crosstalk, low mechanical Q, and technological flexibility are the main features of the piezocomposites. In particular, the gain will be highly sensitive combined with well-damped signals. Additional advantages of the developed ceramic matrix piezocomposites are the possibility of controllable change of main properties in a wide range, compatibility with standard fabrication technologies and processing flexibility.

---

[*] E-mail: harigamypeople@gmail.com.

1. INTRODUCTION

Limited transducer materials are currently available for use in ultrasonic transducers for non-destructive testing (NDT) and medical diagnostics systems, considering combinations of high sensitivity (efficiency), resolution (bandwidth) and operational requirements (working temperature). There are currently no commercially available piezoceramic materials that offer high piezoelectric properties together with low Q, low acoustic impedance, electromechanical anisotropy and high operating temperatures (>250°C) [1]. Those that are available offer either high piezoelectric properties (lead zirconate titanate – PZT, lead magnesium and zinc niobates – PMN, PZN) or high electromechanical anisotropy (lead titanate – PT) and low Q, low acoustic impedance (lead metaniobate – PN) [2].

Therefore, efforts were made to replace lead metaniobate by 1-3 piezocomposite materials. However, to date the production cost of these components well exceeds the common price of a complete standard transducer. Moreover, there are considerable technical limitations with this type of transducer, particularly with respect to the maximum allowable ambient temperature and pressure [3]. Over the past years, considerable advances were made to improve the mechanical properties of ceramics using ceramic composite approaches. Numerous technologies based on incorporation of functional ceramics into structural ones and vice-versa were developed, and novel design ideas were applied in the field of functional ferroelectric ceramics [4, 5]. However, the problem of property trade-off (i.e., the deterioration of electromechanical properties) remains unsolved. Commercialization of composite materials has lead also to the development of new concepts of material and ultrasonic transducers designing [6]. This chapter presents a new family of polymer-free ceramic matrix piezocomposites with properties combining better parameters of PZT, PN type ceramics and 1-3 composites for wide-band underwater ultrasonic transducers.

2. CERAMIC PIEZOCOMPOSITES PZT/α-AL₂O₃

First, a line of chemically, thermally, and technologically compatible ceramic matrix and scattering phase materials were chosen. Different kinds of PZT type piezoceramic powders and various refractory crystals (including stabilized α-Al$_2$O$_3$ powders) were used as initial components for the ceramic composite preparation. The size and shape of the scattering particles were chosen as a compromise between maximal scattering and minimal interfacial area to minimize chemical interactions between these two phases.

Mixing of the powders was carried out using a specially developed method to ensure homogeneous mixture of the components and to prevent destruction and disintegration of powders particles [2]. Powders mixed in different proportions were granulated and cold pressed into cylinders of $\varnothing 23 \times 20$ mm^2. Special pressing and firing regimes as well as porosity agents were used for the formation of microporous piezoceramic matrices. Sintering of the green bodies was carried out at special thermal profiles to prevent cracking because of the different shrinkage and thermal expansion of the composite components. The methods of powder preparation, composite sintering and piezocomposite sample fabrication, as well as technological regimes are described in details in [6, 7].

Density and shrinkage coefficients of the sintered composite bodies were measured by weighing and measuring the volume, as well as by the hydrostatic weighing method. Theoretical density and porosity of the composites were calculated from theoretical densities and volume fractions of the components and the measured density.

For the electrical measurements, Ag-paste was printed on both sides of the grinded elements and then fired at 700°C for 5 min. The poling treatment was carried out in the silicone oil at the same temperature and electrical field as for the corresponding dense piezoceramics. Specimens for optical microscopy were grounded and polished using a specially designed procedure for the composites comprising components with different hardness.

The complex electrical coefficients of the piezocomposite elements were determined by the impedance spectroscopy method using the Piezoelectric Resonance Analysis (PRAP) software [8, 9]. Measurements were made on Solartron Impedance/Gain-Phase Analyzer SL 1260. Pulse-echo and through-transmit measurements of ultrasonic transducers were made using LeCroy digital oscilloscope and Olympus pulser/receiver. The microstructure of polished, chemically etched, and chipped surfaces of composite samples was observed with optical (NeoPphot-21) and scanning electron microscopes (SEM, Karl Zeiss).

3. RESULTS AND DISCUSSION

The composite samples composed of the "soft" PZT matrix and randomly distributed α-Al_2O_3 particles with a mean size ~ 200 µm and volume fractions from 9 up to 26 vol. % [10] were chosen for detailed measurements. Optical micrographs of polished surface of PZT/α-Al_2O_3 samples with 10% and 20% volume fractions of α-Al_2O_3 particles are shown in Figure 1.

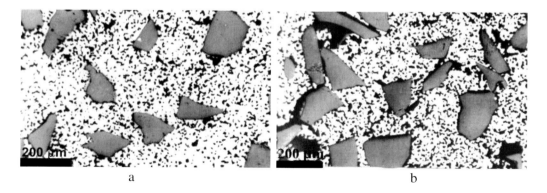

Figure 1. Optical micrographs of polished surface of composite samples with 10 vol. % (a) and 20 vol. % (b) of α-Al_2O_3 particles.

Figures 2 shows impedance spectra and PRAP approximations for thickness modes of PZT/α-Al_2O_3 composite disks at 10% and 20% volume fractions of α-Al_2O_3 particles, respectively.

Figures 2 demonstrates clearly that at low concentrations of scattering particles, the measured impedance spectrum is modulated by individual scattering events with the period corresponding to the average diameter of scattering particles. An increase in scattering particles fraction up to 20 vol. % leads to smoothing of impedance spectra as a result of multiple scattering averaging.

Complex constants of A850L-20 vol.% disk (Ø19.95 × 0.8 mm^2) obtained using PRAP analysis for thickness extensional and radial modes as well as corresponding IEEE Standard results, are summarized in Tables 1 and 2. Additional physical parameters of A850L-20 are as follows: Z_A = 22.8 MRayl, ρ = 6.4 g/cm^3, V_t = 3568 m/s, d_{33}^{quasi} = 300 pC/N, T_C = 340°C.

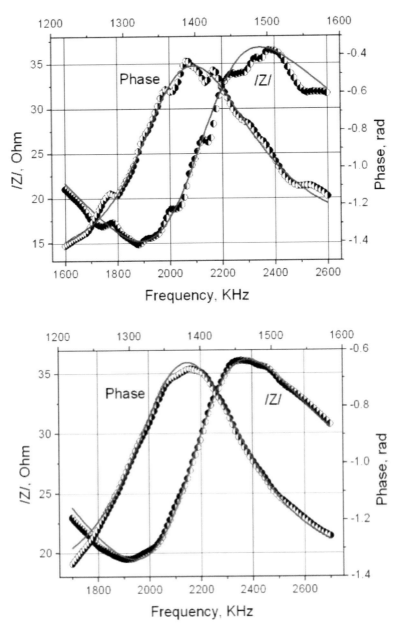

Figure 2. Impedance spectra and PRAP approximations for thickness extensional mode of A850L-10 vol. % and A850L-20 vol. % disks (Ø19.95 × 0.8 mm^2).

Table 1. PRAP Analysis Results for TE Mode of A850L-20 disk

PRAP Parameter	Real	Imaginary	IEEE Standard	% Error
f_p (Hz)	$2.26 \cdot 10^6$	340289	$2.29 \cdot 10^6$	1.3
f_s (Hz)	$2.01 \cdot 10^6$	304601	$2.01 \cdot 10^6$	0.0
k_t	0.492734	-0.00138	0.513653	4.3
$c_{33}{}^D$ (N/m²)	$8.15 \cdot 10^{10}$	$2.52 \cdot 10^{10}$	$8.58 \cdot 10^{10}$	5.3
$c_{33}{}^E$ (N/m²)	$6.17 \cdot 10^{10}$	$1.92 \cdot 10^{10}$	$6.31 \cdot 10^{10}$	2.3
e_{33} (C/m²)	10.6185	1.31335	-	-
h_{33} (V/m)	$1.91 \cdot 10^9$	$3.29 \cdot 10^8$	-	-
$\varepsilon_{33}{}^S$ (F/m)	$5.52 \cdot 10^{-9}$	$-2.64 \cdot 10^{-10}$	-	-

Table 2. PRAP Analysis Results for Radial Mode A850L-20 disk

PRAP Parameter	Real	Imaginary	IEEE Standard	% Error
$f_s{}^1$ (Hz)	105834	1737.27	106037	0.2
$f_p{}^1$ (Hz)	110947	1623.38	111192	0.2
$f_s{}^2$ (Hz)	275536	4522.35	275194	0.1
$S_{11}{}^E$ (m²/N)	$1.72 \cdot 10^{-11}$	$-5.66 \cdot 10^{-13}$	$1.75 \cdot 10^{-11}$	1.7
$S_{12}{}^E$ (m²/N)	$-5.84 \cdot 10^{-12}$	$1.92 \cdot 10^{-13}$	$-6.14 \cdot 10^{-12}$	5.1
$-d_{31}$ (C/N)	$7.35 \cdot 10^{-11}$	$-3.59 \cdot 10^{-12}$	$7.38 \cdot 10^{-11}$	0.4
$\varepsilon_{33}{}^T$ (F/m)	$8.27 \cdot 10^{-9}$	$-2.38 \cdot 10^{-10}$	$8.29 \cdot 10^{-9}$	0.2
k_p	0.339052	-0.00613	0.340468	0.4
σ^P	0.339174	$8.34 \cdot 10^{-6}$	0.351532	3.6
e_{31} (C/m²)	6.46184	-0.10351	6.51219	0.8
$S_{66}{}^E$ (m²/N)	$4.61 \cdot 10^{-11}$	$-1.51 \cdot 10^{-12}$	$4.72 \cdot 10^{-11}$	2.4
$C_{66}{}^E$ (N/m²)	$2.16 \cdot 10^{10}$	$7.10 \cdot 10^8$	$2.12 \cdot 10^{10}$	1.9

Figure 3 shows ultrasonic velocities of thickness V_t and longitudinal $V_1{}^E$ vibration modes of polarized composite elements, measured by the resonance method; longitudinal velocity $V^{isotrop}$ of non-polarized elements, measured by ultrasonic method (1 MHz); as well as theoretical ρ^{theor} and measured ρ^{exper} densities of ceramic composite A850L as a function of volume fraction V, % of the scattering phase.

It is readily observed in Figure 3 that the V_t, $V^{isotrop}$ and $V_1{}^E$ show a different behavior. V_t, $V^{isotrop}$ decrease strongly, whereas $V_1{}^E$ increases slightly with the volume fraction of scattering particles. These variances are caused by the difference in the corresponding elastic moduli ($C_{33}{}^D$, $S_{11}{}^E$ and C^{isotr}) behavior, discussed below. Theoretical density ρ^{theor} calculated from the densities and volume fractions of the composite components decreases linearly with the V, %, whereas the measured density ρ^{exper} decreases more rapidly. The discrepancy between the ρ^{exper} and ρ^{theor} behavior is caused by the fact that the increased scattering particles volume fraction inhibits shrinkage of the composite and leads to microporosity formation during the sintering process. Maximal relative porosity at $V = 26$ vol. % reaches 14%.

Figure 4 shows the dependencies of the real parts of elastic moduli $C'_{33}{}^D$, $C'_{33}{}^E$, $S'_{11}{}^E$ and corresponding mechanical quality factors $Q_M = S'_{11}{}^E / S''_{11}{}^E$, $C'_{33}{}^E / C''_{33}{}^E$ and $C'_{33}{}^D / C''_{33}{}^D$ on scattering phase volume fraction V, %. Figure 13 shows clearly that $S'_{11}{}^E$ is practically constant in measured V, % range, while $C'_{33}{}^D$, $C'_{33}{}^E$ decreases drastically with the V, %.

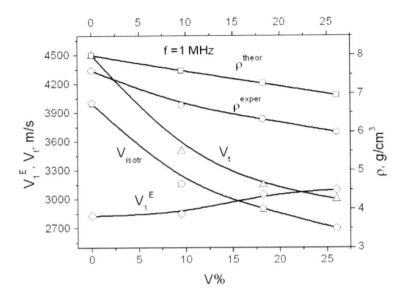

Figure 3. Velocities of thickness V_t and longitudinal $V_1{}^E$ vibration modes, longitudinal velocity V_{isotr}, theoretical ρ^{theor} and measured ρ^{exper} densities versus volume fraction V, % of the scattering particles in PZT matrix.

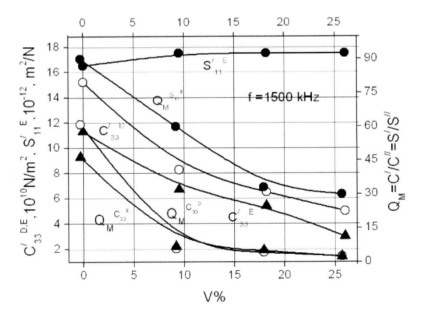

Figure 4. Elastic moduli $C'_{33}{}^D$, $C'_{33}{}^E$, $S'_{11}{}^E$ and corresponding mechanical quality factors $Q_M = S'_{11}{}^E / S''_{11}{}^E$, $C'_{33}{}^E / C''_{33}{}^E$ and $C'_{33}{}^D / C''_{33}{}^D$ versus volume fraction of scattering particles V, % in PZT matrix.

These differences in elastic moduli behavior are related to the inequality of composite structure in radial and thickness directions. For standard piezocomposite disk (Ø20 × 1 mm^2), we have 2-3 rigid scattering particles (Ø200 µm) randomly distributed along the thickness, whereas 50-60 particles are easily packed along the diameter. As a result, the stiffness in thickness direction drops drastically, while compliance in radial direction is practically constant. Mechanical quality factor $Q_M^{rad} = S'_{11}{}^E / S''_{11}{}^E$ decreases from 90 (dense PZT) to 30 (26V, % composite), while $Q_M^{thick} = C'_{33}{}^E / C''_{33}{}^E$ and $C'_{33}{}^D / C''_{33}{}^D$ drop from 45 and 60, respectively, down to 3. This change is caused by the difference in measurement frequency 100 kHz and 1500 kHz for radial and thickness modes, respectively, and as a result, by the change in scattering mechanism from Rayleigh ($\lambda >> D$) to stochastic type ($\lambda \sim D$) and scattering character from multiple scattering on randomly (radial mode) and regularly distributed particles (thickness mode). Scattering on micro pores is negligible in this frequency range.

We can summarize the preceding frequency and volume fraction dependencies of the main composite parameters as follows: the optimal materials for extreme damping at reasonable electromechanical parameters are the ceramic composites with $15 - 25$ vol.% of α-Al$_2$O$_3$ particles in soft PZT matrix at λ/D ratio ~ 10.

In conclusion, the advantages of the ceramic composites were demonstrated with reference to implemented ultrasonic transducers. It was shown that a low-Q and low-Z_A composite element demonstrates high-damped and high amplitude signals even without matching and damper. Notice also that the transducer made of new ceramic composites exhibits a smaller ripple, a larger bandwidth signals, and is free of harmonics, i.e., the spatial resolution for ultrasonic transducers is higher and its dynamic range is broader. Keeping the optimal scatterer size-to-wavelength ratio, a line of wide-band underwater acoustics and NDT ultrasonic transducers with high sensitivity and resolution in frequency range $0.1 - 10$ MHz were fabricated using new ceramic matrix composites.

CONCLUSION

A new family of low-Q ceramic matrix piezocomposites was developed using the "damping by scattering" approach. The main advantages of the new PZT/α-Al$_2$O$_3$ piezocomposites are the high acoustic efficiency, low crosstalk, and low mechanical Q. In particular, the gain will be highly sensitive combined with well-damped signals. Additional advantages of the developed piezocomposites are the possibility of the controllable changes of main properties in a wide range, compatibility with the standard fabrication technologies, and processing flexibility.

ACKNOWLEDGMENTS

Work supported by the RSF grant No. 15-12-00023.

REFERENCES

[1] Rybyanets, A.N. *IEEE Trans. UFFC*. 2011, *vol. 58*, 1492-1507.

[2] Rybyanets, A.N. Ceramic Piezocomposites: Modeling, Technology, and Characterization. In: *Piezoceramic Materials and Devices*, Parinov, I. A. (Ed.). New York: Nova Science Publishers. 2010, Chapter 3, 113-174.

[3] Rybyanets, A.N.; Rybyanets, A. A. *IEEE Trans. UFFC*. 2011, *vol. 58*, 1757-1774.

[4] Rybyanets, A.N. *Feroelectrics*. 2011, *vol. 419*, 90-96.

[5] Rybianets, A.N., Nasedkin, A.V., Turik, A.V. *Integrated Ferroelectrics*. 2004, *vol. 63*, 179-182.

[6] Rybianets, A.N.; Tasker, R. *Ferroelectrics*. 2007, *vol. 360*, 90-95.

[7] Rybyanets, A.N.; Naumenko, A.A.; Shvetsova, N.A. Characterization Techniques for Piezoelectric Materials and Devices. In: *Nano- and Piezoelectric Technologies, Materials and Devices*, Parinov I. A. (Ed.). New York: Nova Science Publishers. 2013, Chapter 1, 275-308.

[8] Rybianets, A.N.; Nasedkin, A.V. *Ferroelectrics*, 2007, *vol. 360*, 57-62.

[9] Rybianets, A.N.; Razumovskaya, O.N.; Reznitchenko, L.A.; Komarov, V.D.; Turik, A.V. *Integrated Ferroelectrics*. 2004, *vol. 63*, 197-200.

[10] Rybianets, A. N. New "damped by scattering" ceramic piezocomposites with extremally low QM values. *Ferroelectrics,* 2007, *vol. 360*, 84-89.

In: Proceedings of the 2015 International Conference ... ISBN: 978-1-63484-577-9
Editors: Ivan A. Parinov, Shun-Hsyung Chang et al. © 2016 Nova Science Publishers, Inc.

Chapter 8

LEAD-FREE PIEZOCERAMIC MATERIALS FOR ULTRASOUND APPLICATIONS

M. V. Talanov[*], L. A. Shilkina and L. A. Reznichenko

Research Institute of Physics, Southern Federal University,
Rostov-on-Don, Russia

ABSTRACT

Ceramics of the system $(1-x-z)NaNbO_3-xKNbO_3-zCuNb_2O_6$ with $x = 0.05-0.50$ and $z = 0.025-0.050$ were prepared by solid state rection synthesis, and then were sintered by conventional ceramic technology. The phase diagram of studied solid solutions was constructed for the first time, and the structure-properties relationship was also established. It was found that the behavior of the major electrophysical parameters of the investigated solid solutions at the change of $KNbO_3$ content correlates with the position of the morphotropic phase boundaries. The prospects of the studied materials application in ultrasound technique were shown.

Keywords: lead-free ceramics, mechanical quality factor, piezoelectric properties, phase diagram, morphotropic phase boundaries

1. INTRODUCTION

Today, ceramic materials based on the binary system $Pb(Zr_{1-x},Ti_x)O_3$ (PZT) are used in the majority of modern piezoelectric devices. It is characterized by the presence of morphotropic phase boundary (MPB) at $x \sim 0.48$, near which the piezoelectric properties are maximal. However, they contain more than fifty percent of the very toxic element lead, which damages human health and the environment. One of the main problems of modern materials science is searching for alternative environmentally friendly piezoelectric materials which do not contain lead, and this searching meets the requirements of a modern legal framework [1].

---

[*] E-mail: tmikle-man@mail.ru.

A prospective basis of such materials are ceramics of the binary system $(Na_{1-x}, K_x)NbO_3$ (KNN) near MPB at $x \sim 0.50$ [2], and parameters of these ceramics (piezoelectric coefficient $d_{33} \sim 80$ pC /N, planar electromechanical coupling factor $K_p \sim 0.36$, etc.) are close to those of some PZT compositions. However, due to the higher values of sound velocity within, they are superior to the latter when used with a microwave technique.

Technological difficulties in obtaining the complex niobium oxides and their solid solutions (SSs), primarily the volatility of the alkali metals at high temperatures and hygroscopicity of starting reagents have been overcome by designing systems with different cuprates (CuO, $K_{5.4}Cu_{1.3}Ta_{10}O_{29}$, $K_4CuNb_8O_{23}$ and $CuNb_2O_6$) [3-11]. Featuring the latest methods which are significantly reduced of optimum sintering temperature, the preservation of the stoichiometry of a given composition increases the relative density of the ceramics.

Corresponding with the foregoing, the aim of the present chapter is to identify the correlation of the dielectric, piezoelectric and mechanical properties of the ceramics system $(1-x-z)NaNbO_3-xKNbO_3-zCuNb_2O_6$ at $x=0.05-0.50$ and $z=0.025-0.075$ with the phase diagram of this system.

2. METHODS

The objects of our study are ceramics of the $(1-x-z)NaNbO_3- xKNbO_3-zCuNb_2O_6$ system with $x=0.05-0.50$ and $z=0.025-0.075$, prepared by solid state reaction synthesis and sintered by conventional ceramic technology [12]. Samples were represented by disks 10 mm in diameter and 1 mm thick with silver-bearing electrodes applied (by double firing) to the flat butt end surfaces.

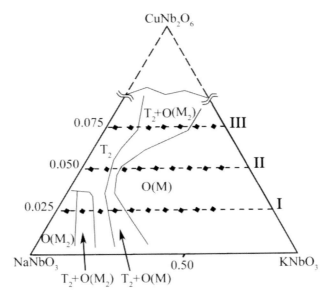

Figure 1. Gibbs triangle of the system $(1-x-z)NaNbO_3-xKNbO_3-zCuNb_2O_6$ with marked compositions studied and phase boundaries on it.

Figure 1 shows a Gibbs triangle of the $(1-x-z)NaNbO_3-xKNbO_3-zCuNb_2O_6$ system with the three quasibinary cuts with $z = 0.025, 0.050$ and 0.075 marked on it and dedicated phase boundaries. The cuts are parallel to the edge of the binary system KNN.

X-ray diffraction studies were performed by the powder diffraction method using a DRON-3 diffractometer ($Co_{K\alpha}$ radiation, Bragg–Brentano focusing). The main electrophysical parameters [dielectric constant of poled samples ($\varepsilon_{33}^{T}/\varepsilon_0$, where ε_0 is dielectric permittivity of the vacuum), planar electromechanical coupling factor (K_p), mechanical quality factor (Q_m), and sound velocity (V_E^1)] were determined by the resonance–antiresonance method using an impedance analyzer Wayne Kerr 6500B. The small-signal piezoelectric coefficient d_{33} was measured using a Berlincourt-type d_{33} meter APC YE2730A.

3. EXPERIMENTAL RESULTS AND DISCUSSION

In cut I at $0 \leq x \leq 0.1$, SSs have a orthorhombic (O) symmetry with a monoclinic (M) perovskite cell, i.e., the phase $O(M_2)$ (the subscript shows the muliplication of the b monoclinic cells parameter). In the range $0.05 < x \leq 0.20$, SSs show tetragonal (T) symmetry with double perovskite cell parameters, i.e., the phase T_2 and in the range $0.15 < x \leq 0.5$, SSs demonstrated O-phase without doubling of the b monoclinic cells parameter, i.e., the phase O(M).

There is the change in symmetry of the unit cell in two morphotropic regions (MRs) which are located in the ranges $0.05 < x \leq 0.10$ and $0.15 < x \leq 0.20$. In the first of these MRs $O(M_2)$ and T_2 phases coexist. In the second MR, T_2 and O(M) phases are coexisting.

In cut II in the range $0.05 \leq x \leq 0.10$ T_2 phase exists. At $x \geq 0.15$ SSs have O(M) symmetry (multiplication of the perovskite cell axis in the O-region is not determined due to the presence of extraneous phase lines on the X-ray pattern). In the range $0.10 < x < 0.15$, there is MR in which T_2 and O (M) phases coexist. In the range $0.15 < x \leq 0.25$, SSs are very inhomogeneous and there is a mixture of O or low-symmetry phases with similar parameters of the unit cell. In the range $0.25 < x \leq 0.35$, two phases, namely O(M) and O(M') with different parameters of the unit cell coexist. At $0.35 < x \leq 0.40$, there is a single-phase O(M) area, and in the range $0.40 < x \leq 0.45$ together with the phase of O(M) exists another phase, whose symmetry could not be determined.

There is the following form of the phase diagram in cut III: T-phase exists in the range of $0.05 \leq x \leq 0.25$, phase O(M) there is in the range $0.10 < x \leq 0.35$. MR in which T and O(M) phase coexist exists in the range $0.10 < x \leq 0.25$. In the range $0.30 < x \leq 0.40$ there is an area in which the two phases O(M) and O(M') coexist.

This diversity of the O-phases in sections II and III can be attributed to the fact that there are three MRs in which O(M)-phases coexist with different multiplications of perovskite cell axes in the binary system $(1-x)NaNbO_3-xKNbO_3$ (range $0.20 < x < 0.43$) [13]. Adding a third component, namely $CuNb_2O_6$ led to the fact that almost all of this MRs transformed into a wide range of MRs with consistently occurring phase transitions. In addition, a certain influence on the phase composition had a stoichiometry violation in the investigated SSs.

Figure 2 shows the major electrophysical parameters of the three cuts of the system studied by the $KNbO_3$ (x) content.

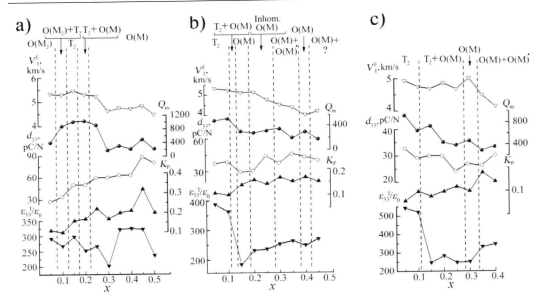

Figure 2. Dependencies of the major electrophysical parameters on the KNbO$_3$ (x) content: cut I (a), cut II (b) and cut III (c).

It is seen that the dependence $\varepsilon_{33}^T/\varepsilon_0$ (x) of SSs in cut I attains extreme value with a diffuse maximum at $x = 0.30$-0.50, i.e., close to the estimated (based on the X-ray diffraction studies) phase boundaries. The maximum values of piezoelectric parameters K_p and d_{33} (0.32 and 88 pC/N, respectively), which, along with the minimum values of V_E^1, confirm the mentioned hypothesis, are observed in SS with $x = 0.45$. Moreover, the fact that in this range of concentrations x, value of Q_m is maximum rather than minimum, as it should be in the vicinity of the MR [14], reflects the priority impact of recrystallization processes during sintering of the ceramics on this parameter. Complex phase diagram with an alternating sequence of structural transitions in the SSs sections II and III as well as in the cut I shows that the amount of impurities determines ambiguous behavior of dielectric, piezoelectric and mechanical properties of the ceramics.

The observed effects (except for the appearance of T$_2$ phase) are associated with introduction of additional component CuNb$_2$O$_6$ in KNN. There are increasing of $\varepsilon_{33}^T/\varepsilon_0$ at $x \leq 0.1$ from ~ 300 (cut I) to ~ 400 (cut II), and to ~ 500-550 (cut III). Moreover, there is a non-monotonic change of Q_m with a sharp initial increase, that may be due to the crystal-chemical characteristics of copper niobate. As mentioned earlier, Cu^{2+} ions can occupy both A and B positions in the perovskite structure.

Occurrence of Cu^{2+} ions in the A-sublattice of KNN would lead to an excess of oxygen, and in the B-sublattice to its deficiency and, consequently, to the appearance of oxygen vacancies.

The latest version is more probable if proceed only from the corresponding ionic radii. In [15, 16] on the base of electron paramagnetic resonance and density functional theory calculations, it was shown that at atmospheric pressure developed just such a scenario. The resulting oxygen vacancies as a source of internal electric fields inhibited movement of domain walls (pinning effect), resulting in the hard properties of ferroelectric ceramics (an increase of Q_m, a decrease of piezoelectric and dielectric parameters) [7, 14], which observed in the SSs cut I.

In addition to crystallochemical substitution, the introduction of $CuNb_2O_6$ leads to an increase of the impurity phase content - $K_4CuNb_8O_{23}$. This compound has a low melting point (1050°C), thereby forming liquid phases during sintering and, therefore, increasing the density of the ceramics [17]. Thus, the electrophysical properties of the studied ceramics at the introduction of $CuNb_2O_6$ are determined by the competitive influence of both processes: crystallochemical substitution (increasing Q_m and decreasing $\varepsilon_{33}^T/\varepsilon_0$, K_p and d_{33}) and formation of glass phases (the increase in average grain size, increased density, Q_m and K_p). Based on the experimental studies, we selected compounds with the best combination of piezoelectric and mechanical properties ($K_p \approx 0.2$, $Q_m \approx 1200$ and $K_p \approx 0.32$, $Q_m \approx 500$) for application in ultrasound technique.

CONCLUSION

The phase diagram of the ternary $(1-x-z)NaNbO_3–xKNbO_3–zCuNb_2O_6$ system with $x=0.05–0.50$ and $z = 0.025-0.050$ has been constructed, and morphotropic phase boundaries have been marked on it. It has been found that the behavior of major electrophysical parameters of SSs being studied at variations of $KNbO_3$ content x correlates with the position of the morphotropic phase boundaries. Principal mechanisms affecting electrophysical properties of the studied ceramics at introduction of $CuNb_2O_6$ have been examined. The selected promising compounds for the application in ultrasound technique are characterized by $K_p \approx 0.2$, $Q_m \approx 1200$ and $K_p \approx 0.32$, $Q_m \approx 500$.

ACKNOWLEDGMENTS

This work was supported by grants of the President of the Russian Federation No. MK-3232.2015.2., Ministry of Education and Science of the Russian Federation (the base and the project of the state. quests: theme number 1927 213.01-2014/012-ВГ, quest No. 3.1246.2014/K) and FTP (agreement No. 14.575.21.0007).

REFERENCES

[1] Directive 2002/95/EC of the European Parliament and of the Council of 27 January 2003. *Official Journal of the European Union*. 2003, *vol. 37*, 19.

[2] Jaffe, B.; Cook, W. R., Jaffe, H. *Piezoelectric Ceramics*. Academic Press: London, New York, 1971.

[3] Park, H.-Y.; Choi, J.-Y.; Choi, M.-K.; Cho, K.-H.; Nahm, S.; Lee, H.-G.; Kang, H.-W. *J. Am. Ceram. Soc.* 2008, *vol. 91*, 2374-2377.

[4] Huang, R.; Zhao, Y.; Zhang, X.; Zhao, Y.; Liu, R.; Zhou, H. *J. Am. Ceram. Soc.* 2010, *vol. 93*, 4018-4021.

[5] Tan, X.; Fan, H.; Ke, S.; Zhou, L.; Mai, Y.-W.; Huang H., *Mater. Res. Bul.* 2012, *vol. 47*, 4472-4477.

[6] Park, B. C.; Hong, I. K.; Jang, H. D.; Tran, V. D. N.; Tai, W. P.; Lee, J.-S. *Mater. Let.* 2010, *vol. 64*, 1577-1579.

[7] Li, E.; Kakemoto, H.; Wada, S.; Tsurumi, T., IEEE Transact. Ultrasonics, Ferroelectrics, and Frequency Control. 2008, vol. 55, 980-987.

[8] Lv, Y.G.; Wang, C.L.; Zhang, J.L.; Zhao, M.L.; Li, M.K.; Wang, H.C., *Mater. Let.* 2008, *vol. 62*, 3425-27.

[9] Hao, J.; Xu, Z.; Chu, R.; Zhang, Y.; Li, G.; Yin, Q., *Mater. Res. Bul.* 2009, *vol. 44*, 1963-1967.

[10] Yang, M.-R.; Tsai, C.-C.; Hong, C.-S.; Chu, S.-Y.; Yang, S.-L., *J. Appl. Phys.* 2010, *vol. 108*, 094103 (1-5).

[11] Yang, M.-R.; Chu, S.-Y.; Chan, I-H.; Yang, S.-L., *J. All. Comp.* 2012, *vol. 522*, 3-8.

[12] Verbenko, I.A.; Shilkina, L.A.; Sadykov, H.A., Phase formation of solid solutions based on alkali metal niobate $CuNb_2O_6$ at the stage synthesis. *Proceedings of the First. Intern. Interdistsiplin. Symp. "Lead-free ferroelectric ceramics and related materials: preparation, properties, and applications (retrospective-modern-forecasts)"* Rostov-on-Don - B. Sochi. 2012, *vol. 1*, 79-84.

[13] Reznichenko, L.A.; Shilkina, L.A.; Razumovskaya, O.N.; Dudkina, S.I.; Gagarina, E.S.; Borodin, A.V., *Inorg. Mat.* 2003, *vol. 39*, 139-50.

[14] Dantsiger, A. Ya.; Razumovskaya, O. N., Reznitchenko, L. A., Sakhnenko, V. P., Klevtsov, A. N., Dudkina, S. I.; Shilkina, L. A.; Dergunova, N. V.; Rybjanets, A. N. *Multicomponent Systems of Ferroelectric Complex Oxides: Physics, Crystallochemistry, Technology. Aspects of Designing Ferroelectric Materials*, Vols. 1 and 2, Rostov University Press: Rostov-on-Don, 2002, pp. 1 – 773 (in Russian).

[15] Erűnal, E.; Jakes, P.; Kőrbel, S.; Acker, J.; Kungl, H.; Elsässer, C.; Hoffmann, M.J.; Eichel, R.-A., *Phys. Rev. B.* 2011. *vol. 84*, 184113(1-11).

[16] Eichel, R.-A.; Erűnal, E., Drahus, M.D., Smyth, D. M., Tol, J., Acker, J., Kungld, H., Hoffmann, M. J., *Phys. Chem. Chem. Phys.* 2009, *vol. 11*, 8698-8705.

[17] Matsubara, M.; Yamaguchi, T.; Sakamoto, W.; Kikuta, K.; Yogo, T.; Hirano, S., *J. Am. Ceram. Soc.* 2005. *vol. 88*, 1190-1196.

In: Proceedings of the 2015 International Conference …   ISBN: 978-1-63484-577-9
Editors: Ivan A. Parinov, Shun-Hsyung Chang et al.   © 2016 Nova Science Publishers, Inc.

Chapter 9

CREATION AND PHYSICAL PROPERTIES OF HIGHLY EFFECTIVE LEAD-FREE SOLID SOLUTIONS

Kh. A. Sadykov, I. A. Verbenko, G. M. Konstantinov, L. A. Reznichenko and A. A. Pavelko*

Research Institute of Physics, Southern Federal University,
Rostov-on-Don, Russia

ABSTRACT

Regularities of changes in the phase content, grain structure and piezoelectric properties of lead-free solid solutions based on $(Na_{1-x}Li_x)NbO_3$ and $Ba_{1-x}Sr_xTiO_3$ systems were established. Based on the results achieved, the ways to control the physical properties of the solid solutions were identified.

Keywords: lead-free ferroelectrics, niobates, BST, phase diagram, grain structure, piezoelectric properties, dielectric properties

1. INTRODUCTION

The development and study of lead-free ferro(piezo)electrics are an urgent task of modern material science. The most promising to get them (with a piezoelectric activity that is comparable to the piezoelectric activity of lead-containing materials) are multicomponent solid solutions based on the $(Na, K)NbO_3$ binary system as more sophisticated and having a great variety of properties [1-3]. Another promising lead-free ceramic system is $Ba_{1-x}Sr_xTiO_3$ (BST), which solid solutions are already widely used in radioelectronics. For example, these materials are used in amplifying techniques. The advantages of these materials include the

---

* Corresponding author: Kh. A. Sadykov. Research Institute of Physics, Southern Federal University, 344090 Rostov-on-Don, Russia. E-mail: hizir-2010@mail.ru.

performance of engineered elements and the use of both fronts of a control pulse to the switching devices based on the BST compared with known power semiconductor and plasma switches.

This chapter presents results of modifying of solid solutions based on the $(Na_{1-x}Li_x)NbO_3$ system of various oxides as well as some features of physical properties of those and BST solid solutions.

2. OBJECTS AND METHODS

Specimens of the $(Na_{1-x}Li_x)NbO_3$ system were prepared by two-stage solid phase synthesis at temperature $T = 850\text{-}870°C$ lasting $\tau_1 = \tau_2 = 6$ h and sintered at $T_s = 1160\text{-}1200°C$ (depending on the composition) for 1 h. BST solid solutions are synthesized by a conventional solid-state reaction in two steps. Synthesis regimes are $T_1 = 1130°C$, $T_2 = 1150°C$, and $\tau_1 = \tau_2 = 4$ h. The sintering conditions are 1375-1500°C depending on the composition. Raw materials of analytical grade used in this study were $BaCO_3$ (99%), $SrCO_3$ (99%), and TiO_2 (99%).

Microstructure of the upper surfaces and chips are evaluated on SEM (Hitachi TM-1000). Dielectric, piezoelectric and elastic parameters of solid solutions were measured using the resonance–antiresonance method [4].

3. RESULTS AND DISCUSSION

Figure 1 shows the modification of non-stoichiometric solid solutions based on the $(Na_{1-x}Li_x)NbO_3$ system. In Figure 2 there are phase diagrams of solid solutions modified by various oxides. Figure 3 contains micrograph cleavage surface of solid solutions modified by either CuO (see Figure 3, a-d) or MnO_2 (see Figure 3, e-h). Figure 4 shows a dependence of electrophysical characteristics of solid solutions of modified NiO (1), MnO_2 (2), CuO (3), CuO + MnO_2 (4), and Bi_2O_3 + Fe_2O_3 (5) on the modifier concentrations at room temperature.

Based on the results, the ways to control the properties of the solid solutions were identified. There are variations of conditions of the phase formation, ways of modifying, the character of input modifiers (simple such as CuO, NiO, MnO_2, and SnO_2, and complex such as CuO + MnO_2 and Bi_2O_3 + Fe_2O_3), and methods for their introduction in the matrix basis.

XRD analysis showed (see Figure 5) that all the samples had a perovskite structure with lack of extraneous phases.

It is found that the phase diagram of $Ba_{1-x}Sr_xTiO_3$ ceramics is as follows: the T phase is in the range of $0 \leq x < 0.4$, in the range of $0.3 \leq x < 0.8$, the PSC phase exists, in the range of $0.8 \leq x \leq 1.0$, there is the C phase. The first morphotropic region (MR1), where the T + PSC phases coexist, is in the range of $0.2 < x < 0.4$, and the second morphotropic region (MR2) associated with the coexisting PSC + C phases is in the range of $0.6 < x < 0.8$.

Also, based on results of the study on $Ba_{1-x}Sr_xTiO_3$ ceramics at $0 \leq x \leq 1$, we conclude that the temperature changes in the manufacture of ceramics and quality of the starting components can lead to a modification of the phase diagram of the system.

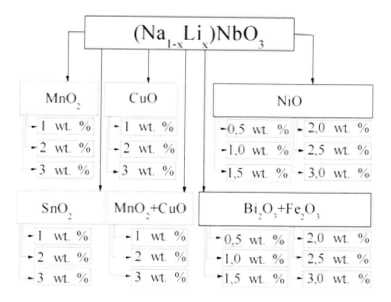

Figure 1. Modification scheme of non-stoichiometric solid solutions.

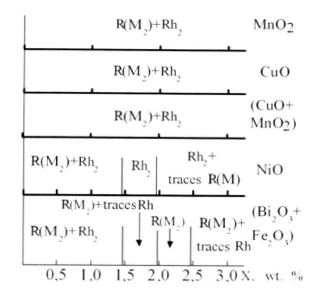

Figure 2. Phase diagrams of solid solutions of the $(Na_{1-x}Li_x)NbO_3$ system modified by various oxides.

The microstructure analysis shows that it is extremely heterogeneous: in some cases, there are traces of submelting. Grain structure of some samples has a clear bimodal nature. The sharp rise in grain occurs in vicinity of two morphotropic regions (at $x = 0.3$ and $x = 0.7$), it is particularly sharp at $x = 0.7$ (Figure 6).

This is a consequence of strengthening the movement of oxygen vacancies in morphotropic region that facilitates the processes of mass transfer.

Figure 7 shows the physical parameters of $BaTiO_3$ and analyzed system ceramics (due to the high electrical conductivity of the last, polarized state of the ones was not achieved).

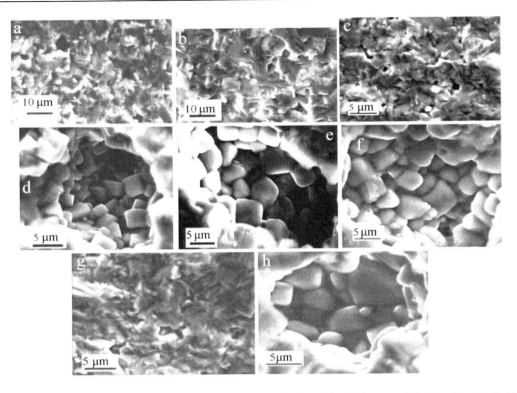

Figure 3. Micrographs of cleavage surfaces of $(Na_{1-x}Li_x)NbO_3$ solid solutions modified by 1 wt.% CuO (a, b), 2 wt.% CuO (c, d), 1 wt.% MnO_2 (e, f), and 2 wt.% MnO_2 (g, h).

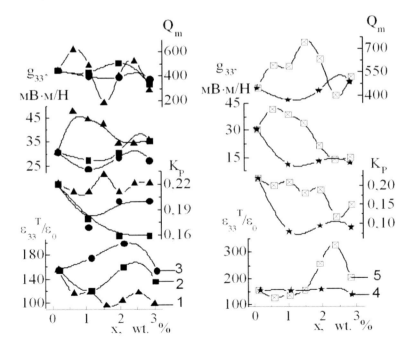

Figure 4. Dielectric and piezoelectric properties of $(Na_{1-x}Li_x)NbO_3$ solid solutions modified by various oxides.

Creation and Physical Properties of Highly Effective Lead-Free Solid Solutions 65

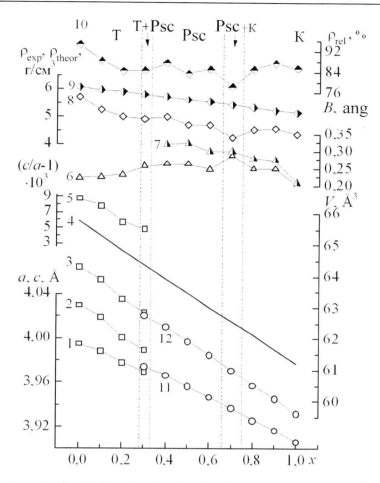

Figure 5. Molar-concentration (x) dependencies of unit-cell parameters a and c, experimental, V_T, V_C and the theoretical V_{theor}, unit-cell volumes of $Ba_{1-x}Sr_xTiO_3$ solid solutions, half-widths of X-ray lines 111, 200, B, and density of ceramic samples ρ: 1–a_m, 2–c_m, 3–V_T, 4–V_{theor}, 5–c_T/a_T, 6–B_{111}, 7–B_{200}, 8–$ρ_{exp}$, 9–$ρ_{theor}$, 10–$ρ_{rel}$, 11–a_c, and 12–V_C.

Figure 6. (Continued).

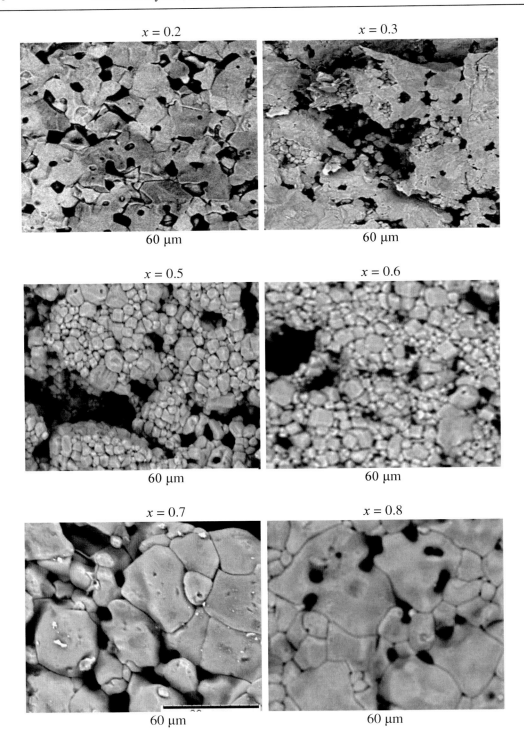

Figure 6. (Continued).

Creation and Physical Properties of Highly Effective Lead-Free Solid Solutions   67

Figure 6. Microstructure of ceramic Ba$_{1-x}$Sr$_x$TiO$_3$ solid solutions at $0 \leq x \leq 1.0$. The actual size of the micrographs is shown thereunder.

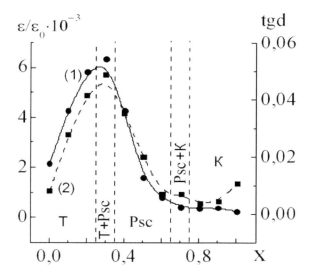

Figure 7. Molar-concentration (x) dependencies of the relative dielectric permittivity $\varepsilon/\varepsilon_0$ (curve 1) and loss tangent tgδ (curve 2) of Ba$_{1-x}$Sr$_x$TiO$_3$ solid solutions.

ACKNOWLEDGMENTS

This work has been performed on the equipment of the Collective Use Center "Electromagnetic, Electromechanical and Thermal properties of solids" of the Institute of Physics, Southern Federal University.

This work was financially supported by the Ministry of Education and Science of the Russian Federation: Grant of the President of the Russian Federation No. MK-3232.2015.2; themes Nos. 1927, 213.01-2014/012-ВГ and 3.1246.2014/K (the basic and project parts of the State task).

REFERENCES

[1] Jaffe, B. et al., *Piezoelectric Ceramics*. Academic Press, New York, 1971.

[2] Kravchenko, O. Yu. et al., *Inorganic Materials*, 2008, vol. 44, 1135-1150.

[3] Reznichenko, L. A. *Crystallography Reports*, 2009, vol. 54, 483-491.

[4] IRE Standards on Piezoelectric Crystals: Determination of the Elastic, Piezoelectric, and Dielectric Constants – the Electromechanical Coupling Factor, *Proc. IRE*, 1958, vol. 46, 764-778.

In: Proceedings of the 2015 International Conference …   ISBN: 978-1-63484-577-9
Editors: Ivan A. Parinov, Shun-Hsyung Chang et al.   © 2016 Nova Science Publishers, Inc.

Chapter 10

DIGITAL PIEZOELECTRIC MATERIAL BASED ON EXTRACOARSE-GRAINED PIEZOCOMPOSITE "POROUS CERAMIC – POLYMER"

*D. I. Makariev**, *A. N. Reznitchenko,*
A. A. Naumenko, E. I. Petrova and A. N. Rybyanets
Institute of Physics, Southern Federal University,
Rostov-on-Don, Russia

ABSTRACT

We have developed extracoarse-grained composite "porous piezoceramic – polymer" suitable for use as an initial material in the manufacture of the piezoelectric elements by the additive technology of binding powder by adhesives. The use of raw piezoceramic particles whose dimensions exceed the final thickness of the piezoelectric element is the main feature of this technology. The presence of horizontal layers of the polymer between the piezoelectric ceramic particles dramatically reduces the effectiveness of standard composites "piezoelectric ceramic – polymer," and our approach allows one to get rid of them. In this chapter, we also provide the electrical characteristics of the developed material.

Keywords: extracoarse-grained composite, composite "porous piezoceramics - polymer," electrical characteristics, porous ceramics

1. INTRODUCTION

In recent years, additive technologies develop rapidly, and can occupy a significant part of the world production in the nearest future. 3D-printing technologies are divided into two large groups: (i) the technologies themselves and devices for 3-D printing, and (ii)

---

* e-mail: dmakarev@rambler.ru.

technologies of initial material production for 3D-printing (such materials are typically called "digital materials"). In terms of 3D-production methods, piezoelectric elements have suitable sizes and shapes for many existing technologies [1–3]. To date, a number of 3D-devices and printing technologies can be adapted to the production of piezoelectric elements. However, the situation becomes complicated by the lack of suitable initial piezomaterials [1, 2, 5]. We have developed extracoarse-grained composite "porous piezoceramic – polymer" suitable for use as an initial material in the manufacture of the piezoelectric elements by the additive technology of "binding powder by adhesives." The use of raw piezoceramic particles whose dimensions exceed the final thickness of the piezoelectric element is the main feature of this technology. The presence of horizontal layers of the polymer between the piezoelectric ceramic particles drastically reduces the effectiveness of standard composites "piezoelectric ceramic – polymer," and our approach allows one to get rid of them. The sample of new composite should be machined by grinding or processing in 3D-miller to adjust the desired size before applying electrodes. The use of porous piezoelectric ceramics [6] instead of the standard dense piezoceramics greatly simplifies such machining and allows to use 3D-miller for plastic processing.

2. MANUFACTURE TECHNOLOGY

The composite elements were made with the assumption that in the future they can be manufactured by an adapted additive technology (binding powder by adhesives). This technology is the most promising for the manufacture of the piezoelectric elements based on the "polymer-piezoceramic" composites. A layer of a powder with a certain thickness is applied to the elevator table of the 3D-printer, then, the adhesive composition or a liquid chemical is applied from the printer cartridge. After bonding or polymerization of the first layer a second layer of powder is applied, which is also glued or polymerized, and so on we obtain final product. In our case, we need to apply only one layer of piezoceramic coarse particles, then bind them by the polymer and grind to the desired thickness as shown in Figure 1.

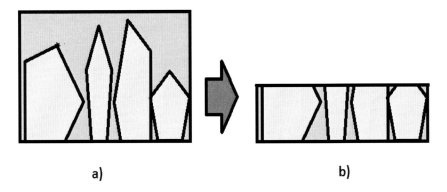

Figure 1. (a) piezocomposite before the one-side machining, (b) piezocomposite after the one-side machining; pink color corresponds to polymer, beige color – to piezoceramics.

We took a mixture of 99 wt.% of PZT-type piezoceramics [6] with a porosity of 40% and 1 wt. % of acrylic polymer as an initial powders. We used a mixture of benzoyl peroxide, an acrylic monomer and N, N-dimethyl-p-toluidine as the liquid curing agent. As the result, we obtained a composite material with 80 vol. % of piezoceramics. Use of a porous piezoceramics instead of a dense piezoceramics greatly simplifies the machining of the obtained material and allows one to use the equipment for plastics processing. We manufactured elements in the form of discs with 15 mm in diameter. The thickness varied from 0.25 mm to 5 mm. We applied a conductive paste on the main surfaces of the discs as the electrodes. Polarization was performed in polyether liquid with an application of a constant electric field with the strength of 3 kV/mm at a temperature of 423 K, the exposure time was 1 h.

3. EXPERIMENT AND DISCUSSION OF THE RESULTS

We have fabricated a few piezoelectric elements from developed piezoelectric composite using the above-specified technology. Each of the elements is sequentially subjected to the one-sided grinding from a thickness $t = 5$ mm to a thickness $t = 0.25$ mm. After grinding to a thickness $t = 3.5$ mm, electrodes were coated on the surface of the piezoelectric element, and electro-physical measurements were conducted. Then the element again subjected to the one-sided grinding, and similar measurements were conducted after each grinding of the piezoelectric element on 0.25 mm. Photographs of all these thin sections of one element are shown in Figure. 2.

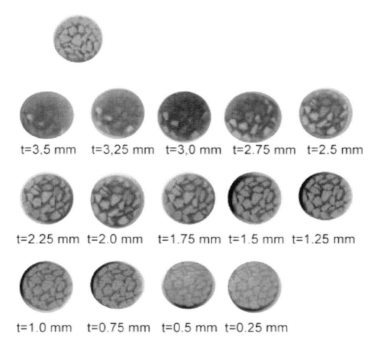

Figure 2. Consecutive thin sections of the extracoarse-grained composite sample made through every 0.25 mm. The upper photograph shows the piezoelectric element surface no subjected to grinding.

The upper photograph shows the piezoelectric element surface no subjected to grinding, but the rest photographs represent thin sections of the opposite surface of the piezoelectric element made through each 0.25 mm in thickness. The photographs show how the fraction of the porous piezoelectric ceramics is gradually grows on the surface of the extracoarse-grained sample (light color in the photographs), and the fraction of the binding polymer is recedes (dark color). As a result, the composite change the properties. The sample practically has not piezoelectric properties at the thickness up to 3.5 mm because of the binding polymer layer are placed between the piezoelectric ceramics and the element electrodes. This polymer layer prevents polarization process of the sample and measurement of the piezoelectric properties. Starting from the thickness of 2.25 mm total surface area of piezoelectrics emerged on the grinded surface of the sample begins to exceed the area occupied by the polymer. From this point, the sample shows noticeable piezoelectric effect measured by the quasi-static method. At the same time, increase in the mechanical quality factor of the sample enables to use dynamic methods of piezoelectric properties measurement. One of the features of the composite is atypical for piezomaterials in general and for composite piezomaterials in particular, namely dependence of composite permittivity on the thickness of the piezoelectric element. To be more exact, the material permittivity depends on the ratio of the average particle size of primary piezoceramics, comprising the ceramic grain size (d) and the piezoelectric element thickness (t). When the ratio of these values becomes close to 1, one can observe a grows of the material permittivity. It is caused by a significant increase in the piezoelectric ceramic fraction on the grinded sample surface.

The dependence of the sample permittivity (ε) on the ratio of the initial size of the piezoceramic grains and the sample thickness (d/t) is shown in Figure 3. It is noticeable that three regions are clearly distinguished: (i) the region of the stably low sample permittivity, (ii) the region of its sharp growth and (iii) the region of stabilization of the permittivity at the new values. Over the range of ratios d/t from 0.33 to 0.5 stably low levels of the composite permittivity are observed, over the range 0.5 to 1.5 a sharp growth of this magnitude from 50 to 375 are observed, and over the range 1.5 to 4, permittivity values are settled down. The origin of this phenomenon becomes clearly apparent by analyzing the photos in Figure 3. For the one-sided grinded extracoarse-grained composite sample, three stages of such machining are distinctly clearly revealed. After the first stage of machining only the largest grains of piezoelectric ceramics emerge on the sample surface, and the polymer compound have the dominant fraction of the sample surface. This stage corresponds to the region of stably low permittivity. After the next stage of grinding, piezoelectric ceramic grains emerge on the grinded surface, thus the piezoelectric ceramics fraction on the surface starts increase sharply, this stage corresponds to the region of sharp rise of the permittivity in Figure 3. Finally, there comes the moment when all piezoelectric ceramics grains, even the smallest, emerge on the grinded surface, and the ratio of polymer and piezoceramics on the grinded surface is stabilized. This stage corresponds to the region of stabilization of the permittivity values at the new level. Some decrease in its value is caused by the fact that most of the piezoelectric ceramic grains have shape of a convex polyhedron and one-sided grinding of the sample to a thickness less than half their original size, increases the polymer fraction in the composite, which leads to decrease of the sample permittivity. It is important to note that the polymer used as a binder is translucent, so visible emergence of the piezoceramic grains on the sample surface does not always corresponds to the real emergence of this grain to the sample surface.

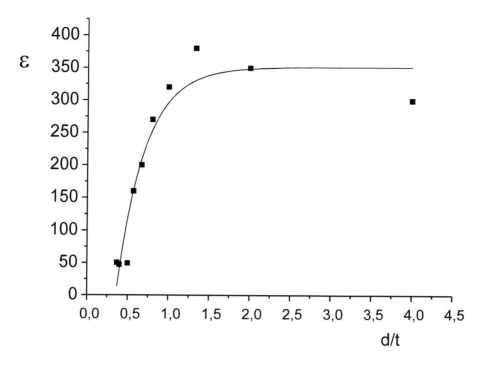

Figure 3. Dependence of the permittivity of the extracoarse-grained "piezoceramic-polymer" composite on the ratio of the initial sample grain size (d) and sample thickness (t).

Measured by quasi-static method piezomodulus d_{33} of the piezoelectric composite was 350 pC/N compared to 400 pC/N for the original PZT porous piezoceramics with a porosity of 30%. Measured by resonance method electromechanical coupling factor k_t of the thickness vibration mode was 0.5, that roughly corresponds to the k_t of the original piezoceramics.

Conclusion

The relatively high values of piezomodulus and piezoelectric electromechanical coupling factor of the composite material create good preconditions for manufacturing of the piezoelectric materials based on the extracoarse-grained "piezoceramic-polymer" composites by additive technologies, in particular by the "binding powder by adhesives" technology. Therefore, we can predicate that the development of a "digital" piezoelectric material based on "polymer-piezoceramics" composites is possible; however, it may require additional machining of the elements prior to application of electrodes.

Acknowledgments

Work supported by the RSF grant No. 15-12-00023.

References

[1] Makarev, D.I.; Rybyanets, A.N.; Mayak, G.M. *Technical Physics Letters.* 2015, *vol. 41,* 317-319.

[2] Rybyanets, A. N.; Rybyanets, A. A. *IEEE Trans. UFFC.* 2011, *vol. 58,* 1757-1773.

[3] Topolov, V.Y.; Filippov, S.E.; Panich, A.E.; Panich, E.A. *Ferroelectrics.* 2014, *vol. 460,* 123-127.

[4] Rybyanets, A.N.; Konstantinov, G.M.; Naumenko, A.A.; Shvetsova, N.A.; Makar'ev, D.I.; Lugovaya, M.A. *Physics of the Solid State.* 2015, *vol. 57,* 527-530.

[5] Rybyanets, A.N.; Naumenko, A.A.; Konstantinov, G.M.; Shvetsova, N.A.; Lugovaya, M.A. *Physics of the Solid State.* 2015, *vol. 57,* 558-562.

[6] Rybyanets, A.N. *IEEE Trans. UFFC.* 2011, *vol. 58,* 1492-1507.

In: Proceedings of the 2015 International Conference ...
Editors: Ivan A. Parinov, Shun-Hsyung Chang et al.

ISBN: 978-1-63484-577-9
© 2016 Nova Science Publishers, Inc.

Chapter 11

NUMERICAL SIMULATION OF FILMS LASER ANNEALING ON THE SAPPHIRE SURFACE

S. P. Malyukov, A. V. Sayenko and Yu. V. Klunnikova[*]

Institute of Nanotechnology,
Electronics and Electronic Equipment Engineering,
Department of Electronic Apparatuses Design,
Southern Federal University, Taganrog, Russia

ABSTRACT

In this chapter we present theoretical and experimental studies of oxide films (TiO_2, Fe_2O_3) laser annealing on the sapphire surface using a solid-state Nd:YAG laser with the wavelength of 1064 nm. We develop a numerical model of films laser annealing based on time-dependent differential equations of heat conduction. It allows one to calculate the temperature on the film surface during laser radiation. The dependencies of temperature on the film surface on the laser power were obtained. The simulation results allow us to determine the optimal modes of films annealing on the sapphire surface.

Keywords: laser annealing, sapphire surface, oxide films, heat conduction, numerical simulation

1. INTRODUCTION

Nowadays laser annealing of films is widely used in various fields of science and technology. The aim of laser annealing is to achieve desired crystalline structure of the films, the orderly particles arrangement in it and to reduce the number of defects in structure [1]. We used sapphire crystals as the substrate for the film during laser annealing. Sapphire crystals have a certain set of physical properties (high melting temperature, chemical and

---

[*] email: yvklunnikova@sfedu.ru.

radiation resistance, high hardness and the laser transparency). Due to these properties, they are used in microelectronics, quantum electronics, optics of high resolution, and nanotechnology [2-5].

In the study of films laser annealing, the most important question is to calculate the temperature on the surface and to determine the heat treatment modes to ensure maximum annealing of defects (vacancies and interstitial atoms), maximal activation of the implanted impurity and efficient film growth [6]. The interaction of laser radiation with material in the spectral range from ultraviolet to infrared occurs at the level of the electronic subsystem [7]. Getting a detailed description of each of these processes is difficult. It is also hard to get a strict mathematical description of the light interaction with a solid even in the particular case. However, for most practically important cases, the time of the processes of excitation, thermalization and recombination of charge carriers is much less than the duration of the light pulse. It allows one to speak about predominantly thermal processes in the material [7].

The transfer of heat energy to material occurs due to the heat-conductivity processes. The process of heat conduction by heat transfer is possible only when the body temperature varies from one point to another. Transfer of heat conductivity in material accompanied by temperature change in both space and time. Thus, a non-stationary temperature field arises during the processing material by laser light. The heat transfer equation sets connection between the values determining the heat transfer [8].

2. THEORY OF NUMERICAL SIMULATION OF LASER ANNEALING

Consider the laser radiation of oxide films with a thickness of 1-2 µm on the sapphire surface (Figure 1).

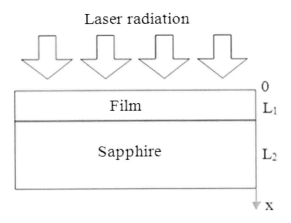

Figure 1. Scheme of film – sapphire structure laser annealing.

The structure of the film – sapphire, which is exposed to pulsed laser radiation (Nd:YAG laser with the wavelength of 1064 nm, laser pulse duration of 84 ns, pulse repetition rate of 10 kHz, and average laser power of 100 W), is isotropic and homogeneous. One-dimensional unsteady heat conduction equation for each layer can be described as [9, 10]:

$$\rho_1 c_1 \frac{\partial T_1(x,t)}{\partial t} = k_1 \frac{\partial^2 T_1(x,t)}{\partial x^2} + F_1(x), \tag{1}$$

$$\rho_2 c_2 \frac{\partial T_2(x,t)}{\partial t} = k_2 \frac{\partial^2 T_2(x,t)}{\partial x^2} + F_2(x), \tag{2}$$

where T_i is the temperature (K), ρ_i is the density (g/cm^3), c_i is the specific heat (J/(g·K)), k_i is the thermal conductivity coefficients (W/(cm·K)) for film and sapphire ($i = 1, 2$), respectively, $F_i(x)$ are the thermal sources (the result of laser radiation absorption) in each material, x is the coordinate, t is the laser exposure time (s).

During heating of different materials by laser radiation the power density distribution of the light flow $I(x)$ in the absorbing medium by the depth x is described by Bouguer-Lambert law in differential form [6]:

$$\frac{dI(x)}{dx} = -\alpha I(x) \tag{3}$$

where α is the absorption coefficient (cm^{-1}).

In the wavelength range from ultraviolet to infrared region, many materials have a large absorption coefficient, which can be considered as constant [8]. In this case, the integral Bouguer-Lambert law (heat source) describes the density change of the light flow in depth. Thus when we heat the film – sapphire structure with laser, the distribution of light flow power density in the absorbing medium by the depth x is determined by the expressions [8]:

$$F_1 = \alpha_1(1 - R_1)I_0 \exp(-\alpha_1 x), \tag{4}$$

$$F_2 = \alpha_2(1 - R_2)(1 - R_1)I_0 \exp\big(-\alpha_2(x - L_1)\big), \tag{5}$$

where I_0 is the power density of the falling laser radiation on the film surface (W/cm^2), R_i is the reflectance of the film and sapphire, respectively.

The solution of Equations (1), (2) is determined by a function that depends on the heat sources $F_i(x)$ and values of the thermal constants ρ_i, c_i, k_i. The uniqueness of the solution is defined by the boundary conditions for each specific task of heating. Therefore, to solve the problem of film–sapphire structure heating, it is necessary to have additional conditions that uniquely determine the task of thermal conductivity (the condition of uniqueness) [8].

We have obtained the boundary and initial conditions for the film – sapphire structure. Environment temperature and the law of heat exchange between the environment and the surface of the processed material are specified. So on the radiated surface (film at $x = 0$) the boundary conditions of the third kind are operating that determine the convection heat transfer (by Newton's law) between environment and surface of the film [8]:

$$-k_1 \frac{\partial T_1(x=0,t)}{\partial x} = \beta(T_1 - T_0), \tag{6}$$

where T_0 is the environment temperature, β is the convective heat transfer coefficient (W/(cm^2·K)), which characterizes the intensity of heat exchange between surface of the films and environment.

On the second boundary of the film – sapphire structure ($x = L_1 + L_2$), the boundary condition of the first kind (no processes with absorption or heat release occur on the body boundary) holds, which gives the temperature distribution on the surface for any time:

$$T_2(x = L_1 + L_2, t) = T_0. \tag{7}$$

The initial condition for the heat equation is the temperature distribution at $t = 0$ at all points of the structure exposed to laser radiation. For film and sapphire, we have:

$$T_1(x, 0) = T_0; \tag{8}$$

$$T_2(x, 0) = T_0. \tag{9}$$

To solve non-stationary heat conduction equations (1) – (5) under laser radiation on the film – sapphire structure with the boundary conditions (6) and (7), we use the numerical method [11], which consists of the partial differential equations approximation by corresponding initial finite differences.

3. RESULTS AND EXPERIMENT

Figure 2 shows the dependence of the temperature on the films surface on the average laser power.

Experimental studies of film formation were conducted on laser equipment LIMO 100-532/1064-U (Nd:YAG laser with the wavelength of 1064 nm) by treatment of $FeCl_3$ solution on the sapphire surface by four passes of the laser beam with power of 80 – 90 W. We process the suspension of particles TiO_2 with a binder by laser power of 70 – 80 W.

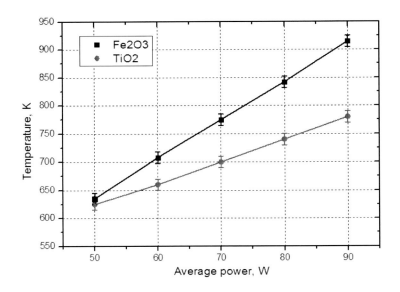

Figure 2. Dependence of temperature on the film surface on the average laser radiation power.

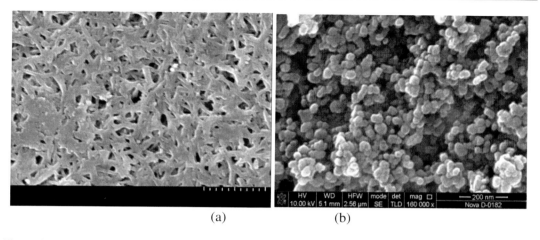

(a)   (b)

Figure 3. Surface morphology of films Fe_2O_3 (a) and TiO_2 (b) on sapphire.

The surface morphology of Fe_2O_3 and TiO_2 films on the sapphire is presented in Figure 3 (a, b).

CONCLUSION

As the result we studied the oxide films laser annealing on sapphire surface using laser equipment (LIMO 100-532/1064-U). The numerical simulation of laser annealing of films Fe_2O_3, TiO_2 on sapphire was carried out. The simulation results show that the temperature dependence of the film surface on the average laser power is linear. It is found that at the average laser power of 80 – 90 W, the temperature of Fe_2O_3 film surface is about 850 – 900 K, which is a prerequisite for film growth on the sapphire surface. It is obtained that at the average laser power of 70 – 80 W, the temperature of TiO_2 film surface is about 650 – 700 K, which is a prerequisite for porous film structure formation. We can vary the morphology of the film structure by changing the laser power and temperature, which allows to reallocate defects in the structure and to improve the quality of the films for their application in microelectronics and thin film optics.

ACKNOWLEDGMENTS

Results were received with the equipment of the Scientific and Educational Center "Laser Technologies," Center of Collective Use and the Scientific and Educational Center "Nanotechnology" in the Institute of Nanotechnology, Electronics and Electronic Equipment Engineering of the Southern Federal University (Taganrog) and the Department of Structural Studies of the Zelinsky Institute of Organic Chemistry (Moscow).

REFERENCES

[1]   Simakin, A. V.; Voronov, V. V.; Shafeev, G. A. Proceedings of the Institute of General Physics in Honour of *A.M. Prokhorov*. 2004, vol. 60, pp. 83-90 (in Russian).

[2]   Dobrovinskaya, E. R.; Lytvynov, L. A.; Pishchik, V. V. Sapphire. Material, Manufacturing, Applications. Springer, New York. 2009. pp. 1-481.

[3]   Cherednichenko, D. I.; Malyukov, S. P.; Klunnikova, Yu. V. In: Sapphire: Structure, Technology and Applications, I. Tartaglia (Ed.), Nova Science Publishers, New York, 2013, pp. 101-118.

[4]   Malyukov, S. P.; Klunnikova, Yu. V., In: Nano- and Piezoelectric Technologies, Materials and Devices, Ivan A. Parinov (Ed.). Nova Science Publishers, New York, 2013, Chapter 5, pp. 133-150.

[5]   Malyukov, S. P.; Klunnikova, Yu. V., In: Advanced Materials Physics, Mechanics and Applications, Series: Springer Proceedings in Physics, Shun-Hsiung Chang, Ivan A. Parinov, Vitaly Yu. Topolov (Eds.). Springer Cham, Heidelberg, New York, Dordrecht, London, 2014, Chapter 6, vol. 152, pp. 55 – 69.

[6]   Dovbnya, A. N.; Efimov, V. P.; Abizov, A. S.; Shapoval, I. I.; Fish, A. V.; Bereznyak, E. P.; Zakutin, V. V.; Reshetnyak, N. G.; Romasko, V. P. *Problems of Atomic Science and Technology*. 2010, vol. 2, pp. 164-167 (in Russian).

[7]   Webb, C. E.; Jones J. D. *Laser Technology and Applications*. Bristol and Philadelphia. 2004, pp. 1-1263.

[8]   Yakovlev, E. B.; Shandybina, G. D. *Interaction of Laser Radiation with Material*. State University of Information Technologies, St. Petersburg. 2011, pp. 1-184 (in Russian).

[9]   Malyukov, S. P.; Sayenko, A. V. *Journal of Russian Laser Research*. 2013, vol. 34(6), pp. 531-536.

[10] Jinjing, F.; Jixiang, Y.; Shouhuan, Z. *Piers online*. 2007, vol. 3(6), pp. 847-850.

[11] Patankar, S. Numerical Methods for Solving Problems of Heat Transfer and Fluid Dynamics. *Energoatomizdat*, Moscow. 1984, pp 1-150 (in Russian).

In: Proceedings of the 2015 International Conference …   ISBN: 978-1-63484-577-9
Editors: Ivan A. Parinov, Shun-Hsyung Chang et al.   © 2016 Nova Science Publishers, Inc.

Chapter 12

NUMERICAL ANALYSIS OF INORGANIC GLASSY DIELECTRIC LASER PROCESSING

S. P. Malyukov, Y. V. Klunnikova and T. H. Bui*
Institute of Nanotechnology,
Electronics and Electronic Equipment Engineering,
Department of Electronic Apparatuses Design,
Southern Federal University, Taganrog, Russia

ABSTRACT

In this chapter we describe the methodology of research of inorganic glassy dielectric laser processing for microsystems using methods of numerical simulation, in particular finite element method. We develop the model of inorganic glassy dielectric laser processing. So we can analyze and determine the temperature distribution and thermal stresses on the surface of borosilicate glass with different laser scanning velocity. We use ANSYS finite element software package for the simulation of laser processing.

Keywords: inorganic glassy dielectric, laser processing, finite element method, laser scanning velocity

1. SOLUTION METHOD AND NUMERICAL RESULTS

Laser technology is experiencing a period of intensive development. Lasers are widely used in materials processing in almost all fields of micro- and nanotechnology. They allow to improve productivity in machining operations and control as well as the quality of manufacturing products and also to ensure the possibility of full automation of technological processes [1].

---

* E-mail: yvklunnikova@sfedu.ru.

Modern methods of mathematical modeling, including progressive finite element method (FEM) allow us to analyze the laser material processing and achieve results similar to the data of experiments [2].

We choose the borosilicate glass ($B_2O_3 - SiO_2 - R_2O - RO$, where R_2O is K_2O, Na_2O, Li_2O and $RO - CaO$, MgO) for solar cells development instead of silicon. The borosilicate glass substrate has high chemical resistance (ISO 719 - 1), high temperature resistance (melting point 1070°C) and radiation resistance. The presence of boron oxide in this type of glass can be used for the doping of the seed layer or an absorbing layer to create solar cells. Thus, the structure of solar cell is free from substrate diffusion barriers and the use of the borosilicate glass substrate allows to reduce the manufacturing cost of solar cells [3 – 5].

The aim of this research is the development of numerical model of inorganic glassy dielectric laser processing. It allows us to analyze the temperature distribution and thermal stress on the surface of substrate with different laser scanning velocity.

The first stage of the numerical analysis of inorganic glassy dielectric laser processing is the simulation of heat source intensity. The different forms of laser beam distribution (Gaussian distribution or rectangular distribution) can be obtained for laser material processing. The Gaussian energy distribution is the most preferred mode for laser material processing, because it has a very small diameter of the focusing spot, resulting in higher power density [6]. The laser beam intensity is defined by formula [7]:

$$I(x,y) = I_0 exp\left[-\frac{\left(x^2 + y^2\right)}{r^2}\right],$$
(1)

where I_0 is the intensity at the center of the Gaussian beam, and r is the radius of laser beam, x and y are the current coordinates.

The laser beam intensity (Nd: YAG laser: the wavelength of 532 nm, pulse duration of 30-45 ns, pulse frequency of 10 kHz) in the range from 10 to 50 MW/m^2 on the surface of substrate is shown in Figure 1. The Gaussian distribution of the laser beam leads to a significant increase in temperature in the area of laser source activity.

The temperature distribution on the surface of the inorganic glassy dielectric is one of the main parameters of laser processing. We use the heat conduction equation for calculation of thermal processes of inorganic glassy dielectric laser processing. It allows us to obtain the temperature dependence of dimensional coordinates and time [8] in the form:

$$\rho C_T \frac{\partial T}{\partial t} = \frac{\partial}{\partial x}\left(\lambda \frac{\partial T}{\partial x}\right) + \frac{\partial}{\partial y}\left(\lambda \frac{\partial T}{\partial y}\right) + \frac{\partial}{\partial z}\left(\lambda \frac{\partial T}{\partial z}\right) + Q,$$
(2)

where ρ is the density, C_T is the thermal conductivity, t is the time, λ is the coefficient of thermal conductivity.

To calculate the heat distribution, in addition to differential equation of heat conduction it is necessary to specify conditions: the initial temperature distribution in the body (the initial condition) and the conditions of heat exchange on the boundary of the body (boundary conditions).

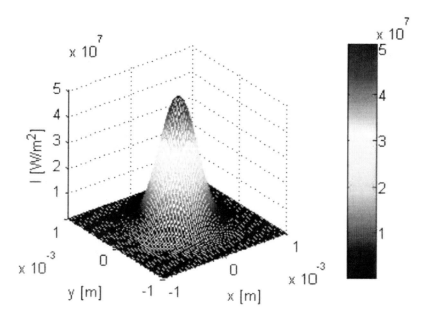

Figure 1. Gaussian distribution of laser beam.

The initial condition is the temperature distribution inside the body at the initial time $t = 0$: $T(x, y, z, 0) = f(x, y, z)$. In practice, the uniform initial temperature distribution $T(x, y, z, 0) = T_0$ is usually used.

The boundary conditions of the third kind hold on the exposed surface of the inorganic glassy dielectric (at $z = 0$). These conditions determine the convection heat transfer between the environment and the surface of substrate [8]:

$$q_0 = \beta(T_1 - T_0), \qquad (3)$$

where β is the convective heat transfer coefficient characterizing the intensity of heat exchange between the glass surface and the environment, T_1 is the current temperature, T_0 is the environment temperature.

We calculate deformations and mechanical (thermal) stresses on the surface of inorganic glassy dielectric material during laser processing. The strain tensor in the Cartesian coordinates for homogeneous and isotropic body is determined by the following formula [8]:

$$\varepsilon_{ij} = \alpha T \delta_{ij}, \qquad (4)$$

where ε_{ij} is the strain, α is the coefficient of thermal expansion, T is the temperature, δ_{ij} is the thermal stresses.

Thermal stresses σ_{ij}, causing additional sample elongation and changes according to the formulae of the classical theory of elasticity have the following form [8]:

$$\varepsilon_{ij} = \frac{1}{2E}\left(\sigma_{ij} - \frac{\mu}{1+\mu}\sigma_{kk}\delta_{ij}\right) + \alpha T\delta_{ij}, \qquad (5)$$

where E is the Young's modulus, μ is the Poisson's coefficient.

In the absence of mass forces, we add the equilibrium condition to the system of equations:

$$\frac{\partial \alpha_{ij}}{\partial x_i} = 0. \qquad (6)$$

In these equations, we apply the Einstein's agreement about summation over repeated indices.

The following assumptions were made for the analysis of inorganic glassy dielectric laser processing:

(i)   the material is isotropic;
(ii) the laser beam is considered as Gaussian distribution at TEM_{00} mode;
(iii) the effect of material evaporation is ignored;
(iv) the heat transfer process is only determined by the effects of conduction and convection.

Transient thermal analysis was used for numerical solution of inorganic glassy dielectric laser processing. It allows studying temperature changes in time [9]. The substrate with dimensions of 10 mm × 10 mm × 1 mm was considered for the simulation of inorganic dielectric laser processing.

Laser processing of materials is based on the fact that the use of laser radiation allows to create high density of the heat flow on the small area of the surface. It is necessary for the intensive heating or melting of virtually any known material. During inorganic glassy dielectric radiation, the portion of the laser beam reflects from it, and the resistance part penetrates inside. The process of heat distribution depends on the intensity of thermal impact and thermal properties of the material [10].

Table 1 shows the properties of inorganic glassy dielectric ($B_2O_3 - SiO_2 - R_2O - RO$) which are used in the simulation.

Table 1. Thermal properties of inorganic glassy dielectric

Specific heat	750 J/kg°C
Density	2500 kg/m^3
Thermal conductivity	1.4 W/m°C
Thermal expansion	$85 - 90 \cdot 10^{-7}$ K^{-1}
Melting point	1070 °C

From simulation results of the laser beam intensity, the temperature distribution on the surface of borosilicate glass with average laser power of 25 W and laser scanning velocity of 10 mm/s at: $a - 0.1$ s; $b - 0.5$ s, $c - 1$s; $d - 5$s is obtained (see Figure 2).

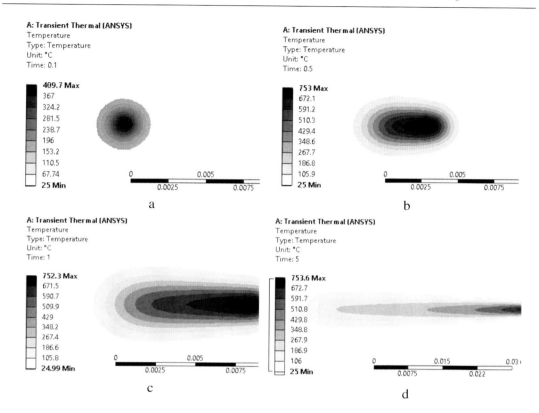

Figure 2. Temperature distribution on the surface of borosilicate glass with average laser power of 25 W and laser scanning velocity of 10 mm/s at: a – 0.1 s; b – 0.5 s, c – 1s; d – 5 s.

We can see in Figure 2 that the maximum temperature is at the center of focusing spot on the sample. The heated zone of thermal influence of laser radiation on the surface of the inorganic glassy dielectric is shown too.

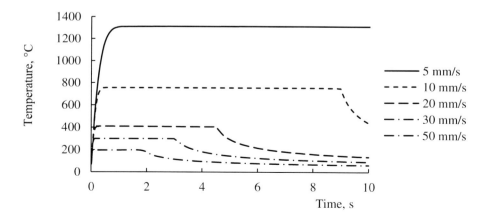

Figure 3. Dependence of the temperature on the surface of inorganic glassy dielectric on time with different laser scanning velocity.

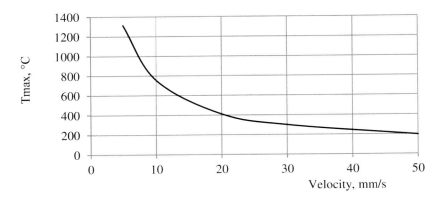

Figure 4. Dependence of the maximum temperature on the glass surface on the laser scanning velocity.

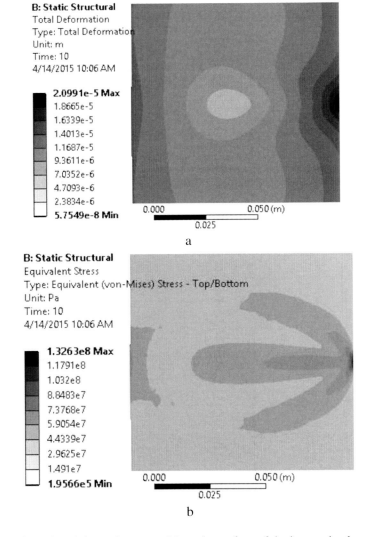

Figure 5. Deformation (a) and thermal stresses (b) on the surface of the inorganic glassy dielectric.

The dependence of the temperature on the surface of inorganic glassy dielectric on time with different laser scanning velocity (5 mm/s, 10 mm/s, 20 mm/s, 30 mm/s, 50 mm/s) is presented in Figure 3. Figure 3 shows the temperature trend on the surface of inorganic glassy dielectric in the laser processing and the start of inorganic glassy dielectric cooling.

The dependence of the maximum temperature on the glass surface on the laser scanning velocity is shown in Figure 4. The possibility of reaching the desired temperature on the surface of the inorganic glassy dielectric by changing laser scanning velocity was obtained.

The calculation results of deformation (Figure 5, *a*) and thermal stresses (Figure 5, *b*) on the surface of the inorganic glassy dielectric caused by the influence of laser processing are presented in Figure 5.

CONCLUSION

We analyzed the processes caused by laser treatment of inorganic glassy dielectric for different laser scanning velocity using the laser system (LIMO 100-532/1064-U, the wavelength of 532 nm). The numerical analysis of laser processing of the inorganic glassy dielectric was performed. The numerical simulation results showed that the temperature at the substrate surface depended on the laser scanning velocity non-linearly. From simulation results, we can conclude that at the average laser power of 25 W with laser scanning velocity of 10 mm/s, the surface temperature is about 750 °C. We calculate the deformations and the thermal stresses on the surface of the inorganic glassy dielectric taking laser processing into account. The obtained results allow to choose the optimal mode of the selected material laser processing, and to improve the surface quality of borosilicate glass used in solar cells manufacture.

ACKNOWLEDGMENTS

Results were obtained by using the equipment of the Scientific and Educational Center "Laser Technologies," Center of Collective Use and the Scientific and Educational Center "Nanotechnology," Institute of Nanotechnology, Electronics and Electronic Equipment Engineering of the Southern Federal University (Taganrog).

REFERENCES

[1]   Klunnikova, Y. V.; Malyukov, S. P.; Saenko, A. V. *Proceedings of the SPbGU LETI.* 2014, *vol. 8*, pp. 15-19 (in Russian).

[2]   Rumyansev, A. V. *Method of Finite Elements for Problems of Heat Conductivity.* Kaliningrad. 1995, pp. 1-170 (in Russian).

[3]   Andra, G.; Plentz, J.; Gawlik, A.; Ose, E.; Falk, F.; Lauer, K. *Proceedings of 22nd Europ. Photovoltaic Solar Energy Conference.* Milan. 2007, pp. 1967-1970.

[4]   Malyukov, S. P.; Kulikova, I. V.; Kalashnikov, G. V. *News of SFU. Technical Science.* 2011, *vol. 7*, pp. 182-187 (in Russian).

[5] Malyukov, S. P. *Glass Dielectrics in Magnetic Heads Production.* TSURE, Taganrog. 1998, pp. 1-181 (in Russian).

[6] Dahotre, N. B.; Harimkar, S. *Laser Fabrication and Machining of Materials.* Springer. 2008, pp. 1-558.

[7] Dubey, A. K.; Yadava, V. *International Journal of Machine Tools and Manufacture.* 2008, *vol. 6*, pp. 609-628.

[8] Yakovlev, E. B.; Shandybina, G. D. *Interaction of Laser Radiation with Material.* State University of Information Technologies, St. Petersburg. 2011, pp. 1-184 (in Russian).

[9] Moaveni, S. *Finite Element Analysis: Theory and Application with ANSYS.* Pearson, India. pp. 1-868.

[10] Grigoranz, A. G.; Shiganov I. N.; Misurov, A. I. *Technological Processes of Laser Treatment.* MGTU, Moscow. 2008, pp. 1-664 (in Russian).

In: Proceedings of the 2015 International Conference ...   ISBN: 978-1-63484-577-9
Editors: Ivan A. Parinov, Shun-Hsyung Chang et al.   © 2016 Nova Science Publishers, Inc.

Chapter 13

FEATURES OF SAPPHIRE AND GLASSY DIELECTRIC JUNCTION FORMATION FOR ELEMENTS OF MICROELECTRONICS

S. P. Malyukov, Yu. V. Klunnikova[*],
A. V. Sayenko and D. A. Bondarchuk

Institute of Nanotechnology, Electronics and Electronic Equipment Engineering,
Department of Electronic Apparatuses Design, Southern Federal University,
Taganrog, Russia

ABSTRACT

The chapter considers features of sapphire and glassy dielectric junction formation. We propose the optimal annealing modes for dielectric films processing on the sapphire substrate surface. The surface of the glassy dielectric film on the sapphire was studied. We also present the scheme of flow route for the manufacturing of sapphire and glassy dielectric junction.

Keywords: glassy dielectric junction, sapphire substrate surface, horizontal directed crystallization (HDC)

1. METHOD AND RESULTS

Nowadays reliable junctions of sapphire and glassy dielectric are used in various areas of microelectronics. The features of junction formation of glassy films with sapphire substrates are particularly interesting for microelectronics [1, 2].

In this chapter, we investigate the features of glassy dielectric junction formation on sapphire substrate. The method of centrifugation makes it possible to fabricate glassy films

---

[*] Corresponding author: E-mail: yvklunnikova@sfedu.ru.

with satisfactory thickness uniformity, acceptable integrity, and good adhesion to the sapphire substrate [3].

We select the method of horizontal directed crystallization (HDC) from various methods for sapphire processing (Verneuil's metod, Czochralski's method, etc.). The HDC method allows to obtain sapphire crystals with large cross-section and to carry out effective removal of detrimental impurities. Table 1 shows the main physicochemical properties of sapphire received by HDC method [4].

We use fusible inorganic glassy dielectric to form the sapphire – glassy dielectric junction. The lead oxide in this compound increases the fusibility of the glass composition [5]. Figure 1 shows the area of glass $PbO - B_2O_3 - ZnO$ formation. We receive the materials diagrams with low softening temperature (T_g) and the necessary indicators of spreading at relatively low temperatures in the system $PbO - B_2O_3 - ZnO$ [6].

Table 1. Main physicochemical properties of sapphire

Al_2O_3 level, %	99.99
Melting temperature, °C	2054
Working temperature, °C	2000
Density, g/cm^3	3.97
Mohs hardness	9
Compression strength, MPa	2450
Tensile strength, MPa	990
Integral optical transparency, %	95
Thermal conductivity, cal/cm·s·°C	0.09
Coefficient of thermal expansion, 1/°K	$5 \cdot 10^{-6}$
Resistivity, Ω·cm	10^{16}
Dielectric capacity	10
Electric strength, V/cm	$4 \cdot 10^5$

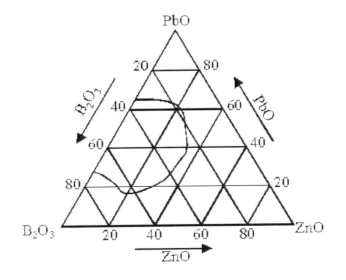

Figure 1. Area of glass $PbO - B_2O_3 - ZnO$ formation.

Table 2 shows the main physicochemical properties of inorganic glassy dielectric PbO – B_2O_3 – ZnO to form the sapphire – glassy dielectric junction.

Glassy dielectric coatings (films) on the substrate surface from the solution (suspension) can be formed by several methods (sol-gel technology): centrifugation, dipping, and diffusion (pulverization). These methods allow us to obtain glassy films with different composition and morphology of the surface without using complex technological equipment. We compare existing methods of film formation in Table 3 [7, 8].

Table 2. Main physicochemical properties of inorganic glassy dielectrics

Material	Required physicomechanical properties
Glass with system PbO – B_2O_3 – ZnO	Coefficient of linear thermal expansion (CLTE) of $85 - 95 \cdot 10^{-7}$ K^{-1} in the temperature interval from 20°C to 500°C
	Spreading temperature does not exceed 600°C
	Microhardness exceeds 350 kg/mm^2
	Absence of crystallization
	Electrical characteristics: $\varepsilon = 10 - 14$; $tg\delta \cdot 10^4 = 20 - 25$; $E_{break} \cdot 10^6 = 5$ V/cm
	Chemical stability corresponds to II dimming class

Table 3. Comparative analysis of film formation methods

Method of film formation	Advantages	Disadvantages
Centrifugation	• suspension durability; • film thickness uniformity; • smaller loss of solution (suspension) for deposition process; • simplicity of film thickness control	• limitations on substrate sizes; • limitations on geometry and form of used substrate
Diffusion (pulverization) figure cen	• possibility of film deposition on surface with various size and form	• film thickness non-uniformity; • bigger loss of solution (suspension) for deposition
Dipping	• possibility of film deposition on surface with various form	• film formation from both sides of substrate; • film thickness non-uniformity; • large quantity of solution for film deposition; • limitations on used substrate sizes

Table 3 shows that glassy films obtained by diffusion (pulverization) and dipping have non-uniform thickness and large quantity of solution is required for film deposition. The centrifugation method allows one to obtain uniform thickness film by solution (suspension) distribution on the substrate surface during rotation. Thus, it is efficient to use the method of centrifugation to obtain small-size glassy dielectric and sapphire junction. It also allows to get a relatively uniform film with thickness from a few nm to tens of microns. The method of centrifugation allows us to easily control the thickness of the applied film by changing speed and duration of rotation.

The suspension solution should be prepared by centrifugation for glassy dielectric and sapphire juncture formation. First of all, the low-melting glass granulate is pulverized to powder with specific surface of 5000 cm^2/g (dry-milling). To prepare the working suspension, we added isobutyl alcohol in the processed powder. We placed the resulting solution in the vibration mill, jasper drum and jasper balls for 24 hours. Low-melting glass of $PbO - B_2O_3 - ZnO$ suspension on sapphire substrate was deposed by centrifugation. It was carried out within 3-5 minutes with a rotor speed of the centrifuge of 7000 rpm. The deposited film is dried in an oven at 50-60°C for 3-5 min. High-temperature annealing was performed in a muffle furnace at $T < 600°C$ during 5-7 min and with isothermal range at $T = 300°C$ during 10 min. Noncrystallizing glassy films of low-melting glass have good adhesion to the substrate materials, acceptable coefficient of linear thermal expansion and consistent temperature of their formation. It provides the most uniform thickness and homogeneous film.

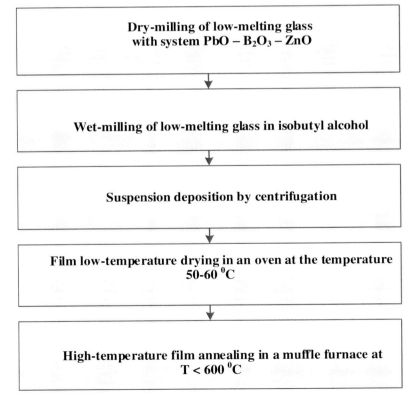

Figure 2. Manufacturing route for sapphire – glassy dielectric junction formation.

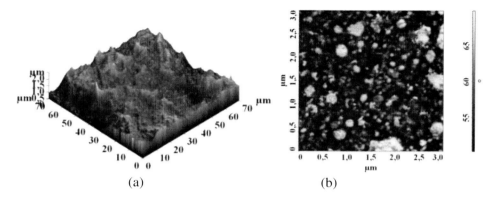

Figure 3. AFM images of films on sapphire substrate.

We develop the manufacturing route for sapphire – glassy dielectric junction formation (see Figure 2).

We use the method of atomic force microscopy (AFM) for quality estimation of the glassy film surface on sapphire substrate. AFM images are presented in Figure 3 (a, b). The film obtained by centrifugation has a thickness of 1 mm, the average height of the peaks is 1.5 mm, the wetting coefficient is 0.7 (which is acceptable), the resistivity of 1.4×10^6 $\Omega \cdot$cm is the minimal one.

CONCLUSION

These experiments allowed us to obtain glassy film with characteristics that are similar to the glass monolith $PbO - B_2O_3 - ZnO$ characteristics.

The study of low-melting glass properties revealed that the dielectrics synthesized in glassy systems that contain oxides of lead, boron, silicon were the most suitable for formation of the sapphire – glassy dielectric junction.

We investigated the low-melting glass of $PbO - B_2O_3 - ZnO$ for junction formation. It had the necessary melting temperature ($T < 600°C$). It allowed us to obtain glassy film with good adhesion to the substrate materials and consistency of linear coefficient of thermal expansion. It gave us the opportunity to obtain the most uniform thickness and homogeneity of the film.

We develop the process of glassy film deposition on sapphire substrates by using centrifugation for 3-5 minutes with a centrifuge rotor speed of 7000 rpm. Low-temperature drying was carried out at 50-60°C for 3-5 minutes. Then the film was annealed at high temperature $T < 600°C$ for 5-7 minutes and with the isothermal exposure at $T = 300°C$ for 10 minutes (we create the manufacturing route). The films morphology (thickness, the average height of the peaks) was studied.

The processing glassy dielectric $PbO - B_2O_3 - ZnO$ is a perspective intermediate stage for the triple sapphire – glassy dielectric – ceramics junction formation for microelectronics.

ACKNOWLEDGMENTS

The results were obtained with the equipment of the Scientific and Educational Center "Laser Technologies," Center of Collective Use and the Scientific and Educational Center "Nanotechnology," Institute of Nanotechnology, Electronics and Electronic Equipment Engineering of the Southern Federal University (Taganrog).

REFERENCES

[1] Savin, JL; Prikhodko, AP; Savin, LS; Makarov, VN; Pshinko, PA; Kononov, DV. *The Structural and Logical Scheme of Protective Covering Formation in GLASS-STEEL System. PGASA, Dnepropetrovsk.*, 2010, pp. 193-197 (in Russian).

[2] Pavlushkin, NM. *Chemical Technology of Glass and Glass Ceramics. Stroyizdat, Moscow.*, 1983, p. 432 (in Russian).

[3] Malyukov, SP. *Glass Dielectrics in Magnetic Heads Production. TSURE, Taganrog.*, 1998, pp. 1 – 181 (in Russian).

[4] Malyukov, SP; Klunnikova, YuV; Nelina, SN. *News of SFU. Technical Science.*, 2010, vol. 7, 210-216 (in Russian).

[5] Malyukov, SP; Kulikova, IV; Kalashnikov, GV. *News of SFU. Technical Science.*, 2011, *vol. 7*, 182-187 (in Russian).

[6] Klunnikova, YuV; Malyukov, SP; Sayenko, AV. *Proceedings of the SPbGU LETI.*, 2014, vol. 8, 15-19 (in Russian).

[7] Malyukov, SP. *News of SFU. Technical Science.*, 2004, *vol. 3*, 175-178 (in Russian).

[8] Sokolov, VI. *Centrifugation. Chemistry.*, 1976, pp. 1 – 209 (in Russian).

In: Proceedings of the 2015 International Conference … ISBN: 978-1-63484-577-9
Editors: Ivan A. Parinov, Shun-Hsyung Chang et al. © 2016 Nova Science Publishers, Inc.

Chapter 14

FEATURES OF TECHNOLOGY OF SAW DEVICES

V. G. Dneprovski and G. Ya. Karapetyan
Vorovich Mathematics, Mechanics and Computer Science Institute,
Southern Federal University, Rostov-on-Don, Russia

ABSTRACT

All the processing steps for the manufacturing of surface acoustic wave (SAW) devices are present at the fabrication of integrated circuits. However, use of piezoelectric single crystals, instead of semiconductor crystals, and specifics of SAW devices, are setting a number of requirements to the known technological processes. In particular, SAW devices require precise alignment of substrates, high resolution of photolithography and uniformity of metal film's thickness. This is very important in the manufacturing of devices in large areas (sometimes over 100 cm^2). Use of manufacturing technology of integrated circuits makes serial production of SAW devices possible. In this chapter, we consider the features of SAW device technology and a number of original methods, which we used in their creation.

Keywords: SAW devices, piezoelectric, IDT, filter, substrate, photomask, technology, film

1. INTRODUCTION

The most common scheme of manufacturing SAW devices includes the following basic operations: production of piezoelectric acoustic line (substrate), preparation of photo master drawing and photomask, metallization of substrate, formation of interdigital transducer (IDT) structures and contact tires, installation in package, and sealing of a device. In the few last years, phototypesetting machines have been used usually to manufacture photomasks [1-3]. They have displaced the photo master drawings.

The main parameters of SAW devices such as operating frequency, bandwidth, insertion loss, thermal stability, distortion due to the effects of the second order, etc. are determined, along with characteristics of piezoelectric substrate material. Therefore, for each design, a choice of the acoustic line material depends on specific characteristics of the device.

2. Piezoelectric Materials for Surface Acoustic Wave Devices

For the manufacturing of the SAW device substrates, both single-crystal and polycrystalline (piezoceramic) materials can be used. The structural perfection single crystals provide small losses in SAW propagation (about $0.1 - 0.5$ dB/cm at frequencies of $1.5 - 2$ GHz). In addition, they are time-stable and have highly reproducible parameters in serial production. Developers use most widely single crystals of quartz, lithium niobate, lithium tantalate and, in some cases, fine-grained piezoceramics as piezoelectric substrates.

Crystallographers grow many piezoelectric single crystals by the Czochralski method in the form of cylindrical boules. Diameter of boules can be up to 150 mm and a length of 200 mm or more. They cut these boules into discs of 0.3–0.5 mm thickness and carefully polish one side. Usually, substrates of quartz, lithium tantalate, and lithium niobate are used in the manufacture of SAW devices. Single-crystal bismuth germanate ($Bi_{12}GeO_{20}$), iodate, lithium silicate, bismuth paratellurid (TeO_2), selenium (Se), as well as films of zinc oxide (ZnO), aluminum nitride (AlN) are materials, promising for use in SAW devices [1, 3, 4]. In recent years, the interest to materials, such as lithium tetraborate ($Li_2B_4O_7$), langasite ($La_3Ga_5SiO_{14}$), langatate ($La_3Ga_{5.5}Ta_{0.5}O_{14}$), gallium orthophosphate ($GaPO_4$), CTGS ($Ca_3TaGa_3Si_2O_{14}$), increased [5, 6].

Perhaps, lithium tantalate ($LiTaO_3$) is the only material, in which high electromechanical coupling coefficient combines with good thermal stability. Therefore, a prime interest exists to use $LiTaO_3$ for thermostable broadband filters.

Bismuth germanate ($Bi_{12}GeO_{20}$) is a suitable material for the delay lines with over long delay due to low SAW propagation velocity and large size of produced crystals. The disadvantage of the material is a high temperature coefficient of delay.

Piezoceramic materials are almost an order of magnitude cheaper than single crystals. Developers control the piezoceramic properties easily by changing of the chemical composition and introducing the modifiers. In addition, the piezoceramics permit to produce blanks for substrates of various configurations, including large-sized. The principal disadvantage of piezoelectric ceramics, as compared to single crystals, is sharp attenuation of SAW with increasing of signal frequency. Besides, porosity of surface sometimes leads to short-circuit or breaking electrodes of interdigital transducer (IDT) of device after metal film deposition and photolithography. These disadvantages are explained by grain structure of piezoceramic. These conditions allow use of piezoceramics for making relatively low-frequency broadband SAW devices, where the values of time instability and reproducibility of properties are not essential. In particular, piezoceramic TV SAW filters successfully produced by enterprises "Horizont" (Minsk, Belarus) and "Murata" (Japan).

3. Preparation of Substrates for Deposition of Films

It should be noted that developers observe variations of SAW propagation velocity within discs of lithium niobate. We observed a variation of SAW velocity and electromechanical coupling coefficient within disk $LiNbO_3$ [7]. If high variation of velocity led to decrease of output percentage of manufacture, we divided substrates (disks) into groups with close SAW velocities. Then we made own photomask for each group of disks. It led to increase of output percentage of manufacture.

Chemical processing of plates occurs before film deposition. Purity of substrates plays the huge role in film deposition. The higher operating frequency of SAW device defines the higher requirements for dust-free surrounded airs. The higher operating frequency of SAW device, the higher requirements for surrounding dust-free air.

We placed the single crystal substrates of lithium niobate in a special cassette and then boiled in toluene and in acetone. We carried out each of the operations within 10 minutes. Then we treated substrates under 323 – 333 K during 15 minutes and thereafter washed them in hot and then cold deionized (or distilled) water. After this, we dried substrates in thermostat at about 423 K [8]. Similarly, we cleaned piezoquartz substrates. But in this case, before purification, we treated them within 10 minutes in $H_2O_2:HCl:H_2O$ composition, prepared in the proportion 1:1:4 at temperature of 333 K.

The described cleaning technique of substrates provides good adhesion and allows obtaining of interdigital transducers (IDT) with high resolution on the surface of piezoelectric crystals after photolithography. Typically we varied the thickness of the aluminum film in the range of 0.1 – 1.0 microns. We varied the thickness of sublayer from vanadium film from 10 to 30 nm. Obviously, the higher operating frequency of device, the smaller thickness of thin metal film.

Adhesion of film increases with additional purification of lithium niobate single crystals by argon ion beam. Immediately, before deposition of film, we made it by argon ion beam at energy of the ions 4–5 keV. In particular, we carried out such treatment of piezoelectric waveguide substrates in vacuum system URM 3.279.026 coupled with the ion source "Radical." Note, that in this case, after substrate cleaning for several tens of minutes, we observed the decrease of electrical resistance of more than an order of magnitude. We explain this effect by selective disruption of bonds of the crystal lattice on the lithium niobate surface during ion bombardment and relaxation processes after its completion.

4. DEPOSITION OF THIN FILMS FOR SAW DEVICES

In most SAW devices, IDTs are made from aluminum thin film, deposited by the vacuum deposition method. Low resistivity, good development of technology, and high chemoresistance of aluminum films to environment, mainly determine selection of the aluminum as the material for electrodes.

As aluminum was relatively low-melting metal, for deposition of films we had used very convenient method of thermal evaporation in vacuum. To ensure good adhesion of aluminum, we condensed 5–30 nm thickness vanadium film as an underlayer. The optimum substrate temperature during deposition of aluminum was 393 ± 10 K and deposition rate was 10 nm/s. Film thickness of aluminum was varied in the range of 80–200 nm, depending on the device operating frequency. Many researchers often also use ion-plasma deposition of films for metallization of piezoelectric substrates.

In SAW device technology, the laser evaporation is also used. At present time, deposition of thin films by laser receives an increased use, particularly for film deposition of nanostructures and different films for sensors of various physical magnitudes [9-12].

The laser evaporation has a number of essential advantages. In particular, they include the possibility of evaporation of any refractory materials and complex compounds, as well as the possibility of deposition of ultrathin films by pulsed laser. The last advantage of the method we used for adjustment of operation frequency of narrow-band acoustoelectronic devices. Furthermore, in the case of pulse mode due to large velocity of condensation, there is possible heating of substrate to very high temperature. This fact provides good adhesion of film without additional external heat source.

To assess the degree of substrate heating at pulse laser deposition we solved one-dimensional task. We supposed that the condensed material is identical to material of the substrate [11]. We considered the power density Q_p, absorbed by condensate, as constant in time and uniform over cross-section, and the substrate, as half-space oriented normally to the flow of steam:

$$\begin{cases} \dfrac{\partial T}{\partial t} = a^2 \dfrac{\partial^2 T}{\partial x^2} - v_c \dfrac{\partial T}{\partial x}, \quad x \geq 0, \quad t \geq 0; \\ \dfrac{\partial T}{\partial x}\bigg|_{x=0} = -\dfrac{Q_n}{K_0}; \\ T(x,0) = T_0, \quad T_s = T(0,t). \end{cases} \qquad (1)$$

In the equation of heat conduction we took into account movement of boundary of vapor–solid phase due to the high condensation speed. The velocity v_c is considered constant. By solving equations (1) and setting $x = 0$, we obtain the following temperature as a function of time:

$$T_s(t) = T_0 + \dfrac{aQ_n}{K_0} \int_0^t \left(\dfrac{e^{-\frac{v_c^2 \tau}{4a^2}}}{\sqrt{\pi \tau}} d\tau + \dfrac{v_c}{2a} Erfc\left(-\dfrac{v_c}{2a}\sqrt{\tau}\right) \right) d\tau. \qquad (2)$$

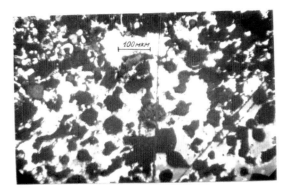

Figure 1. Area of revaporization of rhenium film.

Particularly, for $Q_p = 10^8 \, \dfrac{W}{m^2}$, film surface temperature T_s reaches 1400 K for 10 ms, and for $Q_p = 10^9 \, \dfrac{W}{m^2}$ T_s reaches 4000 K for 1 ms. The values of $10^8 \, \dfrac{W}{m^2} - 10^9 \, \dfrac{W}{m^2}$ are quite achievable. In experiments, we even observed a revaporization (Figure 1) of rhenium films under specific conditions. It should be noted that rhenium is the metal with the highest boiling point (5596°C) [13].

5. PHOTOLITHOGRAPHY

In SAW devices creation one of the challenges is the formation of IDT on chemically active substrates. These substrates were damaged by etchant, during etching target topology of acoustoelectronic device. Sometimes, for solution of this problem, we deposited dielectric thin films (e.g., SiO or GeO [14]) on piezoelectric substrates before their metallization. We improved the quality of thin-film elements and simplified the manufacturing process by eliminating of deposition stage of dielectric thin film. We developed the method for forming IDT on surface of chemically active substrate (CdS). This technique presents substantial interest for planar acoustooptic devices.

We deposited two-layer vanadium-aluminum system on the active surface of CdS substrate. Then, we coated metallized surface of substrate by photoresist, using method of centrifugation. Rotation speed was about 3,000 turnovers per minute. Further, we dried photoresist and exposed it to ultraviolet radiation of mercury lamp. After treatment of photoresist by 0.5% KOH solution, it was subject to strengthening. Then we etched two-layer V-Al system through the protective layer of photoresist. Most of conventional etching agents damages surface of CdS, what is undesirable. Therefore, we etched thin film elements from V-Al through windows, formed in photoresist by special etchant, based on H_3PO_4 and CH_3COOH. This etchant didn't damage the surface of cadmium sulfide. Then we have removed photoresist, and have obtained thin film elements of target topology on the surface of single crystal CdS.

6. ADJUSTMENT OF CENTER FREQUENCY OF RESONATORS AND NARROW-BAND SAW FILTERS

In the manufacture of SAW resonators and other SAW narrow-band devices, there are some inevitable variations in the target frequency values due to imperfection of technology.

In our experiments, relative spreading of the center frequency of narrow-band devices reached up to 10^{-3}. We tuned the center frequency by deposition of thin dielectric film GeO or SiO on the entire operating surface of the SAW device. Thus, by making so, we implemented device passivation.

We condensed films in vacuum by CO_2 laser. This method allowed reducing the central frequency to 2 percents. At that, for example, when Q-factor of the resonator halved, insertion losses increased by 10–20 percents. At technological relative frequency deviation within 10^{-3} the reduction of the Q-factor and increase in insertion losses did not exceed 10 percents.

We carried out experiments [3] for adjustment of the central frequency f_0 of narrowband SAW filters with bandwidths from 0.1% to 0.3% for ST-cut of single crystal quartz. Central frequencies were in the range from 40 to 160 MHz for these filters. In particular, when we adjusted SAW filter with $f_0 = 135$ MHz, we observed deviation of frequency 5.3 kHz/nm.

When the relative change of magnitudes of f_0 was up to 1%, increase of insertion losses did not exceed 3 dB. Relative frequency deviation of ST-cut quartz devices was $(2.5 - 5.0) \times 10^{-5}$ nm^{-1} and was linear at GeO thicknesses up to 200 nm.

When we adjusted frequency of SAW filters based on lithium niobate substrates of YX/I28°-cut, we also observed decrease of f_0 with increase of film thickness of GeO. The relative deviation of f_0 was $(1.5 - 3.5) \times 10^{-5}$ nm−1 for different samples.

The studied method of adjustment hasn't an appreciable effect to the frequency response parameters, except for narrowing of bandwidth filters on substrates of lithium niobate, observed in some cases. Use of laser also allows realization of frequency adjustment in the case of packaged devices, if the package has a window, transparent to ultraviolet, visible or infrared radiation.

7. SEPARATION OF PIEZOCERAMIC PLATES AND PACKAGING OF SAW DEVICES

In the uncut plate, we measured devices parameters, obtained after processes of group photolithography. We carried out screening of devices with the degraded parameters due to changes in the IDT geometry.

Then we scribed or cut plates by diamond disc. After scribing, a plate had cuts with deepness 50–70% of its thickness, and then it was broken to chips.

For production of SAW devices, microcircuit packages are often used. Sealing of such packages was performed by soldering of cover to the base or by laser welding. In some cases, if the frequent connections with special connector are necessary, the special package was developing. As example in Figure 2a, we present the package of the descrambler (SAW decoder) [15] for television channels, which allows frequent connections to antenna socket of a TV receiver and antenna cable. In Figure 2b, we show the bandpass SAW filter in the package of the Terek type.

Sometimes, sealed package with SAW device was mounted in an outer package with tuning elements (in particular, variable inductors and capacities). The availability of these elements allows us more accurately adjust the filter directly into the feeder path, that is not always possible in a stand, especially at high frequencies. They combined both SAW components (such as multi-band filters, antenna duplexsers [16]) and components, which were made by other technology (e.g., BAW, semiconductor IC, MEMS). In particular, in 2011 Murata firm launched the first world mass production of modules combined in a single package SAW filter, power amplifier and magnetic stabilizer. The module does not require any additional matching elements and has sizes $6.6 \times 3.8 \times 1$ mm^3 [17]. Kyocera firm produced miniature module LS-D110S3 with sizes $2.5 \times 2.0 \times 1.1$ mm^3 on the ceramic multilayer substrate, with a high degree of integration for mobile phones and smartphones [18].

Along with simple discrete SAW components (filters, resonators, delay lines etc.), companies produce wide range of integrated components and modules on SAW technology base.

Thus, in the present chapter, we have considered features of SAW devices technology and a number of original methods, which we used at the creation of these devices. Finally, we have discussed cleaning of substrates and laser deposition of thin films. We have considered the technology of interdigital transducers on chemically active substrates, and adjustment of the center frequency of SAW narrowband devices too.

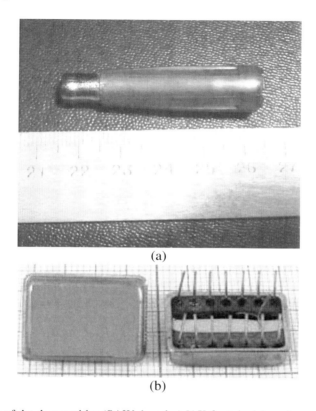

Figure 2. (a) Package of the descrambler (SAW decoder) [15] for television channels developed by Research and Production Enterprise "Piezotron," (b) bandpass SAW filter in the "Terek" type package.

ACKNOWLEDGMENTS

This work was performed with financial support on the theme "Development of Methods of Synthesis, Theoretical and Experimental Study of Nanostructures Based on Zinc Oxide to Create the Elements of Phot–odetectors, Optical Nanoantennas, Piezo- and Chemosensors" No. 16.219.2014K of the project part of the State Task and on the theme "Development and Implementation of Theoretical, Modeling and Experimental Researches of Advanced Piezo-ferroelectrics, Superconductors and Thin Film Devices" No. 213-01-11/2014-25 of the base part of the State Task from the Ministry of Education and Science of the Russian Federation.

REFERENCES

[1] Orlov, V. S,; Bondarenko, V. S. *Surface Acoustic Wave Filters*. Radio and Svyaz, Moscow, 1984, pp. 1 – 272 (in Russian).

[2] Orlikov, L. N. *Technology of Materials and Electronic Products.* TGUSU and RE, Tomsk, 2006, pp. 1 – 364 (in Russian).

[3] Dneprovski, V.G.; Karapetyan, G.Y. *Surface Acoustic Wave Devices.* Southern Federal University, Rostov-on-Don, 2014, pp. 1 – 186 (in Russian).

[4] Ash, E.; Farnell, G.; Gerard, H.; et al. *Surface Acoustic Wave.* A. Oliner (ed.). Springer-Verlag, Berlin, Heidelberg, New York, 1978, pp. 1 – 328.

[5] Morgan, David P. *Surface Acoustic Wave Filters.* 2nd Edition. Elsevier, Amsterdam, Oxford, New York, Tokyo, et al., 2007, pp. 1 – 446.

[6] Balysheva, O.L. *Radio Electronic Journal. Online Journal, No. 6,* 2014. http://jre.cplire.ru/jre/jun14/2/text.html (in Russian).

[7] Dneprovski, V.G.; Karapetyan, G.Y.; Perevoshchikova, T.V. Research of SAW Velocity and Electromechanical Coupling Coefficient Dispersion in $LiNbO_3$ Disk. *Proc. of the Conference "SAW Acoustoelectronic Apparatus of Information Processing."* KPI, Cherkassy, 1990, 427-428 (in Russian).

[8] Dneprovski, V.G.; Baghdasaryan, S.A. Cleaning of Substrates in the Manufacture of Surface Acoustic Wave Devices. *Proc. of the Intern. Conf. "Actual Problems of Radio Engineering and Electronics."* Naukova Dumka, Sevastopol – Kiev, 2004, 17-19 (in Russian).

[9] Zhi Chen; Chi Lu. *Sensor Letters,* 2005, *vol. 3,* 274 – 295.

[10] Nikolaev, A. L.; Karapetyan, G. Ya.; Nesvetaev, D. G.; Lyanguzov, N. V.; Dneprovski, V. G.; Kaidashev, E. M. Chapter 3. Preparation and Investigation of ZnO Nanorods Array Based Resistive and SAW CO Gas Sensors. pp. 27-36. In: *Advanced Materials - Physics, Mechanics and Applications.* Springer Proceedings in Physics, *vol. 152,* Shun-Hsyung Chang, Ivan A. Parinov, Vitaly Yu. Topolov (Eds.). Springer Cham, Heidelberg, New York, Dordrecht, London, 2014, pp. 1 – 380.

[11] Dneprovski, V.G.; Osadin, B.A.; Rusakov, N.V. *Journal of Technical Physics,* 1974, *vol. XLIV(2),* 435-441 (in Russian).

[12] Schuka, A.A.; Dneprovski, V.G.; Dudoladov, A.G. *Foreign Electronic Technique,* 1973, *No. 24(72),* 38 – 65 (in Russian).

[13] Dneprovski, V.G.; Schuka, A.A. Rhenium Films Obtained by Laser Sputtering and Their Properties. In: *Study and Application of Rhenium Alloys.* IMET, USSR Acad. of Sciences, Nauka, Moscow, 1975, 166 -167 (in Russian).

[14] Dneprovski, V.G. *Electronic Technique: Series 6 (Materials),* 1982, *No. 6 (167),* 75-76 (in Russian).

[15] Balakin, V.I.; Voropaev, V.P.; Dneprovski, V.G.; Karapetyan, G.Y.; Perevoshchikova, T.V.; Rozhkov, I.S.; Shikulya, P.I. *Decoder: Industrial Model.* Russian Patent 43226 RF. No. 95500229, 16.02.1997. (in Russian).

[16] Products of TAI-SAW Technology Co., LTD. http://www.taisaw.com/en/product.php.

[17] World's First SAW Filter, Power Amplifier, Magnetic Stabilizer in One Package: Mobile Phone Transmission Module to be Marketed. http://www.murata.com/new/news_release/2011/0913c/index.html.

[18] Developments of Miniature SAW Filter Module "LS-D110S3" for Diversity. http://global.kyocera.com/prdct/electro/news/2011/111031. html.

In: Proceedings of the 2015 International Conference … ISBN: 978-1-63484-577-9
Editors: Ivan A. Parinov, Shun-Hsyung Chang et al. © 2016 Nova Science Publishers, Inc.

Chapter 15

UTILIZATION OF LEONARDITE FOR PRODUCING SLOW-RELEASE FERTILIZER

Petchporn Chawakitchareon[1,], Jittrera Buates[2] and Rewadee Anuwattana[3,†]*

[1,2]Department of Environmental Engineering, Faculty of Engineering,
Chulalongkorn University, Bangkok, Thailand
[3]Thailand Institute of Scientific and Technological Research, Technopolis,
Khlong Luang, Pathumthani, Thailand

ABSTRACT

This chapter was conducted with the objective of producing slow-release fertilizer made from leonardite. The leonadite was obtained as waste from a power generation plant in Thailand and modified by adding certain fertilizer materials to ensure that the nutrient contents closely resembled N:13P:13K:13, which is a commercial formula for slow-release fertilizer. Bentonite was added as a binder at 10%wt. Two series of fertilizer granules were prepared from leonardite. One of the aforementioned was uncoated and the other coated with resin. To evaluate the releases of the N, P and K nutrients, one gram of each fertilizer sample was placed in a flask filled with 50 milliliters of distilled water. Sample solutions were collected after 1, 5, 24 and 48 hours of shaking. According to the chemical analysis after 48 hours, the uncoated sample released 0.1717 percent weight of N, 0.1125%wt of P and 0.1001 percent weight of K. All were higher than the nutrients released from both the coated fertilizer prepared from leonardite and the N:13P:13K:13 commercial slow-release fertilizer formula. After 48 hours, the coated fertilizer sample was found to release N, P and K in amounts of approximately 0.0421, 0.1125 and 0.1001 weight percentage, respectively. Microscopic observation of the coated and uncoated fertilizer granules by Scanning Electron Microscopy (SEM) revealed numerous pores on the surface of the uncoated samples while the coated granule and commercial slow-release fertilizer formula possess rather smooth surfaces. Resin coating helped reduce the

---

[*] Corresponding Author address: E-mail: petchporn.c@chula.ac.th.
[†] E-mail: rewadee_a@tistr.or.th.

pores on the surface of the fertilizer granules, thereby leading to slower release of the nutrients.

Keywords: leonardite, slow-release fertilizer, resin, waste utilization

1. INTRODUCTION

Fertilizer can be defined as any substance containing one or more recognized plant nutrients. It is generally classified by the plant nutrient contents or claims about promoting plant growth. Fertilizers can be composed of multiple fertilizer materials in addition to additives that alter the transformation in the soil, maintain good physical condition, reduce corrosiveness and serve purposes other than providing plant nutrients and micronutrients. The Association of American Plant Food Control Officials (AAPFCO) defines primary nutrients as nitrogen (N), available phosphate (P_2O_5) and soluble potash (K_2O) [1]. At present, fertilizers are regarded an essential input for sustainable development of crop yields. However, a large amount of fertilizer is lost by volatilization into the air and leaching out to the surface and ground water. The consequences are water pollution and environmental problems.

One approach to preventing environmental problems and reducing the nutrient loss is to use slow-release fertilizers (SRFs) capable of gradually releasing nutrients in optimum amounts for plants. Compared to conventional fertilizers, slow-release fertilizers offer many advantages such as decreasing fertilizer loss rate, sustainably supplying nutrients, lowering application frequency and minimizing potential negative effects associated with over-dosage. The slow-release fertilizers are generally classified into the following four types: (i) inorganic materials of low solubility such as metal ammonium phosphates; (ii) chemically or biologically degradable low solubility materials such as urea-formaldehyde; (iii) relatively soluble materials that gradually decompose in soil and (iv) water soluble fertilizers controlled by physical barriers such as coated fertilizers. Coated fertilizers prepared by coating the granules with various materials are the main category of current slow-release fertilizers. The resins often used as coating materials include polysulfone, polyvinyl chloride, polystyrene [2]. Commercial N13:P13:K13 slow-release fertilizer is successfully used in planting because it enhances plant growth with balanced nutrient contents. The N13:P13:K13 commercial formula has been reported as the ideal fertilizer for perennials and slower-growing flowering crops. It is the most appropriate for hanging basket plants [3].

This study aims to produce slow-release N13:P13:K13 fertilizer from leonardite, a natural organic product from the decomposition of lignite for more than 70 million years. Leonardite is an oxidized form of lignite occurring at shallow depths that over lies more compact coal in coalmines [4]. It is undesirable for applications such as fuel due to its high oxygen content at approximately 28-29% wt in comparison with the 19-20%wt of lignite. Leonardite is rich in organic matter and humic acid (50-75%wt and 30-80%wt, respectively) [6]. Humic materials are complex organic molecules that contain a wide variety of functional groups, namely, carboxyl, hydroxyl and carbonyl. Humic acid can be beneficial to plant growth by stimulating the plant to absorb greater quantities of nutrients and inducing greater efficiency in the use of the absorbed nutrients. Hydrophobic experiments have shown low concentrations of humic acid to have a positive effect on nitrate and ammonium uptake in olives (*Oleaeuropea* L. 'Maurino') [7].

2. MATERIALS

The leonardite serving as the raw material to be modified as a slow-release fertilizer in this study was obtained from a power generation plant in Thailand. Table 1 lists some of its physical and chemical characteristics.

The fertilizer used to modify the leonardite contained diammonium phosphate, urea and potassium chloride and was manufactured by YVP Fertilizer Company, Thailand. The bentonite used as a binder was manufactured by TCM Limited Company, China. The commercial slow-release N13:P13:K13 fertilizer formula analyzed in this work was Osmocote® manufactured by Sotus International Company, Nonthaburi, Thailand.

Table 1. Some physical and chemical properties of leonardite

Properties	Units	Value
pH	–	2.65
Ratio of carbon to nitrogen (C/N ratio)	–	63.46
Electrical conductivity (EC)	ds/m	10.71
Cations exchange capacity (CEC)	cmol/kg	53.35
Organic matter (OM)	Percent	25.23
Humic acid	Percent	35.65

3. METHODS

3.1. Determination of Nutrient Contents

The nitrogen content was determined by the Kjeldahl Method, while the phosphorous concentration was estimated colorimetrically by using ammonium molybdate and ammonium metavanadate. The potassium content was determined by using an atomic absorption spectrometer [8].

3.2. Preparation of the Slow-Release Fertilizer

Both the leonardite and fertilizer materials were sifted to obtain particles with a top cut of 0.25 mm. The slow-release fertilizer was prepared by using leonardite as the main component and the nutrient contents of the leonardite were adjusted to closely resemble those of the N13:P13: K13 commercial slow-release fertilizer formula (Osmocote®) by adding diammonium phosphate (18-46-0), urea (46-0-0) and potassium chloride (0-0-60). Moreover, bentonite was used as a binder at 10%wt. All the components of the slow-release fertilizer sample prepared from leonardite were mixed thoroughly before a small amount of water was sprayed over each mixture to form a paste. The slow-release fertilizer from leonardite was prepared in granules sized approximately 4 mm in diameter. The granules were dried for one hour in an air-circulating oven at 105 °C then heated in a muffle furnace for another hour at 200 °C. Some of the slow-release fertilizer granules were coated with resin and re-heated at 50°C for three hours. Figure 1 shows granules of uncoated and coated slow-release fertilizer samples made from leonardite.

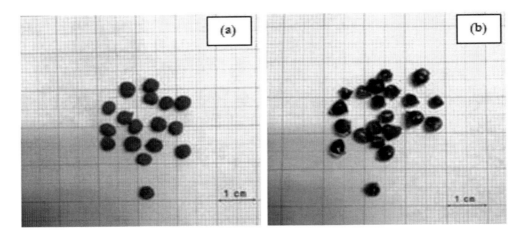

Figure 1. Granules of (a) uncoated and (b) resin-coated slow-release fertilizer made from leonardite.

3.3. Determination of Nutrient Release Contents

To evaluate the nutrient release of the slow-release fertilizer prepared from leonardite, one gram of each sample fertilizer was placed in a 125-ml flask containing 50 ml of distilled water. Then the flasks were placed in an automatic shaker. The sample solutions of each fertilizer formulation were collected after 1, 5, 24 and 48 hours of shaking. The suspended solid matter was separated from the solutions before the nutrient contents were determined for each solution.

4. RESULTS AND DISCUSSION

4.1. Nutrient Contents

Table 2 shows the *nitrogen (N), phosphorus (P) and potassium* (K) nutrients in the leonardite as received. *The leonardite was composed of 0.63%wt of N, 0.10%wt of P and 1.71%wt of K.* Interestingly, leonardite contains rather high levels of primary plant nutrients. Furthermore, the potassium content is much higher than that found in standard organic fertilizers (more than 0.5%wt) [9]. Therefore, leonardite seems to be an appropriate candidate for making fertilizers promoting plant growth [10, 11] as long as the plant nutrients in leonardite can be adjusted at levels comparable to commercially available slow-release fertilizers. Previous studies have reported the use of modified leonardite as a crop growth enhancer in which all of the *N, P and K* nutrients were reported to have been released into the soil [12]. Moreover, the K level in the soil was found to rise higher than standard fertilizers [12].

Table 2. Nutrient contents of leonardite as received

Nutrient Contents	Amount (%wt)
Nitrogen (total N)	0.63
Phosphorus (total P_2O_5)	0.10
Potassium (total K_2O)	1.71

4.2. Surface Morphology

The surface morphology of the N13:P13:K13 commercial slow-release fertilizer and all the slow-release fertilizers prepared from leonardite was investigated microscopically by a Scanning Electron Microscopy (SEM) as illustrated in Figures 2 and 3.

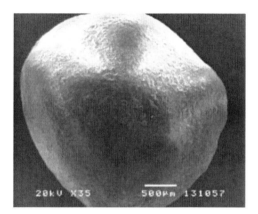

Figure 2. The SEM image shows the rather smooth surface of a N13:P13:K13 commercial slow-release fertilizer granule.

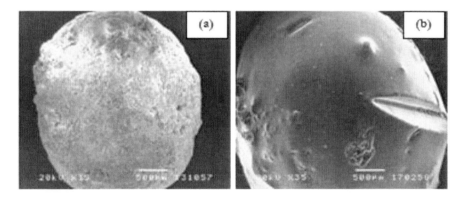

Figure 3. The SEM images show the surfaces of (a) uncoated and (b) resin-coated slow-release fertilizer granules made from leonardite.

The N13:P13:K13 commercial slow-release fertilizer formula in Figure 2 has a rather smooth surface without any large pores, because the N13:P13:K13 commercial slow-release fertilizer formula was coated with resin. The uncoated sample in Figure 3(a) has a rough surface with numerous pores ranging in size from approximately 31 to 188 μm. The resin-coated granule in Figure 3(b) has a smoother surface with fewer pores than the uncoated granule. However, some unevenness remained present in the resin layer of the coat.

4.3. NPK Nutrient Release Contents

Figures 4-6 illustrate the amounts of N, P and K released at various durations for all of the slow-release fertilizers made from leonardite in this study compared with the nutrients released from the N13:P13:K13 commercial slow-release fertilizer formula (Osmocote®). According to the chart, the amounts of N, P and K released dropped significantly with resin coating. The nitrogen amounts released from the uncoated fertilizer after 1, 5, 24 and 48 hours were 0.1210, 0.1626, 0.1756 and 0.1717 weight percentage, respectively. However, the amounts released from the coated fertilizer at identical intervals were much lower at approximately 0.0157, 0.0348 and 0.0421%wt, respectively. Moreover, all of the amounts are similar to the amounts released from the N13:P13:K13 commercial slow-release fertilizer formula (Osmocote®).

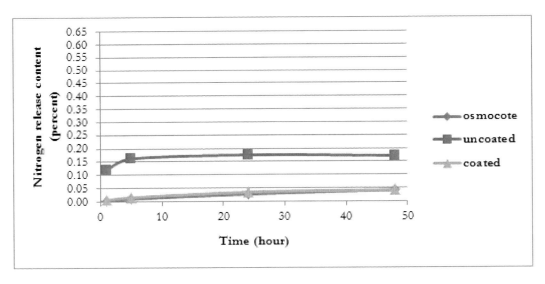

Figure 4. Nitrogen release of Osmocote® in uncoated and coated fertilizer made from leonardite.

For the phosphorus content release illustrated in Figure 5, the uncoated fertilizer made from leonardite released 0.2014, 0.2753, 0.2736 and 0.2706%wt of P after 1, 5, 24 and 48 hours, respectively. Each of the released content amounts was more than double the Preleased from the coated samples at 0.0124, 0.0339, 0.0916 and 0.1125% wt at the same intervals. The N13:P13:K13 commercial slow-release fertilizer formula (Osmocote®) exhibited the slowest release of phosphorus.

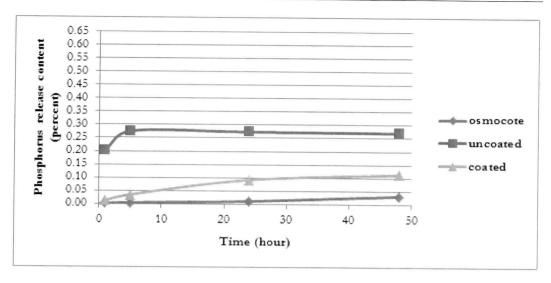

Figure 5. Phosphorus released from the N13:P13:K13 commercial slow-release fertilizer formula (Osmocote®) with samples of uncoated and coated fertilizers made from leonardite.

According to Figure 6, the K release content from the uncoated fertilizer at 1, 5, 24 and 48 hours was exceedingly high at 0.4414, 0.5921, 0.6123 and 0.6277%wt, respectively. According to the chart, the presence of the resin coat dramatically lowered the potassium release rate by nearly twelve-fold after five hours. The drop was near six-fold after 24 and 48 hours. Nevertheless, the coated fertilizer made from leonardite continued to release greater amounts of P than the N13:P13:K13 commercial slow-release fertilizerformula (Osmocote®).

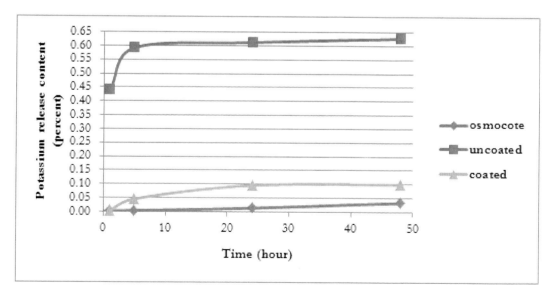

Figure 6. Potassium released from the N13:P13:K13 commercial slow-release fertilizer (Osmocote®) with samples of uncoated and coated fertilizers made from leonardite.

Fertilizer nutrients are released when water permeates into the granules, thereby causing the granules to expand sufficiently to induce the formation of tiny cracks on the surface of the granules [13]. According to the results in Figures 3-6, the presence of the resin coat decelerated all of the N, P and K nutrient release rates. It is indicated, therefore, that the resin coat concealed the numerous pores on the surface of the fertilizer granules. As evident in the SEM micrographs illustrated in Figure 3, the resin coat obstructed the penetration of water and lowered the amounts of nutrients leaching out to plant and soil.

The coated fertilizer made from leonardite had a slow nutrient release rate but slightly higher P and K release rates than those depicted by the N13:P13:K13 commercial slow-release fertilizer (Osmocote®). The desirable slow nutrient release of the Osmocote® might be attributed to its superior smooth surface and thicker resin coat on the surface as shown microscopically in Figure 2.

CONCLUSION

The nutrient release contents of the uncoated fertilizer made from leonardite are higher than the contents from the coated fertilizer and the N13:P13:K13 commercial slow-release fertilizer (Osmocote®). After 48 hours, the N, P and K released from the uncoated fertilizer samples were 0.1717, 0.2706 and 0.6277%wt while the contents from the coated samples were 0.0421, 0.1125 and 0.1001%wt, respectively. The resin coat on the fertilizer surface led to the closure of some pores and subsequent reduction of the amount of nutrients released. The nutrient release rates of the N13:P13:K13 commercial slow-release fertilizer (Osmocote®) were slightly slower than the nutrients released from the coated fertilizer made from leonardite due to the Osmocote's thicker resin coat coupled with superior smoothness.

ACKNOWLEDGMENTS

The authors gratefully acknowledge the financial grant and other support from the TISTR funds for the master's degree student program. Much appreciation also goes to the Graduate School of Chulalongkorn University, Thailand, for its partial financial support. The authors would further like to acknowledge the help and support received from Rochana Tangkoonboribun, PhD, for the use of laboratory instruments.

REFERENCES

[1]   *Background report on fertilizer use, contaminants and regulations. National Program Chemicals,* Division Office of Pollution Prevention and Toxics, U.S. Environmental Protection Agency, Washington D.C., 1999.
[2]   Ni, B.; Liu, M.; Lu, S. *Chemical Engineering Journal*, 2009, vol. 155, 892-898.
[3]   *Fertilizers.* American Plant Products and Services Incorporation, 2010, http://americanplant.com/fertilizers.pdf.
[4]   Stevenson, F. *J. Crops Soils*, 1979, vol. 31, 14-16.
[5]   Baltazzi, E.; Wilip, B.;Wilip, E.; Park, O. *Process for the Halogenation of Leonardite and the Product thereof.* United States Patent No.317,946,1965.

[6]   Akinremi, O. O.; Janzen, R. L.; Lemke, R. L.; Larney, F. J. Can. *J. Soil Sci.*, 2000, vol. 80, 437-443.

[7]   Duval, J. R.; Dainello, F. J.; Haby, V. A.; Earhart, D. R. Hort. *Technology*, 1998, vol. 8, 564-567.

[8]   *Guide to Laboratory Establishment for Nutrient Analysis of Fertilizer*. Agricultural Technology Department, Thailand Institute of Scientific and Technological Research, 2005.

[9]   *Fertilizer Analysis*. Department of Agriculture, Ministry of Agriculture and Cooperatives, Bangkok, 2005.

[10] Sanli, A.; Karadogan, T.; Tonguc, M. Turk. *J. Field Crops*, 2012, vol. 18, 20-26.

[11] Pruitt, N. W. *Controlled Release Composition and Method of Manufacturing Same.* United States PatentNo. 4,975,108, 1990.

[12] Ece, A.; Saltali, K.; Eryigit, N.; Uysal, F. *Journal of Agronomy*, 2007, vol. 6, 480-483.

[13] Scotts Australia, *How Osmocote Works*. 2015, http://www.scotts australia.com.au.

II. Physics

In: Proceedings of the 2015 International Conference … ISBN: 978-1-63484-577-9
Editors: Ivan A. Parinov, Shun-Hsyung Chang et al. © 2016 Nova Science Publishers, Inc.

Chapter 16

PIEZOELECTRIC PERFORMANCE AND FEATURES OF MICROGEOMETRY OF NOVEL COMPOSITES BASED ON FERROELECTRIC ZTS-19 CERAMICS

Petr A. Borzov[1,2], Alexander A. Vorontsov[1], Vitaly Yu. Topolov[2,] and Olga E. Brill[1]*

[1]Scientific Design and Technology Institute 'Piezopribor',
Southern Federal University, Rostov-on-Don, Russia
[2]Department of Physics, Southern Federal University, Rostov-on-Don, Russia

ABSTRACT

Results of experimental studies on piezoelectric properties of the ferroelectric ceramic/polymer composite and features of its microgeometry are discussed at volume fractions of polymer $0.05 \leq m_{pl} \leq 0.50$. An interpretation of experimental data is put forward, and the role of the polymer component and its porosity in forming the piezoelectric response of the composite in a wide volume-fraction range is analyzed. Large values of piezoelectric coefficients $\left| g_{3j}^* \right|$ are achieved in the presence of the polymer component with a high mechanical strength, and this performance is important for sensor applications.

Keywords: ZTS-19 ceramic, F-2ME polymer, volume fraction, piezoelectric coefficient, microgeometry

1. INTRODUCTION

In the last decades, piezocomposites have been of interest due to various effective electromechanical properties and related parameters [1–3], which are not achieved in

---

* Corresponding author: Department of Physics, Southern Federal University, 5 Zorge Street, 344090 Rostov-on-Don, Russia, E-mail: vutopolov@sfedu.ru.

separated piezoelectric components, such as poled ferroelectric ceramics (FCs), single crystals and polymers. The piezocomposites are used in various piezotechnical applications, and knowledge of the performance of these materials in wide volume-fraction ranges is of importance. As is known, effective electromechanical properties of the widespread FC/polymer composites depend [2] on the properties of components, microgeometry, poling conditions, etc. In earlier studies, a link between the microgeometry of composite samples and their effective piezoelectric properties did not discuss in detail, especially at changes in the volume fraction of a piezo-passive polymer component in the wide range. The present chapter is devoted to experimental studies on the piezoelectric properties and microgeometry of a novel FC/polymer composite in a wide volume-fraction range. This composite contains ZTS-19 (PZT-type FC [1, 2]) and F-2ME (polymer with a high mechanical strength, a representative of the fluorine-plastic group). We add that no publications are known that are concerned with a F-2ME-containing piezocomposite.

2. EXPERIMENTAL RESULTS AND INTERPRETATION

The novel composite was manufactured at the Scientific Design and Technology Institute `Piezopribor', Southern Federal University. At the first stage, ZTS-19 FC was prepared using the conventional technology, and the sintered FC samples were crushed into grains. At the second stage, the composite samples were manufactured at mixing the components and then pressed at about 0.1 MPa and 500 K. After cooling, at the third stage, electrodes were put on the sample faces, and then the samples were poled in silicone oil under the electric field $E \approx 3.6$ MV/m. At volume fractions of polymer $0.05 \leq m_{pl} \leq 0.50$, direct and indirect measurements of the following effective piezoelectric coefficients of the composite were performed:

(i) longitudinal d_{33}^* and g_{33}^*,

(ii) transversal d_{31}^* and g_{31}^*,

(iii) hydrostatic $d_h^* = d_{33}^* + 2 d_{31}^*$ and $g_h^* = g_{33}^* + 2 g_{31}^*$.

Based on these data, we evaluate the squared hydrostatic figure of merit $(Q_h^*)^2 = d_h^* g_h^*$ that is often used [1, 2] to characterize a 'signal – noise' ratio.

Microgeometric features (Figure 1) and dielectric and piezoelectric properties of the composite (Figure 2) suggest that changes in the piezoelectric response are caused by different factors. In our opinion, at $0.05 < m_{pl} < 0.10$, the piezoelectric activity of the composite considerably decreases (curves 1 and 2 in Figure 2, b) due to polymer layers and pores, which would weaken an electromechanical interaction in the samples and their dielectric properties (Figure 2, a). At $m_{pl} > 0.10$, several blocks of the poled FC component remain isolated, and therefore, their contribution into the piezoelectric properties is restricted. An increase of $\left| g_{3j}^* \right|$ at $0.40 < m_{pl} < 0.50$ can be associated with a softening of the composite matrix and with a decrease of the dielectric permittivity $\varepsilon_{33}^{*\sigma}$ of the stress-free sample (Figure 2, a). At $0.05 \leq m_{pl} \leq 0.50$, the hydrostatic piezoelectric coefficient g_h^* of the composite is a few times larger than $g_h^{(1)}$ of FC. At $m_{pl} = 0.05$, hydrostatic squared figures of merit of the composite and FC component are linked as $(Q_h^*)^2 \approx 1.4(Q_h^{(1)})^2$, and at $0.10 \leq m_{pl} \leq 0.50$, we observe $(Q_h^*)^2 \approx (Q_h^{(1)})^2$.

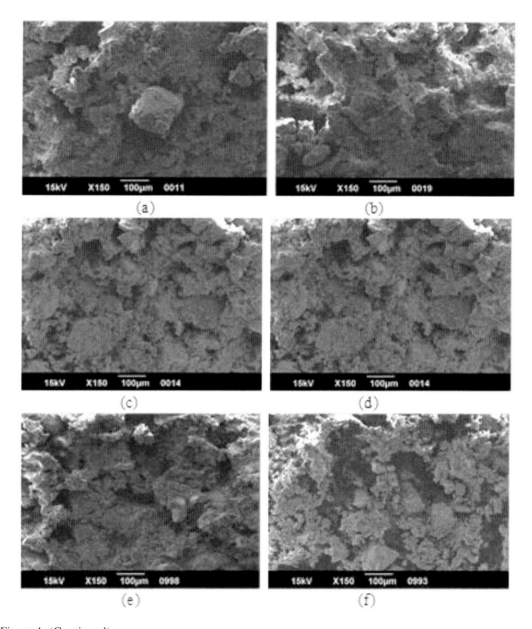

Figure 1. (Continued).

Figure 1. Micrographs of the poled ZTS-19 FC/F-2ME composite at m_{pl} = 0.05 (a), 0.10 (b), 0.15 (c), 0.20 (d), 0.25 (e), 0.30 (f), 0.35 (g), 0.40 (h), 0.45 (i), and 0.50 (j). Micrographs were taken using JEOL JSM-6390LA Analitycal Electron Microscope.

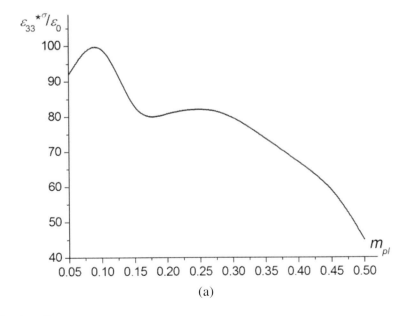

Figure 2. (Continued).

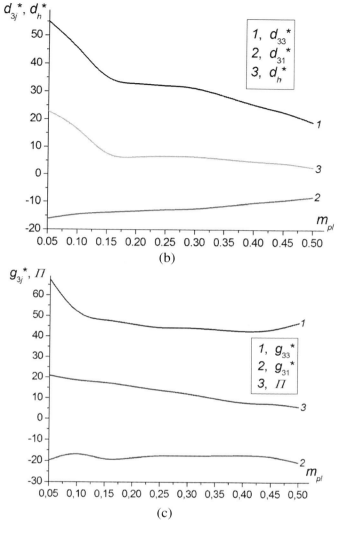

Figure 2. Experimental room-temperature volume-fraction dependencies of the following effective parameters of the composite based on ZTS-19: dielectric permittivity $\varepsilon_{33}^{*\sigma}/\varepsilon_0$ (a), piezoelectric coefficients d_{3j}^* and d_h^* (b, in pC/N) and g_{3j}^* (c, in mV.m/N), and porosity of samples Π (c, in percent).

Now we consider examples of the non-monotonic behavior of the piezoelectric coefficients and related parameters of the studied composite (Figure 2). Such behavior is concerned with a complicated influence of technological, microgeometric factors, poling conditions, and electrodes on the piezoelectric performance in the wide m_{pl} range. A shape of the applied electrodes and their local exfoliating may influence the poling process, remnant polarization and, therefore, piezoelectric parameters of the composite. Various shapes and linear sizes of FC particles (inclusions) may lead to an incomplete matching at FC/polymer interfaces, to pores and non-homogeneous poling characteristics of the studied composite samples.

The piezoelectric coefficients d^*_{3j} of the composite at $0.05 \leq m_{pl} \leq 0.50$ (curves 1 and 2 in Figure 2, b) are approximated as follows:

$$d^*_{33}(m_{pl}) = 78.8 - 578 m_{pl} + 2690 m_{pl}^2 - 5470 m_{pl}^3 + 3870 m_{pl}^4 \text{ (in pC/N)};$$

$$d^*_{31}(m_{pl}) = -19.2 + 82.9 m_{pl} - 452 m_{pl}^2 + 1100 m_{pl}^3 - 888 m_{pl}^4 \text{ (in pC/N)}.$$

Porosity Π of the composite sample (curve 3 in Figure 2, c) can also influence the dielectric permittivity $\varepsilon^{*\sigma}_{33}$ and piezoelectric coefficients g^*_{3j} in the wide m_{pl} range.

The volume-fraction dependence of the piezoelectric coefficient d^*_{3j} can be also interpreted in terms of the model of the 3–0 composite (Figure 3) without pores therein. In this model, the polymer spheroidal inclusions are assumed to be distributed regularly in the composite sample that is poled along the co-ordinate axis OX_3. The spheroidal shape of each inclusion is given by the equation $(x_1/a_1)^2 + (x_2/a_2)^2 + (x_3/a_3)^2 = 1$ in the co-ordinate system $(X_1X_2X_3)$, where a_1, $a_2 = a_1$ and a_3 are semi-axes of the spheroid. We introduce the aspect ratio $\rho = a_1/a_3$ to characterize the shape of the inclusion. The porous polymer matrix is characterized by a regular distribution of air spheroidal pores with the aspect ratio $\rho_{por} = a_{1,por}/a_{3,por}$, and the volume fraction of these pores in the polymer matrix is m_{por}. The shape of each pore is described by the equation $(x_1/a_{1,por})^2 + (x_2/a_{2,por})^2 + (x_3/a_{3,por})^2 = 1$ in the same co-ordinate system. The effective field method [1] is applied to calculate the full set of electromechanical constants of the composite. In this calculation, we use experimental data on ZTS-19 FC [1] and PVDF [3]. Piezo-passive PVDF is regarded as an analog of the F-2ME polymer component. At $0.30 \leq m_{pl} \leq 0.50$, we predict the effective properties of the composite (Table 1) in terms of the aforementioned model, and we see agreement between the calculated and experimental values of two kinds of the piezoelectric coefficients, namely, d^*_{3j} and g^*_{3j}.

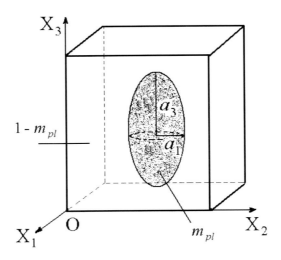

Figure 3. Schematic of the 3–0 composite with spheroidal inclusions. $(X_1X_2X_3)$ is the rectangular co-ordinate system, a_i are semi-axes of the spheroid, m_{pl} is the polymer inclusion volume fraction, and $1 - m_{pl}$ is the FC matrix volume fraction.

Table 1. Calculated and experimental values of piezoelectric coefficients d_{3j}^{*} (in pC/N) and g_{3j}^{*} (in mV·m/N) and relative dielectric permittivity $\varepsilon_{33}^{*\sigma}/\varepsilon_0$ of composites based on ZTS-19 FC

ρ	m_{pl}	d_{33}^{*}	d_{31}^{*}	g_{33}^{*}	g_{31}^{*}	$\varepsilon_{33}^{*\sigma}/\varepsilon_0$
ZTS-19 FC/porous PVDF, $\rho_{por} = 100$, $m_{por} = 0.5$						
0.2	0.5	20.1	−5.2	43.0	−11.2	52.8
0.2	0.4	25.0	−7.4	36.7	−10.8	77.0
0.2	0.3	32.7	−10.7	32.0	−10.5	116
ZTS-19 FC/porous PVDF, $\rho_{por} = 1$, $m_{por} = 0.5$»						
0.2	0.5	20.4	−7.8	39.3	−15.1	58.7
0.2	0.4	25.9	−10.1	34.5	−13.5	84.9
ZTS-19 FC/PVDF						
0.4	0.5	16.3	−6.4	35.3	−13.9	52.4
0.3	0.5	23.3	−9.2	37.8	−14.9	69.8
ZTS-19 FC/F-2ME, experimental data						
–	0.5	19	−8.2	47.7	−20.5	45.0
–	0.4	25	−10.2	42.2	−17.2	66.9
–	0.3	32	−12.7	44.7	−17.7	80.9

CONCLUSION

1. A novel piezo-active composite based on the ZTS-19 FC has been manufactured. In this composite, the second component is a piezo-passive fluorine-containing F-2ME polymer. The effective piezoelectric and dielectric properties of the composite have been studied by taking into account its specific microgeometry and porosity at $0.05 \leq m_{pl} \leq 0.50$.

2. Features of the volume-fraction dependencies $\varepsilon_{33}^{*\sigma}(m_{pl})$, $d_{3j}^{*}(m_{pl})$ and $g_{3j}^{*}(m_{pl})$ have been analyzed for specific ranges within $0.05 \leq m_{pl} \leq 0.50$, and the piezoelectric coefficients $d_{3j}^{*}(m_{pl})$ have been approximated with 4th-degree polynomials. At $0.30 \leq m_{pl} \leq 0.50$, the effective parameters of the studied composite have been interpreted in terms of the model of the 3–0 composite and taking into account porosity.

3. Important advantages of the studied composite consist in large values of $|g_{3j}^{*}|$ that are achieved in the presence of the polymer component with a high mechanical strength.

ACKNOWLEDGMENTS

The authors would like to thank Prof. Dr. A. E. Panich, Prof. Dr. A. A. Nesterov, and Prof. Dr. I. A. Parinov (Southern Federal University, Russia), Prof. Dr. C. R. Bowen

(University of Bath, UK), and Prof. Dr. S.-H. Chang (National Kaohsiung Marine University, Taiwan, ROC) for their constant interest in the research problem. In the present chapters, the results on the research work No.11.1302.2014/K have been represented within the framework of the project part of the state task in the scientific activity area at the Southern Federal University (Russia).

REFERENCES

[1] Topolov, V. Yu.; Bisegna, P.; Bowen, C. R. *Piezo-active Composites. Orientation Effects and Anisotropy Factors.* Berlin and Heidelberg, Springer, 2014.

[2] Topolov, V. Yu; Bowen, C. R. *Electromechanical Properties in Composites Based on Ferroelectrics.* London, Springer, 2009.

[3] Levassort, F.; Lethiecq, M.; Millar, C.; Pourcelot, L. *IEEE Trans. Ultrason.,Ferroelec., a. Freq, Contr.* 1998, vol. 45, 1497-1505.

In: Proceedings of the 2015 International Conference … ISBN: 978-1-63484-577-9
Editors: Ivan A. Parinov, Shun-Hsyung Chang et al. © 2016 Nova Science Publishers, Inc.

Chapter 17

FIGURES OF MERIT AND RELATED PARAMETERS OF MODERN PIEZO-ACTIVE 1–3-TYPE COMPOSITES FOR ENERGY-HARVESTING APPLICATIONS

Vitaly Yu. Topolov[1,], Christopher R. Bowen[2] and Ashura N. Isaeva[1]*

[1]Department of Physics, Southern Federal University,
Rostov-on-Don, Russia
[2]Department of Mechanical Engineering,
University of Bath, Bath, United Kingdom

ABSTRACT

A system of figures of merit is analyzed for 1–3 and 1–3–0 composites based on relaxor-ferroelectric domain-engineered single crystals with high piezoelectric activity. The role of the single-crystal component and its surrounding porous matrix in the 1–3–0 composite is discussed in the context of large values of figures of merit and their anisotropy that is linked with the piezoelectric anisotropy. Examples of volume-fraction dependences of the composite figures of merit are considered in the context of potential energy-harvesting applications.

Keywords: piezo-active composite, figure of merit, relaxor-ferroelectric single crystal, polymer, pore

1. INTRODUCTION

The development of modern piezo-active composites based on relaxor-ferroelectric single crystals (SCs) has a large potential for piezoelectric sensor, actuator, energy-harvesting, and

---

* Corresponding Author address: Department of Physics, Southern Federal University, 5 Zorge Street, 344090 Rostov-on-Don, Russia. Email: vutopolov@sfedu.ru.

other applications. It concerns, for instance, the composites based on SCs of the perovskite-type relaxor-ferroelectric solid solutions such as $(1-x)Pb(Mg_{1/3}Nb_{2/3})O_3 -xPbTiO_3$ (PMN–xPT) with compositions near the morphotropic phase boundary. Interest in the performance and applications of these composites stems from the effective electromechanical properties and related parameters which are concerned with the conversion of electric energy into mechanical energy and vice versa. Figures of merit (FOMs) [1, 2] concerned with electromechanical properties of a piezoelectric material are introduced to characterize its performance and potential for energy-harvesting, transducer, and related applications. To the best of our knowledge, no paper examined the system of FOMs from Refs. 1, 2 in detail for a specific connectivity pattern of a piezo-active composite in a wide volume-fraction range. The aim of the present chapter is to consider the aforementioned FOMs for 1–3-type composites based on PMN–xPT SCs with high piezoelectric activity and to show the potential advantages of these composites over conventional 1–3 composites.

2. FIGURES OF MERIT, MODEL AND METHODS

Hereafter we consider the following FOMs that are used to characterize a 'signal – noise' ratio [3] in piezoelectric transducer systems:

$$F_{V,33}^* = d_{33}^* \, g_{33}^*, \quad F_{q,33}^* = (d_{33}^*)^2, \text{ and } F_{E,33}^* = (k_{33}^*)^4/(Z_{a,33}^*)^2 \tag{1}$$

for the longitudinal piezoelectric effect (33 oscillation mode), and

$$F_{V,3j}^* = d_{3j}^* \, g_{3j}^*, \quad F_{q,3j}^* = (d_{3j}^*)^2, \text{ and } F_{E,3j}^* = (k_{3j}^*)^4/(Z_{a,jj}^*)^2 \tag{2}$$

for the transversal piezoelectric effect (31 and 32 oscillation modes, $j = 1$ and 2). In Equations (1) and (2), d_{3j}^* and g_{3j}^* are piezoelectric coefficients, $k_{3j}^* = d_{3j}^* (\varepsilon_{33}^{*\sigma} s_{jj}^{*E})^{-1/2}$ are electromechanical coupling factors (ECFs), $Z_{a,jj}^* = \rho^* v_j^*$ are values of the specific acoustic impedance on the polar ($f = 3$) and non-polar ($f = 1$ or 2) directions, $\varepsilon_{33}^{*\sigma}$ is dielectric permittivity at mechanical stress σ = const, s_{jj}^{*E} is elastic compliance at electric field E = const, ρ^* is the density of the composite, and v_j^* is the sound velocity along the co-ordinate axis OX_j. In our notations, asterisk (*) is used to show the effective properties and related parameters of the composite.

We analyze FOMs (1) and (2) of the 1–3 SC/polymer and 1–3–0 SC/porous polymer composites. In Figure 1, a, the 1–3 composite based on the [001]-poled SC is shown. In a case of the porous polymer matrix (Figure 1, b), the composite is characterized by 1–3–0 connectivity in terms of Ref. 3. The SC rods are characterized by the square arrangement in the large matrix, and air pores are also distributed regularly therein so that the centers of symmetry of the pores occupy sites of a simple lattice with unit-cell vectors parallel to the

OX_j axes shown in Figure 1, a. It is assumed that the largest semi-axis of each pore is much less than the length of the base of each rod (Figure 1).

Methods used to calculate the effective electromechanical properties of the 1–3-type composite are the matrix method [3], that is applied to the system 'rod – matrix', and the dilute approach [4], that is applied to the system 'air pore – matrix'.

In the matrix method, the effective electromechanical properties of the composite in the co-ordinate system $(X_1X_2X_3)$ are represented as

$$\| C^* \| = \begin{pmatrix} \| s^{*E} \| & \| d^* \|^t \\ \| d^* \| & \| \varepsilon^{*\sigma} \| \end{pmatrix}. \qquad (3)$$

In Equation (3), $\| s^{*E} \|$ is the matrix of elastic compliances at electric field E = const (6 × 6 matrix), $\| d^* \|$ is the matrix of piezoelectric coefficients (3 × 6 matrix), $\| \varepsilon^{*\sigma} \|$ is the matrix of dielectric permittivities at mechanical stress σ = const (3 × 3 matrix), and superscript 't' denotes the transposed matrix.

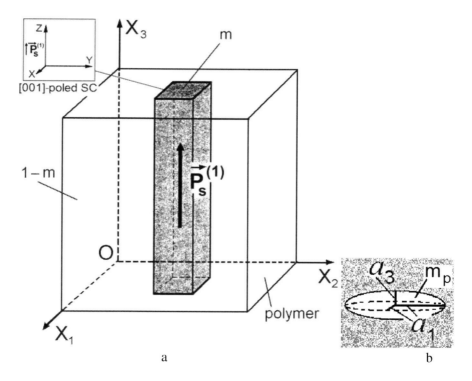

Figure 1. Schematic of the 1–3-type composite based on relaxor-ferroelectric SC. $(X_1X_2X_3)$ is the rectangular co-ordinate system concerned with the composite sample. m and $1 - m$ are volume fractions of SC and polymer, respectively, $P_s^{(1)}$ is the spontaneous polarisation vector of SC (a). The spheroidal pore with semiaxes a_f is shown schematically in graph (b), and m_p is the volume fraction of air in the porous polymer matrix.

The $\| C^* \|$ matrix from Eq. (3) is determined [3] by averaging the electromechanical properties of the components on m and given by

$$\| C^* \| = [\| C^{(1)} \| \cdot \| M \| m + \| C^{(2)} \| (1-m)] \cdot [\| M \| m + \| I \| (1-m)]^{-1}. \tag{4}$$

In Equation (4), $\| C^{(1)} \|$ and $\| C^{(2)} \|$ are matrices of the electromechanical properties of the SC rod and surrounding polymer medium, respectively, $\|M\|$ is concerned with the electric and mechanical boundary conditions at interfaces (Figure 1, a), and $\|I\|$ is the identity 9×9 matrix.

Effective properties of the porous medium with 3–0 connectivity (Figure 1, b) are determined in terms of the dilute approach [4]

$$\| C^{(2)} \| = \| C^{(pol)} \| [\| I \| - m_p ([\| I \| - (1-m_p) \| S \|)^{-1}], \tag{5}$$

where $\| C^{(pol)} \|$ describes the electromechanical properties of the monolithic polymer component, m_p is the volume fraction of the air inclusions therein, and $\|S\|$ contains components of the electroelastic Eshelby tensor [3]. Elements of $\|S\|$ depend on the aspect ratio of the pore $\rho_p = a_1/a_3$ (Figure 1, b) and properties of the polymer component.

Table 1. Maximum values of FOMs from Equations (1) and (2) and related volume fractions of SC m in 1–3-type composites based on PMN–0.33PT SC

ρ_p	max $F^*_{V,33}$, 10^{-10} Pa^{-1}	max $F^*_{E,33}$, 10^{-14} Rayl^{-2}	max $F^*_{V,31}$, 10^{-11} Pa^{-1}	max $F^*_{E,31}$, 10^{-15} Rayl^{-2}
1–3 PMN–0.33PT SC/polyurethane composite				
–	2.42 ($m = 0.109$)	9.57 ($m = 0.087$)	3.45 ($m = 0.138$)	1.85 ($m = 0.097$)
1–3–0 PMN–0.33PT SC/porous polyurethane composite at $m_p = 0.1$				
1	2.86 ($m = 0.088$)	12.8 ($m = 0.073$)	3.74 ($m = 0.106$)	2.07 ($m = 0.077$)
10	4.18 ($m = 0.056$)	20.8 ($m = 0.053$)	2.50 ($m = 0.077$)	1.03 ($m = 0.054$)
100	15.2 ($m = 0.014$)	91.1 ($m = 0.019$)	0.446 ($m = 0.016$)	0.0389 ($m = 0.014$)
1–3–0 PMN–0.33PT SC/porous polyurethane composite at $m_p = 0.2$				
1	3.44 ($m = 0.070$)	17.6 ($m = 0.060$)	4.20 ($m = 0.080$)	2.48 ($m = 0.060$)
10	6.36 ($m = 0.035$)	37.8 ($m = 0.037$)	2.21 ($m = 0.041$)	0.847 ($m = 0.032$)
100	29.4 ($m = 0.008$)	204 ($m = 0.010$)	0.267 ($m = 0.007$)	0.0143 ($m = 0.007$)

Notes. 1. $F^*_{q,33}$ and $F^*_{q,31}$ are monotonically increasing functions of m irrespective of m_p and ρ_p.

2. Room-temperature electromechanical constants of the PMN–0.33PT SC and polyurethane were taken from Refs. 3 and 5, respectively.

3. The [001]-poled PMN–0.33PT SC [3, 6] is characterized by the following FOMs: $F_{V,33} = 1.10 \cdot 10^{-10}$ Pa^{-1}, $F_{q,33} = 7.95 \cdot 10^{-18}$ C^2/N^2, $F_{E,33} = 1.25 \cdot 10^{-14}$ Rayl^{-2}, $F_{V,31} = 0.244 \cdot 10^{-10}$ Pa^{-1}, $F_{q,31} = 1.76 \cdot 10^{-18}$ C^2/ N^2, and $F_{E,31} = 0.106 \cdot 10^{-14}$ Rayl^{-2}.

Conditions $F^*_{V,33} \gg F^*_{V,31}$ and $F^*_{E,33} \gg F^*_{E,31}$ are valid in a wide volume-fraction range, especially for the 1–3–0 composite due to the presence of a system of aligned air pores (see Figure 1, b). The oblate air pores ($\rho_p > 1$) promote the considerable elastic anisotropy of the matrix, and this anisotropy influences the piezoelectric effect so that a more intensive piezoelectric response of the composite along the poling axis OX_3 is achieved in comparison to the response along OX_1 or OX_2. In general, the transversal piezoelectric response of the composite weakens with increasing ρ_p and/or m_p. Data from Table 1 and Figure 2 suggest that the 1–3–0 composite at $\rho_p \gg 1$ provides the best FOM set due to the system of pores in the polymer matrix. Such pores strongly influence the anisotropy of the elastic properties of the matrix and, therefore, the piezoelectric response of the composite as a whole. In this case we select the domain-engineered [001]-poled PMN–0.33PT SC as a piezoelectric component due to its very large piezoelectric coefficient d_{33}: according to experimental data [3, 6], $d_{33} = 2820$ pC/N at room temperature.

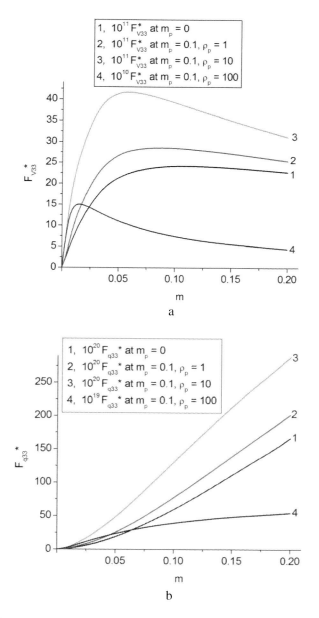

Figure 2. (Continued).

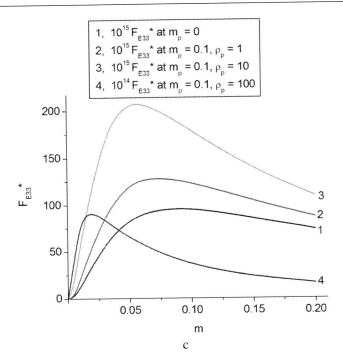

Figure 2. Examples of the volume-fraction dependence of FOMs from Equations (1) for the 1–3 PMN–0.33PT SC/polyurethane composite (at $m_p = 0$) and 1–3–0 PMN–0.33PT SC/porous polyurethane composite (at $m_p = 0.1$).

This performance of the SC component favors large FOM values (see, for instance, data from note 3 in Table 1), and these values are higher than those related to poled PZT-type ceramics [3, 5] and polymers [2]. For example, a highly piezo-active PCR-7M ceramic of the PZT type [3, 5] is characterized by $d_{33} = 760$ pC/N, $d_{31} = -350$ pC/N, $F_{V,33} = 0.131 \cdot 10^{-10}$ Pa^{-1}, $F_{q,33} = 0.578 \cdot 10^{-18}$ C^2/N^2, $F_{V,31} = 0.277 \cdot 10^{-11}$ Pa^{-1}, and $F_{q,31} = 0.123 \cdot 10^{-18}$ C^2/N^2.

According to data from Ref. 2, poled PVDF is characterized by $F_{V,33} = 9.1 \cdot 10^{-12}$ Pa^{-1} and $F_{q,33} = 1.2 \cdot 10^{-21}$ C^2/N^2, and these values are much less than those related to the PMN–0.33PT SC and PCR-7M ceramic. It should be added that values of max $F_{V,33}^*$ from Table 1 are comparable to max $F_{V,33}^* = 6.24 \cdot 10^{-10}$ Pa^{-1} related to a 1–3 PCR-7M/elastomer composite with cylindrical ceramic rods [5].

Maxima of some FOMs of the 1–3–0 composite with $\rho_p \gg 1$ are achieved at relatively small volume fractions of SC m (see Table 1 and curves 4 in Figure 2, a and c). However changes in the FOM values with increasing m are not large (see Figure 2, a and c), and the manufacture of a high-performance composite sample at $m = 0.05$–0.10 is feasible. We add that Choy et al. [7] manufactured a 1–3 ferroelectric ceramic/polymer composite with low volume fractions of ceramic rods (about 0.033, 0.066, *etc.*). Therefore, it is possible, that the technological challenges of forming low volume fractions of SC m can be met in the case of the 1–3-type composite studies in this work.

CONCLUSION

1. The 1–3-type piezo-active composite based on relaxor-ferroelectric PMN–0.33PT SC has been studied to demonstrate its performance with respect to a system of FOMs (1) and (2) in wide volume-fraction ranges.
2. The role of the porous polymer matrix (Figure 1, b) in forming the piezoelectric response and anisotropy of FOMs (conditions $F_{V,33}^* >> F_{V,31}^*$ and $F_{E,33}^* >> F_{E,31}^*$) has been analyzed. Large values of FOMs $F_{V,33}^*$ and $F_{E,33}^*$ in comparison to those of the SC component (see Figure 2 and Table 1) enable us to use the studied 1–3–0 PMN–0.33PT SC/porous polyurethane composite at relatively small porosity ($m_p = 0.1 - 0.2$) in piezoelectric energy-harvesting applications.

ACKNOWLEDGMENTS

The authors would like to thank Prof. Dr. A. E. Panich, Prof. Dr. A. A. Nesterov, and Prof. Dr. I. A. Parinov (Southern Federal University, Russia) and Prof. Dr. S.-H. Chang (National Kaohsiung Marine University, Taiwan, ROC) for their constant interest in the research problem. Applied research has been performed with the financial support from the Ministry of Education and Science of the Russian Federation (project RFMEFI57814X0088) at using the equipment of the Centre of Collective Use 'High Technologies' at the Southern Federal University. Prof. Dr. C. R. Bowen acknowledges funding from the European Research Council under the European Union's Seventh Framework Programme (FP/2007-2013)/ERC Grant Agreement No.320963 on Novel Energy Materials, Engineering Science and Integrated Systems (NEMESIS).

REFERENCES

[1] Priya, S. IEEE Trans. Ultrason., *Ferroelec., a. Freq. Contr.* 2010, vol. 57, 2610-2612.
[2] Sessler, G.; Hillenbrand, *J. Appl. Phys. Lett.* 2013, vol. 103, 122904-4 p.
[3] Topolov, V. Yu.; Bowen, C. R. Electromechanical Properties in Composites Based on Ferroelectrics. London,Springer, 2009.
[4] Dunn, M. L. *J. Appl. Phys.* 1995, vol. 78, 1533-1541.
[5] Topolov, V. Yu.; Turik, A. V. *Tech. Phys.* 2001, vol. 46, 1093-1100.
[6] Zhang, R.; Jiang, B.; Cao, W. *J. Appl. Phys.* 2001, vol. 90, 3471-3475.
[7] Choy, S. H.; Chan, H. L. W.; Ng, M. W.; Liu, P. C. K. *Integr. Ferroelectrics.* 2004, vol. 63, 109-115.

In: Proceedings of the 2015 International Conference ...   ISBN: 978-1-63484-577-9
Editors: Ivan A. Parinov, Shun-Hsyung Chang et al.   © 2016 Nova Science Publishers, Inc.

Chapter 18

PIEZOELECTRIC SENSITIVITY AND ANISOTROPY OF A NOVEL LEAD-FREE 2–2 COMPOSITE

Vitaly Yu. Topolov[1,], Christopher R. Bowen[2], and Franck Levassort[3]*

[1]Department of Physics, Southern Federal University,
Rostov-on-Don, Russia
[2]Department of Mechanical Engineering,
University of Bath, Bath, United Kingdom
[3]F. Rabelais University of Tours, Greman, France

ABSTRACT

Results on the piezoelectric performance of a lead-free 2–2 composite are reported. The role of the piezoelectric component and orientation effect concerned with a rotation of the crystallographic axes are analyzed for a novel 2–2 composite based on a single-domain $KNbO_3$ single crystal. Large values of piezoelectric coefficients $|g_{3j}^*|$ and examples of the large anisotropy of the piezoelectric coefficients d_{3j}^* and electromechanical coupling factors k_{3j}^* are discussed to show advantages of the lead-free 2–2 composite.

Keywords: composite, single crystal, polymer, piezoelectric coefficient, electromechanical coupling factor, anisotropy

---

[*]Corresponding Author address: Department of Physics, Southern Federal University, 5 Zorge Street, 344090 Rostov-on-Don, Russia, E-mail: vutopolov@sfedu.ru.

1. INTRODUCTION

As a rule, piezo-active composites are based on perovskite-type ferroelectrics – either poled ceramics (often of the PZT or $PbTiO_3$ type) [1, 2] or domain-engineered single crystals (SC) such as $(1 - x)Pb(Mg_{1/3}Nb_{2/2})O_3–xPbTiO_3$ and $(1 - x)Pb(Zn_{1/3}Nb_{2/2})O_3–xPbTiO_3$[2–4]. The aforementioned ceramic and SC compositions are lead-based, and their electromechanical properties are chosen in specific ranges to attain optimal parameters in the related composites [2].

However, environmental concerns have led to attempts to eliminate lead-based materials from consumer items, including piezoelectric transducers, sensors and actuators, and this trend is well seen in the last decade. The need for lead-free ferroelectrics gives rise to the challenge of finding an alternative [5, 6] to lead-based piezoelectrics, also in the field of modern piezo-active composites.

Recently, examples of a lead-free 1–3-type SC/polymer composite [6] have been discussed to show its high performance and potential fields of applications.

To the best of our knowledge, laminar lead-free composites with 2–2 connectivity patterns have yet to be discussed in the literature.

The aim of the present chapter is to show the piezoelectric performance and advantages of a novel lead-free 2–2 SC/polymer composite. We choose a piezoelectric component among the ferroelectric perovskite-type SCs. In our opinion, a ferroelectric single-domain $KNbO_3$ SC with the perovskite structure may be of interest as a piezoelectric component in the 2–2 composite.

A full set of electromechanical constants measured on single-domain orthorhombic $KNbO_3$ SC samples at room temperature was published recently [7].

2. MODEL AND EFFECTIVE PARAMETERS OF THE 2–2 COMPOSITE

It is assumed that a regular distribution of layers is observed in the 2–2 parallel-connected SC/polymer composite on the OX_1 direction (Figure 1). The SC and polymer layers are continuously distributed along the OX_2 and OX_3 axes. An orientation of the spontaneous polarization vector P_s of each SC layer is shown in inset 1 of Figure 1, and a rotation of P_s with respect to the co-ordinate axes OX_i is described in terms of the Euler angles shown in the inset of Figure 1. The polymer layer of this composite is piezo-passive and isotropic, therefore, poling of the composite sample as a whole means that the condition P_s = const for each SC layer is achieved in an external electric field E.

To predict the effective electromechanical properties and related parameters of the 2–2 composite at various volume fractions of SC m and Euler angles φ, ψ, and θ, we use full sets of elastic compliances $\| s^{(n),E} \|$ (measured at E = const), piezoelectric coefficients $\| d^{(n)} \|$ and dielectric permittivities $\| \varepsilon^{(n),\sigma} \|$ (measured at mechanical stress σ = const) of components, where n = 1 denotes SC, and n = 2 denotes polymer. Hereafter we take experimental values of the aforementioned constants of the single-domain $KNbO_3$ SC [7] and polyurethane [8]. A rotation of the crystallographic axes of the SC (see inset 2 in Figure 1) is described [9] in terms of the rotation matrix $\|r\|$ (3×3 matrix).

The electromechanical properties of the nth component of the composite in the $(X_1X_2X_3)$ system are represented as

$$\|C^{(n)}\| = \begin{pmatrix} \|s^{(n),E}\| & \|d^{(n)}\|^T \\ \|d^{(n)}\| & \|\varepsilon^{(n),\sigma}\| \end{pmatrix}, \tag{1}$$

where n = 1 and 2, and superscript 'T' denotes the transposed matrix. In Equation (1), $\|C^{(n)}\|$ is the 9×9 matrix.

Averaging the electromechanical properties from Equation (1) is performed along the OX_1 axis by taking into account nine boundary conditions [9] for electric and mechanical fields in the adjacent layers of the composite (Figure 1). For instance, these boundary conditions at x_1 = const involve the continuity of three normal components of the mechanical stress (i.e., σ_{11}, σ_{12} and σ_{13}), three tangential components of the mechanical strain (i.e., ξ_{22}, ξ_{23} and ξ_{33}), one normal component of the electric displacement (i.e., D_1), and two tangential components of the electric field (i.e., E_2 and E_3). Following this procedure for averaging [2, 9], we determine the effective electromechanical properties of the 2–2 composite in the matrix form as follows:

$$\|C^*\| = [\|C^{(1)}\| \cdot \|M\| m + \|C^{(2)}\| (1-m)] [\|M\| m + \|I\| (1-m)]^{-1}. \tag{2}$$

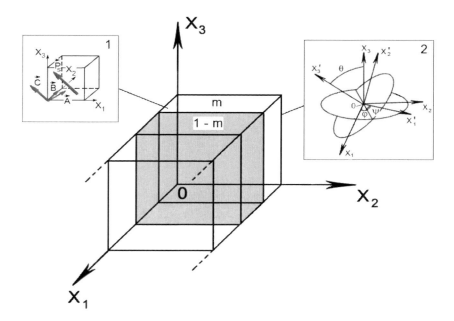

Figure 1. Schematic of the parallel-connected 2–2 SC/polymer composite. m and $1-m$ are volume fractions of SC and polymer, respectively. $(X_1X_2X_3)$ is the rectangular co-ordinate system, The perovskite unit-cell of the single-domain SC layer, its spontaneous polarization

P_s and unit-cell vectors A, B, and C are shown in inset 1, and the Euler angles φ, ψ, and θ, which describe the orientation of the crystallographic axes in each SC layer, are shown in inset 2.

In Equation (2), $\| M \|$ is the 9×9 matrix that describes the aforementioned boundary conditions at $x_1 = \text{const}$, $\| I \|$ is the identity 9×9 matrix, m is the volume fraction of SC, and $\| C^{(n)} \|$ is taken from Equation (1). The full set of effective constants of the 2–2 composite s_{ab}^{*E}, d_{ij}^* and $\varepsilon_{fh}^{*\sigma}$ is determined from matrix (2) by using conventional formulae [10] for piezoelectric media. We add that $\| C^* \|$ from Equation (2) has the form shown in Equation (1). The effective electromechanical properties from Equation (2) are determined in the long-wave approximation [2], i.e., on condition that the wavelength of an external acoustic field is much more than a width of each layer of the composite (Figure 1).

Based on $\| C^* \|$ from Equation (2), we analyze the volume-fraction (m) and orientation (φ, ψ, and θ) dependencies of the piezoelectric coefficients g_{3j}^* (from a relation $d_{3j}^* = g_{3f}^* \varepsilon_{fj}^{*\sigma}$ [10]) and electromechanical coupling factors (ECFs)

$$k_{3j}^* = d_{3j}^* / (\varepsilon_{33}^{*\sigma} s_{jj}^{*E})^{1/2} \tag{3}$$

of the 2–2 composite, where $j = 1$, 2, and 3. Below we consider some examples of the performance of the 2–2 composite based on the single-domain $KNbO_3$ SC.

First, a relatively large piezoelectric sensitivity ($|g_{3j}^*| \approx 100\,\text{mV·m/N}$) and changing sgn g_{32}^* are achieved at variations of the Euler angle φ (Figure 2). Second, a rapid decrease of the piezoelectric sensitivity and changing sgn g_{31}^* are observed with an increase of the Euler angle θ from 0° to 90° (Figure 3). At the same time, a slow decrease of the piezoelectric sensitivity is observed at volume fractions $0.04 < m < 0.1$ (Figure 4).

Third, conditions for the large anisotropy of the piezoelectric coefficients:

$$d_{33}^* / |d_{3f}^*| \geq 5 \tag{4}$$

and ECFs:

$$k_{33}^* / |k_{3f}^*| \geq 5 \tag{5}$$

are valid in volume-fraction ranges $[m_{d1}; m_{d2}]$ and $[m_{k1}; m_{k2}]$, respectively, where $f = 1$ and 2. Data from Figure 5 suggest that the range $[m_{k1}; m_{k2}]$ is wider than $[m_{d1}; m_{d2}]$ at various orientations of the crystallographic axes of SC. This is due to the important role of the elastic anisotropy of the single-domain $KNbO_3$ SC in the orthorhombic phase [7]. We add that ECFs from Equation (3) are highly dependent on the elastic properties of the composite, and therefore, of its components. It should be noted that inequalities (4) and (5) are valid in a restricted range of the Euler angle θ, and the volume-fraction ranges $[m_{d1}; m_{d2}]$ and $[m_{k1}; m_{k2}]$ undergo appreciable changes when varying θ (Figure 5).

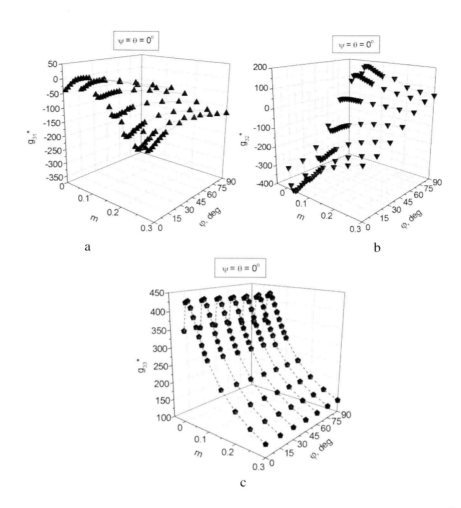

Figure 2. Volume-fraction (*m*) and orientation (*φ*) dependencies of piezoelectric coefficients g_{3j}^* (in mV·m/N) of the 2–2 KNbO$_3$ SC/polyurethane composite.

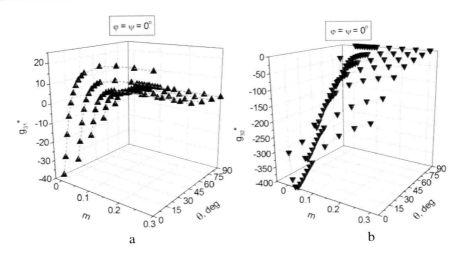
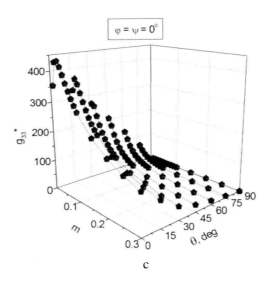

Figure 3. (Continued).

Figure 3. Volume-fraction (m) and orientation (θ) dependencies of piezoelectric coefficients g_{3j}^* (in mV·m/N) of the 2–2 $KNbO_3$ SC/polyurethane composite

This is due to the orientation effect and the key role of the Euler angle θ in forming the piezoelectric response and ECFs (3) which are associated with the OX_3 axis (Figure 1).

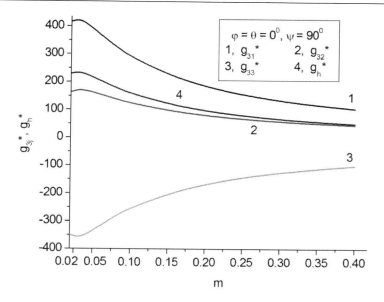

Figure 4. Volume-fraction (m) dependence of piezoelectric coefficients g_{3j}^* and its hydrostatic analog $g_h^* = g_{33}^* + g_{32}^* + g_{31}^*$ (in mV·m/N) of the 2–2 KNbO$_3$ SC/polyurethane composite.

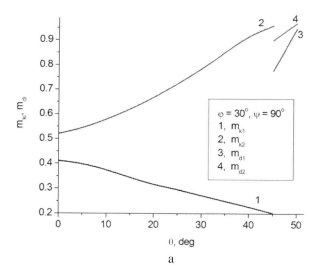

a

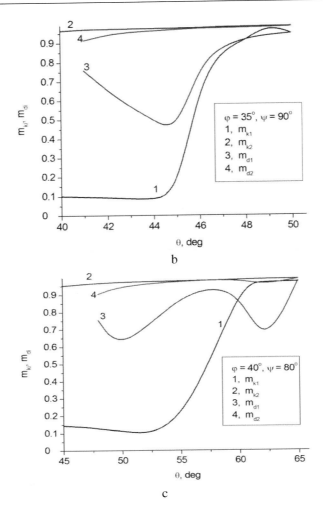

Figure 5. Volume-fraction ranges [m_{d1}; m_{d2}] and [m_{k1}; m_{k2}] where conditions (4) and (5) hold for the 2–2 KNbO$_3$ SC/polyurethane composite at changes in the Euler angle θ.

CONCLUSION

We have shown results on effective parameters of the piezo-active 2–2 composite based on the ferroelectric single-domain KNbO$_3$ SC. We have discussed the important role of the anisotropic SC component and the orientation effect concerned with rotations of the crystallographic axes in each SC layer of the composite.

The studied 2–2 composite based on the single-domain KNbO$_3$ SC is of interest as a lead-free material with the high piezoelectric sensitivity ($|g_{3j}^*|\sim100$ mV·m/N) in the relatively wide volume-fraction (m) range (Figure 4). Conditions (4) and (5) hold at specific orientations of the main crystallographic axes and in specific volume-fraction ranges. Hereby the key role of the Euler angle θ in forming the large anisotropy of the piezoelectric coefficients d_{3j}^* and ECFs k_{3j}^* is shown (Figure 5).

In general, the results discussed in the present chapter suggest that there is the considerable potential for the use of the lead-free 2–2 composite as an active element of piezoelectric sensors, transducers, and energy-harvesting devices.

ACKNOWLEDGMENTS

The authors would like to thank Prof. Dr. A. E. Panich, Prof. Dr. A. A. Nesterov, and Prof. Dr. I. A. Parinov (Southern Federal University, Russia), Prof. Dr. S.-H. Chang (National Kaohsiung Marine University, Taiwan, ROC), and Dr. A. V. Krivoruchko (Don State Technical University, Russia) for their constant interest in the research problem. The results on the research work No.1597 have been represented within the framework of the base part of the state task No.2014/174 in the scientific activity area at the Southern Federal University (Russia). Prof. Dr. C. R. Bowen acknowledges funding from the European Research Council under the European Union's Seventh Framework Programme (FP/2007-2013)/ERC Grant Agreement No.320963 on Novel Energy Materials, Engineering Science and Integrated Systems (NEMESIS).

REFERENCES

[1] Akdogan, E. K.; Allahverdi, M.; Safari A. *IEEE Trans. Ultrason., Ferroelec., a. Freq. Contr.* 2005, *vol. 52*, 746-775.

[2] Topolov, V. Yu.; Bowen, C. R. *Electromechanical Properties in Composites Based on Ferroelectrics*. London, Springer, 2009.

[3] Ren, K.; Liu, Y.; Geng, X.; Hofmann, H. F.; Zhang, Q. M. *IEEE Trans. Ultrason., Ferroelec., a. Freq, Contr.* 2006, *vol. 53*, 631-638.

[4] Wang, F.; He, C.; Tang, Y. *Mater. Chem. Phys.* 2007, *vol. 105*, 273-277.

[5] *Lead-Free Piezoelectrics*, Priya, S., Nahm, S. (Eds.), New York, Dordrecht, Heidelberg, London, Springer, 2012.

[6] Topolov, V. Yu.; Bowen, C.R. *Mater. Lett.* 2015, *vol. 142*, 265-268.

[7] Rouffaud, R.; Marchet, P.; Hladky-Hennion, A.-C.; Bantignies, C.; Pham-Thi, M.; Levassort, F. *J. Appl. Phys.* 2014, *vol. 116*, 194106-7 p.

[8] Gibiansky, L.V.; Torquato, S. *J. Mech. Phys. Solids.* 1997, *vol. 45*, 689-708.

[9] Topolov, V. Yu.; Bisegna, P.; Bowen, C. R. *Piezo-Active Composites. Orientation Effects and Anisotropy Factors*. Berlin, Heidelberg, Springer, 2014.

[10] Ikeda, T. *Fundamentals of Piezoelectricity*. Oxford, New York, Toronto, Oxford University Press, 1990.

In: Proceedings of the 2015 International Conference ...   ISBN: 978-1-63484-577-9
Editors: Ivan A. Parinov, Shun-Hsyung Chang et al.   © 2016 Nova Science Publishers, Inc.

Chapter 19

CHARACTERIZATION TECHNIQUES FOR ADVANCED MATERIALS AND DEVICES

M. A. Lugovaya, E. I. Petrova[1], T. V. Rybyanets[2], G. M. Konstantinov[1] and A. N. Rybyanets[1]*

[1]Institute of Physics, Southern Federal University,
Rostov on Don, Russia
[2]Academy of Biology and Biotechnologies,
Southern Federal University, Rostov-on-Don, Russia

ABSTRACT

Development and wide application of new manufacture technologies, piezoelectric materials, and devices earnestly demand perfection of the existing assessment methods and development of new characterization techniques. Some of these advanced materials are lossy and direct use of the IEEE Standard for the material constant determination leads to significant errors. The accurate description of piezoceramics must include the evaluation of the dielectric, piezoelectric, and mechanical losses accounting for the out-of-phase material response to the input signal. The chapter considers the basic techniques for characterization of piezoelectric materials and devices comprising IEEE Standard on piezoelectricity, impedance spectroscopy, and iterative techniques exploring complex piezoelectric material constants. The chapter presents the basic theory for the methods, modern apparatus setups and software along with the experimental results for a wide range of piezoelectric materials, ultrasonic transducers, and devices. The piezoelectric resonance analysis method and the program (PRAP) were applied to the full set of standard geometries and resonance modes needed to complete complex characterization of a wide range of materials with very high and moderate losses. It was shown that the iterative methods provide a means to accurately determine a complex coefficients accounting for dielectric, mechanical, and electromechanical losses of poled bulk piezoceramics from complex impedance resonance measurements. The results show that the impedance spectroscopy method gives more precise data and in some cases is the only way to obtain correct values for materials with extremely high losses.

---

* e-mail: lugovaya_maria@mail.ru.

Keywords: porous PZT piezoceramics, characterization techniques, resonance modes

1. INTRODUCTION

Demands on piezoelectric and ultrasonic transducer performance have increased in recent years. New monocrystals, piezoceramics and piezocomposite materials are widely used for piezoelectric and ultrasonic transducers with high sensitivity, efficiency and resolution [1-3]. Some of these advanced materials are lossy and direct use of the IEEE Standard for the material constant determination leads to significant errors [4]. The modeling and design of piezoelectric devices by finite element methods, among others, rely on the accuracy of the dielectric, piezoelectric, and elastic coefficients of an active material used which is, commonly, an anisotropic ferroelectric polycrystal. The accurate description of piezoceramics must include the evaluation of the dielectric, piezoelectric, and mechanical losses accounting for the out-of-phase material response to the input signal [5].

Numerous techniques using complex material constants have been proposed [7-10] to take into account losses in low-Q materials and to overcome limitations in the IEEE Standard [6]. Iterative methods [7, 8] provide a means to accurately determine the complex coefficients in the linear range of poled piezoceramics from complex impedance resonance measurements.

The piezoelectric resonance analysis method and the program (PRAP) were proposed for the full set of standard geometries and resonance modes needed to complete complex characterization in a wide range of materials with very high and moderate losses [11, 12].

2. IMPEDANCE SPECTROSCOPY CHARACTERIZATION OF PIEZOELECTRICS

Impedance spectroscopy characterization stems from the relation for impedance Z:

$$\mathbf{Z} = V/\mathbf{I} = \int \mathbf{E} \cdot d\mathbf{x} \bigg/ A \frac{d\mathbf{D}}{dt}, \tag{1}$$

where V is voltage, I is current, and A is the surface area over which current is measured. The piezoelectric equations give the relationship between dielectric displacement D and electric field E through stress T and strain S. For simple one-dimensional resonance modes, many of the electric and mechanical field variables aspire to zero and the sets of piezoelectric equations decouple.

Solving the wave equation for one-dimensional steady-state displacement in the specimen and evaluating S from the derivative of displacement, together with the boundary conditions of an unclamped resonator, the piezoelectric equations can be used to evaluate the relationship between D and E as a function of frequency and material properties. The impedance of a thickness extensional (TE) resonator reduces to [15]:

$$\mathbf{Z} = \frac{\boldsymbol{\beta}_{33}^{S}l + \dfrac{\mathbf{h}_{33}^{2}}{\sqrt{\rho\,\mathbf{c}_{33}^{D}}\,\omega}\tan\left(\dfrac{\omega l\sqrt{\rho}}{2\sqrt{\mathbf{c}_{33}^{D}}}\right)}{i\omega A}, \qquad (2)$$

where ρ is the specimen density, ω is the angular frequency and l is the thickness of the sample in the direction of poling, h is the piezoelectric constant, c is the elastic stiffness at constant D, and β is the inverse permittivity at constant S.

In impedance spectroscopy characterization of piezoelectrics, the material properties of equation (2) can be adjusted until the impedance fits a measured impedance spectrum as a function of frequency.

The common one-dimensional 6mm-symmetry modes include the Length Extensional (LE), Thickness Extensional (TE), Length-Thickness Extensional (LTE), Thickness Shear (TS), Length Shear (LS), and Radial modes. Each of these modes provides electrical, elastic and piezoelectric properties of the material. Table 1 lists the properties of one form of the reduced piezoelectric matrix with the modes that will provide those properties.

Table 1. Properties provided by the common modes

Property	ε^{T}_{11}	ε^{T}_{33}	s^{E}_{11}	s^{E}_{12}	s^{E}_{13}	s^{E}_{33}	s^{E}_{55}	d_{13}	d_{33}	d_{15}
Mode	TS, LS	LE, LTE, Radial	LTE, Radial	Radial		LE	LS, TS	LTE, Radial	LE	TS, LS

By analyzing specimens inducing the TS, LE and Radial modes, all except s^{E}_{13} of the reduced 6mm piezoelectric matrix can be obtained. s^{E}_{13} can be found from the properties determined with the TE, LTE and LE modes for other representations of the piezoelectric linear equations [11], indicating that all modes except the LS mode are required to construct the piezoelectric matrix. Alternatively, using an initial guess for s^{E}_{13}, the reduced matrix can be inverted and c^{D}_{13} compared to that found for the TE mode, eliminating the need for the LTE mode.

In practice, material properties are dispersive (i.e., they vary with frequency). This means that the range of frequency used to fit forms of equation (6) to measured spectra should be limited. It is worth noting that only a few points in the spectrum are required to determine properties. For the TE mode, only three points are required.

The IEEE Standard on Piezoelectricity [6] describes two approaches to obtaining material properties using either the difference between parallel (f_{p}) and series (f_{s}) resonance frequencies, or the frequency ratio between different order resonance peaks. These approaches can be problematic when material properties are dispersive or when a low electromechanical coupling constant results in a small difference between f_{p} and f_{s}.

Smits [9] proposed a general method for the TE, TS, LE and LTE resonators using three points around resonance. This method does not rely on critical frequencies in the spectrum and so does not lose sensitivity for low Q materials. At resonance and anti-resonance, currents can be very large and small and this presents an instrumentation problem. Though less of an issue with low Q materials, Smits's method is less demanding of instrumentation than the

IEEE methods. Finally, because the choice of analysis points is somewhat arbitrary, noisy spectra and samples with mode coupling may still be analyzed using Smits's method.

The IEEE Standard in Piezoelectricity treats all material properties as real, evaluating mechanical and electrical loss independently through Q_M and $\tan\delta$. Losses in the electromechanical coupling constant were assumed zero. This was acceptable for low loss piezoelectric materials such as PZT, but can lead to significant errors when materials have large loss. In real materials, complex impedance results from complex material properties [10]. The impedance is directly associated with dielectric conduction losses [13, 15] and attenuation of the displacement wave in the sample.

The PRAP software [11] analyses impedance spectra to determine complex material properties. This software uses a generalized form of Smits's method to determine material properties for any common resonance mode, and a generalized ratio method for the radial mode [9] valid for all material Q's. By analyzing on each harmonic, PRAP can determine complex material properties as a function of frequency. The software always generates an impedance spectrum from the determined properties to indicate validity of the results.

3. Experimental Scenario

The moderately low-Q porous piezoceramics based on ferroelectrically "hard" PZT composition $PbTiO_3 - PbZrO_3 - PbNb_{2/3}Zn_{1/3}O_3 - PbW_{1/2}Mn_{1/2}O_3 - PbNb_{2/3}Mn_{1/3}O_3$ (trade mark PCR-8) [1] with relative porosity ~20% and average pore size ~50 μm (Figure 1) were chosen for complete complex characterization of material properties.

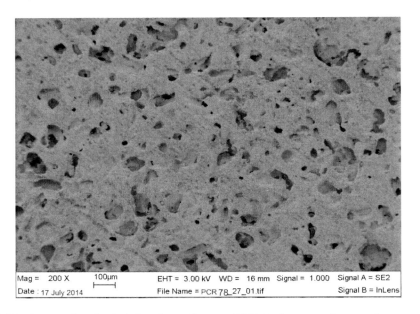

Figure 1. SEM micrographs of grinded surface of porous PCR-8 piezoceramics sample.

Experimental samples of porous piezoceramics were prepared by burning out of organic powders using specially developed technology [1, 3]. An organic salt powder with irregular particle shape, 250°C decomposition temperature and grain-size distribution from 50 to 60 μm were used for producing the pore volumes. The porous ceramic elements with vacuum deposited Cr/Ni electrodes were poled on air at cooling from 380°C to 90°C for 1 min by applying a dc electric field of 1 – 1.5 kV/mm. The following piezoelements were prepared: disks Ø20 × 1 mm (radial and TE modes), rods 1.5 × 1.5 × 6 mm (LTE mode) and squares 0.6 × 0.6 × 6 mm (TS mode). For the experiments, we applied the PRAP automatic iterative method [11] to the full set of standard geometries and resonance modes needed to complete a characterization of porous piezoceramics PCR-8. Measurements of electric parameters were made on the Agilent 4294A Impedance Analyzer.

4. RESULTS AND DISCUSSION

Figures 2 and 3 show measured complex impedance spectra and PRAP approximations for the thickness (TE), radial (RE), length extensional (LE), and thickness shear (ST) modes of PCR-8 porous piezoceramics.

Table 2 summarizes the complex constants of the porous ceramics obtained using PRAP analysis for a full set of standard geometries and resonance modes (Figures 2, 3). The full set of real parameters for dense PCR-8 ceramics is also listed in Table 3 for comparison. Additional physical parameters of PCR-8 porous ceramics measured by ultrasonic and hydrostatic weighting methods are as follows: $Z_A = 18.5$ MRayl, $\rho = 6.25$ g/cm^3, $V_t = 2960$ m/s, $Q_M^t = 150$, $(T_C = 340°C)$ [1]. The results of the complex constant measurements are in good conformity with the previously reported results for other types of piezoceramuics [3, 8, 17]. Experimental and theoretical aspects of porous ceramics modeling and application in ultrasonic transducers and devices are considered in [17-19].

Table 2. 6 mm piezoelectric matrix for PZT piezoceramics

Parameter	Dense	Porous	
	Real	Real	Imaginary
S^E_{11} (m²/N)	$1.25 \cdot 10^{-11}$	$1.95 \cdot 10^{-11}$	$-3.11 \cdot 10^{-14}$
S^E_{12} (m²/N)	$-4.6 \cdot 10^{-12}$	$-6.38 \cdot 10^{-12}$	$2.31 \cdot 10^{-14}$
S^E_{13} (m²/N)	$-5.2 \cdot 10^{-12}$	$-6.05 \cdot 10^{-12}$	$-$
S^E_{33} (m²/N)	$1.56 \cdot 10^{-11}$	$3.03 \cdot 10^{-11}$	$-1.26 \cdot 10^{-14}$
S^E_{55} (m²/N)	$3.53 \cdot 10^{-11}$	$6.25 \cdot 10^{-11}$	$-2.27 \cdot 10^{-12}$
S^E_{66} (m²/N)	$3.42 \cdot 10^{-11}$	$5.19 \cdot 10^{-11}$	$-1.08 \cdot 10^{-13}$
d_{15} (C/N)	$4.1 \cdot 10^{-10}$	$4.69 \cdot 10^{-10}$	$-2.56 \cdot 10^{-11}$
d_{31} (C/N)	$-13 \cdot 10^{-11}$	$-5.92 \cdot 10^{-11}$	$-5.19 \cdot 10^{-14}$
d_{33} (C/N)	$2.9 \cdot 10^{-10}$	$2.54 \cdot 10^{-10}$	$-1.35 \cdot 10^{-12}$
ε^T_{11} (F/m)	$1.22 \cdot 10^{-8}$	$1.13 \cdot 10^{-8}$	$-2.92 \cdot 10^{-10}$
ε^T_{33} (F/m)	$12.39 \cdot 10^{-9}$	$6.44 \cdot 10^{-9}$	$-3.09 \cdot 10^{-12}$

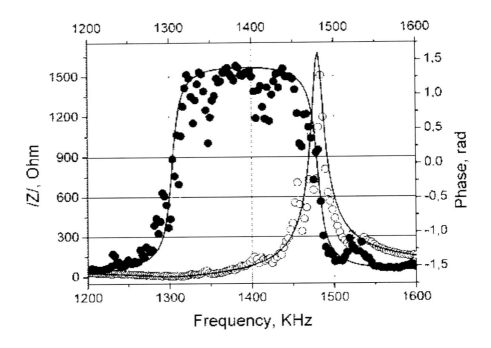
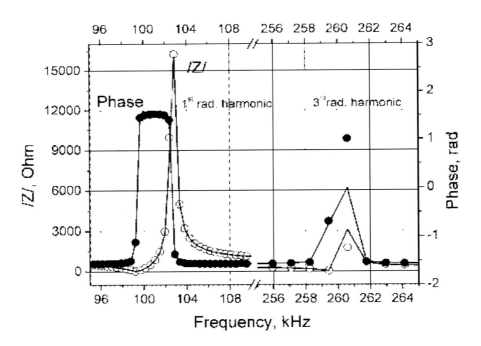

Figure 2. Measured impedance spectra and PRAP approximations for TE and RE modes of porous piezoceramic disk Ø20 × 1 mm.

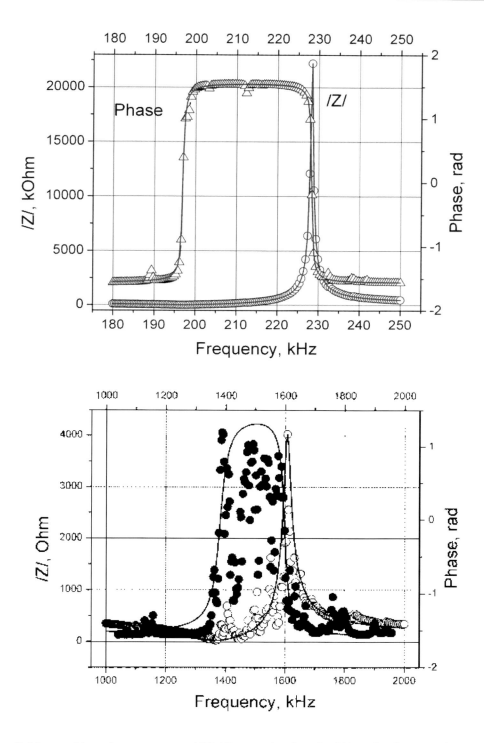

Figure 3. Measured impedance spectra and PRAP approximations for LE-mode of porous piezoceramic rod 1.5 × 1.5 × 6 mm and ST-mode of porous piezoceramic square 0.6 × 0.6 × 6 mm.

CONCLUSION

In this chapter, piezoelectric resonance analysis method for automatic iterative evaluation of complex material parameters of lossy piezoelectrics was described.

Complex sets of material constants for "hard" porous piezoceramics were obtained by PRAP analysis of electric impedance spectra measured for the standard porous piezoceramic elements. By combining the analyzed resonance spectra of the appropriate resonance modes (radial, thickness, length extensional, and shear thickness) the complete 6mm reduced complex piezoelectric matrix of the material was determined.

It was shown that the PRAP method gives accurate results and allows one to take into account elastic, piezoelectric, and dielectric losses of the piezoelectric material.

ACKNOWLEDGMENTS

Work supported by the Southern Federal University (internal grant - 213.01-2014/012-ВГ).

REFERENCES

[1] Rybyanets, A.N.; Naumenko, A.A. Nanoparticles Transport in Ceramic Matrixes: a Novel Approach for Ceramic Matrix Composites Fabrication. In: *Physics and Mechanics of New Materials and Their Applications*, Ivan A. Parinov and Shun-Hsiung Chang (Eds.). Nova Science Publishers Inc., New York. 2013, Chapter 1, 3-18.

[2] Rybyanets, A.N.; Rybyanets, A.A. *IEEE Trans. UFFC*. 2011, *vol. 58*, 1477-1774.

[3] Rybyanets, A.N. *Ferroelectrics*. 2011, *vol. 419*, 90-96.

[4] Rybyanets, A.N.; Naumenko, A.A.; Shvetsova, N.A. Characterization Techniques for Piezoelectric Materials and Devices. In: *Nano- and Piezoelectric Technologies, Materials and Devices*, Parinov, I. A. (Ed.), Nova Science Publishers Inc., New York. 2013, Chapter 1, 275-308.

[5] Rybyanets, A.N. Ceramic Piezocomposites: Modeling, Technology, and Characterization. In: *Piezoceramic Materials and Devices,"* Parinov, I. A. (Ed.), Nova Science Publishers Inc., New York. 2010, Chapter 3, 113-174.

[6] *IEEE Standard on Piezoelectricity*. ANSI/IEEE Std. 1987, 176.

[7] Alguero, M.; Alemany, C.; Pardo, L.; Gonzalez, A.M. *J. Am. Ceram. Soc.* 2004, *vol. 87*, 209-212.

[8] Alemany, C.; Gonzalez, A.M.; Pardo, L.; Jimenez, B.; Carmona, F.; Mendiola, J. *J. Phys. D. Appl. Phys.* 1995, *vol. 28*, 945.

[9] Smits, J.G. *IEEE Trans. Sonics Ultrason.* 1976, *SU-23*, 393-402.

[10] Holland, R. *IEEE Trans. Sonics Ultrason.* 1967, *SU-14*, 18-24.

[11] *PRAP (Piezoelectric Resonance Analysis Program)*. TASI Technical Software Inc. www.tasitechnical.com.

[12] Rybianets, A.N.; Tasker, R. *Ferroelectrics*. 2007, *vol. 360*, 90-95.

[13] Devonshire, A.F. *Philosophical Magazine Supplement,* 1954, 85-130.

[14] Mason, W.P. *Physical Acoustics and the Properties of Solids*. D. Van Nostrand Co. Inc., Princeton, 1958.

[15] Berlincourt, D.A, Curran, D.R., Faffe, H.I. Piezoelectric and Piezomagnetical Materials and Their Function in Transducers. In: *Mason, W.P, eds. Physical Acoustics I. NY: Academic Press,* 1964, Part A, 170-270.

[16] Rybyanets, A.N., Naumenko, A.A. Nanoparticles Transport in Ceramic Matrixes: a Novel Approach for Ceramic Matrix Composites Fabrication. In: *Parinov, I.A, Hsiung-Chang S, eds. "Physics and Mechanics of New Materials and Their Applications," NY: Nova Science Publishers Inc*. 2013, Chapter 1. 3-18.

[17] Rybyanets, A.N. *IEEE Trans. UFFC*. 2011, *Vol. 58*, 1492-1507.

[18] Ramesh, R., Kara, H., Bowen, C.R. *Ultrasonics.* 2005, *Vol. 43*, 173–181.

[19] Bowen, C.R., Kara, H. *Materials Chemistry and Physics*. 2002, *Vol. 75*, 45–49.

In: Proceedings of the 2015 International Conference ...   ISBN: 978-1-63484-577-9
Editors: Ivan A. Parinov, Shun-Hsyung Chang et al.   © 2016 Nova Science Publishers, Inc.

Chapter 20

NEW FUNCTIONAL MATERIALS

K. P. Andryushin, I. N. Andryushina, L. A. Reznichenko, L. A. Shilkina and O. N. Razumovskaya*

Research Institute of Physics, Southern Federal University,
Rostov-on-Don, Russia

ABSTRACT

The present chapter provides information on the effect of changing the content of alkaline-earth elements on dielectric properties of solid solutions based on lead titanate. Two concentration areas with sharply differing nature of manifestations of dispersion phenomena identified. We show that with increasing of introduced modifiers concentration, there is a stabilization of their structure with a gradual decrease in the transition temperature. The influence of thermodynamic history (preparation conditions) and processes associated with time "aging" of the samples in the dielectric properties of the lead titanate ceramics modified with alkaline-earth elements are present.

Keywords: ferroelectric ceramic, lead titanate, alkaline-earth elements, modifiers, solid phase synthesis, dielectric spectra

1. INTRODUCTION

Ferroelectric materials with a high anisotropy of piezoelectric properties, $K_t/K_p > 5$, (K_p and K_t are electromechanical coupling factors at the planar and thickness oscillation modes, respectively) are of great interest for applications in various fields of modern technology (ultrasonic flaw detection, thickness measurement, accelerometer, medical diagnostics, etc.) [1]. Most often, the base of these materials is lead titanate ($PbTiO_3$), which piezoelectric properties and compositions with his participation are studied in sufficient detail [2-4]. On the other hand, on dielectric "behavior" in a wide range of external influences there is quite a bit

---

* E-mail: vortexblow@gmail.com.

of information. This situation restricts the boundaries of possible applications of such materials and makes it relevant to detailed studies of the dielectric spectra with significant variations in the chemical composition, temperature, and frequency measuring alternating electric field. The latter was the goal of this work to continue and develop the previously undertaken research of functional materials for various purposes [5].

2. EXPERIMENTAL

Samples of $(Pb_{1-\alpha1-\alpha2}A_{\alpha1}B_{\alpha2})TiO_3$ solid solutions (where A, B are the alkaline elements, AEs, and their compositions at $0.02 \leq \alpha_1 \leq 0.36$ and $0.0073 \leq \alpha_2 \leq 0.1339$) were prepared by solid phase synthesis followed by sintering according to the conventional ceramic technology (CCT) (solid-state synthesis, sintering without applied pressure). Samples for measuring were prepared in the form of discs (Ø 10 mm × 1 mm or Ø 10 mm × 0.5 mm) with silver electrodes.

Dielectric spectra (dependencies on the relative permittivity $\varepsilon/\varepsilon_0$ on temperature at various frequencies f of the alternating electric field) were studied by the home-made measuring bench using the immitance measuring device LCR-meter Agilent 4980A. The measurements were performed in the temperature range (25–500)°C and frequency range $(25–10^6)$ Hz.

3. RESULTS AND DISCUSSION

Figure 1 shows the temperature dependence of $\varepsilon/\varepsilon_0$ in a wide frequency range (curves a–k) and at a fixed frequency $f = 10^6$ Hz (curve l) for ceramics with different α_1. Figure 2 shows selected temperature dependencies of $\varepsilon/\varepsilon_0$ at different frequencies for ceramics with $\alpha_1 = 0.04$ (a), 0.08 (b), 0.18 (c), 0.22 (d), and 0.36 (e), and dielectric spectra of ceramic with $\alpha_1 = 0.04$ (f), built on the results of experiments, implemented in September 2014 (1) and April 2015 (2). As can be seen from Figure 1, for all compositions, a peculiar characteristic of the $\varepsilon/\varepsilon_0(T)$ dependence with a pronounced maximum at $T = T_c$ is observed. Above T_c after a sharp downturn, $\varepsilon/\varepsilon_0$ grows rapidly, since the temperature T_i is higher then higher the f, while the $T_i(\lg f)$ dependence is almost linear (see insets in Figure 1), and at $f > 10^6$ Hz, the effect of increasing $\varepsilon/\varepsilon_0$ in the studied temperature range absolutely absents.

As can be seen from Figure 1, there are two concentration ranges with sharply differing nature of the manifestations of dispersion phenomena. The first range $(0.02 \leq \alpha_1 \leq 0.24)$ is characterized by a strong dispersion of $\varepsilon/\varepsilon_0$ throughout the studied temperature range and does not allow one to form the maximum of $\varepsilon/\varepsilon_0$ at $T = T_c$ at low frequencies, the second range $(0.24 \leq \alpha_1 \leq 0.36)$ is characterized by a sharp decrease of $\Delta\varepsilon$ in the ferroelectric phase and by its complete disappearance at $\alpha_1 = 0.36$. Such behavior indicates the stabilization of the structure with the introduction of AEs into these solid solutions.

As we noted earlier [6], the observed effect is concerned with increasing conductivity of solid solutions, and as a result, with increasing the contribution into dielectric response mechanisms caused by a migration of mobile charge carriers. The latter is largely due to a change in the valence state of $Ti^{4+} \rightarrow Ti^{3+}$ and the appearance, as a result, of oxygen vacancies under the scheme $PbTi_{1-x}^{4+}Ti_x^{3+}O_{3-x/2}\square_{x/2}$, where $\square$ is a vacancy.

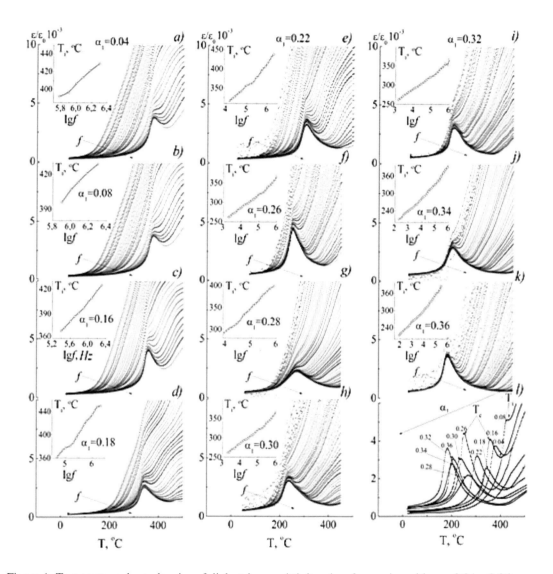

Figure 1. Temperature dependencies of dielectric permittivity $\varepsilon/\varepsilon_0$ of ceramics with $\alpha_1 = 0.04 - 0.36$ at $T = (20 - 500)°C$ and $f = (25 - 10^6)$ Hz (a–l), l is the temperature dependence of $\varepsilon/\varepsilon_0$ at $f = 10^6$ Hz (f measured in Hz).

The decreasing T_c and increasing of $(\varepsilon/\varepsilon_0)_k$ with the introduction of modifiers is associated with decreasing the electro-negativity and polarizing action of the A-cations and, as a consequence, the degree of weakening of covalence A-O bonds, entailing increasing a "ferroelectric soft" performance of solid solutions, and characterized by aforementioned behavior of T_c and $(\varepsilon/\varepsilon_0)_k$ [7-9]. The formation of the two concentration ranges of changes in macro-properties in the studied solid solutions is a consequence of their correlation with the phase diagram, which experiences a transformation in the neighborhood of $\alpha_1 \sim 0.24$, due to a transition from a two-phase to single-phase state. We note that the heterogeneity of the solid solutions with $\alpha_1 < 0.24$ is an additional factor destabilizing their structure.

Analysis of the data (Figure 2) shows a greater dielectric stability of solid solutions with a high AE content (virtually unaffected by T_c and T_s). This is supported by the fact of reproducing the character of dielectric spectra at $\alpha_1 \geq 0.22$ unlike modify profile curves $\varepsilon/\varepsilon_0(T)|_f$ at $\alpha_1 = 0.04$ (Figure 1, e). The splitting of the maxima $\varepsilon/\varepsilon_0 (T)|_f$, obtained on the same samples in April 2015 (i.e., 7 months ago after the initial measurements in September 2014) is the result of the solid-solution degradation, similar to the $PbTiO_3$ composition, probably due to beginning the process of self-destruction ceramics [10].

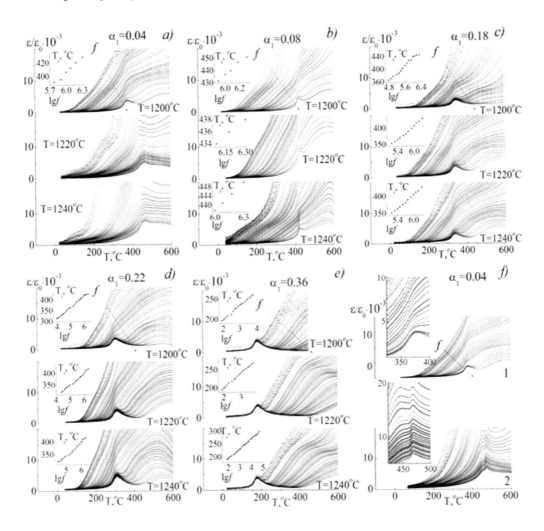

Figure 2. Temperature dependencies $(\varepsilon/\varepsilon_0)|_f$ of ceramics with $\alpha_1=0.04$ (a), 0.08 (b), 0.18(c), 0.22(d), and 0.36 (e), which were obtained at different T_s, and dielectric spectra of ceramic with $\alpha_1= 0.04$ (f), built on the results of experiments, implemented in September 2014 (graph 1) and April 2015 (graph 2). In insets, the $T_i(\lg f)$ dependence is shown (f measured in Hz).

CONCLUSION

The obtained results are expedient to use during the development of functional materials based on the $PbTiO_3$ system promising for ultrasonic testing.

ACKNOWLEDGMENTS

The work was supported by the Russian Ministry of Education and Science: Basic and design of the state job (topic No. 1927, Job No. 3.1246.2014/K, project No. 213.01-2014/012-SH) and FTP (Agreement No. 14.575.21.0007).

REFERENCES

[1] Dantsiger, A. Ya.; Razumovskaya, O. N.; Reznichenko, L. A.; Grineva, L. D.; Devlicanova, R. U.; Dudkina, S. I.; Gavrilyachenko, V. G.; Dergunova, N.V.; Klevtsov, A.N. *High-Efficience Piezoceramic Materials. Handbook.* Kniga, Rostov-on-Don, 1994, pp. 1 – 31 (in Russian).

[2] Chu, S.-Y.; Chen, C.-H. *Sensors and Actuators A.* 2001, *vol. 89*, 210-214.

[3] Te-Yi, Chen; Sheng-Yuan, Chu; Shih-Jeh, Wu; Yung-Der, Juang. *Ferroelectrics*, 2003, *vol. 282*, 37–47.

[4] Reznichenko, L. A.; Razumovskaya, O. N.; Ivanova, L. S.; Dantsiger, A. Ya.; Shilkina, L. A.; Fesenko, E. G. *Inorg. Mater.* 1985, *vol. 21(2)*, 282–285 (in Russian).

[5] Reznichenko, L. A.; Verbenko, I. A.; Andryushina, I. N.; Chernishkov, V. A.; Andryushin, K. P. *Ingenerniy Vestnik Dona.* 2015. *No. 3.* ivdon.Ru/ru/magazine/archive/n2y2015/2860 (in Russian).

[6] Kravchenko, O. Yu.; Reznichenko, L. A.; Gadzhiev, G. G.; Shilkina, L.A.; Kallaev, S. N.; Razumovskaya, O. N.; Omarov, Z. M.; Dudkina, S.I. *Inorg. Mater.* 2008, *vol. 44(10)*, 1265–1281 (in Russian).

[7] Reznichenko, L. A.; Alyoshin, V. A.; Shilkina, L. A.; Talanov, M. V.; Dudkina, S. I. Modification by Barium as Method of Changing Microstructure of Multicomponent ferroelectric ceramics. *Proceedings of the Second Intern. Interdistsiplin. Symp. "Lead-Free Ferroelectric Ceramics and Related Materials: Preparation, Properties, and Applications (Retrospective-Modern-Forecasts)."* 2013, *vol. 2*, 150-157 (in Russian).

[8] Talanov, M. V.; Shilkina, L. A.; Reznichenko, L. A.; Verbenko, I. A. Constructions of Composite Materials. 2014, No. 1, 57-61 (in Russian).

[9] Talanov, M. V.; Shilkina, L. A.; Reznichenko, L. A.; Dudkina, S. I. *Inorg. Mater.* 2014, *vol. 50(10)*, 1154–1160 (in Russian).

[10] Bondarenko, E. I.; Komarov, V. D.; Reznichenko, L. A.; Chernyshkov, V. A. *J. Tech. Phys.*, 1988, *vol. 58(9)*, 1771–1774 (in Russian).

In: Proceedings of the 2015 International Conference …   ISBN: 978-1-63484-577-9
Editors: Ivan A. Parinov, Shun-Hsyung Chang et al.   © 2016 Nova Science Publishers, Inc.

Chapter 21

PIEZOELECTRIC CERAMIC MATERIALS WITH HIGH FERROELECTRIC HARDNESS AND EFFICIENCY

H. A. Sadykov, I. A. Verbenko[*], *S. I. Dudkina and L. A. Reznichenko*

Research Institute of Physics, Southern Federal University,
Rostov-on-Don, Russia

ABSTRACT

The assessment of ferroelectric hardness of ternary solid solutions based on systems of niobate alkali and alkaline earth metals was carried out. PZT-8, PZT-4, and PZT-5H were used as reference materials. The findings should be used in predicting the properties of lead-free functional materials for a variety of applications.

Keywords: ceramic technology, piezoelectric, ferroelectic hardness, trinary systems

INTRODUCTION

At the prediction of properties of functional materials with the aim of identifying the areas of their possible applications, a question about their ferroelectric hardness (FH) often arises.

The concept of FH of ferroelectric (piezoelectric) materials has not still clear definition. Usually, the authors understand that the FH is the totality of the properties associated with more or lesser pliability of the domain structure on the electric and mechanical influences. FH is determined largely from the value of local elastic stresses, which induced by the reorientation of domains. With the growth of these stresses, FH, the coercive field (E_c) and Young's modulus Y_{11}^E increase. The spontaneous deformation or the homogeneous deformation parameter δ can be one of the parameters that determine the level of these

---

[*] E-mail: ilich001@yandex.ru.

stresses [1]. Most of the parameters of the ferro- and piezoelectric ceramics can be associated with the aforementioned δ, E_c, and Y_{11}^E, and it can be used to evaluate FH of materials.

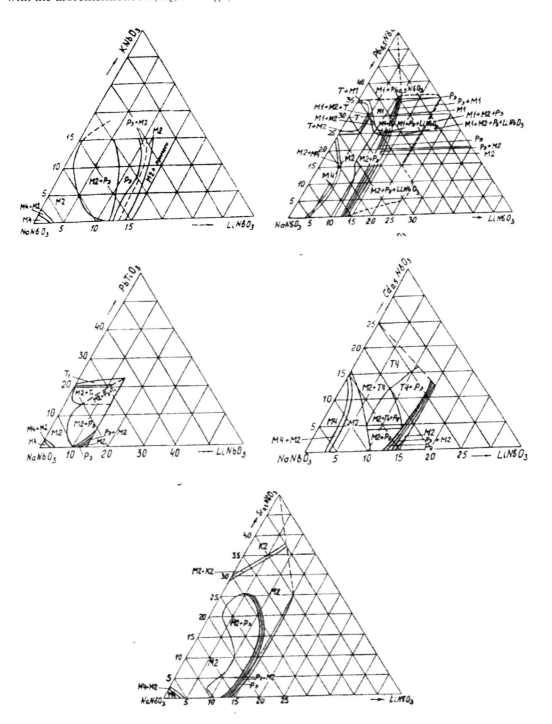

Figure 1. Phase diagrams of ternary systems (Na, Li, K)NbO$_3$ [2], (Na, Li, Pb)(Nb, Ti)O$_3$ [3], (Na, Li, Pb$_{0.5}$)NbO$_3$[4], (Na, Li, Cd$_{0.5}$)NbO$_3$[5], and (Na, Li, Sr$_{0.5}$)NbO$_3$[6], respectively.

Thus, reducing the mobility of domain walls leads to increasing δ and E_c, and, therefore, to decreasing the dielectric loss (tgδ) and mechanical loss (Q_M^{-1}); this means increasing the mechanical q-factor Q_M and dielectric permittivity $\varepsilon_{33}^T/\varepsilon_0$. The sound speed V_1^E changes in the same direction as Y_{11}^E. Thus, with the growth of FH piezomaterials, parameters such as δ that Q_M, V_1^E should grow, but the $\varepsilon_{33}^T/\varepsilon_0$ and tgδ to drop. The electromechanical coupling factor K_p characterizes the effect of converting power in piezoelectric materials and determines their efficiency. K_p and piezoelectric coefficient $|d_{31}|$, because it is dependent on both K_p and $\varepsilon_{33}^T/\varepsilon_0$, tend to decrease with increasing δ.

Due to the toxicity of lead in the widely used piezoelectric ceramics, there is significant interest in developing lead-free piezoelectric ceramics. Especially important is elimination of lead-containing compounds from the composition of special electrotechnical ceramics (piezoceramics).

The objects of the present study are ternary systems of (Na, Li, K)NbO$_3$, (Na, Li, Pb$_{0.5}$)NbO$_3$, (Na, Li, Pb)(Nb, Ti)O$_3$, (Na, Li, Cd$_{0.5}$)NbO$_3$, and (Na, Li, Sr$_{0.5}$)NbO$_3$ (see phase diagrams in Figure 1). On the base of medium values of the parameters of morphotropic solid solutions corresponding of the section $z = 5$; 10; 15, (here z is PbTiO$_3$, KNbO$_3$, Pb$_{0.5}$NbO$_3$, Cd$_{0.5}$NbO$_3$, or Sr$_{0.5}$NbO$_3$ in mol. %), FH and the efficiency of the studied ternary systems were estimated.

Table 1. Evaluation results on ferroelectric stiffness systems based on alkali metal niobate

$z = 5$	Electrophysical parameters	hard	medium	soft	$z = 10$	hard	medium	soft	$z = 15$	hard	medium	soft
Cd	Q_M	0.30	**0.60**	4.62	Cd	0.35	**0.69**	5.31	Cd	0.26	**0.52**	4.03
	$(\varepsilon_{33}^T/\varepsilon_0)^{-1}$	**2.12**	2.77	7.23		**1.87**	2.45	6.42		**1.25**	1.63	4.25
	K_P	**0.36**	0.31	0.28		**0.29**	0.25	0.22		**0.52**	0.45	0.40
	$|d_{31}|$, pC/N	**0.23**	0.17	0.08		**0.20**	0.15	0.07		**0.52**	0.39	0.18
Pb	Q_M	0.40	**0.80**	6.15	Pb	0.35	**0.70**	5.38	Pb	0.28	**0.55**	4.23
	$(\varepsilon_{33}^T/\varepsilon_0)^{-1}$	**4.35**	5.65	1478		**3.64**	4.72	1236		**2.27**	2.96	7.72
	K_P	**0.44**	0.38	0.34		**0.38**	0.33	0.29		**0.40**	0.34	0.31
	$|d_{31}|$, pC/N	**0.15**	0.11	0.05		**0.17**	0.13	0.06		**0.22**	0.16	0.07
Sr	Q_M	**0.68**	7.35	10.38	Sr	0.36	**0.72**	5.54	Sr	0.28	**0.56**	4.31
	$(\varepsilon_{33}^T/\varepsilon_0)^{-1}$	**5.88**	7.65	20.0		**5.26**	6.84	1789		**3.33**	4.33	1133
	K_P	**0.28**	0.24	0.22		**0.29**	0.25	0.22		**0.39**	0.33	0.30
	$|d_{31}|$, pC/N	**0.10**	0.07	0.03		**0.11**	0.08	0.04		**0.34**	0.17	0.06

Commentary: values close to unity are shown in bold.

The coefficient P uses for the evaluations, that allows one to judge about the degree FH and efficiency of solid solutions [7].

The coefficient represents the attitude of the parameters of the systems as compared to values of this parameter of industrial analogs of PZT-8, PZT-4, PZT-5H [8], which are examples of a hard material, medium hard materials, and a soft material, respectively. The parameters for characterizing FH are Q_M and $(\varepsilon_{33}{}^T/\varepsilon_0)^{-1}$, which increase with increasing FH. Efficiency was estimated by K_p and $|d_{31}|$. Evaluation results on the ferroelectric stiffness systems, based on alkali metal niobate, are present in Table 1.

Our analysis of experimental data shows that the parameters $(\varepsilon_{33}{}^T/\varepsilon_0)^{-1}$, K_p, $|d_{31}|$ allow us to classify systems as "hard," and Q_M as systems with an intermediate "hardness." The parameters of these lead-free materials have a higher dispersion than solid solutions based on the PZT system.

The obtained data can be used for the prediction of the properties, development and creation of novel functional materials based on niobates alkaline and alkaline-earth elements.

ACKNOWLEDGMENT

This work was financially supported by the Ministry of Education and Science of the Russian Federation (Basic and Design State Tasks Nos. 1927 (213.01-11/2014-21), 213.01-2014/012-SH, and 3.1246.2014/K) and President Grant № MK-3232.2015.2.

REFERENCES

[1] Fesenko, E. G. Family of Perovskite and Ferroelectricity. Atomizdat, Moscow. 1972, pp. 1 – 248 (in Russian).

[2] Reznichenko, L. A.; Razumovskaya, O. N.; Shilkina, L. A.; Dudkina, S. I.; Borodin, A. V. *Technical Physics Letters*, 2002, vol. 28(2), 83 - 86.

[3] Reznichenko, L. A.; Shilkina, L. A.; Tais'eva, V. A.; Kupriyanov, M. F.; Turik, A. V. *Inorganic Materials*, 1980, vol. 16(11), 2002 - 2004.

[4] Reznichenko, L. A.; Razumovskaya, O. N.; Dantsiger, A. Ya. Phase transitions in system (Na, Li, $Pb_{0,5}$)NbO_3. Proceedings of the International scientific and practical conference "Piezoelectric technology - 97." Obninsk. 1997, 197 - 207 (in Russian).

[5] Reznichenko, L. A.; Razumovskaya, O. N.; Shilkina, L. A.; Pozdnyakova, I. V.; Servuli, V. A. *Technical Physics*, 2001, vol. 46(1), 24 - 28.

[6] Reznichenko, L. A.; Razumovskaya, O. N.; Shilkina, L. A.; Pozdnyakova, I. V.; Servuli, V. A. *Technical Physics*, 2000. vol. 45(11), 1432 - 1436.

[7] Fesenko, E. G.; Dantsiger, A. Ya.; Razumovskaya, O. N.; Gavrilyachenko, S. V.; Dergunova, N. V. *Ferroelectrics*, 1994, vol. 167, 197 - 204.

[8] Jaffe, H.; Berlincourt, D. A. Piezoelectric Transducer Material. *Proc. IEEE*, 1965, vol. 53, 1552.

In: Proceedings of the 2015 International Conference ...   ISBN: 978-1-63484-577-9
Editors: Ivan A. Parinov, Shun-Hsyung Chang et al.   © 2016 Nova Science Publishers, Inc.

Chapter 22

EXPERIMENTAL STUDY OF HIGH TEMPERATURE POROUS PIEZOCERAMICS

I. A. Shvetsov, M. A. Lugovaya, N. A. Shvetsova, G. M. Konstantinov and A. N. Rybyanets*

Institute of Physics, Southern Federal University, Rostov-on-Don, Russia

ABSTRACT

A manufacture technology of porous ferroelectric ceramics on the base of high temperature compositions of pure and modified lead titanate and lead metaniobate compositions, as well as lead free sodium-potassium niobate solid solutions was developed. Special attention was given to the microstructure studying of porous ceramics with different porosity types. The porosity dependencies of elastic, dielectric and piezoelectric properties of the specified compositions were measured in 0-40% relative porosity range. It was shown that the porous ceramics technology allows one to produce stable in time piezoelements of such "technologically difficult" ceramics as lead titanate and lead metaniobate with excellent and reproducible properties.

Keywords: porous piezoceramics, microstructure, porosity, $PbTiO_3$, $PbNb_2O_6$

1. INTRODUCTION

To date, PZT-type compositions are mainly used as initial materials for porous ferroelectric ceramics manufacture [1]. Other ceramic systems were not studied as porous ferroelectric ceramics because of high initial anisotropy, low piezoelectric activity, and technological difficulties inherent for ceramic compositions of these systems.

With the growing demands to piezoceramics for advanced piezoelectric applications alongside with rising operational and environment requirements [2, 3], lead titanate ($PbTiO_3$) and lead metaniobate ($PbNb_2O_6$), as well as lead free porous ferroelectric ceramics are of

---

* e-mail: wbeg@mail.ru.

considerable interest in both scientific research and practical applications. Due to unique properties, first of all high Curie temperature, high d_{33}/d_{31} ratio, and low mechanical quality factor Q_M, $PbNb_2O_6$ and $PbTiO_3$ ceramics are optimum for application in the broadband ultrasonic transducers working in extreme operational conditions. Unfortunately, pure $PbNb_2O_6$ and $PbTiO_3$ ceramics as a materials are very difficult to produce because of the inherent internal stresses resulting in cracking and destruction of piezoceramic samples. In the same time, available on the market modified piezoceramics based on $PbNb_2O_6$ and $PbTiO_3$ are in part deprived the advantages inherent for pure ceramic compositions. Lead free compositions based on $Na(K_x,Nb_{1-x})O_3$ and $Na(Li_x,Nb_{1-x})O_3$ solid solutions are also characterized by unique set of properties (low acoustic impedance, high electromechanical anisotropy and sound velocity, and high Curie temperature) and can be used in high temperature ultrasonic transducers.

In this work porous piezoceramics based on $PbTiO_3$, $PbNb_2O_6$, $(Pb_x,K_{1-x})Nb_2O_6$, $(Pb_x,Ca_{1-x})TiO_3$ and $Na(K_x,Nb_{1-x})O_3$ compositions with excellent and reproducible properties were fabricated and studied. Standard powders of piezoceramic compositions prepared using routine ceramic technology (solid phase synthesis) were chosen as the starting materials. For the porous ceramics preparation specially developed technology, supposing controllable formation of microporous structures were used. Special attention was given to the microstructure studying of porous ceramics with different porosity types.

The porosity dependencies of elastic, dielectric and piezoelectric properties of the high temperature porous piezoceramics were measured in $0-40\%$ porosity range using impedance spectroscopy and ultrasonic methods. Temperature dependencies of piezoelectric and dielectric parameters were measured in the $0-600$ °C range. It was shown that the porous ceramics technology allows one to produce stable in time piezoelements of such "technologically difficult" high temperature ceramics as lead titanate and lead metaniobate with excellent and reproducible properties.

A line of high temperature ultrasonic transducers and sensors for NDT and ultrasonic applications with high sensitivity and resolution was manufactured and tested. Additional advantages of the developed porous ceramics are the possibility of controllable change of main parameters in a wide range, compatibility with standard fabrication technologies, and processing flexibility.

2. POROUS CERAMIC PREPARATIONS

Standard powders composed of $PbTiO_3$, $PbNb_2O_6$, $(Pb_x, K_{1-x})Nb_2O_6$ and $(Pb_x, Ca_{1-x})TiO_3$, as well as $Na(K_x,Nb_{1-x})O_3$ prepared using routine ceramic technology (solid phase synthesis) were chosen as the starting materials. For the preparation of porous ceramics, specially developed methods [4, 5] were used, that enabled the controllable formation of micro- (intracrystalline), mesa- (intercrystalline) and macro-porous structures.

The macro-porous ceramics were prepared by burning out organic powders. Mixtures of organic crystal powders with different particle shapes, decomposition temperatures and grain-size distribution (from 5 to 150 μm) were used for producing the pore volumes. For intracrystalline micro-porosity formation (capture of initial pores during crystalline growth) special sintering and heat treatment regimes were used. For intercrystalline (mesa) porosity creation, the inhibitors of densification and shrinkage (presintered fine grain ceramic powder of the same composition, or passive fine-grained refractory powders) were added to the initial ceramic powder before green body compacting.

The porous ceramics samples with vacuum deposited Cr/Ni electrodes were poled in silicone oil at $140 - 160°C$ by applying a dc field of $4 - 6$ kV/mm for 30 min. Micrographs of polished and thermally etched surfaces of PFC samples were made using high resolution optical and scanning electron microscopes.

3. METHODS OF MEASUREMENTS

The basic techniques for finding material constants of piezoelectric materials are outlined in the IEEE Standard on Piezoelectricity (1987) [6]. The IEEE Standard describes two approaches to obtaining material properties using either the difference between parallel (f_p) and series (f_s) resonance frequencies, or the frequency ratio between different order resonance peaks. These approaches can be problematic when material properties are dispersive or when a low electromechanical coupling constant results in a small difference between f_p and f_s.

The IEEE Standard works for many of the most widely used commercial piezoceramics based on lead-titanate-zirconate (PZT) compositions that are high-Q_M and high-coupling coefficient piezoelectric materials. However, there is a general agreement that their use in many new piezoelectric materials such as porous ceramics, piezoelectric polymers or piezoelectric composites may lead to significant errors. Furthermore, the current IEEE Standard does not comprehensively account for the complex nature of material coefficients, as it uses only the dielectric loss factor (tan δ) and the mechanical quality factor (Q_M) to account for loss.

Numerous techniques using complex material constants have been proposed to take into account losses in low-Q_M materials and to overcome limitations in the IEEE Standard [7-11]. Iterative methods [7, 8] provide a means to accurately determine the complex coefficients in the linear range of poled piezoceramics from complex impedance resonance measurements.

Smits [9] proposed a general method for the thickness extensional (TE), thickness shear (TS), length extensional (LE), and length-thickness extensional (LTE) resonators using three points around resonance. This method does not rely on critical frequencies in the spectrum and so does not lose sensitivity for low Q materials. At resonance and anti-resonance, currents can be very large and small and this presents an instrumentation problem. Though less of an issue with low Q materials, Smits's method is less demanding of instrumentation than the IEEE methods. Finally, because the choice of analysis points is somewhat arbitrary, noisy spectra and samples with mode coupling may still be analyzed using Smits's method.

The PRAP software [11] analyses impedance spectra to determine complex material properties. This software uses a generalized form of Smits's method to determine material properties for any common resonance mode, and a generalized ratio method for the radial mode [10] valid for all material Q's. By analyzing on each harmonic, PRAP can determine complex material properties as a function of frequency. The software always generates an impedance spectrum from the determined properties to indicate validity of the results.

Measurements were carried out on standard samples of porous piezoceramics fabricated in the same technological regimes. For each porosity value, not less than 10 samples of each composition were measured. Measurement error for porosity (density) was not exceeding ± 1.5%. Dielectric, piezoelectric and elastic parameters were measured using the Solartron Impedance/Gain-Phase Analyzer SL 1260 and Agilent 4291A Impedance Analyzer according IEEE Standard [6] and PRAP software [11, 12] methods.

The porosity dependencies of elastic, dielectric and piezoelectric properties of the specified porous piezoceramics were measured in a 0 – 40% relative porosity range.

The microstructure of polished, chemically etched, and chipped surfaces of composite samples was observed with optical (NeoPhot-21).

Sound velocity and attenuation of longitudinal waves in porous ceramic samples were measured by pulse-echo and through-transmit ultrasonic methods in the frequency range 1 – 5 MHz using a LeCroy Wave Surfer 422 digital oscilloscope and Olympus 5800, 5077 pulser/receivers using standard Olympus transducers.

4. RESULTS AND DISCUSSIONS

Figure 1 shows a typical optical micrograph of polished and thermally etched surfaces of $PbNb_2O_6$, $(Pb, K)Nb_2O_6$ and $PbTiO_3$ porous ceramics with different types of porosity. An optimized micro-, mesa- and macro-porosity combination in all $PbNb_2O_6$ porous ceramics samples prevents excessive crystalline growth and cracking. For $PbTiO_3$ porous ceramics, mixed macro- and mesa-porosity prevent ceramic cracking and destruction on grain boundaries.

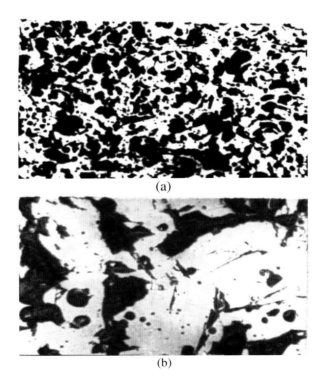

(a)

(b)

Figure 1. (Continued).

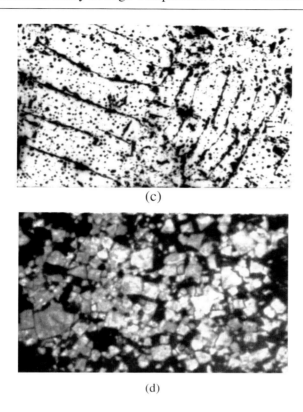

Figure 1. Optical micrograph of polished and thermally etched surfaces of different porous ceramics: (a) PbNb$_2$O$_6$ porous ceramics with mixed macro-, mesa- and micro porosity ($p = 15\%$) - magnification 115X; (b) PbNb$_2$O$_6$ porous ceramics with mixed micro- and mesa porosity ($p = 15\%$) - magnification 735X; (c) (Pb,K)Nb$_2$O$_6$ porous ceramics with optimized micro porosity ($p = 15\%$). - magnification 300X; (d) PbTiO$_3$ porous ceramics with mixed macro- and mesa porosity ($p = 25\%$) - magnification 740X.

Table 1 and 2 summarize the results of elastic, dielectric and piezoelectric constants measurement for PbTiO$_3$, PbNb$_2$O$_6$, and Na(K$_x$,Nb$_{1-x}$)O$_3$ porous ceramics with optimal porosities.

Table 1. Properties of PbTiO$_3$ and PbNb$_2$O$_6$ porous piezoceramics

Parameter/Material	PbNb$_2$O$_6$ Porosity 15%	(Pb, K)Nb$_2$O$_6$ Porosity 15%	PbTiO$_3$ Porosity 25%	(Pb, Ca)TiO$_3$ Porosity 25%
k_p	0	0	0	0
k_t	0.4	0.42	0.50	0.56
d_{33} (10^{-12}), C/N	65	70	50	75
$\varepsilon_{33}^T/\varepsilon_0$	160	160	150	80
$\tan\delta$, %	0.6	0.8	2	2
Q_M^t	< 10	< 10	< 10	< 10
ρ (10^3), g/cm^3	5.7	5.6	6.0	5.6
T_C, °C	560	550	490	325
N_t, kHz·mm	1300	1380	1400	1300

Table 2. Properties of Na(K$_x$,Nb$_{1-x}$)O$_3$ porous piezoceramics

Material/ Parameter	$\varepsilon_{33}^T/\varepsilon_0$	$tg\delta$, %	d_{33}, pC/N	$|d_{31}|$, pC/N	k_t	k_p	V_t, m/sec	Q_M^t	ρ, g/cm^3	Z_A, Mrayl
PCR-34*	310	2.50	102	32	0.52	0.09	3250	20	2.75	8.93
PCR-34**	200	2.75	98	56	0.51	0.13	4500	35	3.50	15.75

Note: * - relative porosity $p = 0.4$; ** - relative porosity $p = 0.2$.

It was found that for PbTiO$_3$, PbNb$_2$O$_6$ and Na(K$_x$,Nb$_{1-x}$)O$_3$ porous ceramics (in contrast to PZT based porous ceramics [2, 5]) piezoelectric constant d_{33} and electromechanical coupling factor k_t decrease slightly as a function of relative porosity.

It was also shown that the porous ceramics technology results in increasing of piezoelectric anisotropy, removal of internal mechanical stress, increasing of mechanical durability, preventing of cracking, and finally, allows production of stable in time elements of such "technologically difficult" ceramics as lead titanate and lead metaniobate with excellent and reproducible properties [6, 7].

CONCLUSION

In this work, porous piezoceramics based on PbTiO$_3$, PbNb$_2$O$_6$, (Pb$_x$,K$_{1-x}$)Nb$_2$O$_6$, (Pb$_x$,Ca$_{1-x}$)TiO$_3$ and lead free Na(K$_x$,Nb$_{1-x}$)O$_3$ compositions with high Curie temperature and reproducible properties were fabricated and studied. The porosity dependencies of elastic, dielectric, and piezoelectric properties of the high temperature porous piezoceramics were measured in 0 – 40% porosity range using impedance spectroscopy and ultrasonic methods. Temperature dependencies of piezoelectric and dielectric parameters were measured in the 0 – 600 °C range. It was shown that the porous ceramics technology allows one to produce stable in time piezoelements of such "technologically difficult" high temperature ceramics as lead titanate and lead metaniobate with excellent and reproducible properties. A line of high temperature ultrasonic transducers and sensors for NDT and ultrasonic applications with high sensitivity and resolution was manufactured and tested. Additional advantages of the developed porous ceramics are the possibility of controllable change of main parameters in a wide range, compatibility with standard fabrication technologies, and processing flexibility.

ACKNOWLEDGMENTS

Work supported by the RSF grant no. 15-12-00023.

REFERENCES

[1]   Rybyanets, A.N. *IEEE Trans. UFFC*. 2011, *vol. 58*, 1492-1507.
[2]   Rybyanets, A.N.; Rybyanets, A.A. *IEEE Trans. UFFC*. 2011, *vol. 58*, 1757-1774.
[3]   Rybyanets, A.N. *Ferroelectrics*. 2011, *vol. 419*, 90-96.

[4] Rybyanets, A.N.; Naumenko, A.A.; Shvetsova, N.A. In: *Nano- and Piezoelectric Technologies, Materials and Devices,* Parinov I. A. (Ed.), Nova Science Publishers Inc., New York, 2013. Chapter 1, 275-308.

[5] Rybyanets, A.N. Ceramic Piezocomposites: Modeling, Technology, and Characterization. In: *Piezoceramic Materials and Devices.* Parinov I. A. (Ed.), Nova Science Publishers Inc., New York. 2010, Chapter 3, 113-174.

[6] *IEEE Standard on Piezoelectricity.* ANSI/IEEE Std. 1987, 176.

[7] Alguero, M.; Alemany, C.; Pardo, L.; Gonzalez, A.M. *J. Am. Ceram. Soc.* 2004, *vol. 87,* 209-212.

[8] Alemany, C.; Gonzalez, A.M.; Pardo, L.; Jimenez, B.; Carmona, F.; Mendiola, J. *J. Phys. D. Appl. Phys.* 1995, *vol. 28,* 945.

[9] Smits, J.G. *IEEE Trans. Sonics Ultrason.* 1976, *SU-23,* 393-402.

[10] Holland, R. *IEEE Trans. Sonics Ultrason.* 1967, *SU-14,* 18-24.

[11] *PRAP (Piezoelectric Resonance Analysis Program).* TASI Technical Software Inc. *www.tasitechnical.com.*

[12] Rybianets, A.N.; Tasker, R. *Ferroelectrics.* 2007, *vol. 360,* 90-95.

[13] Rybyanets, A.; Naumenko, A. In: *Physics and Mechanics of New Materials and Their Applications.* Parinov I. A., Shun-Hsiung Chang, (Eds.), Nova Science Publishers Inc., New York, 2013. Chapter 1, 3-18.

In: Proceedings of the 2015 International Conference …   ISBN: 978-1-63484-577-9
Editors: Ivan A. Parinov, Shun-Hsyung Chang et al.   © 2016 Nova Science Publishers, Inc.

Chapter 23

STRUCTURAL PECULIARITIES OF POROUS FERROELECTRIC CERAMICS

G. M. Konstantinov[*], Y. B. Konstantinova, N. A. Shvetsova, N. O. Svetlichnaya and A. N. Rybyanets
Southern Federal University, Rostov-on-Don, Russia

ABSTRACT

The porous ferroelectric ceramic materials have found a wide application due to a favorable combination of their physical parameters. This paper offers an experimental study of interrelations between fabrication methods of porous ceramics based on CTS-19 ferroelectric material, their microstructure, crystal structure, and set of physical properties.

Keywords: porous ferroelectric ceramics, microstructure, crystal structure, nanopores, morphotropic transition region, rhombohedral phase, tetragonal phase, pore size, intragranular porosity, macrostructural inhomogeneities, microstructural inhomogeneities, ceramic grain size

1. INTRODUCTION

Development of fundamentally new functional materials with controlled physical properties (by means of regulation of nanopores system and alloying materials with nanoparticles of different substances) and study of the physical processes inside nanopores present relevant scientific and technical problems.

The porous ferroelectric materials are becoming more widely used, thanks to a useful combination of physical properties (e.g., [1, 2]). Today, we have some experience in the development of technologies making porous ferroelectrics based on the system of solid

---

[*] e-mail: georgy.konstantinov@yandex.ru.

solutions of lead zirconate titanate (PZT) with compositions corresponding to the morphotropic transition region (MTR).

Meanwhile, so far there is no experimental proof of the possibility to obtain porous ferroelectrics with intragranular porosity and pores sizes of a few micrometers. Obviously, the creation of technologies forming ferroelectric materials with controlled porosity scale of several micrometers allows one to have a different look at the control of the properties of these materials by means of their targeted doping with micro- and nanoparticles of various materials.

This paper describes the first manufactured and studied porous ferroelectric PZT-based materials with pores in individual crystallites (pore size is about several micrometers). The CTS-19 porous ceramics is the object of the investigation. Experiments were conducted to study the relations between the microstructure and crystal structure of the material obtained under various processing modes. When obtaining the porous ferroelectrics, the sintering time and temperature, as well as the type and amount of blowing agent were varied.

2. EXPERIMENTAL RESULTS

The study of material microstructure was conducted using the Scanning Electron Microscope (SEM) HITACHI TM 1000 (with increases from 200x to 5000x). X-ray diffraction tests of porous ferroelectrics were made using DRON-6.0 diffractometer and C_0 K_α radiation. The electrical and physical parameters of ceramic samples were determined by standard methods.

With the use of a special selection of processing methods, a CTS-19 porous ferroelectric material was obtained with intragranular pore size of 3-10 micrometers. In the photo of Figure 1, an arrow shows an intragranular pore size of about 3 microns. As one can see this pore is inside a large crystallite size of about 18 micrometers.

Figure 1. Intragranular pore size of about 3 microns in porous ceramics CTS-19.

Figure 2 shows a photo of an intragranular pore size of about 10 microns. The pore is located inside a crystallite size of about 40 micrometers. At the same photo, arrows show the nucleus of the intragranular pore size of 5-10 micrometers, as well as a large intercrystalline pore size of about 80 microns and smaller intrercrystalline pores. This pattern of intercrystalline pores is typical for porous PZT-based ferroelectrics.

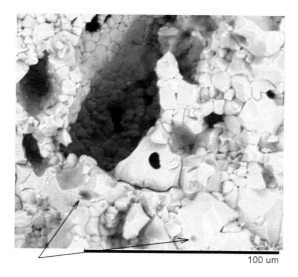

Figure 2. Intragranular pore size of about 10 microns and nucleus of the intragranular pore size of 5-10 micrometers in porous ceramics CTS-19.

Figure 3. Typical microstructure of porous ceramics CTS-19 with a developed intercrystalline porous structure.

It should be noted, that observed intragranular pores (Figures 1 and 2) were located in the center of large crystallites (ceramic grains) of irregular geometric shape. Apparently, this is due to the fact that the process of pore formation is caused, on the one hand, by the size and shape of particles of a blowing agent, and on the other hand, by the ratio of its particles size and granules of the obtained ferroelectrics. Probably in the ceramic grains of small size (a few micrometers), the process of pore formation is difficult.

The photo of Figure 3 shows a typical microstructure of porous ceramics with a developed intercrystalline porous structure. In the center of the photo, one can observe a large pore size of about 50 micrometers, with an average ceramic grain size of 5-10 microns.

Using the X-ray diffraction experiment, we could find out a significant difference in the structural parameters of the CTS-19 porous ferroelectrics obtained under different process conditions. The CTS-19 ferroelectrics, which were processed by hot-pressing, characterized by the presence of only tetragonal phase of perovskite structure in its composition at room temperature. We established the coexistence of rhombohedral and tetragonal ferroelectric phases at the room temperature in the CTS-19 porous materials, which is typical for PZT compositions in the region of morphotropic transition [3, 4]. According to the ratio of the integrated intensities of diffraction reflections with hkl 200 of the perovskite structure, we defined the concentration relation of these phases. The composition of CTS-19 porous materials prepared under different process conditions consists of 30-40% of rhombohedral phase and 70-60% of tetragonal phases.

It is well known (e.g., [3-5]) that the relation of phases concentrations coexisting in the MTR of ceramic systems of PZT-based solid solutions is determined not only by the composition of the solid solution, but also by structural inhomogeneities of ceramics. Such inhomogeneities can be divided into two groups:

i) macrostructural inhomogeneities associated with the defects in the grain structure of ceramics, and a nonuniform distribution of the solid solution components in ceramic grains (crystallites);

ii) microstructural inhomogeneities associated with the defects in the crystal lattice, and a nonuniform distribution of the solid solution components inside the crystallites of ceramics.

It is obvious that the formation of pores in the porous ferroelectric ceramics is another kind of inhomogeneities affecting the uniform distribution of the components in a solid solution. Apparently, the observed change in the concentration relation of the ferroelectric rhombohedral and tetragonal phases is a proof of all the inhomogeneities mentioned above.

It should be noted, that the reasons of the process conditions of preparing ferroelectrics impact on uniformity of distributing PZT solid solution components in the volume of crystallites are in a significant difference of temperatures, required for the formation of $PbTiO_3$-based and $PbZrO_3$-based PZT solid solutions. The lead titanate forms from PbO and TiO_2 oxides at temperatures 450-500°C, while the lead zirconate synthesizes at higher temperatures (above 700°C). These results consist in appearance of two most possible solid solutions in the volume of any PZT-based synthesized material. One of the compounds associates with the previously formed solid solution with a predominant content of $PbTiO_3$, the other - with the later formed solid solution with the predominant content of $PbZrO_3$ that prepares at higher temperatures.

Changes in the process conditions of preparing ceramics result in the change of uniform distribution of the PZT solid solution components inside ceramics crystallites, which has an impact on the concentration relation of ferroelectric phases in the region of morphotropic transition.

Thus, the present paper provides an experimentally established fact that by choosing processing methods of preparation of PZT-based porous ferroelectric ceramics, it is possible to obtain ceramics with the intragranular pores size of several micrometers. Varying process conditions causes changes in the composition of rhombohedral and tetragonal phases coexisting in MTR.

On the base of experimental results, we can state the selection of methods to prepare porous ferroelectric ceramics. It is possible to receive intentionally porous ceramic structures based on the solid solutions system with periodic deployment of intragranular pores size of a few micrometers. When having such structures, it is possible

i) to carry out doping of ferroelectrics with nanoparticles of various materials and thus to create new composite structures;
ii) to create specific PZT- based ferroelectric porous structures.

ACKNOWLEDGMENTS

Work supported by the Ministry of Education and Science of RF (base and project parts of state order theme No. 213.01-11/2014-21 and 3.1246.2014/K), Southern Federal University (internal grant - 213.01-2014/012-ВГ) and FCP (ГК No. 14.575.21.0007).

REFERENCES

[1] Rybyanets, A.N. *IEEE Trans. UFFC*. 2011, *vol. 58(7)*, 1492-1507.

[2] Bunin, M.A.; Rybyanets, A.N.; Naumenko, A.A.; Suchomlinov, D.I.; Fedorovskiy, A.E. *Ferroelectrics*. 2015, *vol. 475*, 96-103.

[3] Bogosova, Ya.B.; Konstantinov, G.M. *J. Neorg. Mat.* 1995, *vol. 31(4)*, 563-566.

[4] Konstantinov, G.M.; Bogosova, Ya.B.; Abdulvahidov, K.G.; Kupriyanov, M.F. *Izv. RAN, Ser. Phys.* 1995, *vol. 59(9)*, 89-92.

[5] Konstantinov, G.M.; Bogosova, Ya. B. *Technical Physics*. 1995, *vol. 65*, 93-102.

In: Proceedings of the 2015 International Conference ...   ISBN: 978-1-63484-577-9
Editors: Ivan A. Parinov, Shun-Hsyung Chang et al.   © 2016 Nova Science Publishers, Inc.

Chapter 24

COMPLEX PARAMETERS OF POROUS PZT PIEZOCERAMICS MEASURED FOR DIFFERENT VIBRATIONAL MODES

M. A. Lugovaya[*], A. A. Naumenko, E. I. Petrova and A. N. Rybyanets
Southern Federal University, Rostov-on-Don, Russia

ABSTRACT

The accurate description of piezoceramics must include the evaluation of the dielectric, piezoelectric and mechanical losses, accounting for the out-of-phase material response to the input signal. Numerous techniques using complex material constants have been proposed to take into account losses in low-Q_M materials and to overcome limitations in the IEEE Standard. Iterative methods provide a means to determine accurately the complex coefficients in the linear range of poled piezoceramics from complex impedance resonance measurements. This chapter presents the results of experimental study of porous PZT piezoceramics with the composition of $Pb_{0.95}Sr_{0.05}Zr_{0.53}Ti_{0.47}O_3$ + 1wt.% Nb_2O_5, and relative porosity 35%. Both the IEEE Standard and the Piezoelectric Resonance Analysis Method (PRAP) were used to determine material constants corresponding to thickness and length-thickness extensional modes of piezoelectric plate vibrations. Complex sets of elastic, dielectric and piezoelectric coefficients of the porous ceramic piezoelements were measured by impedance spectroscopy method using (PRAP) software. Reference data for real parts of material constants were measured by IEEE Standard method. Material constants obtained by the different methods were compared. The results show that the PRAP method gives results that are more accurate and allow one take into account elastic, piezoelectric, and dielectric losses.

Keywords: porous PZT piezoceramics, elastic, dielectric and piezoelectric coefficients

---

[*] E-mail: lugovaya_maria@mail.ru.

1. INTRODUCTION

The basic techniques for finding material constants of piezoelectric materials are outlined in the IEEE Standard on Piezoelectricity (1987) [1]. These methods work for many of the most widely used commercial piezoceramics based on lead-zirconate-titanate (PZT) compositions that are high-Q_M and high-coupling coefficient piezoelectric materials. However, there is a general agreement that their use in many new piezoelectric materials such as porous ceramics, piezoelectric polymers or piezoelectric composites may lead to significant errors. Furthermore, the current IEEE Standard does not comprehensively account for the complex nature of material coefficients, as it uses only the dielectric loss factor (tan δ) and the mechanical quality factor (Q_M) to account for loss.

Numerous techniques using complex material constants have been proposed to take into account losses in low-Q_M materials and to overcome limitations in the IEEE Standard [2-6]. Iterative methods [7-10] provide a means to determine accurately the complex coefficients in the linear range of poled piezoceramics from complex impedance resonance measurements. The piezoelectric resonance analysis method and program (PRAP) has been proposed for the full set of standard geometries and resonance modes needed to complete complex characterization in a wide range of materials with very high and moderate losses [11, 12].

In this Chapter, porous PZT piezoceramics with the composition of $Pb_{0.95}Sr_{0.05}Zr_{0.53}Ti_{0.47}O_3$ + 1wt.% Nb_2O_5, and relative porosity 35% were studied. Both the IEEE Standard and the Piezoelectric Resonance Analysis Method (PRAP) were used to determine material constants corresponding to thickness and length-thickness extensional modes of piezoelectric plate vibrations. Complex sets of elastic, dielectric and piezoelectric coefficients of the porous ceramics piezoelements were measured by impedance spectroscopy method using (PRAP) software. Reference data for real parts of material constants were measured by IEEE Standard method. Material constants obtained by the different methods were compared. The results show that the PRAP method gives results that are more accurate and allow one take into account elastic, piezoelectric, and dielectric losses. Wide-band ultrasonic transducers for underwater and NDT applications were fabricated using studied PZT piezoceramics.

2. IMPEDANCE SPECTROSCOPY CHARACTERIZATION OF PIEZOELECTRICS

Impedance spectroscopy characterization stems from the relation for impedance Z:

$$\mathbf{Z} = V/\mathbf{I} = \int \mathbf{E} \cdot d\mathbf{x} \Big/ A \frac{d\mathbf{D}}{dt},$$

(1)

where V is voltage, I is current, and A is the surface area over which current is measured. The piezoelectric equations give the relationship between dielectric displacement D and electric field E through stress T and strain S. For simple one-dimensional resonance modes, many of the electric and mechanical field variables aspire to zero and the sets of piezoelectric equations decouple. Solving the wave equation for one-dimensional steady-state displacement in the specimen and evaluating S from the derivative of displacement, together with the boundary conditions of an unclamped resonator, the piezoelectric equations can be used to evaluate the relationship between D and E as a function of frequency and material properties. The impedance of a thickness extensional (TE) resonator reduces to [9, 10]:

$$Z = \frac{\beta_{33}^{S} l + \dfrac{\mathbf{h}_{33}^{2}}{\sqrt{\rho \mathbf{c}_{33}^{D}} \omega} \tan\left(\dfrac{\omega l \sqrt{\rho}}{2 \sqrt{\mathbf{c}_{33}^{D}}}\right)}{i \omega A}, \tag{2}$$

where ρ is the specimen density, ω is the angular frequency and l is the thickness of the sample in the direction of poling, h is the piezoelectric constant, c is the elastic stiffness at constant D, and β is the inverse permittivity at constant S.

In impedance spectroscopy characterization of piezoelectrics, the material properties of Equation (2) can be adjusted until the impedance fits a measured impedance spectrum as a function of frequency.

The common one-dimensional 6mm-symmetry modes include the Length Extensional (LE), Thickness Extensional (TE), Length-Thickness Extensional (LTE), Thickness Shear (TS), Length Shear (LS), and Radial modes. Each of these modes provides electrical, elastic and piezoelectric properties of the material. Table 1 lists the properties of one form of the reduced piezoelectric matrix with the modes that will provide those properties.

Table 1. Properties provided by the common modes

Property	ε_{11}^{T}	ε_{33}^{T}	s_{11}^{E}	s_{12}^{E}	s_{13}^{E}	s_{33}^{E}	s_{55}^{E}	d_{13}	d_{33}	d_{15}
Mode	TS, LS	LE, LTE, Radial	LTE, Radial	Radial		LE	LS, TS	LTE, Radial	LE	TS, LS

By analyzing specimens inducing the TS, LE and Radial modes, all except s_{13}^{E} of the reduced 6mm piezoelectric matrix can be obtained. s_{13}^{E} can be found from the properties determined with the TE, LTE and LE modes for other representations of the piezoelectric linear equations [11], indicating that all modes except the LS mode are required to construct the piezoelectric matrix. Alternatively, using an initial guess for s_{13}^{E}, the reduced matrix can be inverted and c_{13}^{D} compared to that found for the TE mode, eliminating the need for the LTE mode.

In practice, material properties are dispersive (i.e., they vary with frequency). This means that the range of frequency used to fit forms of equation (6) to measured spectra should be limited. It is worth noting that only a few points in the spectrum are required to determine properties. For the TE mode, only three points are required.

The IEEE Standard on Piezoelectricty [1] describes two approaches to obtaining material properties using either the difference between parallel (f_p) and series (f_s) resonance frequencies, or the frequency ratio between different order resonance peaks. These approaches can be problematic when material properties are dispersive or when a low electromechanical coupling constant results in a small difference between f_p and f_s.

Smits [9] proposed a general method for the TE, TS, LE and LTE resonators using three points around resonance. This method does not rely on critical frequencies in the spectrum and so does not lose sensitivity for low Q materials. At resonance and anti-resonance, currents can be very large and small and this presents an instrumentation problem. Though less of an issue with low Q materials, Smits's method is less demanding of instrumentation than the IEEE methods. Finally, because the choice of analysis points is somewhat arbitrary, noisy spectra and samples with mode coupling may still be analyzed using Smits's method.

The IEEE Standard in Piezoelectricity treats all material properties as real, evaluating mechanical and electrical loss independently through Q_M and $\tan\delta$. Losses in the electromechanical coupling constant were assumed zero. This was acceptable for low loss piezoelectric materials such as PZT, but could lead to significant errors when materials had large loss. In real materials, complex impedance results from complex material properties [13]. The impedance is directly associated with dielectric conduction losses [14, 15] and attenuation of the displacement wave in the sample.

The PRAP software [11] analyses impedance spectra to determine complex material properties. This software uses a generalized form of Smits's method to determine material properties for any common resonance mode, and a generalized ratio method for the radial mode [9] valid for all material Q's. By analyzing on each harmonic, PRAP can determine complex material properties as a function of frequency. The software always generates an impedance spectrum from the determined properties to indicate validity of the results.

3. EXPERIMENTAL SCENARIO

Porous PZT piezoceramics with the composition of $Pb_{0.95}Sr_{0.05}\,Zr_{0.53}Ti_{0.47}O_3 + 1wt.\%$ Nb_2O_5, relative porosity 35%, and average pore size ~50 μm were chosen for the complex characterization of material properties.

Experimental samples of porous piezoceramics were prepared by burning out of organic powders using specially developed technology [2, 3, 16]. An organic salt powder with irregular particle shape, 250 °C decomposition temperature and grain-size distribution from 50 to 60 μm were used for producing the pore volumes. The porous ceramics elements with vacuum deposited Cr/Ni electrodes were poled on air at cooling from 380 °C to 90 °C for 1 min by applying a dc electric field of $1 - 1.5$ kV/mm. The following piezoelements were prepared and studied: rectangular plates $29.5 \times 24 \times 1$ mm^3. For the experiments, we applied the PRAP automatic iterative method [11] to the set of resonance modes needed to complex characterization of porous piezoceramics plates. Measurements of electric parameters were made on the Agilent 4291A Impedance Analyzer.

4. RESULTS AND DISCUSSION

Figures 1 and 2 show measured complex impedance spectra and PRAP approximations for the thickness (TE) and length thickness extensional (LTE) modes of porous piezoceramics plates.

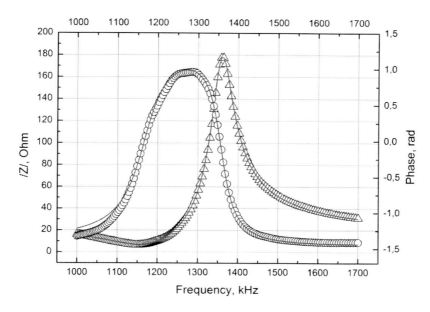

Figure 1. Complex impedance spectra and PRAP approximations for the thickness extensional (TE) mode of porous piezoceramics plates.

Table 2 summarizes the complex constants of porous ceramics obtained using PRAP analysis of complex impedance spectra (Figures 1 and 2). Additional physical parameters of the porous ceramics measured by ultrasonic and hydrostatic weighting methods are as follows: $Z_A = 12.4$ MRayl, $\rho = 5350$ g/cm^3, $V_t = 2320$ m/c., $Q_M^t = 18$, $T_C = 290$°C.

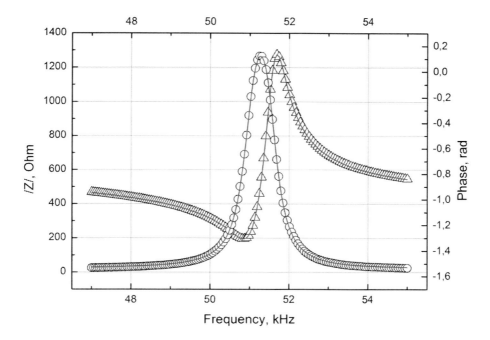

Figure 2. Complex impedance spectra and PRAP approximations for the length thickness extensional (LTE) mode of porous piezoceramics plates.

Table 2. Complex constants of $Pb_{0.95}Sr_{0.05}Zr_{0.53}Ti_{0.47}O_3$ + 1wt.% Nb_2O_5 porous ceramics

Parameter	Real	Imaginary	IEEE Standard
Length Thickness Extensional (LTE) Mode			
f_p (Hz)	51582.2	281.772	51572.2
f_s (Hz)	50958.7	287.521	50977.2
k_{31}	0.172209	-0.0012571	0.168267
s_{11}^E (m²/N)	$2.06 \cdot 10^{-11}$	$-2.33 \cdot 10^{-13}$	$2.07 \cdot 10^{-11}$
ε_{33}^T (F/m)	$9.22 \cdot 10^{-9}$	$-2.96 \cdot 10^{-10}$	-
d_{31} (C/N)	$7.51 \cdot 10^{-11}$	$-2.18 \cdot 10^{-12}$	-
g_{31} (Vm/N)	0.008153	$2.56 \cdot 10^{-5}$	-
Thickness Extensional (TE) Mode			
f_p (Hz)	$1.35 \cdot 10^6$	27032.6	$1.36 \cdot 10^6$
f_s (Hz)	$1.16 \cdot 10^6$	47656.4	$1.16 \cdot 10^6$
k_t	0.551805	-0.0305506	0.563025
c_{33}^D (N/m²)	$3.95 \cdot 10^{10}$	$1.57 \cdot 10^9$	$3.96 \cdot 10^{10}$
c_{33}^E (N/m²)	$2.75 \cdot 10^{10}$	$2.43 \cdot 10^9$	$2.71 \cdot 10^{10}$
e_{33} (C/m²)	8.52861	-0.856122	-
h_{33} (V/m)	$1.41 \cdot 10^9$	$4.12 \cdot 10^7$	-
ε_{33}^S (F/m)	$6.02 \cdot 10^{-9}$	$-7.83 \cdot 10^{-10}$	-

CONCLUSION

In this Chapter, piezoelectric resonance analysis method for automatic, iterative evaluation of complex material parameters of lossy piezoelectrics was described.

Complex sets of material constants for "hard" porous piezoceramics were obtained by PRAP analysis of electric impedance spectra measured for the standard porous piezoceramic elements. By combining the analyzed resonance spectra of the appropriate resonance modes (radial, thickness, length extensional, and shear thickness), the complete 6mm reduced complex piezoelectric matrix of a material were determined.

It was shown, that the PRAP method gives accurate results and allows one to take into account elastic, piezoelectric, and dielectric losses of the piezoelectric material.

ACKNOWLEDGMENTS

Work supported by the Southern Federal University (internal grant - 213.01-2014/012-ВГ).

REFERENCES

[1] IEEE Standard on Piezoelectricity. *ANSI/IEEE Std*. 1987, 176.
[2] Rybyanets, A.N. *IEEE Trans. UFFC*. 2011, *vol. 58*, 1492-1507.
[3] Rybyanets, A.N., Rybyanets, A.A. *IEEE Trans. UFFC*. 2011, *vol. 58*, 1757-1774.

[4]   Rybyanets, A.N. *Ferroelectrics.* 2011, *vol. 419*, 90-96.

[5]   Rybyanets, A.N., Naumenko, A.A., Shvetsova, N.A. Characterization Techniques for Piezoelectric Materials and Devices. In: *Nano- and Piezoelectric Technologies, Materials and Devices*, Parinov, I. A. (Ed.), Nova Science Publishers Inc., New York, 2013, Chapter 1, 275-308.

[6]   Rybyanets, A.N. Ceramic Piezocomposites: Modeling, Technology, and Characterization. In: *Piezoceramic Materials and Devices,* Parinov, I. A, (Ed.), Nova Science Publishers Inc., New York, 2010, Chapter 3, 113-174.

[7]   Alguero, M.; Alemany, C.; Pardo, L.; Gonzalez, A. M. *J. Am. Ceram. Soc.* 2004, *vol. 87,* 209-212.

[8]   Alemany, C.; Gonzalez, A.M.; Pardo, L.; Jimenez, B.; Carmona, F.; Mendiola, J. *J. Phys. D: Appl. Phys.* 1995, *vol. 28,* 945.

[9]   Smits, J. G. *IEEE Trans. Sonics Ultrason.* 1976, SU-23, 393-402.

[10]   Holland, R. *IEEE Trans. Sonics Ultrason.* 1967, SU-14, 18-24.

[11]   *PRAP (Piezoelectric Resonance Analysis Program).* TASI Technical Software Inc. www.tasitechnical.com.

[12]   Rybianets, A.N.; Tasker, R. *Ferroelectrics.* 2007, *vol. 360*, 90-95.

[13]   Devonshire, A.F. *Philosphical Magazine Supplement.* 1954, *vol. 3*, 85-130.

[14]   Mason, W.P. *Physical Acoustics and the Properties of Solids.* D. Van Nostrand Co. Inc., Princeton, 1958.

[15]   Berlincourt, D.A.; Curran, D.R.; Jaffe, H.I. Piezoelectric and Piezomagnetical Materials and Their Function in Transducersn. In: *Physical Acoustics I.* Mason W. P. (Ed.), Academic Press, New York, 1964, Part A, 170-270.

[16]   Rybyanets, A.N.; Naumenko, A.A. Nanoparticles Transport in Ceramic Matrixes: a Novel Approach for Ceramic Matrix Composites Fabrication. In: *Physics and Mechanics of New Materials and Their Applications*, Parinov, I.A, Shun-Hsiung Chang (Eds.), Nova Science Publishers Inc., New York 2013, Chapter 1, 3-18.

In: Proceedings of the 2015 International Conference … ISBN: 978-1-63484-577-9
Editors: Ivan A. Parinov, Shun-Hsyung Chang et al. © 2016 Nova Science Publishers, Inc.

Chapter 25

EXPERIMENTAL STUDY OF ULTRASONIC ATTENUATION AND DISPERSION RELATIONSHIPS FOR CERAMIC MATRIX COMPOSITES

E. I. Petrova[*], A. A. Naumenko, S. A. Shcherbinin and A. N. Rybyanets

Institute of Physics, Southern Federal University, Rostov-on-Don, Russia

ABSTRACT

The chapter presents a verification of the generalized relationships between the ultrasonic attenuation and dispersion for materials with high scattering losses. The basic techniques for characterization of lossy piezoelectric materials comprising impedance spectroscopy and iterative methods exploring complex material constants are considered. Experimental frequency dependencies of attenuation and ultrasonic velocities for different ceramic matrix composites with high scattering losses (porous piezoceramics, composites ceramics/crystals) are compared with the theoretical results obtained using approximate forms of general Kramers-Kronig relations.

Keywords: lossy piezoelectric materials, ultrasonic attenuation, dispersion, scattering losses, Kramers-Kronig relations

1. INTRODUCTION

Quantitative relationships between attenuation and the frequency dependence of the phase velocity (dispersion), the validity of which depends only upon the properties of linearity and causality of the system under investigation, have proven useful in a number of

---

[*] E-mail: harigamypeople@gmail.com.

settings. Examples include the Kramers-Kronig relationships [1-4] connecting the in-phase and out-of-phase components of the appropriate susceptibility in electromagnetic and acoustic phenomena and the Bode relationship [5] connecting the gain and phase shift in amplifier circuits.

Historically, the subject of the mechanisms of acoustic wave propagation has developed such that the topics of absorption and scattering are usually dealt, separately. Such a distinction between propagation phenomena is, however, in some senses artificial and may become difficult to maintain. As pointed out in [6, 7] one grey area lies within a general discussion of acoustic loss mechanisms, particularly in relation to speed dispersion. There is no sharp distinction between the various loss mechanisms; a local absorption of energy may be thought of as one limit of a scattering event. Furthermore, phenomenological descriptions exist which define speed dispersion as arising only from the presence of a frequency dependent attenuation coefficient, which may be due to any of the loss mechanisms. Discussion of the general causal attenuation–dispersion relationships is considerably extended in [8], where various theories for a variety of attenuating media were compared.

The chapter presents a verification of the generalized relationships between the ultrasonic attenuation and dispersion for materials with high scattering losses. The basic techniques of attenuation and dispersion estimation for lossy piezoelectric materials comprising impedance spectroscopy and iterative methods exploring complex material constants [9, 10] are considered. Experimental frequency dependencies of attenuation and ultrasonic velocities for different ceramic matrix composites with high scattering losses are measured and compared with the theoretical results obtained using approximate forms of general Kramers-Kronig relations.

2. KRAMERS-KRONIG RELATIONSHIP BETWEEN ULTRASONIC ATTENUATION AND PHASE VELOCITY

Detailed derivation of the general Kramers-Kronig relations linking the attenuation and dispersion for a linear acoustic system was proposed in [6, 7]. Approximate, nearly local forms for these general relations, derived under the conditions that the attenuation and the dispersion do not vary rapidly as functions of frequency, were also present. The validity of these approximate relationships hinges only on the properties of causality and linearity, and does not depend upon details of the mechanism responsible for the attenuation and dispersion. The use of these approximate relationships for a wide range of acoustic systems exhibiting substantially different physical properties was illustrated and the implications of the existence of these relations concerning investigation of the physical mechanisms responsible for the observed attenuation were examined in [6, 7].

The dispersion $dC(\omega)$ and the attenuation coefficient $\alpha(\omega)$ are connected by the following nearly local generalized ultrasonic attenuation-dispersion relations [6]:

$$\alpha(\omega) = \frac{\pi\omega^2}{2C^2(\omega)}\frac{dC(\omega)}{d\omega}. \tag{1}$$

$$\frac{1}{C_0} - \frac{1}{C(\omega)} = \frac{2}{\pi}\int_{\omega_0}^{\omega}\frac{\alpha(\omega')}{\omega'^2}d\omega', \tag{2}$$

where C_0 is the sound velocity at fixed frequency ω_0. The magnitude of the dispersion is usually small, so that these Kramers-Kronig relations can be further simplified to the approximate forms:

$$\alpha(\omega) = \frac{\pi \omega^2}{2 C_0^2} \frac{dC(\omega)}{d\omega},$$ (3a)

$$\Delta C = C(\omega) - C_0 = \frac{2 C_0^2}{\pi} \int_{\omega_0}^{\omega} \frac{\alpha(\omega')}{\omega'^2} d\omega',$$ (3b)

where $C(\omega)$ is written as $C_0 + \Delta C(\omega)$ with $\Delta C(\omega) \ll C_0$, and only terms of order $\Delta C(\omega)$ are retained. In the next section, the validity of Equations (3) in several different acoustic systems is discussed.

3. VERIFICATION OF THE NEARLY LOCAL RELATIONSHIPS FOR ACOUSTIC SYSTEMS WITH HIGH SCATTERING LOSSES

The multiphase ceramic matrix composites are very complex objects for theoretical modeling, NDT (non-destructive testing), and ultrasonic measurements [11]. The accurate description of piezocomposites must include the evaluation of the dielectric, piezoelectric and mechanical losses, accounting for the out-of-phase material response to the input signal [12, 13]. It was shown [14] that pulse-echo measurements of frequency dependencies of elastic properties for dispersive and lossy ceramic composites are inaccurate and ambiguous. In its turn, piezoelectric resonance measurements (PRAP) can give accurate and reproducible results that well agree with the results of 3D finite-difference simulations. Different methods were proposed for theoretical modeling and evaluation of ceramic composites properties [15-20]. However, physical mechanisms and interrelations of elastic losses and dispersion in complex objects, such as, porous ceramics and ceramic matrix piezocomposites remain difficult to deep understanding and they would require a dedicated study.

Complex elastic, dielectric and piezoelectric coefficients of porous ceramics and ceramic matrix piezocomposite elements were determined by impedance spectroscopy method using PRAP software [9, 19]. Measurements were made using Agilent 4291A Impedance Analyzer. The microstructure of polished and chipped surfaces of composite samples was observed with optical (NeoPhot-21) and scanning electron microscopes (SEM, Karl Zeiss).

The following two types of ceramic composites [10-12] with high losses were chosen as model samples for simulation of ultrasonic wave propagation and comparison with PRAP and ultrasonic measurements (Figure 1):

(i) ceramic matrix composites A850L (Figure 1a) consisting of soft PZT matrix with randomly distributed α-Al_2O_3 crystals with a mean particle diameter ~ 200 μm and volume fraction from 9 up to 26 vol.%;

(ii) porous PZT piezoceramics PCR-1 (porosity 18%, average pore size ~ 20 μm) shown in Figure 1b.

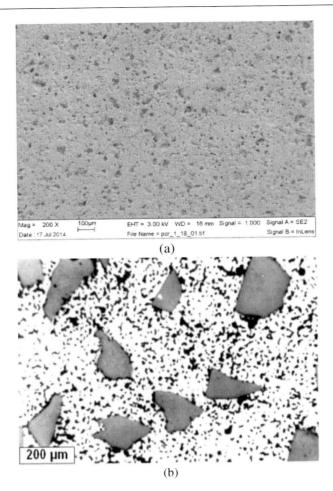

Figure 1. Optical and SEM micrographs of porous PCR-1 piezoceramics (a) and A850L composites (b) samples.

Figure 2 shows the frequency dependencies of ultrasonic velocities $V_t^D = \sqrt{\frac{c_{33}^{/D}}{\rho}}$ and $V_t^E = \sqrt{\frac{c_{33}^{/E}}{\rho}}$ and corresponding attenuation coefficients $\alpha(V_t^D) = \frac{c_{33}^{//D}\omega_0}{2c_{33}^{/D}}$ and $\alpha(V_t^E) = \frac{c_{33}^{//E}\omega_0}{2c_{33}^{/E}}$ obtained from PRAP results for the higher harmonics of the TE mode of PCR-1 porous ceramic disks with different fundamental resonant frequencies. The following standard notations were used above: $c_{33}^{/D}$, $c_{33}^{/E}$ is the real part and $c_{33}^{//D}$, $c_{33}^{//E}$ is the imaginary part of elastic stiffness, ρ is the density, ω_0 is the cyclic frequency.

The least squares approximations of the frequency dependencies of Figure 3 have shown that in the frequency range corresponding to Rayleigh scattering of high frequency ultrasonic waves on pores ($\lambda \gg D$, where D is the average pores diameter, λ is the wavelength), attenuation coefficients $\alpha(V_t^D)$ and $\alpha(V_t^E)$ grow with the frequency approximately as f^4. At further frequency growth scattering mechanism changes from Rayleigh type to stochastic one ($4 \leq \lambda/D \leq 10$) and frequency dependencies of $\alpha(V_t^D)$ and $\alpha(V_t^E)$ are approximated by f^2 function. The corresponding frequency dependence of ultrasonic velocity V_t^E, defined by the elastic modulus C_{33}^E, is approximated by f^3 function (normal dispersion type) in Rayleigh range and by linear f function that agrees with the theoretical predictions for the dispersion (Equations 3). In its turn, ultrasonic velocity V_t^D defined by the elastic modulus C_{33}^D, decreases with the frequency (anomalous dispersion) as result of the electromechanical contribution to C_{33}^D according the equation $C_{33}^D = C_{33}^E/(1-k_t^2)$ (note that effective value of k_t measured on higher harmonics drops drastically with the harmonics number as $k_{eff.n}^2 = \frac{8k_t^2}{((2n+1)\pi)^2}$ [11]).

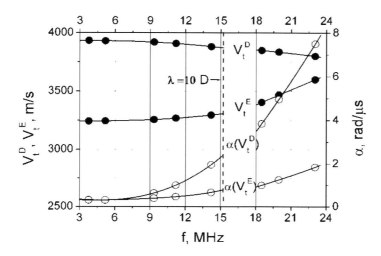

Figure 2. Frequency dependencies of ultrasonic velocities V_t^D and V_t^E and corresponding attenuation coefficients $\alpha(V_t^D)$ and $\alpha(V_t^E)$ obtained from PRAP results for higher harmonics of the TE mode of PCR-1 porous ceramic disks with different fundamental frequencies; vertical dashed line depict boundary between Rayleigh and stochastic scattering regions.

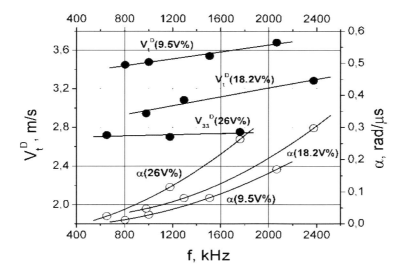

Figure 3. Frequency dependencies of ultrasonic velocities V_t^D and corresponding attenuation coefficients $\alpha(V_t^D)$ for ceramic composites PZT/α-Al$_2$O$_3$ (A850L) at different volume fractions of α-Al$_2$O$_3$.

Figure 3 shows the frequency dependencies of ultrasonic velocities $V_t^D = \sqrt{\frac{c_{33}^{/D}}{\rho}}$ and corresponding attenuation coefficients $\alpha(V_t^D) = \frac{c_{33}^{//D} \omega_0}{2 c_{33}^{/D}}$ for ceramic composites PZT/α-Al$_2$O$_3$ with different volume fractions of α-Al$_2$O$_3$ measured on standard disk samples by piezoelectric resonance (PRAP) method. The results of complex material constants evaluation agree well with the previously reported results for lossy piezoelectrics [10-13].

The least squares approximations of the frequency dependencies of Figure 3 have shown that in all measured frequency ranges and for all concentration of α-Al$_2$O$_3$ particles, attenuation coefficients $\alpha(V_t^D)$ and $\alpha(V_t^E)$ grow with the frequency as f^2 and corresponding frequency dependence of ultrasonic velocity V_t^D is approximated by linear f functions. This behavior corresponds to stochastic type of scattering ($4 \leq \lambda/D \leq 10$, D is the average α-Al$_2$O$_3$ particle diameter) and is described well by the approximate dispersion relations (Equations (3a) and (3b)). It can be mentioned that unlike from V_t^D behavior measured on higher harmonics for porous ceramics PCR-1 the dispersion of ultrasonic velocity V_t^D measured on fundamental resonant frequency has normal character, because of constancy of electromechanical contribution (k_t) to C_{33}^D.

4. DISCUSSION

The validity of the nearly local, approximate Kramers-Kronig relations was tested in a number of acoustic systems possessing a range of loss and dispersion mechanisms. One of the most important results of this work is verification of the approximate relationships for different ceramic matrix composites with strong spatial dispersion and high scattering losses. It was shown that depending on the frequency range and scattering particles sizes, the scattering mechanism for ceramic matrix composites and porous ceramics can changes from Rayleigh ($\lambda \gg D$) to stochastic type ($\lambda \sim D$) that leads to corresponding changes in the dispersion character, however for all frequency range attenuation and dispersion are well described by the approximate relations. The results of theoretical analysis and experiments presented clearly show that the approximate relations represent an accurate description of acoustic propagation in several systems, which do not exhibit rapid variations (resonances) with frequency over the range of interest.

ACKNOWLEDGMENTS

Work supported by the RSF grant No. 15-12-00023.

REFERENCES

[1]   Kronig, R. *J. Opt. Soc. Am.* 1926, *vol. 12*, 547-551.
[2]   Kronig, R.; Kramers, H.A. *Zeits f. Phys.* 1928, *vol. 48*, 174-178.

[3]   Landau, L. D.; Lifshitz, E.M. *Statistical Physics*. Addison-Wesley, Reading, MA, 1958, pp. 1 – 392.

[4]   Mangulis, V. *Acoust. Soc. Am.* 1964, *vol. 36*, 221-222.

[5]   Bode, W. *Bell Syst. Tech. J.* 1940, *vol. 19*, 421-425.

[6]   O'Donnell, M., Jaynes, E.T., Miller, J.G. *J. Acoust. Soc. Am.* 1978, *vol. 63*, 1935-1938.

[7]   O'Donnell M., Jaynes, E.T., Miller, J. *J. Acoust. Soc. Am.* 1981, *vol. 69*, 696-701.

[8]   Szabo, T.L. *J. Acoust. Soc. Am.* 1995, *vol. 97*, 14-24.

[9]   Rybianets, A.N.; Tasker, R. *Ferroelectrics*. 2007, *vol. 360*, 90-95.

[10]   Rybyanets, A.N.; Naumenko, A.A.; Shvetsova, N.A. Characterization Techniques for Piezoelectric Materials and Devices. In: *Nano- and Piezoelectric Technologies, Materials and Devices*, Parinov I.A. (Ed.). Nova Science Publishers Inc., New York, 2013, Chapter 1, 275-308.

[11]   Rybyanets, A.N. Ceramic Piezocomposites: Modeling, Technology, and Characterization. In: *Piezoceramic Materials and Devices*, Parinov I. A. (Ed.), Nova Science Publishers Inc., New York, 2010, Chapter 3, 113-174.

[12]   Rybyanets, A.N. *IEEE Trans. UFFC*. 2011, *vol. 58*, 1492-1507.

[13]   Rybyanets, A.N.; Rybyanets, A.A. *IEEE Trans. UFFC*. 2011, *vol. 58*, 1757-1774.

[14]   Rybyanets, A.N. *Feroelectrics*. 2011, *vol. 419*, 90-96.

[15]   Rybianets, A.N.; Nasedkin A.V.; Turik, A.V. *Integrated Ferroelectrics*. 2004, *vol. 63*, 179-182.

[16]   Rybianets, A.N. *Ferroelectrics*. 2007, *vol. 360*, 84-89.

[17]   Rybianets, A. *Proc. 2007 IEEE Ultrason. Symp*. 2007, 1909-1912.

[18]   Rybianets, A.N.; Nasedkin, A.V. *Ferroelectrics*. 2007, *vol. 360*, 90-95.

[19]   Rybianets, A.N.; Razumovskaya, ON.; Reznitchenko, L.A.; Komarov, V.D.; Turik, A.V. *Integrated Ferroelectrics*. 2004, *vol. 63*, 197-200.

[20]   Rybianets, A.N.; Tasker, R. *Ferroelectrics*. 2007, *vol. 360*, 90-95.

In: Proceedings of the 2015 International Conference …   ISBN: 978-1-63484-577-9
Editors: Ivan A. Parinov, Shun-Hsyung Chang et al.   © 2016 Nova Science Publishers, Inc.

Chapter 26

CORRELATION BONDS: STRUCTURE – ELECTROPHYSICAL PROPERTIES OF THE BINARY PZT SYSTEM

I. N. Andryushina[*], L. A. Reznichenko, L. A. Shilkina, S. I. Dudkina, K. P. Andryushin and O. N. Razumovskaya

Research Institute of Physics, Southern Federal University,
Rostov-on-Don, Russia

ABSTRACT

Electrophysical parameters of solid solutions of the system $PbZr_{1-x}Ti_xO_3$ were studied in a wide range of concentrations of components. It was found that changes in the parameters have non-monotonic, jump character, caused by the periodic changes of the phase composition, based on the real structure of the objects and on the different contributions from the spontaneous deformation and domain reorientations into their formation.

Keywords: PZT, intraphase transitions, defective state, electrophysical properties

1. INTRODUCTION

The binary $(1 - x)PbZrO_3–xPbTiO_3$ system is of great practical importance and is currently the basis for the development of new high-performance ferroelectric (FE) materials due to their unique properties: wide isomorphism defining the complete solubility of the components in the range $0.00 \leq x \leq 1.00$ [1–3] and an almost unlimited ability to "absorb" double and triple Pb-containing oxides that actually exist (perovskites) or are hypothetical [4, 5], which makes it possible to purposefully vary the ratio of the components; the rich filling phase, consisting in the presence of an antiferroelectric (AFE)–FE transition near

---

[*] E-mail: futur6@mail.ru.

PbZrO$_3$ and two morphotropic region (MR$_{1,2}$) (with a complex sequence of transformations in one of them; [6] and references therein) and anticipating their clustering patterns [7], which divide the system into fundamentally different (in terms of properties) fields; the presence in the chemical compositions of ions of variable valence (Pb, Ti), contributing to the development of a defective situation in the system and, as a consequence, the emergence in one-symmetry fields of multiple phase states and their areas of coexistence with the local extrema properties [8]; as a consequence of the foregoing, the formation of regions with different combinations of parameters of solid solutions (SSs), which are suitable for different piezotechnical applications, overlapping almost all applied branches. All this makes it possible to create multi-component compositions with the number of components in excess of two on the basis of this system. As shown in work [9], further increases in the complexity of systems are not advisable due to the lower piezoelectric parameters that are associated with disordered structure (the development of crystallochemical disorder) when introduced into the crystal lattice of the alien ions. Since the macroscopic properties (dielectric, piezoelectric, ferroelastic) of such materials are, as we have shown, dependent on the position of the SS on the phase diagram (PD) of the respective systems, knowledge of the correlations between these characteristics and the crystal structure of substances is required. The condition of the structure may be characterize by the integral parameter: homogeneous deformation parameter, δ, which describes the magnitude of the spontaneous deformation of oxides with perovskite structure, as well as the parameter η, characterizing the degree of domain reorientations other than $180°$ committed in the process of polarized ceramics. They are introduced in works [4, 10] to compare the properties of objects with different symmetries. However, analysis of the literature showed that in the binary system of correlation between the structural parameters and macro-responses, SSs were considered either in the limited area of concentration of components near MR [11] or with a wide concentration step [12]. In a few works [13–17], such correlations were established in detail only with regard to the characteristics of the grain structure of ceramics. In more complex systems, SSs were studied almost exclusively in the MR and its surrounding area, with a large concentration step, and the correlations of the properties were determined only with parameter δ [10, 18–20]. It is obvious that the known data are insufficient for correctly establishing the relationships between composition, structure, and properties in such systems and, above all, in the PZT matrix system. In this regard, the aim of this work was to establish the correlation between the structural and dielectric, piezoelectric, and ferroelastic properties of a SS of this system in the whole concentration range with a small step. For the realization of this aim it was necessary to carry out a detailed analysis of the dependency of the electrophysical characteristics on the structural parameters of the SS located in different areas of the PD of the PZT system, which became the subject of the present study.

2. EXPERIMENTAL

The objects of our study were the SS composition $(1-x)$PbZrO$_3$–xPbTiO$_3$ ($0.00 \leq x \leq 1.00$) studied with a research concentration step $\Delta x = 0.01$ in the range of $0.00 \leq x \leq 0.04$, $0.08 \leq x \leq 0.12$, $0.30 < x \leq 0.42$, and $0.52 \leq x \leq 0.57$; in the range of $0.04 < x < 0.06$ the concentration step $\Delta x = 0.0025$; in the range of $0.42 < x < 0.52$ the concentration step $\Delta x =$

0.005; in the range of $0.12 < x \leq 0.30$ the concentration step $\Delta x = 0.02$; and in the range of $0.60 \leq x \leq 1.00$ the concentration step $\Delta x = 0.025$. The samples were prepared from stoichiometric mixtures of the oxides by two-stage solid-phase synthesis at temperatures of $T_1 = T_2 = 870\ °C$: the duration of isothermal exposure was $\tau_1 = \tau_2 = 7h$, and this was followed by sintering at temperatures of $T_S = 1220 - 1240\ °C$ (depending on composition), $\tau_s = 3h$.

Radiographic studies were conducted by the method of powder diffraction diffractometry using a Dron-3 instrument with $Co_{K\alpha}$ radiation. The parameters of the perovskite cell were calculated by the standard method [4]. Relative measurement errors of the structural parameters have the following values: linear: $\delta a = \delta b = \delta c = \pm 0.05\%$; angle: $\delta \alpha = \pm 5\%$; volume: $\delta V = \pm 0.07\%$. A homogeneous deformation parameter, δ, was determined by the formula $\delta \approx \cos\alpha$ for rhombohedral (Rh) phase, and $\delta \approx 2/3(c/a-1)$ for tetragonal (T) phase (c, a, and α are parameters of the perovskite cell) [4]. The following dielectric, piezoelectric, and ferroelastic parameters were determined in accordance with OST 11 0444-87 using a Wayne Kerr 6500B precision impedance analyser and a YE2030A d_{33} meter: the relative dielectric constant before and after the polarization of the samples ($\varepsilon/\varepsilon_0$ and $\varepsilon_{33}^{T}/\varepsilon_0$, respectively), piezoelectric moduli ($|d_{31}|$, d_{33}), the electromechanical coupling coefficient of planar oscillation modes (K_p), the mechanical quality factor (Q_M), and Young's modulus (Y_{11}^{E}). The relative errors in the determination of parameters of the SS were $\delta\varepsilon = \pm 1.0\%$; $\delta K_p = \pm 2.0\%$; $\delta d_{31} = \pm 4.0\%$; $\delta d_{33} = \pm 5.0\%$, $\delta Q_M = \pm 12\%$; and $\delta Y_{11}^{E} = \pm 0.7\%$.

3. RESULTS AND DISCUSSION

Figure 1 shows the concentration dependences of the electrophysical characteristics $\varepsilon_{33}^{T}/\varepsilon_0$, K_p, $|d_{31}|$, d_{33}, Q_M, and Y_{11}^{E} of the SS system $PbZr_{1-x}Ti_xO_3$ in the interval $0.00 < x \leq 0.80$ (when $x > 0.80$, it disrupted the integrity of the ceramic samples due to their self-destruction due to large internal mechanical stresses [21]). Figure 1 also shows the field phase states and their coexistence on the PD, whose transcript is presented in the figure caption. Extreme values are achieved at the transition from one phase to another. The maximum values of the main characteristics $\varepsilon_{33}^{T}/\varepsilon_0$, K_p, $|d_{31}|$, and d_{33} correspond to MR ($Rh \rightarrow T$). The electrophysical parameters change slightly in the transition from one phase state to another. These changes are non-monotonic uneven character. All of this provides a strong "irregularity" of the concentration dependences of the electrophysical characteristics.

As noted above, the relationships of the electrophysical parameters of SS multicomponent systems with δ are established in work [10]. Thus, with the increase of δ, due to the reduced mobility of domain walls $\varepsilon_{33}^{T}/\varepsilon_0$, dielectric loss, tg$\delta$, and mechanical loss, $1/Q_M$, should decrease while Q_M, Y_{11}^{E} and the speed of sound, V_1^{E}, should increase; K_p, which characterizes the energy conversion in piezomaterial more fully, tends to decrease with increasing δ, and the piezoelectric modulus $|d_{31}|$ is also reduced, because it depends on K_p and $\varepsilon_{33}^{T}/\varepsilon_0$ $|d_{31}| \sim K_p \sqrt{\varepsilon_{33}^{T}}$).

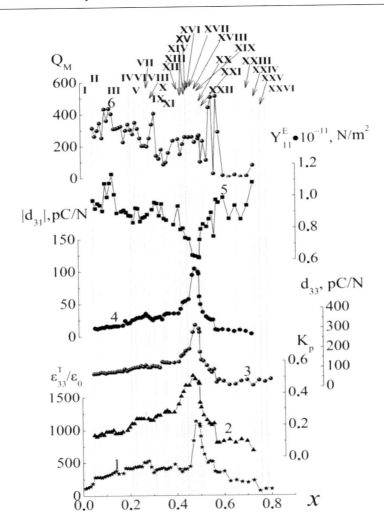

Figure 1. Dependencies of electrophysical characteristics of solid solutions of the system $PbZr_{1-x}Ti_xO_3$ on concentration x in the range $0.00 < x \leq 0.80$: $\varepsilon_{33}^T/\varepsilon_0$ (1), K_p (2), d_{33} (3), $|d_{31}|$ (4), Y_{11}^E (5), Q_M (6). The dotted lines in the figure illustrate different phase states. Interpretation of phases: I –R ($0.00 \leq x \leq 0.04$); II – $R + Rh_1$ ($0.04 < x \leq 0.065$); III – Rh_1 ($0.065 < x \leq 0.20$); IV – $Rh_1 + Rh_2$ ($0.20 < x \leq 0.22$); V – Rh_2 ($0.22 < x \leq 0.24$); VI – $Rh_2 + Rh_3$ ($0.24 < x \leq 0.26$); VII – Rh_3 ($0.26 < x \leq 0.28$); VIII – $Rh_3 + Rh_4$ ($0.28 < x \leq 0.30$); IX – Rh_4 ($0.30 < x \leq 0.34$); X – $Rh_4 + Rh_5$ ($0.34 < x \leq 0.35$); XI – Rh_5 ($0.35 < x \leq 0.39$); XII – $Rh_5 + Rh_6$ ($0.39 < x \leq 0.41$); XIII – Rh_6 ($0.41 < x \leq 0.425$); XIV – $Rh_6 + Rh_7$ ($0.425 < x \leq 0.44$); XV – Rh_7 ($0.44 < x \leq 0.445$); XVI – $Rh_7 + Psk_1$ ($0.445 < x \leq 0.45$); XVII – $Rh_7 + Psk_1 + Psk_2$ ($0.45 < x \leq 0.455$); XVIII – $Rh_7 + Psk_1 + Psk_2 + T_1$ ($0.455 < x \leq 0.48$); XIX – $Psk_2 + T_1$ ($0.48 < x \leq 0.49$); XX – T_1 ($0.49 < x \leq 0.50$); XXI – $T_1 + T_2$ ($0.50 < x \leq 0.515$); XXII – T_2 ($0.515 < x \leq 0.65$); XXIII – $T_2 + T_3$ ($0.65 < x \leq 0.725$); XXIV – T_3 ($0.725 < x \leq 0.75$); XXV – $T_3 + T_4$ ($0.75 < x \leq 0.775$); XXVI – T_4 ($0.775 < x \leq 0.925$).

We observe exactly such changes of the parameters in the T-phase of this binary system (Figure 1). It should be noted that these changes have some peculiarities: a stepwise increase in δ and decrease of $\varepsilon_{33}^T/\varepsilon_0$. In Figure 2a, we see the field of constancy of δ and $\varepsilon_{33}^T/\varepsilon_0$ in the region of the coexistence phase states (region XXIII), as was observed inside the MR [22], and within one phase state (region XXII), which indicates the occurrence of some transformation inside the phase, most likely due to changes in the real structure of the SS. The strong irregularity of the concentration dependences of the parameters $\varepsilon_{33}{}^T/\varepsilon_0$, Q_M, and Y_{11}^E (without the formation of areas of their constancy) and the symbate dependence of $\varepsilon_{33}{}^T/\varepsilon_0$ and δ (Figure 2b, dashed line) are the specifics of the Rh-region.

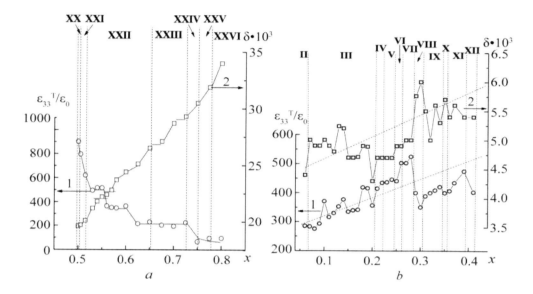

Figure 2. Dependence of $\varepsilon_{33}{}^T/\varepsilon_0$ (1) and δ (2) of the solid solution of the system PbZr$_{1-x}$Ti$_x$O$_3$ on concentration x in T (a) and Rh (b) phases.

The first is caused by a large number (N) of possible directions of the vector of spontaneous polarization in the Rh-phase compared to the T-phase (N = 8 and 6, respectively, [10]) and by the low value of δ, which determines the increased mobility (inertia-free) and the ferroelectric softness of Rh-compositions and, as a result, abrupt changes of the parameters $\varepsilon_{33}{}^T/\varepsilon_0$, Q_M, and Y_{11}^E (Figures 1 and 2b).

The second is determined by the dependence of the electrophysical properties not only on the spontaneous deformation of crystalline cells (δ) but also on the degree of domain reorientations that are different from 180° that occur in the process of polarization of the ceramic, which is expressed by the structural parameter η, which is the maximum in this phase (dashed curve $\eta(x)$ in Figure 3 according to [10] for SSs of multicomponent systems based on PZT). The dependence of $\Delta\varepsilon/\varepsilon(x)$ ($\Delta\varepsilon/\varepsilon = (\varepsilon/\varepsilon_0 - \varepsilon_{33}{}^T/\varepsilon_0)\varepsilon_0/\varepsilon$) is an indirect reflection of the behaviour of η and characterizes changes in the dielectric constant of the investigated SS in the process of polarization (the dashed curve $\Delta\varepsilon/\varepsilon(x)$ in Figure 3 virtually reproduces the dependence of η on x), which was used in the analysis of the properties of other SS systems [25, 26]. For small δ (Rh phase), the contribution of η to the formation of electrophysical properties becomes crucial, and that fact that $\eta \sim \delta$ explains why $\varepsilon_{33}{}^E/\varepsilon_0$ and $\delta(x)$ show the same behaviour.

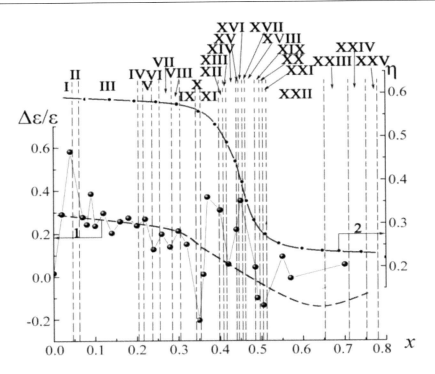

Figure 3. Dependence of $\Delta\varepsilon/\varepsilon$ of solid solutions of the system $PbZr_{1-x}Ti_xO_3$ on the concentration x (1) and the schematic dependence of $\eta(x)$ (2) (values according to [10]).

CONCLUSION

The concentration dependencies of the dielectric, piezoelectric, and ferroelectric parameters of the SS system $PbZr_{1-x}Ti_xO_3$ were studied. It is shown that their changes are non-monotonic with an uneven character associated with the periodic change of the phase composition. It is established that the behaviour of the electrophysical properties of the SS is associated with different contributions to their formation: the spontaneous deformation is dominant in the T-region, while domain reorientations that are different from 180° are prevalent in the Rh phase.

ACKNOWLEDGMENTS

The work was supported by the Russian Ministry of Education and Science: Basic and design of the state job (topic No. 1927, Job No. 3.1246.2014/K, project No. 213.01-2014/012-SH) and FTP (Agreement No. 14.575.21.0007).

REFERENCES

[1] Shirane G.; Suzuki K. *J. Phys. Soc. Jpn.* 1952, *vol. 7*, 333–337.

[2] Sawaguchi, E. *J. Phys. Soc. Japan.* 1953, *vol. 8*, 615–629.

[3] Jaffe, B.; Cook, U.; Jaffe, H. *Piezoelectric Ceramics*, Mir, Moscow, 1974, pp. 1 – 288.

[4] Fesenko, E.G. *The Family of Perovskite and Ferroelectricity*, Atomizdat, Moscow, 1972. pp. 1 – 248 (in Russian).

[5] Fesenko, E.G.; Kupriyanov, M.F.; Davlikanova, R.U.; Filip'iev, V.S. *Crystallography*, 1974, *vol. 19(1)* 118–122 (in Russian).

[6] Noheda, B.; Cox, D.E.; Shirane, G.; Gonzalo, J.A.; Cross, L.E.; Park, S-E. *Appl. Phys. Lett.* 1999, *vol. 74(14)*, 2059–2061.

[7] Titov, S.V.; Shilkina, L.A.; Reznichenko, L.A.; Dudkina, S.I.; Razumovskaya, O.N.; Shevtsova, S.I.; Kuznetsova, E.M. *Tech. Phys. Lett.* 2000, *vol. 26(18)* 9–16 (in Russian).

[8] Titov, S.V.; Shilkina, L.A.; Razumovskaya, O.N.; Reznichenko, L.A.; Vlasenko, V.G.; Shuvaev, A.T.; Dudkina, S.I.; Klevtsov, A.N. Inorg. Mater. 2001, *vol. 37*, 849–856 (in Russian).

[9] Dantsiger, Y.A.; Dudkina, S.I.; Kupriyanov, M.F.; Razumovskaya, O.N.; Reznichenko, L.A. *Izv. of the USSR Academy of Sciences. Ser. Physics.* 1995, *vol. 59(9)*, 104–105 (in Russian).

[10] Fesenko, E.G.; Dantsiger, A.Ya.; Razumovskaya, O.N. *New Piezoceramic Materials*, Rostov State University Press, Rostov-on-Don, 1983, pp. 1 – 156 (in Russian).

[11] Berlincourt, D.A.; Cmolic, C.; Jaffe, H. *Proc. Institute Radio Engineers,* 1960, *vol. 48(2)* 220–229.

[12] Dantsiger, A. Ya.; Fesenko, E.G. *J. Phys. Soc. Jpn.* 1970, *vol. 28 Suppl.* 325–327 (*Proc. Second Int. Meeting Ferroelectricity*, 1969).

[13] Andryushina, I.N.; Reznichenko, L.A.; Alyoshin, V.A.; Shilkina, L.A.; Titov, S.V.; Titov, V.V.; Andryushin, K.P.; Dudkina, S.I. *Ceram. Int.* 2013, *vol. 39(1)*, 753–761.

[14] Andryushina, I.N.; Reznichenko, L.A.; Shilkina, L.A.; Andryushin, K.P.; Dudkina, S.I. *Ceram. Int.* 2013, *vol. 39(2)*, 1285–1292.

[15] Andryushina, I.N.; Reznichenko, L.A.; Shilkina, L.A.; Andryushin, K.P.; Dudkina, S.I. *Ceram. Int.* 2013, *vol. 39(3)*, 2889–2901.

[16] Andryushina, I.N.; Reznichenko, L.A.; Shmytko, I.M.; Shilkina, L.A.; Andryushin, K.P.; Yurasov, Yu. I.; Dudkina, S.I. *Ceram. Int.* 2013, *vol. 39(4)*, 3979–3986.

[17] Andryushina, I.N.; Reznichenko, L.A.; Shilkina, L.A.; Andryushin, K.P.; Yurasov, Yu. I.; Dudkina, S.I. *Ceram. Int.* 2013, *vol. 39(7)*, 7635–7640.

[18] Bogdanov, Ya.S.; Dantsiger, A.Ya.; Razumovskaya, O.N.; Fesenko, E.G. *Izv. of the USSR Academy of Sciences. Inorgan. Mater.* 1981, *vol. 17(12)*, 2243–2247 (in Russian).

[19] Dantsiger, A.Ya.; Razumovskaya, O.N.; Reznichenko, L.A.; Sakhnenko,V.P.; Klevtsov, A.N.; Dudkina, S.I.; Shilkina, L.A.; Dergunova, N.V.; Rybyanets, A.N. *Multicomponent Systems Ferroelectric Complex Oxides: Physics, Chemistry, Technology. Design Aspects Ferroelectric Materials*, Rostov State University Press, Rostov-on-Don, 2001, *vol. 1*, pp. 1 – 437 (in Russian).

[20] Dantsiger, A. Ya.; Razumovskaya, O.N.; Reznichenko, L.A.; Sakhnenko, V.P.; Klevtsov, A.N.; Dudkina, S.I.; Shilkina, L.A.; Dergunova, N.V.; Rybyanets, A.N. *Multicomponent Systems Ferroelectric Complex Oxides: Physics, Chemistry, Technology. Design Aspects Ferroelectric Material*s, Rostov State University Press, Rostov-on-Don, 2002, *vol. 2*, pp. 1 – 365 (in Russian).

[21] Bondarenko, E.I.; Komarov, V.D.; Reznichenko, L.A.; Chernyshkov, V.A. *J. Tech. Phys.* 1988, *vol. 58(9)*, 1771–1774 (in Russian).

[22] Turik, A.V. *Crystallography*, 1981, *vol. 26(1)*, 171–173 (in Russian).

[23] Reznichenko, L.A.; Shilkina, L.A.; Razumovskaya, O.N.; Yaroslavtseva, E.A.; Dudkina, S.I.; Demchenko, O.A.;Yurasov, Yu.I.; Esis, A.A.; Andryushina, I.N. *Phys. Solid State.* 2009, *vol. 51(5)*, 1010–1018.

[24] Rao, C.N.R.; Gopalakrishnan, J. *New Directions in Solid State Chemistry*, 2nd ed., Cambridge University Press, London, 1997, pp. 1 – 568.

[25] Reznichenko, L.A.; Razumovskaya, O.N.; Ivanova, L.S.; Dantsiger, A.Ya.; Shilkina, L.A.; Fesenko, E.G. *Inorg. Mater.* 1985, *vol. 21(2)* 282–285 (in Russian).

[26] Fesenko, E.G.; Reznichenko, L.A.; Ivanova, L.S.; Razumovskaya, O.N.; Dantsiger, A. Ya.; Shilkina, L.A.; Dergunova, N.V. *J. Tech. Phys.* 1985, *vol. 55(3)*, 601–606 (in Russian).

In: Proceedings of the 2015 International Conference … ISBN: 978-1-63484-577-9
Editors: Ivan A. Parinov, Shun-Hsyung Chang et al. © 2016 Nova Science Publishers, Inc.

Chapter 27

REFINEMENT OF THE $R3c \rightarrow R3m$ TRANSITION LINE IN THE PHASE DIAGRAM OF PZT SOLID SOLUTIONS

A. A. Pavelko[*], L. A. Shilkina, K. P. Andryushin, I. N. Andryushina, S. I. Dudkina, L. A. Reznichenko and O. N. Razumovskaya

Research Institute of Physics, Southern Federal University,
Rostov-on-Don, Russia

ABSTRACT

The phase-transition line $R3c \rightarrow R3m$ on the phase x-T diagram ($T = 25\text{-}350°C$) of $PbZr_{1-x}Ti_xO_3$ was refined using results on temperature studies of the resonance frequency of the radial oscillation modes of poled samples. Piezoelectric properties of solid solutions of $PbZr_{1-x}Ti_xO_3$ were studied in a wide temperature range. It is shown that their behavior with increasing x depends on features of the phase diagram.

Keywords: ferroelectrics, PZT, phase diagram, piezoelectric properties

1. INTRODUCTION

The traditional phase diagram (PD) of the $PbZr_{1-x}Ti_xO_3$ (PZT) system [1, 2] contains rhombic (R), rhombohedral (Rh), tetragonal phases, and the morphotropic phase transitions between them. It is known that the dielectric, piezoelectric and mechanical properties of solid solutions of the PZT system significantly affected by the type and concentration of defects in the crystal lattice. Moreover, Meitzler [3] states that the presence of any significant concentration of crystallographic planes shift type ranked planar defects within its structure plays the major role in changing the properties of electroactive ceramics. Therefore, it is of interest to consider the PD of the PZT system, which would take into account the real defect

---

[*] Corresponding Author address: E-mail: aapavelko@sfedu.ru.

structure of the solid solutions. Such a phase x-T-diagram in the temperature range 25-700°C and the whole concentration range was built by the authors of the present chapter on the results of detailed studies of the structure and electrical properties of solid solutions with a small concentration step [4, 5]. However, the construction of PD caused some difficulties in determining the position of the $R3c{\rightarrow}R3m$ phase transition, as X-ray diffraction superstructure reflections are not detected. It should be noted that the $R3c{\rightarrow}R3m$ transition studied by various experimental techniques. The first experimental evidence of a phase transition between these phases were represented by Barnett [6], who found anomalies in the temperature dependencies of the dielectric loss, pyroelectric charge and thermal expansion of the $PbZr_{0.94}Ti_{0.06}O_3$ sample, doped with 1 mol.% Nb_2O_3. In [2] single crystals were investigated, and the position of the line of this transition have established unequivocally by a single point of inflection in the perovskite cell parameters temperature dependence. In this study we investigated polycrystalline objects that are very different from the single-crystal complex hierarchical structure, so depending on the temperature of the cell parameter in the Rh phase sometimes had several inflection points and anomalies in the form of permanence.

Analysis of published data showed that in the $R3c{\rightarrow}R3m$ transition area most significant changes occur in ferroelastic properties. Thus, the study of some elastic and electrical properties of Nb and Sn modified PZT sample, with $x = 0.07$ [7], resonance frequency, f_r, temperature dependence demonstrated the greatest change in the PT point. In [8], where the acoustic, dielectric properties, and f_r of $PbZr_{0.65}Ti_{0.35}O_3$ ceramics were investigated, the longitudinal sound velocity exhibits the highest sensitivity to the PT. The above statement also confirmed in [9] on the basis of measurements of the dielectric, piezoelectric and ferroelastic properties of the two PZT compounds, doped with 1 mol.% Nb_2O_3, with $x = 0.05$ and $x = 0.25$. In addition, the authors note the $R3c{\rightarrow}R3m$ transition nature changing with increasing x, reflected in the disappearance of the dielectric and piezoelectric anomalies of the Ti-rich sample in the PT temperature range.

Unfortunately, the data presented in the literature cover only the elected compounds of the system, and therefore, the aim of this study was to clarify the position of the $R3c{\rightarrow}R3m$ phase transition line on the PD using additional methods.

2. OBJECTS AND METHODS

Solid solutions of the $PbZr_{1-x}Ti_xO_3$ system ($0.00 \leq x \leq 0.32$) were studied with a research concentration step $\Delta x = 0.0025$-0.025, depending on the position of the solid solution on the PD in the Rh region. The samples were obtained from a stoichiometric mixture of oxides by two-stage solid-phase synthesis at temperatures $T_1 = T_2 = 870$°C, isothermal extracts $\tau_1 = \tau_2 = 7h$, followed by sintering at temperatures $T_s = 1220$-1240°C (depending on composition), $\tau_s = 3h$. Test samples were prepared as flat discs (diameter 10 mm, thickness 1 mm).

High-temperature studies were carried out by powder X-ray diffraction diffractometer ADP-1 (Bragg-Brentano focusing) using filtered $Co_{K\alpha}$ radiation. The accuracy of the temperature stabilization was $\pm$ 1°C, the rate of temperature rise was arbitrary, isothermal holding 10 min, in some cases, it varied from 0 to 30 min. Perovskite cell parameters were calculated by the standard method [10]. The relative error of measurement of structural parameters had the following values: linear, $\delta a = \delta b = \delta c = \pm 0.05\%$; angular, $\delta \alpha = \pm 5\%$; volume, $\delta V = \pm 0.07\%$.

The dielectric, piezoelectric, and ferroelectric properties of solid solutions in the temperature range 30-400°C were investigated using the resonance - antiresonance technique (IEEE Standard on Piezoelectricity, 1987) on an Agilent 4980A precision LCR-meter. We determined the dielectric permeability of polarized samples ($\varepsilon_{33}^T/\varepsilon_0$), the piezoelectric modulus ($|d_{31}|$), and the elastic constant (S_{11}). The resonance (f_r) and antiresonance frequencies were calculated using specially developed software based on the ALGLIB (nonlinear least_squares) algorithms with equivalent circuit modeling.

3. RESULTS AND DISCUSSION

X-ray studies aimed to clarifying the $R3c \rightarrow R3m$ transition line did not allow us to identify uniquely its position in the PD. It is related, as noted above, with a non-monotonic temperature dependence of the perovskite cell parameters: dome-shaped dependencies of parameter a, and the volume, V; dependence of $\alpha(T)$ also changes smoothly without a clear inflection point. Hence, there was a need to involve other research methods.

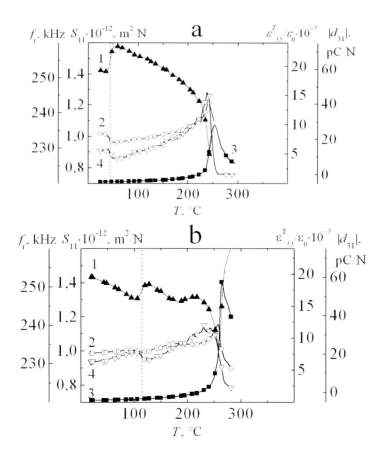

Figure 1. (Continued).

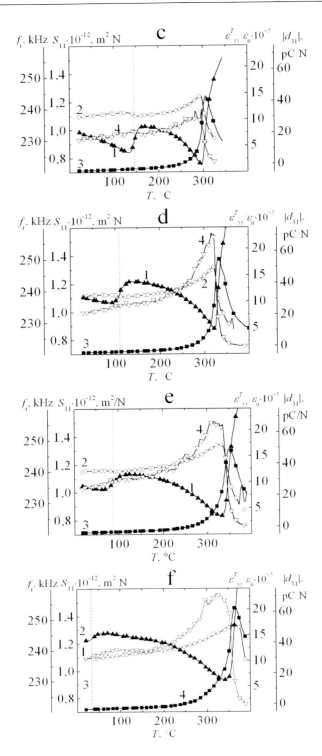

Figure 1. (*1*) Resonance frequency f_r, (2) elastic constant S_{11}, (3) relative dielectric permeability $\varepsilon_{33}^T/\varepsilon_0$, measured at a frequency of 1 kHz; and (4) piezoelectric modulus $|d_{31}|$ of PZT solid solutions as a function of temperature at various *x*: (a) *x* = 0.05; (b) 0.11; (c) 0.15; (d) 0.20; (e) 0.22; (f) 0.28.

Figure 1 shows the temperature dependencies of $f_r(T)$, $S_{11}(T)$, $\varepsilon_{33}^T/\varepsilon_0(T)$, $|d_{31}|(T)$ for the various PZT compounds. Figure 1 shows clearly that the temperature dependencies of all piezoelectric properties exhibit anomalies at the $R3c \to R3m$ transition only up to $x = 0.15$. Above this value of concentration, the PT is clearly effect on the temperature dependencies of ferroelastic parameters (f_r, S_{11}), only. It should be noted that the greatest change at the PT area have shown the temperature dependence of f_r, which we will discuss in more detail.

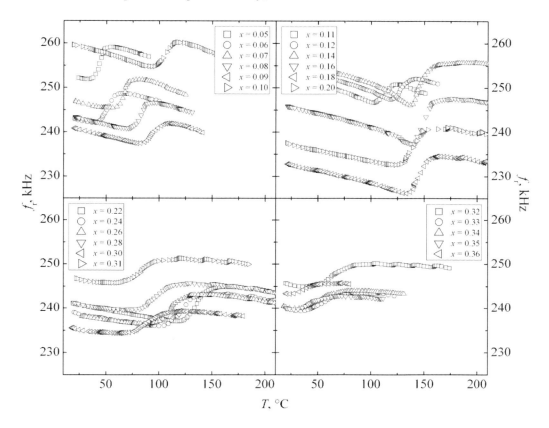

Figure 2. $f_r(T)$ dependencies of Pb(Zr$_{1-x}$Ti$_x$)O$_3$ solid solutions, measured in R3c $\to$ R3m transition (measurement in heating mode).

The $f_r(T)$ dependencies of all investigated solid solutions are shown in Figure 2. All of them Attention is called to gradual smoothing and broadening of $f_r(T)$ anomaly with increasing x, which is apparently due to the blurring of $R3c \to R3m$ transition. Behavior of $f_r(T)$ is caused by the position of the solid solutions in the PD. The abrupt change of $f_r(T)$ in the PT area have solid solutions with $0.07 \leq x < 0.18$ from the single-phase region, which is preserved up to paraelectrics (PE) phase transition. When $x \geq 0.20$ isosymmetrical states (IS) in single-phase regions, characterized by differing behavior of structural and electrophysical parameters, as well as regions of their coexistence (ISCR) with constant unit cell volumes are formed, replacing each other with increasing temperature, which leads to the diffusion of the PT. In this regard, as a PT point on the PD we chose a one corresponding to the mid-course of its intended range.

Fragment of the phase x-T diagram ($25 \leq T \leq 350°C$) of PZT, constructed according to our data in [5] and refined in the present work is shown in Figure 3. Specification of PD was performed using data of additional high-temperature X-ray diffraction studies of some solid solutions in the Rh phase (region with $x = 0.28$ and $x = 0.32$), as well as investigations of $f_r(T)$ dependencies in $R3c \to R3m$ phase transition area (empty circles). Confidence interval corresponds to the width of the latter anomaly of $f_r(T)$ dependencies. Different shading on the PD highlighted areas other than phase states.

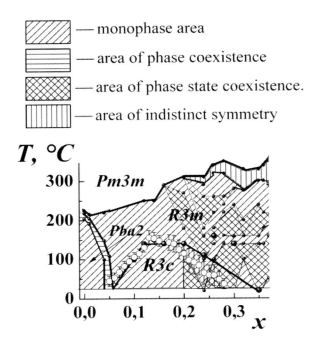

Figure 3. Fragment of the phase x-T diagram ($25 \leq T \leq 350°C$) of PZT, adjusted according to the high-temperature measurements of the piezoelectric parameters.

CONCLUSION

The data obtained in the experimental studies confirmed the effectiveness of the resonance frequency method in the localization of $R3c \to R3m$ transition temperature position in PZT ceramic solid solutions. This method enabled us to identify more accurately the $R3c \to R3m$ PT boundary, as well identified a number of features that require further explanation.

According to the piezoelectric resonance study the correlation of the $R3c \to R3m$ transition nature with the position of solid solutions on the PD was established: a sharp transition corresponds to single phase region, continuing up to PE phase transition, and diffuse transition corresponds to areas of the PD, where periodical changes of IS and ISCR occur with temperature rise.

ACKNOWLEDGMENTS

This work was financially supported by the Ministry of Education and Science of the Russian Federation: Grant of the President of the Russian Federation No. MK-3232.2015.2; Agreement No. 14.575.21.0007 (Federal Target Program); themes Nos. 1927, 213.01-2014/012-ВГ and 3.1246.2014/К (the basic and project parts of the State task).

REFERENCES

[1] Jaffe, B.; et al., *Piezoelectric Ceramics*, Academic Press, New York, 1971.
[2] Eremkin, V. V. et al., *Fizika Tverdogo Tela*, 1989, *vol. 31*, 156 (in Russian).
[3] Meitzler, A. H. *Ferroelectrics*, 1975, *vol. 11*, 503.
[4] Andryushina, I. N. et al., *Ceramics International*, 2013, *vol. 39*, 1285.
[5] Andryushina, I. N. et al., *Ceramics International*, 2013, *vol. 39*, 2889.
[6] Fesenko, E. G. *Perovskite Family and Ferroelectricity*, Atomizdat, Moscow, 1972 (in Russian), pp. 1-248.
[7] Barnett, H. M. *J. Appl. Phys.*, 1962, *vol. 33*, 1606.
[8] Berlincourt, D. et al., *J. Phys. Chem. Solids*. 1964, *vol. 25*, 659.
[9] Krause, J. T., O'Bryan, H. M. Jr., *J. Am. Ceram. Soc*. 1972, *vol. 52*, 497.
[10] Dong, X. L., Kojima, S. *J. Phys. Condens. Matter*. 1997, *vol. 9*, L171.

In: Proceedings of the 2015 International Conference …   ISBN: 978-1-63484-577-9
Editors: Ivan A. Parinov, Shun-Hsyung Chang et al.   © 2016 Nova Science Publishers, Inc.

Chapter 28

ABSORPTION SPECTRA OF MICROWAVE ENERGY OF PIEZOELECTRIC CERAMICS SYNTHESIZED UNDER DIFFERENT SINTERING TEMPERATURE

E. N. Sidorenko[1,], V. G. Smotrakov[2], V. V. Eremkin[2], I. I. Natkhin[1], M. E. Agarkova[3], A. A. Naumenko[2] and E. I. Petrova[2]*

[1]Southern Federal University, Rostov on Don, Russia
[2]Research Institute of Physics, Southern Federal University,
Rostov on Don, Russia
[3]Branch of the St. Petersburg Hydrometeorological State University,
Rostov-on-Don, Russia

ABSTRACT

Energy absorption spectra were studied for ferro-electrically-soft piezoceramics ZTP-19 in the frequency range of 3.1–12 GHz. The studied samples differed by the technology of ceramics production and formed two basic sets. The first set was obtained by ordinary ceramic technology using solid-phase synthesis, and the second one was produced from the powder by chemical co-precipitation of metal hydroxides with deaglomeration in a planetary mill. It has been found that shape of absorption spectra depends on the technology of ceramics production, and polarization applied to samples as well. Large (up to 40 dB) sharp maxima existing in the spectra probably result from piezoresonant oscillations of wedge-type domains in crystallites of ceramics. This assumption is confirmed by the structure of central part of the absorption maximum, which consists of many narrowband peaks at frequencies changing continuously in time.

Keywords: piezoceramics, microwave energy absorption, crystallite grains, domains, dielectric permittivity, ZTP-19, PCR-1

---

[*] E-mail: ensidorenko@sfedu.ru.

1. INTRODUCTION

Search for new radio absorbing substances includes research of ferroelectric materials [1], and little knowledge of microwave properties of the latter has stimulated the study of absorbing ability of ZTP-19 ceramics in centimeter wavelength band. In this band, dispersion of dielectric permittivity is known [2] to be observed for all poly-domain ferroelectrics.

2. OBJECTS AND METHODS

We have studied ceramic samples of $Pb_{0.95}Sr_{0.05}Zr_{0.53}Ti_{0.47}O_3$ + 1% mass fraction of Nb_2O_5 (ZTP-19) of two types obtained by different techniques, and at different temperatures. Piezoceramics of type *A* with various porosity was obtained by standard ceramic techniques using solid-phase synthesis at the sintering temperature T_s = 1245°C. Ceramic samples of type *B* were sintered of ZTP-19 powder [3] at temperatures from 900 to 1250°C. The powders were obtained by chemical co-precipitation of metal hydroxides and deaglomerated in a planetary mill. Phase composition of powders corresponds to morphotropic phase boundary. The powder had high concentration of big hard agglomerates due to 3.5% mass fraction of free PbO. The size of agglomerates affecting the density of preliminary forms at forming and sintering processes decreases after 2 hours of grinding in a planetary mill. Ceramics of this type has been synthesized for production of piezoactuators and has improved characteristics: higher values of dielectric permittivity, piezomoduli, electromechanic coupling ratio, and lower values of dielectric loss tangent and size of crystallite grains [4].

Polarized and non-polarized samples had cylindrical shape with a diameter of 10 mm and a height of 1 mm. Using microwave equipment we studied energy absorption spectra of all piezoelectric ceramic samples (without electrodes). The experimental set-up worked in the "transmission" mode of electromagnetic wave, and comprised three sweep frequency generators covering 3.2–12 GHz band, and a VSWR/attenuation indicator [5]. As a measuring cell, we used a broadband microstrip line with test samples laid on its surface.

3. RESULTS

We obtained the following results of our measurements. For polarized and non-polarized samples of *A*-ceramics with porosity of 30% we observed two large (more than 40 dB) peaks in absorption spectra at frequencies in the intervals of 3.5–4.5 GHz and 5.5–6 GHz (Figure 1, curves 1 and 3). At higher frequencies (8–12 GHz) the absorption was practically flat varying from 5 to 10 dB. Polarization of samples of this ceramics shifts the absorption maxima to higher frequencies (Figure 1, curves 2 and 4), which is also typical for ferro-electrically-soft piezoceramics PCR-1 [5].

For the same sample of high-density piezoceramics of 1% porosity Figure 2 shows four absorption curves in the area of one peak in more detail. Energy absorption graphs $L(f)$ represent non-polarized sample (curve 1), polarized sample (curve 2), sample depolarized by heating over the temperature of phase transition and measured at once after cooling (curve 3), and the sample depolarized, and measured 6 month later (curve 4).

The shape of the spectra is practically independent of ceramic porosity (Figure 1, curves 1 and 3, Figure 2, curves 1 and 2). However, for 30%-porosity ceramics, along with clear sharp peaks (Figure 1, curves 1, 3, 4), for some samples in the interval from 4 to 6 GHz we observed large absorption maxima with rugged tops (Figure 1, curve 2). We found that those maxima and even large sharp peaks had complex fine structure. Changing the frequency limits of the sweep generator made it possible to expand the absorption peak and examine it in more detail in a narrower bandwidth.

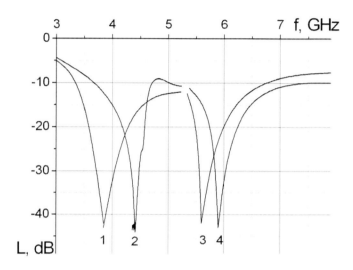

Figure 1. Energy absorption spectra of *A*-type 30%-porosity piezoceramics for non-polarized (curves 1 and 3), and polarized (curve 2 and 4) samples.

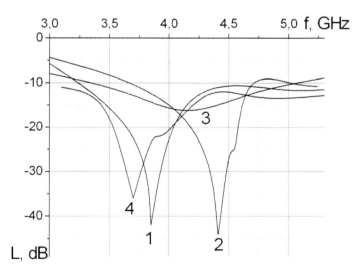

Figure 2. Energy absorption spectra in more detail: 1 – non-polarized sample; 2 – polarized sample; 3 – sample depolarized by heating and measured at once after cooling; 4 – depolarized sample 6 months later.

Figure 3 shows oscillograms of the spectra for different values of bandwidth $\Delta f = 2.5$, 0.7, and 1.13 GHz. It appeared (Figure 3) that absorption maximum consisted of multiple narrowband peaks with natural frequencies, which were not stable but varied in time thus forming the dynamic picture.

For non-polarized samples of B-type ceramics, we observed approximately uniform energy absorption at levels from 2 to 10 dB without distinct maxima in the whole frequency band. Polarization of these samples changed the absorption spectra drastically (Figure 4). For all samples, it was found that absorption maxima appeared in the interval from 3.5 to 4.5 GHz. The peaks became higher and less blurred with the growth of sintering temperature. Some samples had spectra with split maxima. For samples with high T_s, the additional one or two maxima appeared at higher frequencies with energy absorption of up to 15 dB. For piezoceramics with lower values of T_s, the absorption peaks were severely blurred or absent.

Large sharp peaks really indicate the absorption of energy and have nothing in common with reflection of electromagnetic waves from the sample laid on the microstrip line. To check this, we inverted the mode of our experimental set-up from "transmission" to "reflection" of electromagnetic waves. This procedure resulted in low VSWR minima of 1.05-1.15 at the frequencies of absorption peaks. In some cases, we used liquid electrodes to study the effect of gap between the sample and central conductor of microstrip line, and found that character of spectra did not change.

4. DISCUSSION AND CONCLUSION

Sharp peaks that we have found in energy absorption spectra may be attributed to resonant oscillations of groups of small wedge-type domains in piezoceramic crystallites [7]. Spectrum of a sample can have several absorption maxima. The frequencies of spectral maxima are determined by mean values of piezoresonance frequencies of the groups of wedge-type domains. The values of piezoresonance frequencies depend on thickness and type of wedge-type domains. For a piezoresonance frequency $f_0 \approx 10$ GHz the domain thicknesses are about tenths of a micrometer [6].

There is no common opinion in literature on which direction the center of dielectric dispersion shifts when a ferroelectric is polarized. The absorption spectra of polarized and non-polarized A-type piezoceramics ZTP-19 (Figures 1 and 2) definitely indicate that absorption maxima and therefore the center of microwave permittivity dispersion shift towards higher frequencies.

Crystallite grains of B-type ceramics are small in size [4], and structure of wedge-type domains cannot realize or domains are mechanically gripped. Therefore energy absorption is low for non-polarized samples of this ceramics. It is known [8] that for ZPT-19 ceramics at frequencies of about 10^{10} Hz the dielectric dispersion ratio k is about 0.5, where $k = (\varepsilon^{LF} - \varepsilon^{UHF})/\varepsilon^{LF}$, ε^{LF} and ε^{UHF} are values of ceramics permittivity at frequencies of about 10^3 and 10^{10} Hz, respectively. Therefore, the fact of small absorption of the energy for this ceramics indicates that the center of permittivity dispersion is probably below 3 GHz.

Polarization of B-type piezoceramic samples changes their absorption spectra, which now feature absorption maxima in $L(f)$ dependencies. These maxima are due to rebuilding of domain structure, and probably to formation of groups of wedge-type domains in crystallite grains. It is known [4] that with T_s decreasing for *B*-ceramics from 1250 до 900°C, the size of crystallite grains decreases from 2.1 to 0.76 micrometers, respectively. Therefore the concentration and mobility of the domains seems to decrease with decreasing of T_s, which results in spectra blurring and lowering of the absorption peaks or even their full vanishing.

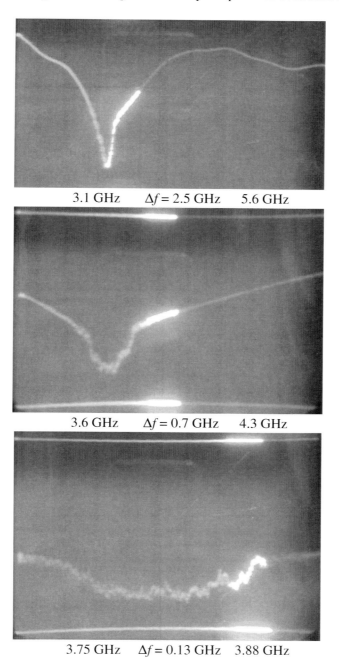

Figure 3. Change of absorption maximum appearance with frequency span Δf decreasing by 3.5 and 20 times for *A*-type ceramics.

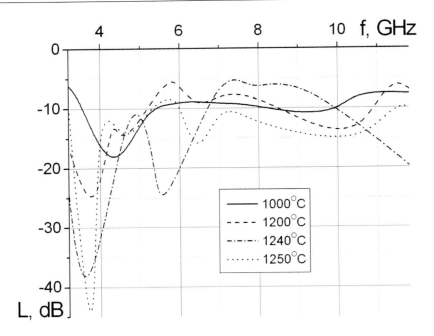

Figure 4. Absorption spectra for polarized ceramics of type B with different sintering temperatures.

Figure 3 shows that large sharp absorption maximum has fine structure consisting of many narrowband maxima. Maximum energy absorption in spectrum is caused by group of domains with piezoresonance natural frequencies in the range of 80–100 MHz.

Both a- and c-domain wedges can resonate in alternating electric field. When these domains oscillate, mechanical stresses and electric field strength change due to piezoeffect, which leads to changing of piezoresonance frequencies of the domains. For this reason, the shape of the top of a peak in absorption spectra looks rugged and is not static.

The obtained results indicate the resonant character of dielectric permittivity dispersion in ZTP-19 piezoceramics. Using the model of a harmonic oscillator, it is possible to describe a resonance dispersion spectrum by Drude-Lorenz equation, and get equations for frequency dependencies of real ε' and imaginary ε'' parts of complex dielectric permittivity. From the dependence $\varepsilon' = \varepsilon'(\omega)$ for small values of Γ ($\Gamma < 1$), the value of ε'' at maximum is found [2]:

$\varepsilon''_{max} = \dfrac{\varepsilon_o - \varepsilon_\infty}{\Gamma}$. Here ω_0 is the natural frequency of the oscillator, $\Gamma = \dfrac{\Delta\omega}{\omega}$ is damping constant of the oscillator, $\Delta\omega$ is 3dB bandwidth of the loss. Thus, having a big maximum in an absorption spectrum, it is possible to calculate the value of damping constant, and to estimate maximum value ε''_{max} of imaginary part of permittivity and dielectric loss tangent tg δ.

From the calculation of parameters for polarized B-ceramics obtained at T_s = 1250°C, for $\varepsilon_0 = \varepsilon_{LF} = 2000$, $\varepsilon_\infty = \varepsilon_{UHF} = 1000$, it was found that $\Delta f_1 = 1.75$ GHz at $f_0 = 3.6$ GHz, $\Gamma = 0.49$, $\varepsilon''_{max} = 2040$, tg$\delta$ = 1. The similar calculation for A-ceramics showed that $\Gamma = 0.25$, $\varepsilon''_{max} = 3600$, tg δ = 2.

Thus, the obtained results show that the ability of piezoceramics ZTP-19 to absorb microwave energy depends on the technology of its production, and large values of ε''_{max} and tg δ derived from absorption spectra indicate the resonant mechanism of UHF dispersion of permittivity in piezoceramics.

REFERENCES

[1] Petrov, V. M. et al. Izv. RAN, *Ser. Neorg. Matter.* 2001, vol. 37, 135 (in Russian).

[2] Poplavko, Yu. M.; Pereverzeva, L. P.; Rayevsky, I. P. Physics of Active Dielectrics. Southern Federal University Press, Rostov-on-Don. 2009, pp. 1 − 478 (in Russian).

[3] Khazhinskiy, M. Yu. et al. Russian Patent RU2042629 (in Russian).

[4] Eremkin, V. V. et al. *Proceedings of the International Scientific and Technical Conference* "Intermatic-2011." Moscow. 2011, Pt. 2, 117 (in Russian).

[5] Sidorenko, E. N. et al. *Ferroelectrics.* 2003, vol. 286, 131.

[6] Sidorenko, E. N. Abstracts of the International Symposium on Physics and Mechanics of New Materials and Underwater Applications (PHENMA-2014), March 27-29, 2014, Khon-Kaen, Thailand, p. 81.

[7] Sidorenko, E. N. et al. Electromagnetic Waves and Electronic Systems. 2013, vol. 18, 51 (in Russian).

[8] Turik, A. V. et al. *Journal of Technical Physics.* 1980, vol. 50, 2146 (in Russian).

In: Proceedings of the 2015 International Conference …   ISBN: 978-1-63484-577-9
Editors: Ivan A. Parinov, Shun-Hsyung Chang et al.   © 2016 Nova Science Publishers, Inc.

Chapter 29

PYROELECTRIC ACTIVITY AND DIELECTRIC RESPONSE OF THE $PbZr_{1-x}Ti_xO_3$ SYSTEM IN THE RANGE OF $0.37 \leq X \leq 0.57$

*Yu. N. Zakharov, A. A. Pavelko**, *A. G. Lutokhin, V. G. Kuznetsov, L. A. Shilkina and L. A. Reznichenko*
Research Institute of Physics, Southern Federal University,
Rostov-on-Don, Russia

ABSTRACT

Temperature dependences of pyroelectric and dielectric properties of ceramics of solid solutions of the PbZr1-xTixO3 system were studied in $0.37 \leq x \leq 0.57$ range. It has been established, that on the isothermal sections corresponding to the R3m$\leftrightarrow$P4mm phase transition and the phase transformations of pseudo-cubic states and also in the region of indistinct symmetry, one can observe the anomalies of pyroelectric and dielectric properties. Weak relaxor properties not registered earlier for binary PZT compositions were revealed in the region of indistinct symmetry.

Keywords: ferroelectrics, PZT, phase diagram, pyroelectric properties

1. INTRODUCTION

This chapter presents the results of studying the pyroelectric and dielectric properties of the PZT solid-solution (SS) ceramics, the analysis of which is performed from the positions of their real structure. Actuality of such studies is due to the fact that the modern highly effective PZT-based ferro-piezoelectric materials are manufactured, mainly, in the form of ceramics, which have a compicated hierarchical structure with different types of imperfections.

---

* Corresponding author: E-mail: aapavelko@sfedu.ru.

2. OBJECTS AND METHODS

The objects for studying were prepared by means of the common ceramic technology from high purity oxides: PbO, TiO$_2$ and ZrO$_2$.

Discs of 10 mm diameter and 1 mm thickness coated with Ag electrodes were polarized in a dc ielectric field $E = (5 - 7) \cdot 10^6$ V·m^{-1} at $T \sim 140°C$ for 40 min with the following cooling down to 80°C. Registration of $i_{dyn}(T)$ was conducted at the frequency of 6.5 Hz by the sinusoidal modulation of the infrared radiation flow, and $i_{qu.st.}(T)$ – at heating and cooling at the linear rate of 0.03-0.1 ks^{-1} synchronously for each sample. Simultaneously the temperature dependences of the relative dielectric permittivity, $\varepsilon(T)$, were determined ($\varepsilon(T) = \varepsilon_{33}^T/\varepsilon_0$, where ε_{33}^T the dielectric permittivity of the poled sample and ε_0 is the vacuum permittivity) by the ac bridge E7-20. The technique of studying included, also, a thermocycling of the samples according to the following scheme: heating from T_{room} to $T_1 \rightarrow$ stabilization of T_1 in time $t_1 \rightarrow$ cooling down to $T_2 < T_1 \rightarrow$ stabilization of T_2 for t_2. Each following cycle was accompanied by the increase of T_1 to the values not breaking the poled state. Information registering and processing, as well the control of temperature were performed by means of the personal computer equipped by the system of data collection and the "L-CARD" software.

3. RESULTS AND DISCUSSION

Figure 1 shows the temperature dependence $i_{dyn}(T)$ for the PbZr$_{0.63}$T$_{0.37}$O$_3$ sample characterizing all the stages of thermally induced phase transitions (PT) R3c→R3m→P4mm for the given x concentration established in [1] by the x-ray studies. The PT boundaries are indicated by the dashed lines fixed at the $i_{dyn}(T)$ maxima. No anomalies were observed on the dependences $\varepsilon(T)$ and $i_{qu.st.}(T)$.

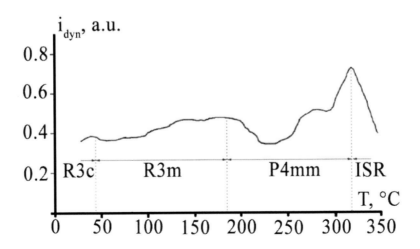

Figure 1. Temperature dependence of $i_{dyn}(T)$ for the PbZr$_{0.63}$T$_{0.37}$O$_3$ samples.

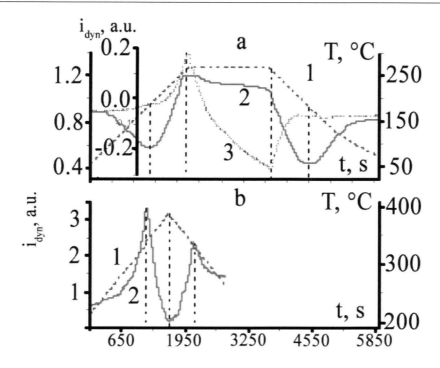

Figure 2. The plot of temperature variation: $T(t)$-1, $i_{dyn}(t, T)$, $i_{qu.st.}(T)$ -3 in the case of thermocycling of the sample with $x = 0.37$ to $T = 250°C$ (a) and $T = 380°C$ (b).

For the purpose of determining a repetition of the obtained results the repeated thermocycling with a fixation of temperature of the chosen isothermal section for 5-10 min was carried out. Figure 2a presents the fragments of the process of thermocycling of the sample for $x = 0.37$ with the recording of $i_{qu}(T)$ and $i_{dyn}(T)$. Upon heating and cooling $i_{dyn}(T)$ shows the stable minima at $T \sim 180°C$ while on approaching $T \sim 250°C$ it relaxes slowly with increasing amplitude. The i_{qu} increases with increasing temperature, at $T \sim$ const it inverts the sign for 5 min and for 30 min it reaches the maximum value exceeding the initial one by a factor of 1.5 in modulus. The decrease of $i_{dyn}(T)$ with time at $T \sim$ const is due to depolarization, whereas the behavior of $i_{qu.st.}(T)$ cannot be related only to the pyroelectric effect of ferroelectric polarization of the sample. Figure 2,b shows a disappearance of the dynamic pyroelectric activitity at $T = 380°C$ and its recovery at the cooling.

In accordance with Figure 2,b, in this temperature range a study of dispersion of the relative dielectric permittivity was carried out. Results of the study are presented in Figure 3. Figure 3 shows the relaxor behavior of the samples of the PZT SS system at $x = 0.37$ as evidenced by the decreased magnitude of the relative dielectric permittivity maximum, its shift to the high-temperature range by 4°C and, also, the PT diffusion with increasing frequency of the measuring field. Similar studies were conducted for the samples with the increased x-concentration. An example of such a study is shown in Figure 4 for the $PbZr_{0.53}Ti_{47}O_3$ sample, where $\varepsilon(T)$, $i_{qu.st.}(T)$ and $i_{dyn}(T)$ are indicated by 1, 2 and 3, respectively. As follows from the analysis of Figure 4, in vicinity of T_{1mL}, there exist the anomalies of the above mentioned dependencies, the manifestation of which will be discussed below.

The phase diagram of the binary PbZr$_{1-x}$Ti$_x$O$_3$ system presented in Figure 5 follows from the x-ray analysis data, according to work [2].

In all the SS, with the increase of temperature from room to a certain (specific for each SS`) temperature T_L the distortions of noncubic (Rh, T) cells typical of $T > T_{LO}$ decrease ($c/a \to 1$; $a \to 90°$) and at $T = T_L$ the cells become, practically, cubic (C) (diffraction reflections are close to single ones, without any splittings peculiar to the low-symmetry phases). However, at this temperature the $\varepsilon/\varepsilon_0$ maxima are not reached.

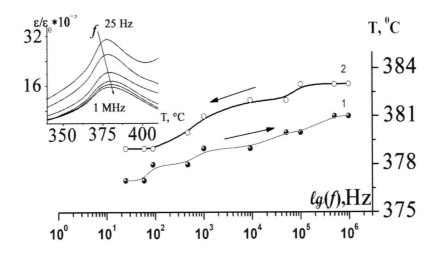

Figure 3. Temperature of $\varepsilon/\varepsilon_0(T)$ maxima in the vicinity of the Curie temperature versus the measuring field frequency for the samples with x = 0.37; 1 - the heating, 2 - the cooling. The inset shows a general view of the $\varepsilon/\varepsilon_0(T)$ dependence dispersion.

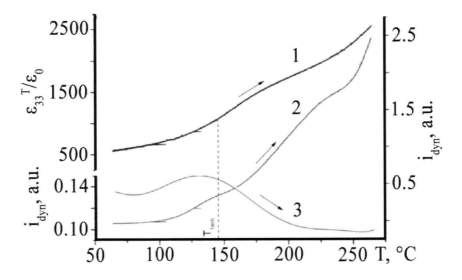

Figure 4. Temperature dependences $\varepsilon(T)$ -1, $i_{qu.st.}(T)$-2 and $i_{dyn.}(T)$ - 3 at the heating of the PbZr$_{0.63}$T$_{0.37}$O$_3$ samples. T_{1mh} is the temperature interval of manifestation of the anomalies in the above dependences.

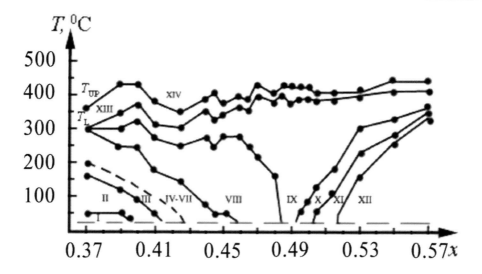

Figure 5. Phase *x-T* diagram of the system in the component concentration range under study. The phases are denoted as follows:

I: Rh_1
II: $Rh_1 + Rh_2$
III: Rh_2
IV: $Rh_2 + Rh_3$
V: Rh_3
VI: $Rh_3 + PSC_1$
VII: $Rh_3 + PSC_1 + PSC_2$
VIII: $Rh_3 + PSC_1 + PSC_2 + T_1$
IX: $PSC_2 + T_1$
X: T_1
XI: $T_1 + T_2$
XII: T_2
XIII: ISR
XIV: C

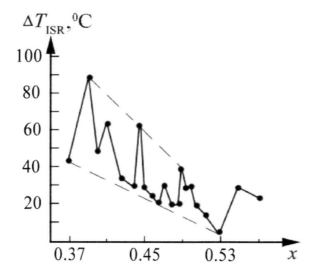

Figure 6. Width of the indistinct-symmetry region, ΔT_{ISR}, as a function of *x*.

At the elevated temperatures (below the Curie temperature) the x-ray analysis hase revealed a region in with the crystal lattice parameters could not be clearly determined; it was called an indistinct-symmetry region. Figure 6 presents this region width as a function of the *x* concentration. Analysis of the data reported in [2] and presented in Figures 5 and 6 shows that, at $0.37 \leq x \leq 0.57$, for SS of the binary PZT system, on the isothermal section of a rhombohedral (Rh) – tetragonal (T) PT there exist two intermediate phases indenfied (because of weak splittings) as pseudocubic (P_sC_1 and P_sC_2), and also the indistinct-symmetry region the boundaries of which become closer as x increases. As a result, at $0.37 \leq x \leq 0.51$, on the isothermal sections the regions with different phase states of the PZT crystal lattice may exist separately or coexist. Variations in temperature and also the keeping at the direct current may lead to a breaking of the existing balance and a setting of a new energetic one in such regions. This process is considerably complicated by the point defects (vacancies) caused by the heterophase compositions and the extended ones (crystallographic shear planes, imperfections connected with the infinitely adaptive TiO_2 structure, and spatial inhomogeneity of the ceramics manifesting themselves, especially, in the regions of phase transformations. Thus the dependencies $i_{qu.st.}$ (t, T) (Figure 2,a), integrating a variation of all kinds of polarization and thermostimulated currents, will require special methods for their interpretation. They may be the lower steps of cyclic heating, the longer periods of keeping at the direct current or others but they will be possibly used in another study.

In Figure 4, the dependencies $\varepsilon(T)$, $i_{dyn}(T)$ and $i_{qu.st}(T)$ clearly illustrate a transition between phase states VIII into phase state IX in accordance with the phase diagram [2] or a R3c $\rightarrow$ P4mm phase transition for the PD [1]. For $i_{dyn}(T)$ it is a pronounced maximum, and for $\varepsilon(T)$ and $i_{qu.st.}(T)$ it is the smooth step-like kinks on curves 1 and 2.

CONCLUSION

To summarize, we can make the following conclusion. A study of the pyroelectric effect in the dynamic and quasistatic regimes of measurement enables one to determine from with high precision in the samples of ferroelectric materials the isothermal sections, on which there occur the phase variations and transformations and, also, to observe the relaxation and other physical phenomena caused by them. As a result, it becomes possible to get additional information about the state of ferroelectric polarization obtained by the X-ray methods.

ACKNOWLEDGMENTS

This work was financially supported by the Ministry of Education and Science of the Russian Federation, theme No. 1880 (the basic part of the State task).

REFERENCES

[1] Eremkin, VV; et al., *Fizika Tverdogo Tela*, 1989, vol. 31, 156 (in Russian).
[2] Reznichenko, LA; et al., *Physics of the Solid State.*, 2008, vol. 50, 1527.

In: Proceedings of the 2015 International Conference ...   ISBN: 978-1-63484-577-9
Editors: Ivan A. Parinov, Shun-Hsyung Chang et al.   © 2016 Nova Science Publishers, Inc.

Chapter 30

POLYCRYSTALLINE SOLID SOLUTIONS BASED ON ALKALI METAL NIOBATES MODIFIED WITH GLASS-FORMING ADDITIVES

I. A. Verbenko[*], L. A. Reznichenko, A. G. Abubakarov, L. A. Shilkina, S. I. Dudkina and A. A. Pavelko

Research Institute of Physics, Southern Federal University,
Rostov-on-Don, Russia

ABSTRACT

Polycrystalline solid solutions based on the $NaNbO_3$-$LiNbO_3$ system were prepared. Their dielectric, piezoelectric and elastic properties were measured. The most effective modifier was found adding of which leads to a substantial reduction in sintering temperature and increase of the K_t/K_p ratio, which can be used in practice.

Keywords: ceramic technology, piezoelectric, glass ceramics, environmental problems

1. INTRODUCTION

Among the global issues [1], one of the most difficult is the problem of environmental safety. The difficulty of solving this problem consists in the lack of natural financial advantage to encourage international manufacturers to seek appropriate "green" technologies. The work on the analysis of existing scientific developments and their potential development needs to create the legal framework. One major cause of environment pollution is wide use of lead compounds, which are present as a key component in most commercial piezoelectric ceramics. Therefore, one of the challenges for modern materials research is to find a safe alternative to lead-containing ceramics. In recent years, alternative materials have been the

---

[*] E-mail: ilich001@yandex.ru.

subject of intense attention. Solid solutions based on alkali metal niobate are most promising. However, there are a number of technological challenges: high temperature of sintering, narrow range of optimum sintering temperature, and fragility of ceramics and inhomogeneous materials.

2. OBJECTS AND METHODS

Glass-modified solid solutions of the $NaNbO_3$-$LiNbO_3$ system [$Na_{0.875}Li_{0.125}NbO_3$ (A), $Na_{0.850}Li_{0.122}W_{0.028}NbO_3$ (B) and $Na_{0.86625}Li_{0.12375}Sr_{0.01}Nb_{0.998}O_3$ (C)] were prepared by the conventional fabrication technique. Powders of Li_2CO_3 (99.99%), $NaHCO_3$ (99.8%), $SrCO_3$ (99.5%), WO_3 (99.99%) and Nb_2O_5 (99.9%) were used as raw materials. Glass forming additives include oxides in the ratio of Na_2O - 7,97 wt.%; Nb_2O_5 - 34.1 wt.%; WO_3 - 44.7 wt.%; SiO_2 - 3.9 wt.%; B_2O_3 - 9.0 wt.%; Al_2O_3 - 0.33 wt.% (α); Na_2O - 9.9 wt.%; Nb_2O_5 - 53 wt.%; SrO - 20.6 wt.%; SiO_2 - 5.0 wt.%; B_2O_3 - 11.1 wt.%; Al_2O_3 - 0.4 wt.% (β); Na_2O - 9.1 wt.%; Nb_2O_5 - 48.6 wt.%; SrO - 18.9 wt.%; SiO_2 - 4.6 wt.%; H_3BO_3 - 18.4 wt.%; Al_2O_3 - 0.4 wt.% (γ). All of them were introduced in an amount of (0.75-5) wt.% above the stoichiometric composition (Tables 1 and 2).

The study of the crystal structure of the ceramics was performed by an X-ray diffraction (XRD) patterns using CoK_α radiation in the θ-2θ scan mode (DRON-3, Russia). The density of the sintered sample was measured by the Archimedes method (in octane). The piezoelectric ceramics were poled at 140°C for 15 min in an electric field of 6 kV/mm. Dielectric, piezoelectric and elastic parameters of solid solutions were measured using the resonance-antiresonance method [2]. The following parameters were determined: the relative dielectric permittivity of poled ($\varepsilon_{33}^T/\varepsilon_0$) and unpoled ($\varepsilon/\varepsilon_0$) samples, the dielectric loss in a low electric field (loss tangent, tgδ), the piezoelectric modulus ($|d_{31}|$, absolute value), the electromechanical coupling factor of planar (K_p) and transversal (K_t) vibration modes, the mechanical quality factor (Q_M), Young's modulus (Y_{11}^E), and the sound velocity (V_R). The accuracy of these measurements was $\leq \pm$ 1.5% for $\varepsilon_{33}^T/\varepsilon_0$, $\leq \pm$ 2.0% for K_p, $\leq \pm$ 4.0% for $|d_{31}|$, $\leq \pm$ 12% for Q_M, and $\leq \pm$ 1.0% for Y_{11}^E [3]. DC resistivity (ρ_v) was measured with a high resistance meter Agilent 4339B at a voltage of 1 V.

3. RESULTS AND DISCUSSION

The main phase of all the studied ceramics exhibits a perovskite structure. Diffraction peaks related to impurities in modified glass-additives solid solution are not observed. However, it is shown that in the glass-ceramics of C-α-system, the concentration of the rhombohedral phase is growing with the increasing content of the glass.

Our analysis of the influence of different glass-forming additives on the sintering temperature (T_{sint}) of $Na_{0.875}Li_{0.125}NbO_3$ shows the following. In the A-α system, a decrease of T_{sint} is observed at all the concentrations of glass. The most significant change in T_{sint} (100°C) is recorded in the area with a low content of a modifier (less than 1.5 wt.%). In the range of 1.5-3 wt.% of glass, T_{sint} decreases by (25-30)°C. However, increasing the amount of α-glass in the system more than by 3 wt.% leads to a slight change of T_{sint} only. The influence of β-

and γ- glass-forming additives on T_{sint} is less pronounced. With the introduction of glass-forming additives, the density of ceramics in all cases increases.

It is found that the nature of changes of electrophysical properties of glass-ceramics is weakly depends on the chemical composition of the initial solid solutions. At that small additions of glass (up to 1.5 wt.%) leads to sharp changes of $\varepsilon/\varepsilon_0$ and $\varepsilon_{33}^T/\varepsilon_0$, increase tg$\delta$ and reduce ρ_v, noticeable deterioration of piezoelectric properties (reduction of $|d_{31}|$, K_p, K_t). Increase of their concentration (1.5-5wt.%) contributes to the stabilization of certain characteristics (values of $|d_{31}|$, K_p, and K_t remain virtually unchanged) and insignificant changes of $\varepsilon/\varepsilon_0$, $\varepsilon_{33}^T/\varepsilon_0$, ρ_v, tgδ. Parameters Q_M, V_R and Y_{11}^E of glass ceramics containing various (α, β, γ) glasses exhibit somewhat different behavior: increasing the number of α-glass results in a linear increase in these parameters, but their changes in the A-β and A-γ systems were nonmonotonic. Thus, with small additions of glass (< 0.75 wt.% for β-glass and < 1.5 wt.% for α-glass), V_R and Y_{11}^E increase, but Q_M decreases in the range of $(0.75-1.5)$ wt% (β-glass) and $(1.5-3)$ wt.% (γ-glass), on the contrary, V_R and Y_{11}^E reduced, but Q_M increases. If $\beta > 1,5$ wt.% and $\gamma > 3$ wt.%, then V_R and Y_{11}^E increase, and Q_M increases in the case of β-glass and decreases in the case of γ-glass. An interesting feature of all glass-ceramics is a heterogeneous rate of K_p and K_t decreasing, which is why at a certain concentration of glass, K_t/K_p ratio is considerably higher than the corresponding value in the original solid solution ($K_t/K_p \sim 1$). Thus, the K_t/K_p ratio reaches values of 2.5-2.69 at the concentration of glass 1.5 – 3 wt.% with the highest value (2.69) implemented in the A-α system with 3 wt.% of glass.

Table 1. The composition of modifying glasses

Composition of glass / Glass-ceramics	Na_2O	Nb_2O_5	WO_3	SiO_2	B_2O_3	Al_2O_3	SrO	H_3BO_3
A-α	7.97[*]	34.1	44.7	3.9	9.0	0.33	-	-
	20.3[**]	20.3	29.7	9.4	20.3	0.50		
A-β	9.9	53.0	-	5.0	11.1	0.4	-	-
	19.9	24.9		9.45	19.9	24.9		
A-γ	9.1	48.6	-	4.6	-	0.4	18.9	18.4
	16.49	20.53		8.6		4.49	20.48	33.44
B-β	9,9	53.0	-	5.0	11.1	0.4	20.6	-
	19.9	24.9		9.95	19.9	4.98	24.9	
C-α	7.97	34.1	44.7	3.9	9.0	0.33	-	-
	20.3	20.3	29.7	9.4	20.3	0.50		

[*] wt.%.
[**] mol.%.

Table 2. The composition of modifying glasses and its contents in the original solid solutions

Composition of glass / Glass-ceramics	The content of glass in the original solid solutions			
	wt.%			
	0.75	1.5	3.0	5.0
	Mol.%			
A-α	1.20	2.41	4.76	7.85
A-β	0.89	1.86	3.73	6.19
A-γ	1.039	2.08	4.123	6.89
B-β	0.95	1.86	3.75	6.37
C-α	1.10	2.21	4.65	7.25

CONCLUSION

The comparative analysis of the obtained experimental data revealed the composition and concentration of input glass-forming additives, provides the most effective change in the parameters of the original solid solutions. The most effective modifier is α-glass, adding 3 wt.% of which in the A-base leads to a substantial reduction in T_{sint} and an increase in the K_t/K_p ratio, which can be used in practice.

ACKNOWLEDGMENT

The work was supported by the Russian Ministry of Education and Science: FTP (Agreement No. 14.575.21.0007).

REFERENCES

[1] Directive 2002/95/EC of the European Parliament and of the Council of 27 January 2003 on the Restriction of the Use of Certain Hazardous Substances in Electronic Equipment, *Official Journal of the European Union*, 2003, vol. 37, 19-23.

[2] IRE Standards on Piezoelectric Crystals: Determination of the Elastic, Piezoelectric, and Dielectric Constants - the Electromechanical Coupling Factor, *Proc. IRE*, 1958, vol. 46, 764-778.

[3] Dantsiger, A. Y., Razumovskaya, O. N., Reznitchenko, L. A., Sakhnenko, V. P., Klevtsov, A. N., Dudkina, S. I., Shilkina, L. A., Dergunova, N. V., Rybyanez, A. N., *Multicomponent Systems of Ferroelectric Complex Oxides: Physics, Crystallochemistry, Technology. Aspects of Designing Ferroelectric Materials*, Rostov University Press, Rostov-on-Don, 2001-2002, vol. 1-2 (in Russian).

In: Proceedings of the 2015 International Conference ...   ISBN: 978-1-63484-577-9
Editors: Ivan A. Parinov, Shun-Hsyung Chang et al.   © 2016 Nova Science Publishers, Inc.

Chapter 31

EFFECT OF ZR AND (TI, ZR) DOPING ON FERROELECTRIC AND MAGNETIC PHASE TRANSITIONS IN $Pb(Fe_{1/2}Nb_{1/2})O_3$

I. P. Raevski[1,], V. V. Titov[1], S. I. Raevskaya[1], V. V. Laguta[2,3], M. Marysko[2], S. P. Kubrin[1], H. Chen[4], C.-C. Chou[5], M. A. Malitskaya[1], A. V. Blazhevich[1], D. A. Sarychev[1], L. E. Pustovaya[6], I. N. Zakharchenko[1], E. I. Sitalo[1] and V. Yu. Shonov[1]*

[1]Research Institute of Physics and Physical Faculty, Southern Federal University, Rostov-on-Don, Russia
[2]Institute of Physics, AS CR, Prague, Czech Republic
[3]Institute for Problems of Materials Science NASU, Kiev, Ukraine
[4]Institute of Applied Physics and Materials Engineering, Faculty of Science and Technology, University of Macau, Macau, China
[5]National Taiwan University of Science and Technology, Taipei, Taiwan
[6]Don State Technical University, Rostov-on-Don, Russia

ABSTRACT

We carried out structural, dielectric, magnetization and Mössbauer studies of the $(1-x)PbFe_{1/2}Nb_{1/2}O_3-xPbZr_{1/2}Ti_{1/2}O_3$ (PFN$-x$PZT) and PFN$-x$PbZrO$_3$ (PFN$-x$PZ) solid solution ceramics. Though, in contrast to PFN$-x$PbTiO$_3$ (PFN$-x$PT), in both these systems the lattice parameter increases with x and the symmetry remains rhombohedral at $x < 0.25$, their magnetic x, T-phase diagrams appeared very similar to that of PFN-xPT. We suppose that anomalies of composition dependence of magnetic phase transition temperatures at $x \approx 0.05 - 0.1$ in both PFN$-x$PZ and PFN$-x$PZT systems are due to the percolation phase transition.

---

[*] E-mail: igorraevsky@gmail.com.

Keywords: lead iron niobate, multiferroic, phase transition, x, T-phase diagram

1. INTRODUCTION

Lead iron niobate $PbFe_{1/2}Nb_{1/2}O_3$ (PFN) is a complex perovskite multiferroic oxide possessing both ferroelectric and magnetic properties [1-6]. Recent burst of interest to PFN-based materials was initiated by a report that some PFN-Pb(Zr, Ti)O_3 compositions possess a large room-temperature magnetoelectric response [7]. However, the data on the ferroelectric and magnetic phase transitions in PFN-based solid solutions are rather fragmentary and contradictory. Such scatter of data is, to a large extent, caused by a difficulty of fabrication of highly-resistive PFN-based ceramics [5]. On cooling, PFN undergoes a sequence of phase transitions: from the cubic (C) paraelectric to tetragonal (T) ferroelectric phase at $T_{CT} \approx 380$ K, then to the rhombohedral (Rh) ferroelectric phase, at $T_{TM} \approx 360$ K, and, finally, to the G-type antiferromagnetic (AFM) phase [1-6]. At $T_g \approx 10$-20 K a spin glass state appears which coexists with the AFM one [1, 6, 8]. The temperature of the AFM phase transition (the Neel temperature, T_N) determined in different works varies for PFN from 130 to 200 K [1-6]. The possible reasons of such scatter in T_N values seem to be local compositional ordering and/or clustering of Fe^{3+} and Nb^{5+}ions [3, 9].

The most studied PFN-based solid solution system is $(1 - x)PFN - xPbTiO_3$ (PFN $-$ xPT) exhibiting high piezoelectric and pyroelectric responses at the morphotropic phase boundary (MPB) between the Rh and T phases [2, 10, 11]. This MPB is strongly curved and at room temperature, it locates at $x \approx 0.06 - 0.07$ [2, 10] while at low temperatures it is at about $x \approx 0.1$ (Figure 1a). Magnetic x, T-phase diagram of this system, determined in [2, 5, 8] has some peculiar features at $x \approx 0.1$. Ti-doping leads to lowering of T_N, but above $x \approx 0.1$ fast lowering of the T_N determined from the Mössbauer studies stops and a new magnetic state becomes stable in a rather wide compositional range (Figure 1a, curve 3). On the other hand, magnetization (M) measurements have shown [2, 8] that T_g at first increased with x, but above $x \approx 0.1$ it gradually decreased (Figure 1a, curve 4). The maximum in the $T_g(T)$ dependence was ascribed in [2] to the competing action of two factors. Decreasing of the unit cell volume in the PFN $-$ xPT system with increasing x (Figure 2a), promotes the magnetic coupling leading to increase of T_g while dilution of the Fe subsystem leads to the opposite effect. One more possible origin of the observed peculiarities is the change of the lattice symmetry from Rh to T which occurs at low temperatures in the vicinity of $x \approx 0.1$ (Figure 1a).

The scope of the present work was to test these assumptions. For this purpose, we undertook structural, dielectric, magnetization and Mössbauer studies of the $(1 - x)PFN - xPbZrO_3$ (PFN $-$ xPZ) system in which the room-temperature symmetry remains Rh up to $x \approx 0.9$ [12] and $(1 - x)PbFe_{1/2}Nb_{1/2}O_3 - xPbZr_{1/2}Ti_{1/2}O_3$ (PFN $-$ $xPZT$) system in which Rh-T phase boundary shifts to much higher x values as compared to PFN-PT. In both these systems, the lattice parameter increases with x.

Effect of Zr and (Ti, Zr) Doping on Ferroelectric and Magnetic Phase ...   227

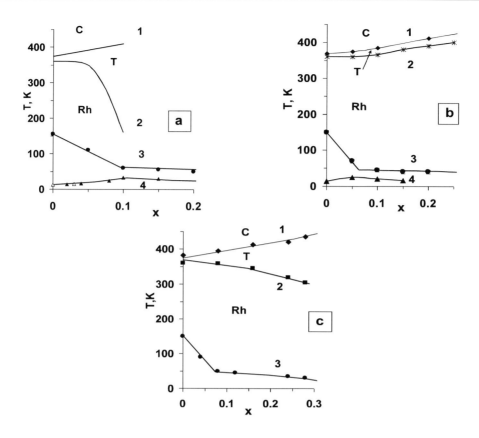

Figure 1. Composition dependencies of the temperatures of the C-T (1) and T-Rh (2) phase transitions, the temperature of the $\eta(T)$ anomaly (3) and the maximum of the $M(T)$ dependence in the ZFC mode (4) for PFN – xPT (a), PFN – xPZ (b) and PFN – xPZT (c) ceramics; the data for PFN – xPT system were taken from Refs. [2, 5, 8].

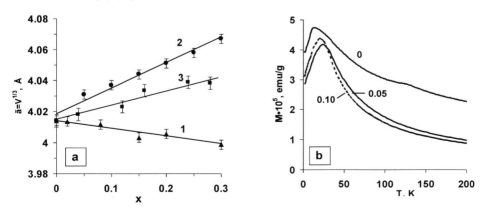

Figure 2. (a) Composition dependencies of the pseudocubic lattice parameter $\tilde{a} = V^{1/3}$ for PFN – xPT (1), PFN-xPZ (2) and PFN – xPZT (3) ceramics studied; (b) temperature dependencies of magnetization M measured under magnetic field of 1 kOe in ZFC mode for PFN – xPZ ceramics; numbers at the curves correspond to x values.

2. Methods

Ceramic samples of PFN $-$ xPZ ($x = 0 - 0.30$) and PFN $-$ xPZT ($x = 0 - 0.28$) solid solution compositions doped with 1 wt% of Li_2CO_3 in order to reduce their conductivity [5, 10] were fabricated by a one-step sintering method [13]. The sintering was performed at 1050-1100°C for 2 hours in a closed alumina crucible filed with $PbZrO_3$ powder to reduce PbO losses. The density of ceramics obtained was about 85-95% of the theoretical one. The mean grain size was $5 - 15$ μm. X-ray phase analysis was performed using DRON-3 diffractometer and Co-K$_\alpha$ radiation. Dielectric studies of the samples with fired Ag electrodes were carried out in the $10^2 - 10^6$ Hz range in the course of both heating and cooling at a rate of $2 - 3$°C/min with the aid of a Wayne Kerr 6500B impedance analyzer. Mössbauer spectra were recorded with the aid of the MS-1104E rapid spectrometer and analyzed using the original computer program UNIVEM. Magnetic measurements were performed in the temperature range $2 - 300$ K using the SQUID magnetometers MPMS-5S and MPMS-XL (Quantum Design).

3. Experimental Results and Discussion

Room-temperature X-ray diffraction studies have shown that all the fabricated Li-doped PFN $-$ xPZ and PFN-xPZT ceramic compositions are single-phase and have a perovskite structure. The pseudocubic lattice parameter ã $= V^{1/3}$ in these systems increases with x (Figure 2a). Tetragonal distortion of the unit cell appears in the PFN $-$ xPZT system at $x \approx 0.25$.

Figure 3 shows the temperature dependencies of the real part of dielectric permittivity ε' measured at 1 kHz for some of Li-doped PFN $-$ xPZT ceramic compositions. While temperature T_m of the $\varepsilon'(T)$ maximum increases as x grows, the step on the $\varepsilon'(T)$ curve corresponding to the Rh-T phase transition shifts to lower temperatures. As the latter step lowers and becomes more diffused with increase of x, for precise determination of the phase transition temperature the minimum in the temperature dependence of the resonance frequency f_r was used (Figure 3b). Similar to Li-doped PFN ceramics [5, 10], no frequency shift of T_m was observed for all the PFN $-$ xPZT and PFN $-$ xPZ compositions studied.

Dependence of magnetization M on temperature for PFN measured on heating after zero field cooling (ZFC) exhibits a maximum at $T_g \approx 12$ K and a bump at $T_N \approx 130$ K (Figure 2b). For PFN $-$ xPZ compositions with $x \geq 0.05$ only the low-temperature maximum is observed in the $M(T)$ curves (Figure 2b). This is typical for solid solutions of multiferroics [2, 8]. The temperature T_g of the $M(T)$ maximum for PFN $-$ 0.05PZ is higher than that for PFN, however in the $x > 0.05$ compositional range T_g decreases gradually with x. Thus $T_g(x)$ dependence for the PFN $-$ xPZ compositions has a maximum at $x \approx 0.05 - 0.07$. As was already mentioned above, a similar increase of T_g with x growing was observed previously for the PFN $-$ xPT ceramics in the $x < 0.1$ composition range [2, 8]. This increase was attributed to a decrease of the lattice parameter â with x and a subsequent increase of the magnetic coupling [2]. However, in both PFN $-$ xPZ and PFN $-$ xPZT systems â increases with x (Figure 2a) and such explanation is inappropriate. It is worth noting that a maximum in the $T_g(x)$ dependence was observed for the PFN $-$ $BaFe_{1/2}Nb_{1/2}O_3$ [8] and $PbFe_{1/2}Ta_{1/2}O_3 - PbTiO_3$ [13] solid solutions where â also increases with x. We suppose that the observed increase of T_g in all these systems at small x values is a result of the increase of the average size of the confined percolation clusters, with x, because of the decrease of the strength of the infinite cluster [8].

Temperatures of magnetic phase transitions were determined from the abrupt drop in the temperature dependence of the magnitude η of Mössbauer spectra intensity within the 0 – 1.2 mm/s velocity range normalized to its value at 300 K [3]. This method was successfully used previously to determine values of T_N in several multiferroics and their solid solutions and the results obtained were very similar to the data obtained by traditional methods such as the magnetization measurements (see Refs. [3, 9, 13] and references therein).

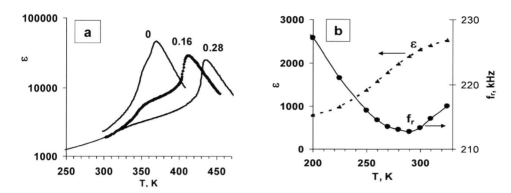

Figure 3. (a) Temperature dependencies of permittivity ε' measured at 1 kHz for some PFN – xPZT compositions; numbers at the curves correspond to x values; (b) anomalies of the $\varepsilon'(T)$ and resonance frequency corresponding to Rh-T phase transition for PFN – 0.28PZT composition.

The results of dielectric, Mössbauer and magnetization studies for PFN – xPZ and PFN – xPZT solid solution system are summarized in Figures 1b and 1c. One can see that the results obtained are very similar to those for PFN – xPT system (Figure 1a). According to Mössbauer data, at low x values T_N rapidly decreases as x grows. Lowering of T_N with x in both systems is quite expectable due to dilution of the magnetic subsystem. However, above a certain compositional threshold ($x \approx 0.05 - 0.1$), fast lowering of T_N with the increase of x stops and a new magnetic state with comparatively high (~50 K) transition temperature becomes stable in a rather wide compositional range. According to a large difference between $M(T)$ curves measured in the ZFC and FC modes, magnetic state stable at $x > 0.05 - 0.1$ seems to be a spin-glass-like one. In all the systems studied T_g values for compositions with $x > 0.05 - 0.1$ are lower by about 20 K than those determined from the Mössbauer studies. This difference seems caused by the fact that the upper limit of the spin relaxation rates in these samples is above the characteristic Mössbauer time [14]. Similar behavior was already reported for $PbFe_{12-x}Cr_xO_{19}$ [15] and for $PbFe_{1/2}Ta_{1/2}O_3 - PbTiO_3$ solid solutions [13].

CONCLUSION

Highly-resistive Li-doped PFN – xPZ and PFN – xPZT ceramics were fabricated. Dilution of Fe sublattice by Zr or (Ti, Zr) lowers the Neel temperature T_N. However, above a certain composition threshold ($x \approx 0.05 - 0.1$) fast lowering of T_N stops and a new magnetic state stable in a rather wide compositional range arises. Large difference between the zero-field cooled and field-cooled magnetization curves as well as between the temperature of magnetic phase transition determined from Mössbauer studies and the temperature T_g of the ZFC magnetization curve maximum for compositions with $x > 0.05 - 0.1$ implies that this state is a spin glass.

We suppose that both the maximum of the $T_g(x)$ dependence and a change of the slope of $T_N(x)$ dependence at $x \approx 0.05 - 0.1$ are due to the percolation phase transition in the PFN – xPZ, PFN – xPZT and PFN – xPT solid solution systems. Similar transition was revealed recently in the $PbFe_{1/2}Ta_{1/2}O_3 - PbTiO_3$ solid solution system [13].

ACKNOWLEDGMENT

The work was partially supported by Russian Foundation for Basic Research (grant 14-02-90438_Ucr_a), research project 3.1137.2014K of the Ministry of Education and Science of Russian Federation, and Research Committee of the University of Macau under Research and Development Grant for Chair Professor No RDG007/FST-CHD/2012.

REFERENCES

[1] Kleemann, W.; Shvartsman, V.V.; Borisov, P.; Kania, A. *Phys Rev Letters.* 2010, *vol. 105*, 257202-1-257202-4.

[2] Singh, S.P.; Yusuf, S.M.; Yoon, S.; Baik, S.; Shin, N.; Pandey, D. *Acta Mater.* 2010, *vol. 58*, 5381-5392.

[3] Raevski, I.P.; Kubrin, S.P.; Raevskaya, S.I.; Stashenko, V.V.; Sarychev, D.A.; Malitskaya, M.A.; Seredkina, M.A.; Smotrakov, V.G.;Zakharchenko, I.N.; Eremkin, V.V. *Ferroelectrics.* 2008, *vol. 373*, 121-126.

[4] Laguta, V.V.; Rosa, J.; Jastrabik, L.; Blinc, R.; Cevc, P.; Zalar, B.; Remskar, M., Raevskaya, S.I.; Raevski, I.P. *Mater Res Bull.* 2010, *vol. 45*, 1720-1727.

[5] Raevski, I.P.; Kubrin, S.P.; Raevskaya, S.I.; Prosandeev, S.A.; Malitskaya, M.A.; Titov, V.V.; Sarychev, D.A.; Blazhevich, A.V.; Zakharchenko, I.N. *IEEE Trans. Ultrason. Ferroelect. Freq. Contr.* 2012, *vol. 59*, 1872-1878.

[6] Falqui, A.; Lampis, N.; Geddo-Lehmann, A.; Pinna, G. *J Phys Chem B.* 2005, *vol. 109*, 22967-22970.

[7] Sanchez, D.A.; Ortega, N.; Kumar, A.; Sreenivasulu, G.; Katiyar, R.S.;Scott, J.F.; Evans, D.M.; Arredondo-Arechavala, M.; Schilling, A.; Gregg, J.M. *J Appl Physics.* 2013, *vol. 113*, 074105-1-074105- 7.

[8] Laguta, V.V.; Glinchuk, M.D.; Maryško, M.; Kuzian, R.O.; Prosandeev, S.A.; Raevskaya, S.I.; Smotrakov, V.G.; Eremkin, V.V.; Raevski, I.P. *Phys Rev B.* 2013, *vol. 87*, 064403-1- 064403-8.

[9] Raevski, I.P.; Kubrin, S.P.; Raevskaya, S.I.; Sarychev, D.A.; Prosandeev, S.A.; Malitskaya, M.A. *Phys Rev B.* 2012, *vol. 85*, 224412-1-224412 -5.

[10] Sitalo, E.I.; Raevski, I.P.; Lutokhin, A.G.; Blazhevich, A.V.; Kubrin, S.P.; Raevskaya, S.I.; Zakharov, Yu.N.; Malitskaya, M.A.; Titov, V.V.; Zakharchenko, I.N. *IEEE Trans Ultrason Ferroelect Freq Control.* 2011, *vol. 58*, 1914-1918.

[11] Zakharov, Yu.N.; Raevskaya, S.I.; Lutokhin, A.G.; Titov, V.V.; Raevski, I.P.; Smotrakov, V.G.; Eremkin, V.V.; Emelyanov, A.S.; Pavelko, A.A. *Ferroelectrics.* 2010, *vol. 399*, 20-26.

[12] Marbeuf, A.; Bavez, J.; Demazeaw, G. *Rev. Chim. Miner.* 1974, *vol. 11*, 198-206.

[13] Raevski, I.P.; Titov, V.V.; Malitskaya, M.A.; Eremin, E.V.; Kubrin, S.P.; Blazhevich, A.V.; Chen, H.; Chou, C.-C.; Raevskaya, S.I.; Zakharchenko, I.N.; Sarychev, D.A.; Shevtsova, S.I. *J. Mater. Science.* 2014, *vol. 49*, 6459-6466.

[14] Campbell, I.A. *Hyperfine Interactions.* 1986, *vol. 27*, 15-22.

[15] Albanese, G.; Watts, B.E.; Leccabue, F.; Diaz, C. S. *J. Magn. Magn. Mater.* 1998, *vol. 184*, 337-343.

In: Proceedings of the 2015 International Conference …
Editors: Ivan A. Parinov, Shun-Hsyung Chang et al.
ISBN: 978-1-63484-577-9
© 2016 Nova Science Publishers, Inc.

Chapter 32

INFLUENCE OF MnO_2 ON DIELECTRIC CHARACTERISTICS, GRAIN AND CRYSTAL STRUCTURE OF $PbFe_{0.5}Nb_{0.5}O_3$ CERAMICS

N. A. Boldyrev[1], A. V. Pavlenko[1,2], L. A. Reznichenko[1] and L. A. Shilkina[1,]*

[1]Research Institute of Physics, Southern Federal University,
Rostov-on-Don, Russia
[2]Southern Scientific Center of the Russian Academy of Sciences,
Rostov-on-Don, Russia

ABSTRACT

Dielectric and ceramic characteristics of pure (PFN) and modified by 1 wt.% (PFNM1), 2 wt.% (PFNM2) and 3 wt.% (PFNM3) of MnO_2 lead iron niobate $PbFe_{0.5}Nb_{0.5}O_3$ (PFN) were investigated. Explanation of observed effects was suggested considering crystalochemical specification of an introduced modifier.

Keywords: ceramics, lead iron niobate, multiferroics

1. INTRODUCTION

Currently great attention in material science is paid to a creation and detailed research of properties of functional structures based on multiferroics, which demonstrate the interaction of the magnetic and electrical subsystems in the wide temperature range. Series of compounds containing Fe Mn Ni with the perovskite structure and common composition $AB'_xB''_{1-x}O_3$ belong to them. Among them, PFN attracts specific attention. PFN is often used as a various purpose functional structures component [1] in modified ceramics. The aim of the present

---

[*] Corresponding Author address: E-mail: huckwrench@gmail.com.

chapter is to establish regularities of the MnO_2 influence on structure and dielectric characteristics of the formation processes in PFN ceramics.

2. TECHNIQUES OF SAMPLES OBTAINING AND INVESTIGATING

The objects of our study were pure PFN and PFN with stoichiometric MnO_2 additives additives in an amount of 1 wt.% (PFNM1), 2 wt.% (PFNM2) and 3 wt.% (PFNM3). The samples were synthesized by method of solid phase reactions from oxides PbO, Fe_2O_3, Nb_2O_5 MnO_2 of high purity grade by the two stage calcination with intermediate milling. Test samples were prepared as discs with a diameter of 10 mm and a thickness of 1 mm. The metallization is carried out by heating the two-time silver-containing paste.

X-ray research were performed by powder X-ray diffraction using a diffractometer DRON-3 (CoKα radiation, Bragg - Brentano focusing). The calculation of perovskite unit cell parameters (linear a, angular α, volume V, and homogeneous deformation parameter δ) was carried out by standard procedures [2]. Complex permittivity $\varepsilon^* = \varepsilon' - i\varepsilon''$ (ε' and ε'' are the real and imaginary parts of ε^*, respectively) in the temperature [(300–700) K] and frequency [(10– 10^5)Hz] ranges was measured using a measuring bench based LCR-meter HIOKI 3522-50 and Warta thermostat TP 703.

Measurements of the unpolarized specimens dependence $\varepsilon'/\varepsilon_0(E)$ were carried out at room temperature and at frequency of the alternating electric field measuring 1 kHz by measuring bench based on LCR-meter Agilent 4263V. Measurements were made with reversible changing of electrostatic field intensity constant in the range from −1 kV/mm to +1 kV/mm. The dielectric controllability K was calculated using the formula $K = (\varepsilon'/\varepsilon_0(E = 0) - \varepsilon'/\varepsilon_0(E))/\varepsilon'/\varepsilon_0(E = 0)$.

3. EXPERIMENTAL RESULTS AND DISCUSSION

The X-ray diffraction analysis has shown that obtained samples are unalloyed and monophase at room temperature. PFN, PFNM1, PFNM3 have rhombohedral structure, PFNM2 has pseudocubic structure (crystal structure couldn't be determined because of low distortion of perovskite cell). Increment of modificator concentration had led to increasing of parameters a, V and α (Table 1). It is at variance with logic of changes of V at localization of Mn^{4+} cations in B-position [2] of the perovskite structure (Pb^{2+} – 1.26 Å, Fe^{3+} – 0.67 Å, Nb^{5+} – 0.66 Å, Mn^{4+} – 0.52 Å).

Probably, part of introduced modificators takes up both regular and nonregular (including interstices) positions forming interstitial solid solutions, which demonstrate increment of crystal lattice volume [3]. Moreover low temperature (which can be lower in multicomponent composition) of a transformation in "chain" $Mn^{(4+)}O_2 \xrightarrow{873 K} Mn_2^{(3+)}O_3 \xrightarrow{1173K} Mn_3^{(2+)+(3+)}O_4 \xrightarrow{1573K} Mn^{(2+)}O$ [4], can be possible in the presence of manganese cations with the different oxidation level (Mn^{4+}, Mn^{3+}, Mn^{2+}) in observed materials. Introduction of Mn^{3+} (0.70 Å) and Mn^{2+} (0.91 Å) in B-positions of perovskite structure also can lead to observed increment of parameters a and V.

Fragments of cleavage image of studied ceramics are shown in Figure 1.

Table 1. Results of subjects of inquiry X-ray diffraction analysis

	PFN	PFNM1	PFNM2	PFNM3
Phase composition	unalloyed			
Symmetry	Rh	Rh	PsC	Rh
a, Å	4.014	4.0144	4.0153	4.0171
α, degree	89.92	89.93	-	89.94
V, Å^3	64.68	64.69	64.74	64.83

A sufficiently dense structure formed of crystals of irregular polygon form is shaped always. Introduction of modificators have led to increment of ceramic's average grain size. Cleavage in PFN passes through grains and its borders (Figure 1a). That evidences of closeness durabilities of grain boundary and grains. But cleavage in PFNM (1-3) passes almost only through grains (Figure 1b-d).

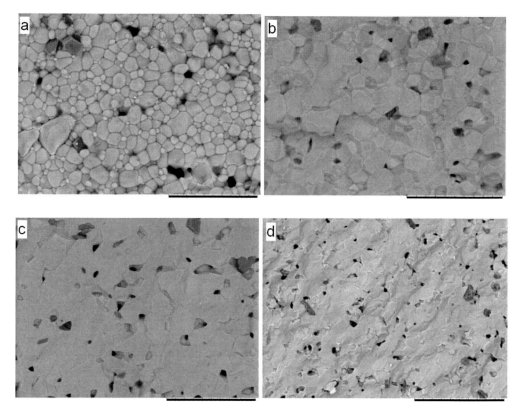

Figure 1. Micrographs of fragments of cleavage of ceramics of PFN (a), PFNM1 (b), PFNM2 (c), PFNM3 (d). (Markers are $3 \cdot 10^{-5}$ m long).

It shows that modification have led to increment of grain borders strength property. It may be explained by change of sintering nature from solid-phase sintering to liquid phase sintering. The source of liquid phase is MnO$_2$, which has low temperature of melting.

Moreover, changes in sintering nature have led to strong decrement of melting temperature of investigated ceramics. Absence of obvious signs of liquid phase existence can be excited by demonstration of "cementing action" of Mn-containing liquid phases, which strongly bands grains and increases demonstrated during machining ceramics durability.

Figure 2 shows dielectric spectra of investigated ceramics.

Maximums of $\varepsilon'/\varepsilon_0$ at the temperature $T = T_C = 371$ K (T_C is Curie temperature that corresponds to maximum $\varepsilon'/\varepsilon_0(T)$) are due to the phase transition from the ferroelectric phase to paraelectric phase [5]. Maximum of $\varepsilon'/\varepsilon_0$ in PFN ceramics at the temperature $T > T_C$ is related to dielectric relaxation which is due to results of the Maxwell-Wagner (interlayer) polarization [7]. An introduction of modificators have led to decreasing T_C and blurring of the phase transition from the ferroelectric phase to the paraelectric one. Moreover in PFN and PFNM1, Curie temperature doesn't depend on alternating electric field frequency (Figure 2 a, b). Such behavior is usually observed in classical ferroelectrics with smeared phase transitions. But in PFNM2 and PFNM3 forming of frequency-dependent moving to the high temperature range maximums are observed. But one should point out that no acceptable T_m dependence approximation using the Vogel – Fulcher expression $f = f_0 \exp(E_{act}/(k(T_m-T_f)))$ was obtained (f_0 is the frequency of attempts to overcome the potential barrier E_{act}, k is Boltzmann constant, and T_f is Vogel – Fulcher temperature).

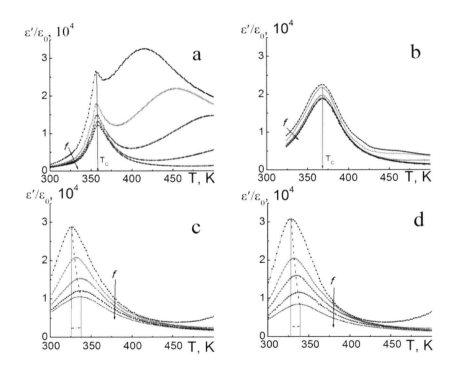

Figure 2. Dielectric spectra in the temperature range (300–500) K and frequency range (10–10⁵) Hz in PFN (a), PFNM1 (b), PFNM2 (c) и PFNM3 (d) (Pointer shows the f increment direction).

Such an evolution of dielectric spectra and demonstration of relaxor properties increasing with increasing the molar concentration of MnO_2 can be related to growth of crystal chemical disorder and composition fluctuation in a material volume due to a localization of different valent Mn-cations in a various crystallographic structure positions of PFN. Observed changing of ferroelectric properties character in samples is clearly demonstrated in investigating of the reversible dielectric nonlinearity (Figure 3).

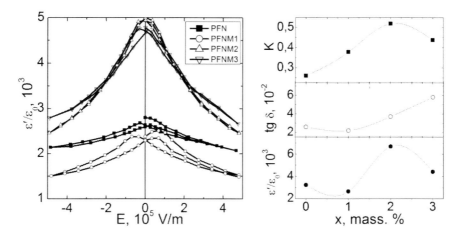

Figure 3. Dependences $\varepsilon'/\varepsilon_0(E)$ of PFN, PFNM1, PFNM2, and PFNM3 ceramics at $f = 1$ kHz. On the right – dependence of the controllability coefficient K(x), $\varepsilon'/\varepsilon_0(x)$ and tg$\delta(x)$.

Dependences $\varepsilon'/\varepsilon_0(E)$ in PFN and PFNM1 have peculiar to ferroelectric with blurring phase transition form of "butterfly" [8] and hysteresis effects. In PFNM2 and PFNM3 "dome-shaped" anhysteretic curves are implemented. Controllability coefficient K in these samples is more a half times K in PFN. Also it reaches a maximum value in PFNM2 characterized by $\varepsilon'/\varepsilon_0$ maximum and low tgδ (Figure 3). $\varepsilon'/\varepsilon_0(E)$ measurements at $E = (0\text{-}10^6)$ V/m shows that controllability in investigated materials reaches $K = 0.8$.

CONCLUSION

The obtained results indicate that modifying the PFN ceramic enables us to effectively control its dielectric characteristics and to form structures with high dielectric controllability.

ACKNOWLEDGMENTS

The work was supported by Ministry of Education and Science of the Russian Federation (Federal program: contract No. 14.575.21.0007; contract No. 3.1246.2014/K, pr. No. 1927, theme No. 213.01-2014/012-IG).

REFERENCES

[1] Kantselson, L.M.; Bokov, G.A.; Kuznetsova, T.K.; Sytnik L.P. *The Semiconductor Ceramic Thermistor Material* Russian patent No. 2066077, 1996 (in Russian).

[2]   Fesenko, E.G. *The Family of Perovskite and Ferroelectricity*. Atomizdat, Moscow, 1972, pp. 1 – 248 (in Russian).

[3]   Umanskiy, I. S. *Radiography of Metals and Semiconductors*. Metallurgia, Moscow, 1969. pp. 1 – 496 (in Russian).

[4]   *Chemical Encyclopedia*. Knuniants, I. L. (Ed.), Soviet Encyclopedia, Moscow, 1988, pp. 1 – 623.

[5]   Pavlenko, A.V.; Shilkina, L.A.; Reznichenko, L.A. *Crystallography*. 2012, *vol. 56(4)*, 729–734.

[6]   Reznichenko, L.A.; Razumovskaya, O.N.; Klevtsov, A.N. About manufacturability manganese ferroelectric ceramics *Proceedings of the International Scientific and Practical Conference "Piezotechnics-99."* Rostov-on-Don. 1996. pp. 268- 275.

[7]   Pavlenko, A.V.; Turik, A.V.; Reznichenko, L.A.; Shilkina, L.A.; Konstantinov, G.M. *Solid State Physics*. 2011, *vol. 53(9)*, 1773-1776.

[8]   Talanov, M.V.; Turik, A. V.; Reznichenko, L.A. *Technical Physics Journal*. 2013, *vol. 83(11)*, 60-66.

In: Proceedings of the 2015 International Conference …   ISBN: 978-1-63484-577-9
Editors: Ivan A. Parinov, Shun-Hsyung Chang et al.   © 2016 Nova Science Publishers, Inc.

Chapter 33

EFFECT OF DYNAMIC FATIGUE ON DIELECTRTIC AND PYROELECTRIC PROPERTIES OF $(1 - x)$ PBFE$_{1/2}$NB$_{1/2}$O$_3$ – XPBTIO$_3$ CERAMICS

A. F. Semenchev[1], I. P. Raevski[1], S. I. Raevskaya[1,], T. A. Minasyan[1], S. P. Kubrin[1], H. Chen[2], C-C Chou[3], M. A. Malitskaya[1] and V. V. Titov[1]*

[1]Research Institute of Physics and Physics Department,
Southern Federal University, Rostov on Don, Russia,
[2]Faculty of Science and Technology,
University of Macau, Macau, China
[3]National Taiwan University of Science and Technology,
Taipei, Taiwan, ROC

ABSTRACT

Electric dynamic fatigue (EDF) - a gradual degradation of switchable polarization under electrical cycling of ferroelectrics, limits greatly the applications of ferroelectrics. In the present work, we study the effect of EDF on polarization loops, pyroelectric and dielectric responses of $(1 - x)\mathrm{Pb}(\mathrm{Fe}_{1/2}\mathrm{Nb}_{1/2})\mathrm{O}_3 - x\mathrm{PbTiO}_3$ ceramics. We suppose that dynamic fatigue in these ceramics is due to space-charge effects. Free charge carriers screen external electric field at domain boundaries. As a result, some domains are excluded from the process of polarization switching, and the number of such pinned domains increase in the course of cycling.

Keywords: electric dynamic fatigue (EDF), pyroelectric and dielectric properties, $(1 - x)\mathrm{Pb}(\mathrm{Fe}_{1/2}\mathrm{Nb}_{1/2})\mathrm{O}_3 - x\mathrm{PbTiO}_3$

---

* Corresponding author: E-mail: igorraevsky@gmail.com.

1. INTRODUCTION

Electric dynamic fatigue (EDF) of ferroelectric materials comes out at multiple polarization switching by the action of bipolar electric field exceeding the coercive field E_c [1-14]. EDF results in decrease of macroscopic polarization P, and changes of thermal, dielectric and electromechanical properties. This kind of instability imposes significant limitations on the ability of practical application of ferroelectrics in devices based on the effect of the polarization switching (for example, non-volatile memory cells, shift registers) and other practically important devices based on ferroelectric materials [2-7]. The effect of fatigue can be triggered by numerous factors, including both internal characteristics of materials and external conditions. Internal characteristics are the composition and microstructure of the bulk material or thin films; external conditions include state of the surface, electrodes, and the temperature as well as the frequency, intensity and type of the applied field [5-8]. Various mechanisms are suggested to explain EDF in ferroelectric (single-crystal, ceramic, thin film) materials however, a comprehensive understanding of this phenomenon has not been reached, so far. At the same time, it is generally accepted that EDF is a result of a complex interaction of electrical, mechanical and electrochemical processes [2-8]. Space-charge processes lead to screening of the external electric field in the regions, close to electrodes, and at grain boundaries, weakening external fields acted in the bulk of material and thereby preventing the polarization switching. As a result, a system of non- switched, so-called "opposite" domains forms gradually in the material. Their walls are inclined under some angle with respect to the direction of the vector of spontaneous polarization P_s and consequently such domain walls are charged. Free charges screen the domain walls and stabilize them, leading to the formation of the "frozen" domains. With increasing of the number of polarization switching cycles the number of "frozen" domains increases and the value of switching polarization decreases [6-10].

Electromechanical effects are the consequences of significant local mechanical stresses, arising because of incompatibility of piezoelectric strain at the boundaries of the electrodes, grains and "frozen" domains. These effects generate point and extended defects in the crystal structure. Interaction of these defects leads to formation of micro-cracks, which can grow during cycling to macro-cracks [2, 6-13]. In Ref. [14] microscopic a-domains at the side walls and micro-cracks at the front walls of "frozen" domains were identified in fatigued $PbTiO_3$ single crystals, which clearly indicates the presence of electromechanical effects. In the bulk polycrystalline ferroelectric materials, such effects should appear, mostly, because of disoriented grains. Micro-cracking leads to a decrease of electric field in the bulk of ceramics, complicating polarization switching, whereby, with the increasing of the number of micro-cracks, the value of polarization decreases.

Electrochemical processes lead to formation of clusters of point defects [2]. Their role in the EDF seems to be the same as that of the space-charge processes.

In this paper, we present the results of EDF studies for some of $(1-x)$ Pb $(Fe_{1/2}Nb_{1/2})$ O_3 $-x PbTiO_3$ (PFN $-x$ PT) solid solution ceramic compositions.

2. EXPERIMENTAL

Ceramic samples of PFN – xPT solid solution compositions ($x = 0$ and $x = 0.08$) were prepared by the solid-state reaction route using high-purity Fe_2O_3, Nb_2O_5, TiO_2 and PbO. Both compositions were doped with 1 wt. % of Li_2CO_3. This additive promotes formation of the perovskite phase and reduces dramatically the conductivity of PFN-based ceramics [15]. The synthesis was carried out at 900°C for 2 hours. The samples for the final sintering were pressed at 1000 kg/cm^2 using 3% water solution of the polyvinyl alcohol as a plasticizer. The final sintering was performed at 1020 – 1070°C for 2 hours in the closed alumina crucible. Weight losses during sintering appeared to be less than 1 wt.%. The sintered ceramics had a high density (about 93-96% of the theoretical one) and their mean grain size was 6 – 10 μm.

X-ray phase analysis was performed using DRON-3 diffractometer and Co-K$_\alpha$ radiation. The electrodes for dielectric measurements were deposited on the grinded disks of 9 mm in diameter and of 0.9 mm in height by firing on the silver past at 500°C. Dielectric studies were carried out in the $10^2 – 10^6$ Hz range in the course of both heating and cooling at a rate of 2 – 3°C/min with the aid of a Wayne Kerr 6500B impedance analyzer. Pyroelectric current was measured in a quazistatic regime on heating at a constant rate of 3°C/min.

Fatigue was estimated from the rate of degradation of the remnant polarization P_r, determined from the dielectric hysteresis loops measured using the Sawyer–Tower oscilloscope method at 50 Hz. The samples were placed between two spherical metal clamps serving as electrical contacts. The setup was immersed in silicone oil to avoid arcing and to enhance heat dissipation from the sample during cycling. Fatigue cycling was carried out by application of a sinusoidal ac electric field having a frequency of 50 Hz. The cycling field amplitude was steadily increased from zero to the maximum value within 3 s and was decreased in a similar fashion and at a similar rate after the respective number of switching cycles N. In all the experiments, the maximal cycling field amplitude was 2 times larger than the value of the coercive field E_c of the initial hysteresis loop.

3. RESULTS AND DISCUSSION

Room-temperature X-ray diffraction studies have shown that all the fabricated Li-doped PFN – xPT ceramic compositions are single-phase and have a perovskite structure. The symmetry of PFN – 0.0PT ceramics was rhombohedral while that of PFN-0.08PT composition was tetragonal in good agreement with the literature data [15].

Figure 1 illustrates the change in the shape of dielectric hysteresis loops for the PFN – xPT samples with different content of lead titanate $PbTiO_3$ after $6 \cdot 10^6$ switching cycles. One can see that while the hysteresis loop for PFN – 0.0PT composition only slightly reduces in height, staying saturated, the loop for PFN – 0.08PT composition changes dramatically, indicating a decrease in remnant polarization P_r. This difference seems to be due to rhombohedral structure of PFN at room temperature, with relatively small value of homogeneous spontaneous deformation parameter (a relative elongation of the prototype unit cell along the polar axis) [16] $\delta_{Rh} \approx 1.5 \cdot 10^{-3}$ in comparison with $\delta_T \approx 5 \cdot 10^{-3}$ for tetragonal PFN – 0.08PT composition. Small value of homogeneous deformation parameter is one of the signs of soft ferroelectric ceramics [16]. This difference in δ values may explain why polarization switching occurs easier in rhombohedral ferroelectric compositions than in the tetragonal ones.

Figure 2 shows relative remnant $P_r(N)/P_r(0)$ polarization dependencies on the number N of switching cycles. As one can see in Figure 2, for both PFN – xPT compositions studied dynamic fatigue does not occur until the number of switching cycles reaches $\approx 10^5$, while on further increase of N the values of P_r, begin to decrease. Interestingly, both PFN – xPT compositions appear more stable with respect to dynamic fatigue during polarization switching as compared with the soft PZT-based piezoelectric ceramics [12, 13].

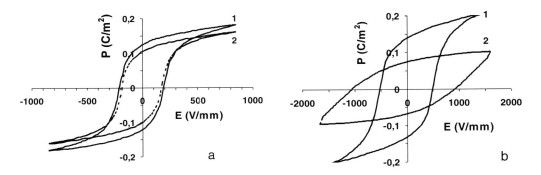

Figure 1. Effect of dynamic fatigue on dielectric hysteresis loops measured at 50 Hz for (a) PFN – 0.0PT and (b) PFN – 0.08PT ceramics: (1) before cycling, (2) after $6 \cdot 10^6$ switching cycles.

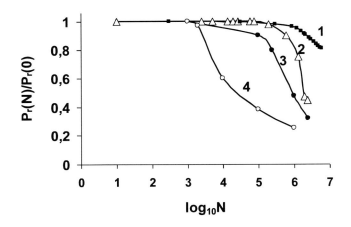

Figure 2. Evolution of normalized remnant polarization determined from the dielectric hysteresis loop during bipolar electric cycling for several piezoelectric ceramics. The ratio of the amplitude of the applied ac electric field E_m and the coercive field E_c for each composition are shown in parentheses: (1) PFN – 0.0PT ($E_m = 2E_c$); (2) PFN – 0.08PT ($E_m = 2E_c$); (3) Pb$_{0.99}$[Zr$_{0.45}$Ti$_{0.47}$(Ni$_{0.33}$Sb$_{0.67}$)$_{0.08}$]O$_3$ ($E_m = 2E_c$) [12]; (4) PCR-7M ($E_m = 1.5E_c$) [13].

Figure 3 illustrates the effect of fatigue on the temperature dependencies of dielectric permittivity ε and pyroelectric current for PFN – 0.08PT ceramics. One can see that both pyroelectric and dielectric properties also exhibit degradation in the course of cycling. In contrast to the data of Ref. [12], grinding off the 0.1 mm-thick surface layers from the fatigued PFN – 0.08PT sample did not lead to the recovering of the initial $\varepsilon(T)$ curve. This fact implies that the fatigued regions of the sample are not located near the electrodes. Such behavior seems to correspond to the fatigue mechanism, according to which the domain walls are fixed by space charge injected from the electrode, forming the, so-called, frozen domains. The number of these frozen domains increases with switching cycles. In order to evaluate the fraction of the sample's volume distorted by the local electric fields arising during the cycling experiments we tried to use Mössbauer spectroscopy, as the quadrupole splitting of doublet in Mössbauer spectrum may be influenced by such fields.

At room temperature Mössbauer [57]Fe spectra of both initial and fatigued PFN – 0.08PT samples were doublets. The values of both quadrupole splitting ($\approx$ 0.4 mm/s) and isomer shift ($\approx$ 0.4 mm/s) relative to metallic iron corresponded to the Fe^{3+} ions located inside the oxygen octahedra of the perovskite lattice [17]. This result corroborates the data of the X-ray photoelectron studies showing for PFN single crystals and ceramics the presence of only trivalent iron within the sensitivity of the method [18]. The value of quadrupole splitting of doublet in Mössbauer spectrum appeared practically the same for both initial and fatigued PFN – 0.08PT samples indicating that local electric fields arise only in a very small part of the sample's volume.

4. SUMMARY

Dynamic fatigue effect was studied for PFN – 0.0PT and PFN – 0.08PT ceramic compositions. We found out that in the samples studied fatigue does not occur until ~10^5 switching cycles. We suppose that dynamic fatigue in PFN – PT ceramics is due to space-charge effects. Free charge carriers screen external electric field at domain boundaries of non-switched domains, stabilizing them. As a result, some domains are excluded from the process of polarization switching, and the number of such pinned domains increase in the course of cycling.

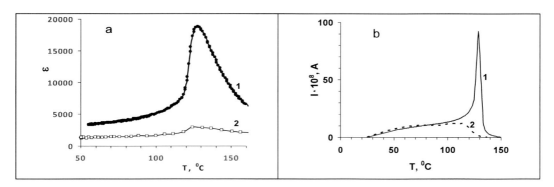

Figure 3. Impact of dynamic fatigue (4.4·10^6 switching cycles) on (a) dielectric permittivity and (b) pyroelectric current for PFN – 0.08PT ceramics: (1) before cycling, (2) after 6·10^6 switching cycles.

As compared with the soft PZT-based piezoelectric ceramics, the PFN – xPT compositions appear more stable with respect to dynamic fatigue during polarization switching.

ACKNOWLEDGMENT

The work was partially supported by the Ministry of Education and Science of Russian Federation (research projects 2132 and 3.1137.2014K), Southern Federal University (grant No. 213.01-2014/012-VG), and Research Committee of the University of Macau under Research and Development Grant for Chair Professor No RDG007/FST-CHD/2012.

REFERENCES

[1] Burfoot, JC; Taylor, GW. *Polar Dielectrics and Their Applications*. The Macmillan Press LTD, N.-Y. 1979.

[2] Lupascu, DC. *Fatigue in Ferroelectric Ceramics and Related Issues*. Springer, Heidelberg, 2004.

[3] Duiker, HM; Beale, PD; Scott, JF; Paz de Araujo, CA; Melnick, BM; Cuchiaro, JD; McMillan, LD. *J. Appl. Phys.*, 1990, 68, 5783-5791.

[4] Ozgul, M; Takemura, K; Trolier-McKinstry, S; Randall, CA. *J. Appl. Phys.*, 2001, vol. 89, 5100-5106.

[5] Robels, U; Arlt, G. *J. Appl. Phys.*, 1993, vol. 73, 3454-3460.

[6] Cao, H; Evans, AG. *J. Am. Ceram. Soc.*, 1994, vol. 77, 1783–1786.

[7] Kim, SJ; Jiang, Q. *Smart Mater. Struct.*, 1996, vol. 5, 321–326.

[8] Zhu, T; Fang, F; Yang, W. *J. Mater. Sci. Let.*, 1999, vol. 18, 1025-1029.

[9] Nuffer, J; Lupascu, DC; Rödel, J. *J. Eur. Ceram. Soc.*, 2001, vol. 21, 1421–1423.

[10] Nuffer, J; Lupascu, DC; Rödel, J. *Acta Mater.*, 2000, vol. 48, 3783-3794.

[11] Zhang, Y; Lupascu, DC; Aulbach, E; Baturin, I; Bell, A; Rödel, J. *Acta Mater.*, 2005, vol. 53, 2203-2213.

[12] Verdier, C; Morrison, FD; Lupascu, DC; Scott, JF. *J. Appl. Phys.*, 2005, vol. 97, 024107-1 – 024107-6.

[13] Gavrilyachenko, VG; Kuznetsova, EM; Semenchev, AF; Sklyarova, EN. *Phys. Solid State.*, 2006, vol. 48, 1149–1152.

[14] Semenchev, AF; Gavrilyachenko, VG; Kuznetsova, EM; Sklyarova, EN. *Phys. Solid State.*, 2006, vol. 48, 1146–1148.

[15] Sitalo, EI; Raevski, IP; Lutokhin, AG; Blazhevich, AV; Kubrin, SP; Raevskaya, SI; Zakharov, Yu. N; Malitskaya, MA; Titov, VV; Zakharchenko, IN. *IEEE Trans Ultrason Ferroelect Freq Control.*, 2011, vol. 58, 1914-1918.

[16] Fesenko, EG. Perovskite Family and Ferroelectricity. *Atomizdat, Moscow*, 1972 (In Russian).

[17] Raevski, IP; Kubrin, SP; Raevskaya, SI; Stashenko, VV; Sarychev, DA; Malitskaya, MA; Seredkina, MA; Smotrakov, VG; Zakharchenko, IN; Eremkin, VV. *Ferroelectrics.*, 2008, vol. 373, 121-126.

[18] Kozakov, AT; Kochur, AG; Googlev, KA; Nikolsky, AV; Raevski, IP; Smotrakov, VG; Yeremkin, VV. *J. Electron Spectrosc. Related Phenom.*, 2011, vol. 184, 16-23.

In: Proceedings of the 2015 International Conference …   ISBN: 978-1-63484-577-9
Editors: Ivan A. Parinov, Shun-Hsyung Chang et al.   © 2016 Nova Science Publishers, Inc.

Chapter 34

INFLUENCE OF LI$_2$CO$_3$ ON DIELECTRIC AND CERAMIC CHARACTERISTICS OF PBFE$_{0.5}$NB$_{0.5}$O$_3$

N. A. Boldyrev[1], A. V. Pavlenko[1,2] and L. A. Shilkina[1]*

[1]Research Institute of Physics, Southern Federal University,
Rostov-on-Don, Russia
[2]Southern Scientific Center of the Russian Academy of Sciences,
Rostov-on-Don, Russia

ABSTRACT

Unalloyed ceramics samples of pure and modified by 1, 2 and 3 wt.% of Li$_2$CO$_3$ lead iron niobate PbFe$_{0.5}$Nb$_{0.5}$O$_3$ (PFN) and PFNL1, PFNL2, PFNL3 are obtained by convensional ceramics technology. An influence of the concentration of various introducing modificators on sintering kinetics, grain and crystal chemical structure and dielectric characteristics was investigated.

Keywords: ceramics, lead iron niobate, multiferroics

1. INTRODUCTION

Lead iron niobate, PbFe$_{0.5}$Nb$_{0.5}$O$_3$ (PFN), combining ferroelectric and antiferromagnetic properties below (140 – 170) K is a representative of functional materials group that demonstrates the interaction of the magnetic and electrical subsystems. We have shown [1] that the introduction of Li$_2$CO$_3$ in amount of 1% (over stoichiometry) on the stage of synthesis led to the stabilization of dielectric properties and increase of piezoelectric characteristics. The aim of this chapter is to establish regularities of influence of Li$_2$CO$_3$ on ceramic and dielectric characteristics of lead iron niobate in wider range of concentrations (1–3 mass. %).

---

* Corresponding Author address, E-mail: huckwrench@gmail.com.

2. TECHNIQUES OF SAMPLES OBTAINING AND INVESTIGATING

The objects of investigation were pure PFN and PFN with stoichiometric Li_2CO_3 additives in an amount of 1 wt. % (PFNL1), 2 wt. % (PFNL2) and 3 wt. % (PFNL3). The samples were synthesized by method of solid phase reactions from oxides PbO, Fe_2O_3, Nb_2O_5 Li_2CO_3 of high purity grade by the two stage calcination with intermediate milling at temperatures $T_1 = T_2 = 1123$ K and holding times $\tau_1 = \tau_2 = 4$ h. Ceramics was sintered at $T_{si} = (1333 - 1353)$ K for 2.5 h.

X-ray research was performed by the powder X-ray diffraction using a diffractometer DRON-3 ($CoK\alpha$ radiation, Bragg - Brentano focusing). The calculation of perovskite unit cell parameters (linear - a, angular - α, volume - V, homogeneous deformation parameter - δ) carried out by standard procedures [2]. Complex permittivity $\varepsilon^* = \varepsilon' - i\varepsilon''$ (ε' and ε'' are the real and imaginary parts of ε^*, respectively) in the temperature [(300–700) K] and frequency [(10–10^5)Hz] ranges was measured using a measuring bench based LCR-meter HIOKI 3522-50 and Warta thermostat TP 703.

Dielectric hysteresis loops in the temperature range (300 – 450) K at frequency 50 Hz were obtained by the oscillographic Sawyer – Tower method.

3. EXPERIMENTAL RESULTS AND DISCUSSION

The introduction of Li_2CO_3 has led to decrease of T_{si} from 1373 K (PFN) to 1353 K (PFNL1) and 1333 K (PFNL2 and (PFNL3).

The X-ray diffraction analysis has shown that obtained samples are unalloyed (excepting PFNL3 samples that have have marks of PbO on radiograph) and monophase at room temperature. All samples have rhombohedral structure. An increase of the modificator concentration had led to increasing of parameters a, V and α (Table 1).

Table 1. X-Ray diffraction results on PFN, PFNL and PFNM

Comp.	Sym.	a, Å	α, ang. deg.	V, Å^3	δ	P_{exp} g/cm^3	ρ_{re}, %
PFN	Rh	4.0141	89.92	64.68	0.00216	8.07	95.0
PFNL1	Rh	4.0149	89.93	64.72	0.00200	7.76	91.0
PFNL2	Rh	4.0167	89.92	64.80	0.002004	7.79	91.3
PFNL3	Rh	4.0166	89.92	64.80	0.002215	7.57	88.7

Fragments of cleavage images of investigated ceramics are shown in Figure 1. Previously we supposed [1] that the introduction of modificators could lead to a change of sintering nature from solid-phase sintering to liquid phase sintering. The strong T_{si} decrease in PFNL2 and PFNL3 and increasing the number of intergranular interlayers with increasing the modificator concentration confirm our suggestion. The source of the liquid phase is Li_2CO_3 [3] with a low temperature of melting.

A strong increase of the average grain size can be explained by the fact that the Li-containing liquid phases are the transport medium promoting a mass transfer and diffusion during recrystallization sintering. The increase of the modificator concentration has led to the decrease of the average grain size that can be explained by the growth rate of the new-grain formation on the synthesis stage. The increase of the Li$_2$CO$_3$ concentration has also led to the increase of the number of intergranular layers and to the appearance of impurity inclusions.

The increase of the linear size of the unit cell (Table 1) can be explained by an incorporation of Li^{1+} ions into the base compound structure and by a possibility of their localization in irregular lattice positions. The Li^{1+} ions usually can incorporate into the B-positions of the structure [4] (instead of Fe^{3+} or Nb^{5+}), but in the perovskite structure they also can incorporate into the A-positions and interstice. The unit-cell volume, according to the logic of ionic radius change (Pb^{2+} – 1.26 Å, Fe^{3+} – 0.67 Å, Nb^{5+} – 0.66 Å, and Li^{1+} – 0.68 Å), should be preserved, however we don't observe it (Table 1). Probably introduced modificators take up both regular and nonregular (including interstices) positions forming interstitial solid solutions, which demonstrate the increase of the crystal lattice volume. It can be suggested that the Li^{1+} cations incorporate into the additional tetrahedral and octahedral positions formed by the shift of crystallographic planes in niobates with the complicated block structure. Aliovalent impurities introducing an additional distortion in the structure can be also attracted by planar defects.

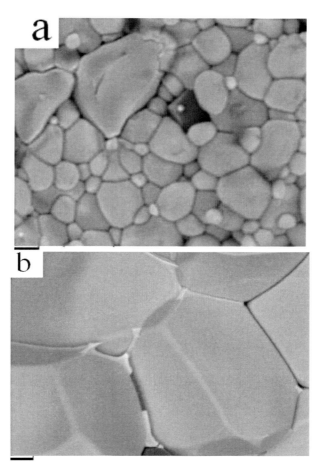

Figure 1. (Continued).

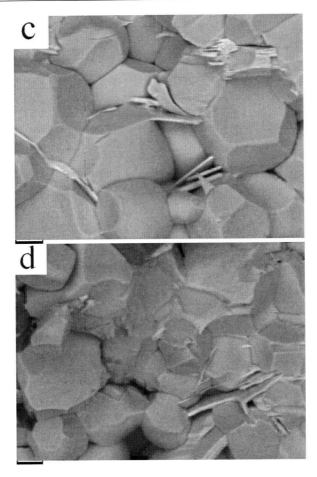

Figure 1. Fragments of cleavage images of the following ceramics: PFN (a), PFNL1 (b), PFNL2 (c), and PFNL3 (d). The marker length equals $3 \cdot 10^{-6}$ m.

Figure 2 shows the results of dielectric hysteresis investigation. Saturated loops being typical of classical ferroelectrics are formed. Modifications have led to a small increase of coercive field from $5 \cdot 10^5$ V/m in PFN to $7 \cdot 10^5$ V/m in PFNL3 (it may be explained by a presence of impurities and intercrystalline layers with the localized Li^{1+} ions that prevents polarization processes) and the remanent polarization from 0.18 C/m^2 in PFN to 0.2 C/m^2 in PFNL1 and PFNL2 (it may be due to the increase of the average grain size which would lead to a domain reorientation). The decrease of the remanent polarization observed in PFNL3 can be explained by strong smearing of the phase transition and its shift to a lower temperature range.

The temperature dependences of the complex permittivity real and imaginary parts in the temperature [(300–600) K] and frequency [(10^2–10^5) Hz] ranges are shown in Figure 3.

Introducing the modificator has led to a decrease of Curie temperature T_C from 371 K (PFN) до 351 K (PFNL3). Temperatures related to max($\varepsilon'/\varepsilon_0$) in PFN, PFNL1, and PFNL2 do not depend on the alternating electric field frequency, but in PFNL3, frequency-dependent maxima of $\varepsilon'/\varepsilon_0$ with a strong dispersion increase near T_C are observed.

Imperceptible in other samples and related to features of the real (defect) structure, additional extrema are formed on curves of $\varepsilon'/\varepsilon_0(T)$ in PFN at $T > T_C$. The strong dispersion of $\varepsilon''/\varepsilon_0(T)$ and the monotonic increase at $T > 400$ K (it can be explained by increasing the conductivity of the studied ceramics) are observed in all the samples.

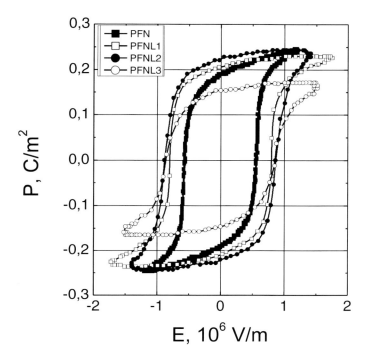

Figure 2. Dielectric hysteresis loops of PFN, PFNL1, PFNL2, PFNL3 ceramics.

An analysis of width values $\varepsilon'/\varepsilon_0$ spectra ΔT calculated for frequency $f = 10^5$ Hz and taken on the half the height of the peak value have shown that the increase of the introduced modificator concentration has led to the smearing of the phase transition. Introducing the modificator has also led to the increase of the resistivity of the studied ceramics. Maximal value of the specific resistivity ρ is observed in PFNL1 ($2.6 \cdot 10^{10}$ $\Omega \cdot$m). A subsequent increase of the Li_2CO_3 concentration has led to a decrease of the resistivity.

CONCLUSION

The effects observed in our study can be related to the localization of Li^{1+} ions in regular and non-regular lattice positions which have led to interstitial solid-solution forming. It can be confirmed by the increase of the unit-cell volume, decrease of T_{si} decrement, and blurring of the phase transition increment in modified samples.

We suggest that the optimal Li_2CO_3 concentration is 1–2 wt. %. Such samples have high values of the remanent polarization, resistivity and relatively weak smearing of the phase transition. A further increase of the modificator concentration has led to the increase of the conductivity and to the appearance of impurity phases.

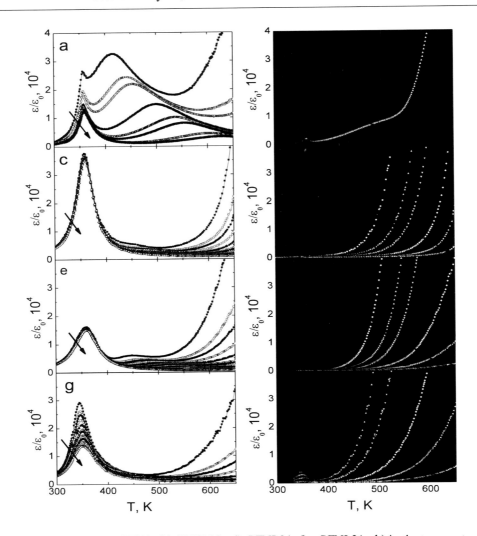

Figure 3. Dielectric spectra PFN(a, b), PFNL1(c, d), PFNL2(e,f) и PFNL3(g, h) in the temperature range (300 – 650) K and frequency range (100 – 105) Hz. The arrow indicates the direction of f growth.

Acknowledgments

The work was supported by Ministry of Education and Science of the Russian Federation (Federal program: contract No. 14.575.21.0007; contract No. 3.1246.2014/K, pr. No. 1927, theme No. 213.01-2014/012-IG).

References

[1] Pavlenko, A.V.; Boldyrev, N.A.; Reznichenko, L.A.; Verbenko, I.A.; Konstantinov, G.M.; Shilkina, L.A. *Inorganic Materials*, 2014, *vol. 50(7)*, 750-756.

[2] Fesenko, E.G. *The Family of Perovskite and Ferroelectricity*. Atomizdat, Moscow, 1972, pp. 1 – 248.

[3] Okadzaki, K. *Ceramic Dielectrics Technology*. Energia, Moscow, 1976, pp. 1 – 336.

[4] Bochenek, D.; Surowiak, Z. *Phys. Status Solidi A*, 2009, *vol. 206(12)*, 2857–2865.

In: Proceedings of the 2015 International Conference … ISBN: 978-1-63484-577-9
Editors: Ivan A. Parinov, Shun-Hsyung Chang et al. © 2016 Nova Science Publishers, Inc.

Chapter 35

DIELECTRIC CHARACTERISTICS OF $(BA_{0.5}SR_{0.5})NB_2O_6$ CERAMICS AT TEMPERATURES FROM $-250°C$ TO $180°C$

A. G. Abubakarov[1,], A. V. Pavlenko[1,2], L. A. Reznichenko[1] and A. S. Nazarenko[2]*

[1]Research Institute of Physics, Southern Federal University,
Rostov-on-Don, Russia
[2]Southern Scientific Center of the Russian Academy of Sciences, Russia

ABSTRACT

Solid solutions of niobates of barium-strontium $[(Ba_{1-x}Sr_x)Nb_2O_6$, BSN] fall into to a class of ferroelectric connections with structure of tetragonal tungsten bronze [1 – 3]. In this binary mixture solid solutions with $x = 0.5$, structure holds almost central position of solubility of extreme components that, along with its localization on border of morfotropic transition from rhombic to a tetragonal phase, defines high processibility of the object. The present chapter concerns with studying grain structure and regularities of formation of dielectric characteristics of $(Ba_{0.50}Sr_{0.50})Nb_2O_6$ ceramics in a wide interval of external influences.

Keywords: niobates of barium-strontium, dielectric characteristics, grain structure

1. OBJECTS AND METHODS TO STUDY

Object of research was the ceramics of a solid solution of $(Ba_{0.50}Sr_{0.50})Nb_2O_6$. Regulations of synthesis and baking of ceramics are submitted century. Research of a zyorenny structure of objects was conducted by means of the color laser scanning 3D microscope of KEYENCE VK-9700. Temperature dependencIes of the relative permittivity,

---

* E-mail: abubakarov12@mail.ru.

$\varepsilon/\varepsilon_0$ (ε_0 is the permittivity of vacuum) and loss teangent tg δ in a temperature range (-250 –180)°C at frequencies from a range (10–10^5) Hz were studied by means of the measuring set-up by using the LCR meter of HIOKI 3522-50 and the temperature regulator Warta TP 703. Loops of a dielectric hysteresis in a temperature range (25–130)°C at 50 Hz were obtained by means of the oscillographic Sawyer – Tower method. It enabled us to estimate a remanent polarization and coercive field of the studied samples.

2. EXPERIMENTAL RESULTS AND DISCUSSION

In work [4], it is shown that the obtained ceramic samples are single-phase, neat and have tetragonal structure at room temperature. Their grain structure is uniform, almost nonporous, and grains have sizes (5–10) μm (Figure 1).

Figure 1. BSN-50 ceramics chip micrograph.

Loops of the dielectric hysteresis have the extended form (Figure 2), the reference on ferroelectric-relaxor properties is peculiar to this composition. To reach "saturation" of $P(E)$ at $T = 25$°C, the electric field $E = 3.2 \cdot 10^6$ V/m is applied, and in this sample, values of $P_R = 0.024$ C/m^2 and $E_C = 7.2 \cdot 10^5$ V/m are achieved. When heating the sample, the loop of the dielectric hysteresis becomes more narrow, it is less on amplitude and disappears at $T \sim 117$°C that testifies to a phase transition from a polar state to a non-polar state.

Dependencies of $\varepsilon/\varepsilon_0(T)$ and tg$\delta(T)$ of BSN-50 ceramics in the considered temperature and frequency range are shown in Figures 3–5. With increasing T from –255 to –175°C, $\varepsilon/\varepsilon_0$ and tg δ almost do not change and have values ~ 420 and 0.015, respectively. A further temperature increase is accompanied by $\varepsilon/\varepsilon_0$ growth (inclination change is observed on the curve $(\varepsilon/\varepsilon_0)^{-1}(T)$, see Figure 2) and formation of diffused maximum of the tg$\delta(T)$ which relates to change of incommensurate material structure according to work [5]. At room temperature and $f = 10^5$ Hz, this material is characterized by $\varepsilon/\varepsilon_0 \sim 800$ and tg $\delta \sim 0.05$. With a rise in temperature there is a growth of $\varepsilon/\varepsilon_0$ at $T > 30$°C and formation of "hump" (obvious when f is low) at $T \sim 82$°C. Its position does not depend on f. Then within the range of maximum values of $T = 80 - 110°C$, these values of T_m shift to the higher temperature range since the increase of f corresponds to the FE → paraelectric (PE) phase transition (P4bm → P4/mbm) [4].

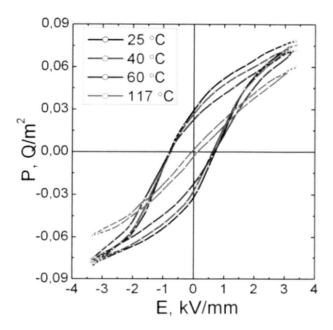

Figure 2. $P(E)$ dependencies of BSN-50 ceramics.

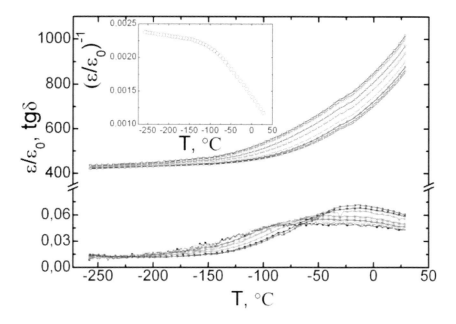

Figure 3. Temperature dependencies of $\varepsilon/\varepsilon_0$ and tg δ of BSN-50 ceramics at $T = (-260–30)°C$ and $f = (10^2–10^5)$ Hz (the direction of increasing f is shown by arrows).

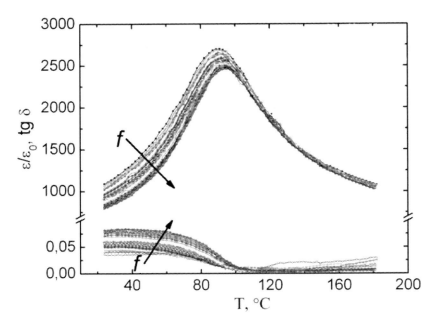

Figure 4. Temperature dependencies of $\varepsilon/\varepsilon_0$ and tg δ of BSN-50 ceramics at $T = (30–180)°C$ and $f = (10–10^5)$ Hz (the direction of increasing f is shown by arrows).

On the curves of tg$\delta(T)$ as the temperature rises, we can see a decrease of tg δ and formation of minimum values at $T = 85 – 110°C$. At $T > 110°C$, there is a decrease of $\varepsilon/\varepsilon_0$ growing with increase of dispersion temperature and exists a continuous increasing in the case of tg δ corresponding to the growth of electrical conductivity of the material at these temperatures.

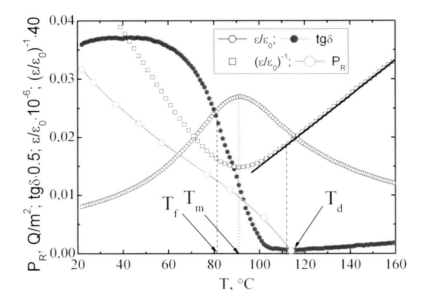

Figure 5. Temperature dependencies of $\varepsilon/\varepsilon_0$, tg δ, $(\varepsilon/\varepsilon_0)^{-1}$, and P_R of BSN-50 ceramics at $T = (30–180)°C$.

At $T_m < T < T_d$ (T_d is the Burns temperature, at lower values of which the polar nanodomains appear) the $\varepsilon/\varepsilon_0(T)$ dependence can be well approximated (Figure 6) by formula:

$$1/\varepsilon - 1/\varepsilon_m = (T - T_m)^\gamma / C_1, \qquad (1)$$

where γ is the parameter of difussion of phase transition, ε_m and T_m is the high level and temperature of $\varepsilon/\varepsilon_0(T)$ maximum, respectively. During approximation of $T_m(f)$ dependence, the best results were achieved when the Vogel-Fulcher equation was used (Figure 6):

$$f = f_0 \exp(E_{act}/(k(T_m - T_f))), \qquad (2)$$

where f_0 is the frequency of attempts to overcome a potential barrier E_{act}, k is the Boltzmann constant, T_f is the Vogel-Fulcher temperature.

The specified high value of γ is ~ 1.99; Curie–Weiss law is applicable only at $T \geq T_d$ (see Figure 5) and convergence $P_r \to 0$ is only at $T \sim T_d$, that indicates high level of diffusion of ferroelectric-paraelectric phase transition. The calculated values of E_{act} and f_0 (0.015 eV and 10^{11} Hz, respectively) are close to the ones proper for relaxor ferroelectrics.

The defined value of T_f = 355 K allowes one to relate the detected anomalies on the $\varepsilon/\varepsilon_0(T)$ and tg $\delta(T)$ curves in this temperature range to the transition of BSN-50 ceramics from relaxor to macrodomain state. The obtained results point that a diffusion of the "paraelectric – ferroelectric" phase transition takes place in BSN-50 ceramics in the condition of relaxor state in the temperature range T = 80 – 110°C. The phase transition diffusion in $Ba_{1-x}Sr_xNb_2O_6$ solid solutions is caused by fluctuation of chemical composition due to misalignment of Ba и Sr atoms in the structure. The features determining material properties change from one microsection to another that leads to enlargement of phase transition area.

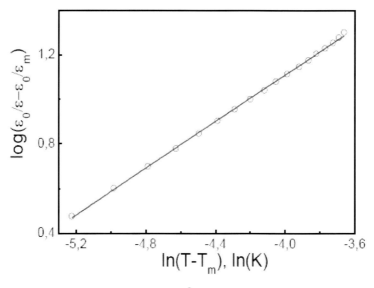

a

Figure 6. (Continued).

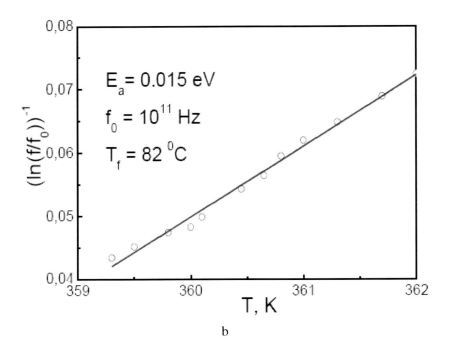

Figure 6. (a) Dependence of $\ln(\varepsilon_0/\varepsilon - \varepsilon_0/\varepsilon_m)$ on $\ln(T)$ for BSN ceramics; the continuous line is an approximation on Eq. (1), and (b) dependence of $(\ln(f/f_0))^{-1}(T_m)$. Straight line is a result of calculation according to the Vogel-Fulcher equation.

ACKNOWLEDGMENTS

The work was performed with financial support from the Ministry of Education and Science of the Russian Federation (basic and project parts of the state task of the topic No. 213.01-11/2014-21, 213.01-2014/012-ВГ, the task 3.1246.2014/К), Federal Special Purpose Program (SC No.14.575.21.0007) and SD-1689.2015.3.

REFERENCES

[1] Kuzminov, Yu. S. *Ferroelectric Crystals for Control of Laser Radiation*, Nauka, Moscow, 1982, pp. 1 – 400 (in Russian).
[2] Layns, M.; Glass, A. *Ferroelectric and Congenerous to them Materials*, Mir, Moscow, 1981. pp. 1 – 736 (in Russian).
[3] Jamicson, P. B.; et al. *J. Chem. Phys.* 1968, *vol. 48(11)*, 5048-5057.
[4] Abubakarov, A. G.; Verbenko, I. A.; Pavlenko, A. V.; et al. *News of the Russian Academy of Sciences. Physical Series.* 2014, *vol. 78(8)*, 943-947.
[5] Huang, T. G.; Xu, Z.; Viehland, D.; Neurgaonkar. R. R. *J. Appl. Phys.* 1995, *vol. 77(4)*, 1677.

In: Proceedings of the 2015 International Conference …   ISBN: 978-1-63484-577-9
Editors: Ivan A. Parinov, Shun-Hsyung Chang et al.   © 2016 Nova Science Publishers, Inc.

Chapter 36

STRUCTURE, MICROSTRUCTURE AND COMPOSITION OF $Bi_{1-x}Pr_xFeO_3$ CERAMICS

S. V. Titov, I. A. Verbenko, L. A. Reznichenko, L. A. Shilkina, V. A. Aleshin, V. V. Titov, S. I. Shevtsova and A. P. Kovtun*

Research Institute of Physics, Rostov State University,
Rostov-on-Don, Russia

ABSTRACT

Ceramics with the nominal composition of $Bi_{1-x}Pr_xFeO_3$ at $0.00 \leq x \leq 0.20$ were obtained by solid-phase synthesis. We studied their chemical composition and crystalline structure. Multifractal parameters of the grain structure were determined. Sequence of changes in the ceramics' structure was studied as a function of Pr content. New possibilities to create ceramic ferromagnetic materials based on $BiFeO_3$ doped by rare-earth ions were discussed.

Keywords: bismuth ferrite, rare-earth ions, solid solution, ceramics, multiferroics, magneto-electric materials, X-ray diffraction, microstructure, multifractal analysis

1. INTRODUCTION

Bismuth ferrite $BiFeO_3$ (BFO) is a multiferroic with high ferroelectric Curie temperature (1123 K) and antiferromagnetic Neel temperature (643 K) [1, 2]. It is a promising base for magneto-electric materials with high efficiency. Multiferroics are free from different drawbacks known for materials, used currently in magnetoelectric converters. In particular, such drawbacks are: eddy currents and resistive losses, heating and burnout of devices, difficulties of manufacturing. The substitution of Bi-ions by rare-earth ions leads to

---

* E-mail: svtitov@sfedu.ru.

significant improvements in physical properties of $BiFeO_3$-based solid solutions (SSs). Wherein, the increasing of rare-earth ions concentration leads to the sequence of phase changes. Better electromecanical properties are observed in the vicinity of phase boundaries of studied SSs [3, 4]. The goal of the current study is to reveal the mechanism of structural reorganization in BFO ceramics at different scales, caused by increasing of rare-earth ions content in solid solution.

2. EXPERIMENT

$Bi_{1-x}Pr_xFeO_3$ SSs at $0.00 < x \leq 0.20$ (with $\Delta x = 0.010$ for $0.09 \leq x \leq 0.15$ and $\Delta x = 0.025$ for the rest of the interval) were obtained by conventional ceramic technology using the solid-phase synthesis with subsequent sintering. The phase composition, perovskite unit-cell parameters, microstrain $\Delta d/d$, and density of ceramics have been determined by X-ray powder diffraction studies of solid solutions with the help of DRON-3 diffractometer using Bragg-Bretano geometry and Cobalt K-alpha radiation. The study of the elemental composition of objects was performed on the scanning electron microscope "Camebax-micro" with WDS analytical system. Microstructure of thermal etched samples was studied by using optical microscope with magnification from ×200 to ×1050 using regular and polarized light. We prepared the fine ceramics grain's pattern images for multifractal analysis using digital treatment.

Images of ceramics grain's pattern were analyzed by methods, described in [5]. Multifractal parameters $f(\alpha(q))$ - spectra, Dq - spectra of dimensions (Renyi's entropy spectra), f_∞ - homogeneity parameter of measurement medium and Δ_∞ - parameter of order in the range of q (exponent) from -40 up to $+40$ have been calculated.

3. RESULTS AND DISCUSSION

We revealed by using X-ray analysis that impurity-free samples were obtained only for praseodymium concentrations $x \geq 0.12$. All the remaining samples contained intermediate compositions, $Bi_{25}FeO_{40}$ and $Bi_2Fe_4O_9$. Generally, when we used other rare-earth elements (La, Nd, Sm, Eu, Gd) for doping $BiFeO_3$, the quantities of $Bi_{25}FeO_{40}$ and $Bi_2Fe_4O_9$ compositions decrease gradually while the concentration of rare-earth element increased. In the $Bi_{1-x}Pr_xFeO_3$ ceramics, the amount of impurity phases decreased in a non-monotonic way. The lowest quantity of impurities were detected in solid solutions with $x = 0.075$ and $x = 0.09$.

The impurities in the pictures visually differ in colors and forms of grains (Figure 1).

An analysis of the diffraction reflections has shown that with increasing of x, the definite consequence of phases observed (Figure 2). For $0 \leq x < 0.05$, only rhombohedral (R) phase (R3c group) exists. For $0.05 < x \leq 0.075$, the R-phase and two orthorhombic phases, namely, O_2 of the $PbZrO_3$ type (Pnam group) and O_1 of the $GdFeO_3$ type (Pnma group), coexist.

It is worth noting that the concentration of both the orthorombic phases does not exceed 10%. For compositions with $0.075 < x \leq 0.09$, the pure R-phase appears again. For $0.09 < x \leq 0.20$, three phases R, O_2, and O_1 co-exist, and in this concentration range, the content of the R-phase is the highest up to $x = 0.20$.

Structure, Microstructure and Composition of $Bi_{1-x}Pr_xFeO_3$ Ceramics

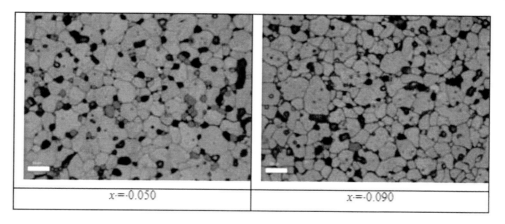

Figure 1. Grain's pattern image of ceramics. Grey grains of rectangular-like shape are $Bi_{25}FeO_{40}$ and $Bi_2Fe_4O_9$ impurities.

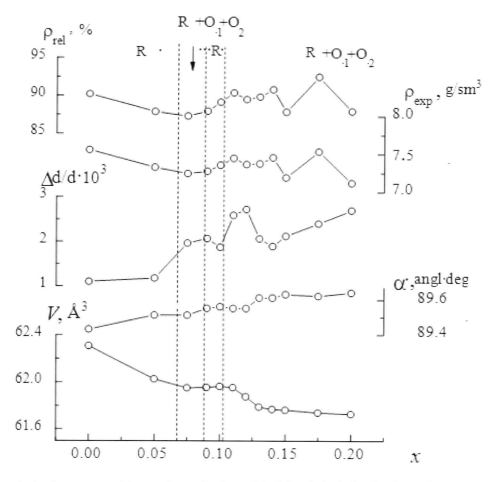

Figure 2. Angle parameter (α), experimental volume (V) of rhombohedral cell, microdeformation ($\Delta d/d$), experimental density, (ρ_{exp}), and relative density, (ρ_{rel}), of $Bi_{1-x}Pr_xFeO_3$ solid solution ceramics are shown depending on x.

Both the concentration and ratio of the O_2 and O_1 phases in this range are unstable and change from minimal at $x = 0.13$ up to the maximal at $x = 0.20$.

The appearance of the R-phase in the multiphase concentration range correlates with data, described in works [6, 7]. In Refs. 6 and 7, the dependencies of the sequences of structural phase transitions as a function of both temperature and rare-earth ions concentration in $Bi_{1-x}La_xFeO_3$ [6] and $Bi_{1-x}Pr_xFeO_3$ [7] SSs were studied. Phase transitions $R+O_2 \rightarrow R \rightarrow O_1$ were observed for the compositions with $x = 0.16$ and $x = 0.185$ with increasing temperature [6]. For $x = 0.16$, the R-phase exists in the temperature range 350–530°C, while for $x = 0.185$, it exists in the temperature range 450–500°C. The decreasing of the R-phase quantity with temperature in two-phase (R+O_2) SS with $x = 0.125$ was revealed [7], while in a narrow temperature range (380–410°C), the three-phase state $R+O_2+O_1$ was observed in this composition.

Previously we assumed the same origin of concentrational and temperature phase transitions in SSs of $BiFeO_3$ with lanthanites caused by real (defect) structure of objects [8]. The sequence of concentration phase transitions, which appear similar to the sequence of temperature phase transitions [6] revealed in the current study, confirms the validity of our assumption.

Electron probe microanalysis showed that ceramics with $0 < x \leq 0.050$ are SSs with Pr contents close to the nominal. Coexistense of grains with Pr content from 4 to 7% was observed in ceramics with $x = 0.075$ (Figure 3, Table 1). At $x = 0.090$, praseodymium is evenly distributed in ceramics' grains, concentration value is 6–7%. At $x = 0.10$, we observed coexistence of grains again with the Pr content from 5% to 9% and small regions of $Bi_{0.70}Pr_{0.30}FeO_3$. Further, for $0.010 < x \leq 0.20$, all is the same. The interval of coexistence of different Pr concentrations in SS is shifting to higher concentrations. At $x = 0.20$, the areas of $Bi_{0.5}Pr_{0.5}FeO_3$ appear.

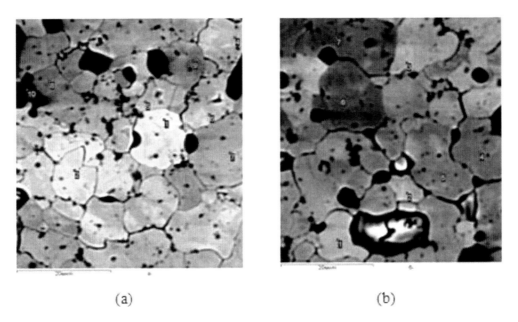

(a)   (b)

Figure 3. SEM-picture of $Bi_{0.935}Pr_{0.075}FeO_3$ ceramics in reflected electrons. "Dots" are fields of analysis (see Table 1).

Structure, Microstructure and Composition of $Bi_{1-x}Pr_xFeO_3$ Ceramics

Table 1. Compositions of analized areas

Figure 3, area	%, ат.				Normalized by (Bi + Pr) = 1
	Bi	Pr	Fe	O	
a,7	19.4	0.8	18.7	61.1	$(Bi_{0.96}Pr_{0.04})Fe_{0.94}O_{3.1}$
a,3	18.7	0.9	18.5	62.0	
b,2	18.7	0.9	18.4	62.0	
a,4	18.9	0.9	19.1	61.1	
b,7	18.6	0.9	18.8	61.7	
a,5	19.0	1.0	18.4	61.6	
b,5	18.9	1.0	18.4	61.8	
a,6	18.9	1.0	18.8	61.3	
b,3	18.3	1.0	18.2	62.5	
a,8	19.6	1.0	19.2	60.2	
b,6	19.0	1.0	19.4	60.6	
a,1	18.2	1.1	18.3	62.4	
b,4	18.8	1.1	19.3	60.8	
a,9	18.6	1.4	19.1	61.0	$(Bi_{0.93}Pr_{0.07})Fe_{0.94}O_{3.1}$
a,2	18.9	1.4	18.8	60.9	
b,1	18.1	1.4	18.5	62.0	

Thus, in the nominal composition area of $Bi_{1-x}Pr_xFeO_3$ ($0.05 < x \le 0.20$), there is no continuous formation of a substitutional SS.

Considering the structural data, given above, we can assume that doping bismuth ferrite by Pr leads to formation of solid solutions according the schema showed in Figure 4. By our opinion, the formation of interstitial solid solutions SS_1 occurs in compositions with $0 < x \le 0.050$ and R-symmetry of BFO cell is not changes. Starting at $x = 0.075$, the forming of stable solid solution or composition of approximate $Bi_{0.93}Pr_{0.07}FeO_3$ (SS_2) with R-symmetry occurs. Then, this output and SS1 produce the solid solutions SS3 with O-symmetry. The next stable point is compostiton of $Bi_{0.70}Pr_{0.30}FeO_3$ (SS_4). This composition produces with previous ones the solid solutions SS_5 with O-symmetry. Next special point of $Bi_{0.5}Pr_{0.5}FeO_3$ (SS_6) produces with previous ones the solid solutions SS_7 with O-symmetry.

With the increase of the Pr content, changes of microstrain and the unit cell volume as well as the reorganization of the structure at the mesoscopic and macroscopic levels take place. Behavior of multifractal parameters is shown in Figure 5.

In the single-phase areas, we observe the decreasing of f_∞ parameter, caused by high mechanical stresses at microstructure level of ceramics. The decomposition of the system into three phases leads to the relaxation of mechanical stresses. Both f_∞ and Δ_∞ parameters change periodically as x grows further.

Such behavior is due to the changes in the value of ratio of rhombohedral and orthorhombic phases in ceramics. The $\Delta d/d(x)$ dependence for the R-phase has two maxima at $x = 0.09$ and $x = 0.12$, with the small intermediate minimum.

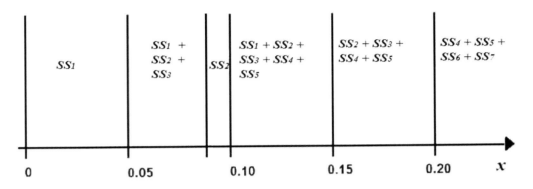

Figure 4. Schematic of forming SSs in $Bi_{1-x}Pr_xFeO_3$ ceramics.

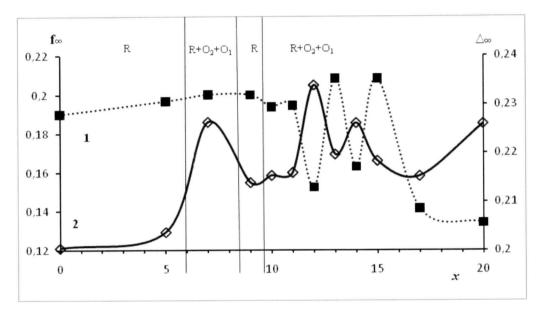

Figure 5. Compositional dependence of multifractal parameters Δ_∞ (1) and f_∞ (2) for $Bi_{1-x}Pr_xFeO_3$ ceramics.

Weak correlation between $\Delta d/d$ and f_∞ parameters is due to the multi-phase character of the system and the presence of $Bi_{25}FeO_{40}$ and $Bi_2Fe_4O_9$ impurities. Probably, the same reason can explain small influence of Δ_∞ parameter on phase changes for the compositions with $x \geq 0.12$. Behavior of multifractal parameters and the $\Delta d/d$ dependence for $x > 0.11$ shows that increase of the quantity of the R-phase in SSs leads rather to the formation of stresses in grains' structure than to the increase of the R-phase cell microstrain. The behavior of the multifractal parameters' shows that optimal state of samples' microstructure is observed in composition ranges $0.07 \leq x \leq 0.08$, $0.12 \leq x \leq 0.13$, and $0.14 \leq x \leq 0.15$.

There is a large probability that phases coexist as micro-regions within one grain in compositions with $0.07 \leq x \leq 0.08$ and $0.12 \leq x \leq 0.13$. The probability of its spatial separation becomes higher at $0.14 \leq x \leq 0.15$, thus, the latter composition range is less promising to obtaining the maximal dielectric response.

CONCLUSION

SS ceramics compositions of $Bi_{1-x}Pr_xFeO_3$ with $0.05 < x < 0.20$, are usually complex three-phase systems. We have revealed two compositional areas of $Bi_{1-x}Pr_xFeO_3$ SSs with $0.07 \leq x \leq 0.08$ and $0.12 \leq x \leq 0.13$ which are promising for obtaining maximal dielectric response. Compositions from these ranges are multiphase and contain minimal amount of impurities. The revealed mechanisms of the Pr influence on the structure of bismuth ferrite open the new possibilities to control the fabrication of novel magnetoelectric materials.

ACKNOWLEDGMENT

This study has been performed using equipment of Collective use center "Electromagnetic, Electromechanical, and Thermal Properties of Solids" of Institute of Physics, Southern Federal University and supported by the MES of Russia (Basic and Design State Tasks Nos. 1927 (213.01-11/2014-21), 213.01-2014/012-SH, and 3.1246.2014/K) and the Federal Program (GC № 14.575.21.0007).

REFERENCES

[1]   Wong, I.; Neaton, I. B.; Zheng, H. *Science*, 1965, vol. 299, 52-69.
[2]   Zvezdin, A. K.; Pyatakov, A. P. *Sov. Phys. Usp.,* 2004, vol. 47, 416–421.
[3]   Rusakov, D. A.; Abakumov, A. M.; Yamaura, K.; Belik, A. A.; VanTendeloo, G.; Takayama-Muromachi, E. *Chem. Mater.,* 2011, vol.23, 285-292.
[4]   Troyanchuk, I. O.; Karpinsky, D. V.; Bushinsky, M. V.; Mantytskaya, O. S.; Tereshko, N. V.; Shut, V. N. *J. Am. Ceram. Soc.,* 2011, vol. 94, 4502-4506.
[5]   Vstovsky, G. V.; Kolmakov, A. G.; Bunin, I. Zh. Introduction in Multi-Fractal Parametrization of Material Structures, NITs "Regul. Khaot. Din.," Izhevsk, 2001 (in Russian).
[6]   Karpinsky, D. V.; Troyanchuk, I. O.; Mantytskaja, O. S.; Chobot, G. M.; Sikolenko, V. V.; Efimov, V.; Tovar, M. *Physics of the Solid State,* 2014, vol. 56(4). 701-706.
[7]   Karpinsky, D. V.; Troyanchuk, I. O.; Sikolenko, V. V.; Efimov, V.; Efimova, E.; Silibin, M. V.; Chobot, G. M.; Willinger, E. *Physics of the Solid State,* 2014, vol. 56(11). 2263-2268.
[8]   Reznichenko, L. A.; Shilkina, L. A.; Razumovskaya, O. N.; Yaroslavtseva, E. A.; Dudkina, S. I.; Demchenko, O. A.; Yurasov, Yu. I.; Esis, A. A.; Andryushina, I. N. *Physics of the Solid State,* 2008, vol. 50(8), 1527-1533.

In: Proceedings of the 2015 International Conference …   ISBN: 978-1-63484-577-9
Editors: Ivan A. Parinov, Shun-Hsyung Chang et al.   © 2016 Nova Science Publishers, Inc.

Chapter 37

DIELECTRIC CHARACTERISTICS OF CERAMIC SOLID SOLUTIONS OF $(1-x)BiFeO_3 - xPb(Fe_{0.5}Nb_{0.5})O_3$ $(x = 0.5-1.0$ AND $\Delta x = 0.05)$

A. V. Pavlenko[1,2], L. A. Reznichenko[2], V. S. Stashenko[2], A. V. Markov[3] and N. A. Boldyrev[2,]*

[1]Southern Scientific Center of the Russian Academy of Sciences,
Rostov-on-Don, Russia
[2]Research Institute of Physics, Southern Federal University,
Rostov-on-Don, Russia
[3]Department of Physics, Southern Federal University,
Rostov-on-Don, Russia

ABSTRACT

Dielectric and magnetic characteristics of ceramic samples of $(1 - x)BiFeO_3 - xPb(Fe_{0.5}Nb_{0.5})O_3$ solid solutions $(x = 0.5 - 1.0, \Delta x = 0.05)$ were studied. The system phase diagram based on obtained results was found.

Keywords: ceramics, bismuth ferrite, lead iron niobate, multiferroics

1. INTRODUCTION

Due to the fact that the effects of characterizing the interaction between electric and magnetic subsystems in high temperature multiferroics $BiFeO_3$, $BiMnO_3$, $PbFe_{0.5}Nb_{0.5}O_3$ (PFN) appear weakly great attention in material science is paid to the preparation and investigating properties of solid solutions based on these materials. One of them is solid solution system $(1 - x)BiFeO_3 - xPbFe_{0.5}Nb_{0.5}O_3$.

---

[*] E-mail: tolik_260686@mail.ru.

According to work [1], compositions with $0 \leq x \leq 0.25$ have rhombohedral structure, compositions with $0.40 \leq x \leq 1.00$ have pseudocubic, and above phases coexist in the range $0.25 < x < 0.40$. We have shown in [2] that the "actual" phase diagram of the system is more complicated. Solid solutions from concentration range $x = 0.50 - 1.00$ are of great interest because of its small scrutiny and possibility of practical apply.

The aim of this work was establishing appropriateness of the dielectric characteristics formation of the investigated solid solution in the concentration range $x = 0.50 - 1.00$.

2. OBJECTS, TECHNIQUES OF SAMPLES OBTAINING AND STUDY

The objects of our study were solid solutions of the binary system of $(1 - x)BiFeO_3 - xPb(Fe_{0.5}Nb_{0.5})O_3$ with molar concentrations in the range of $x = 0.50 - 1.00$ The samples were prepared by solid phase synthesis followed by sintering by conventional ceramic technology.

A study of the actual nature of the change $(\varepsilon'/\varepsilon_0)$ and imaginary $(\varepsilon''/\varepsilon_0)$ relative parts of the complex permittivity $\varepsilon^*/\varepsilon_0 = \varepsilon'/\varepsilon_0 - i\varepsilon''/\varepsilon_0$ (ε_0 is the dielectric constant of free space) at $T = (300 - 700)$ K in the frequency range $f = (2 \cdot 10^2 - 10^6)$ Hz was carried out using a measuring bench based on the LCR-meter Agilent E4980A thermostat and Warta TP 703.

Mössbauer spectra were investigated at the Institute of Physics by means of the MS1104Em spectrometer (built at the Southern Federal University, Russia) with a source of gamma radiation ^{57}Co in the Cr matrix.

A model transcript of spectra was carried out using the program UnivemMS. The area of the magnetic phase transition is determined by analyzing the temperature dependence of the line intensities of the Mössbauer spectra.

3. EXPERIMENTAL RESULTS AND DISCUSSION

It is shown [2] that the studied solid solutions in all concentration range (excepting $x = 0.90$ and $x = 0.95$) are unalloyed and have perovskite structure. Figure 1 shows the $\varepsilon'/\varepsilon_0(T)$ and $\varepsilon''/\varepsilon_0(T)$ dependencies at $T = (300 - 700)$ K and $f = (2 \cdot 10^2 - 10^6)$ Hz, and these dependencies were obtained in the cooling mode.

In the neighbourhood of a phase transition from a ferroelectric to paraelectric phase at $T > 300$ K, ferroelectric-relaxor behavior with forming blurring $\varepsilon'/\varepsilon_0(T)$ maxima decreased and shifted to the high temperature range with increasing frequency. Also the $\varepsilon'/\varepsilon_0(T)$ dispersion is observed at $T < T_m$ (T_m is the temperature that relates to max $\varepsilon'/\varepsilon_0$). In the concentration range $0.50 < x < 0.85$, increasing x have led to smooth decrement of T_m which have minimum ($T_m = 323$ K.) in pseudocubic and rhombohedral phases coexistence region at $x = 0.85 - 0.90$. It is typical of solid solutions from morphotropy area. T_m increases in the concentration range $0.90 < x < 1.00$ and reaches value 371 K in PFN.

Best approximation results of dependencies $T_m(f)$ have been obtained by means of Vogel – Fulcher relations (Figure 2) $f = f_0 \exp(E_{act}/(k(T_m - T_f)))$ (f_0 is the frequency of attempts to overcome the potential barrier E_{act}, k is Boltzmann constant, and T_f is Vogel–Fulcher temperature).

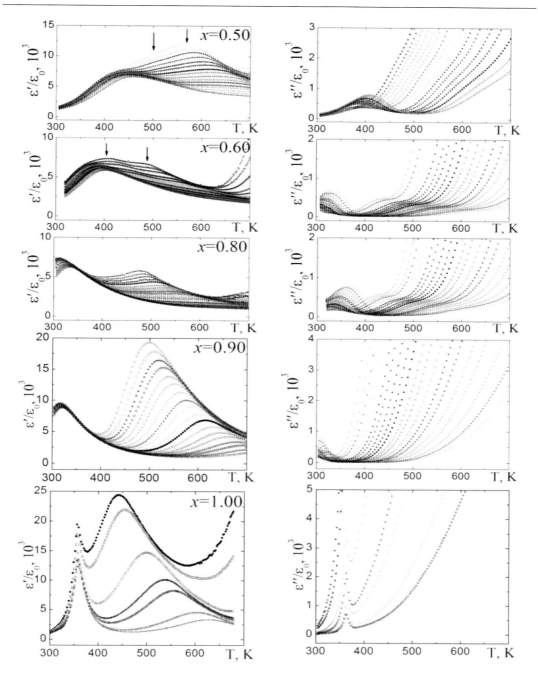

Figure 1. Temperature dependencies of $\varepsilon'/\varepsilon_0$ and $\varepsilon''/\varepsilon_0$ for the solid-solution system of $(1-x)\mathrm{BiFeO_3}-x\mathrm{Pb(Fe_{0.5}Nb_{0.5})O_3}$ in the concentration range $x = 0.50 - 1.00$, temperature range $T = (300 - 700)$ K and frequency range $f = (2 \cdot 10^2 - 10^6)$ Hz.

It is evident that at first dependence $\varepsilon'/\varepsilon_0(T)$ maximum was spreading but then it was narrowing. Also values of $\varepsilon'/\varepsilon_0(T = T_m)$ at first were decreasing but then they are increasing. It may be explained by consecutive changing of phase transition blurring from ferroelectric to paraelectric phase.

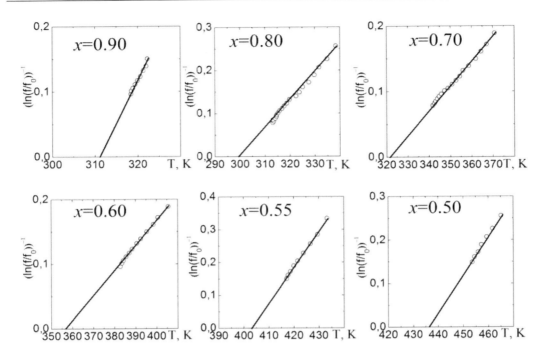

Figure 2. Illustration of fulfilment of Vogel – Fulcher relation for $(1-x)BiFeO_3 - xPb(Fe_{0.5}Nb_{0.5})O_3$ system at $x = (0.50 - 0.90)$.

Ferroelectric – relaxor properties manifestation may be associated with chemical composition volume fluctuations presence in samples.

This fluctuations are due to random redistribution of cations Pb^{2+} and Bi^{3+} to the A-positions and Nb^{5+} and Fe^{3+} to the B-positions of perovskite structure. Increasing the PFN concentration leads to increment of the ions Pb^{2+} part ($R = 1.26$ Å) in A-position instead Bi^{3+} ($R = 1.20$ Å) and ions Nb^{5+} part in B-positions instead Fe^{3+}.

At first crystal-chemical disorder was intensify but then it was decreasing at $x = 0.75 - 0.85$ because of intensification of the PFN role in forming solid solutions properties. At $T > T_m$, additional peaks appeared in curves $\varepsilon'/\varepsilon_0(T)$ and $\varepsilon''/\varepsilon_0(T)$.

Double peak/twin peaks for solid solutions in pseudocubic phase area unresolved by temperature from main peak, exceeding in value (at $x = 0.50 - 0.60$) or approximating (at $x = 0.60 - 0.70$) strong relaxation. For solid solution at $x = 0.80 - 1.00$ single peak stands separate with significant dispersion of $\varepsilon'/\varepsilon_0(f)$ and $\varepsilon''/\varepsilon_0(f)$.

The observed evolution of dielectric spectra in the range $T > T_m$ can be explained by changing the prevalent mechanism (or change of type) giving contribution to resultant dielectric response.

In consideration of difficult phase filling at room temperature in investigated concentration range, it can be suggested that observed in the samples temperature-frequency behavior of the dielectric characteristics may be related with complex sequence of phase transitions and actual (defect) solid solutions structure changing.

It can be associated with variable valence of ions Fe and Nb (presence of certain amount of Fe^{2+} and Nb^{4+} ions).

Dependencies $(\varepsilon'/\varepsilon_0)^{-1}$ $(T, f = 10^5$ Hz) are shown for example in Figure 3. It illustrates that dependencies $(\varepsilon'/\varepsilon_0)^{-1}$ in all investigated solid solutions are linear beginning from value of $T_b > T_m$ (T_b is Burns temperature) (Figure 4).

Figure 5 shows joint dependencies T_m and T_N (T_N is Neel temperature) and based on them phase diagram in concentration range $0.50 \leq x \leq 1.00$ for the investigated solid solutions system.

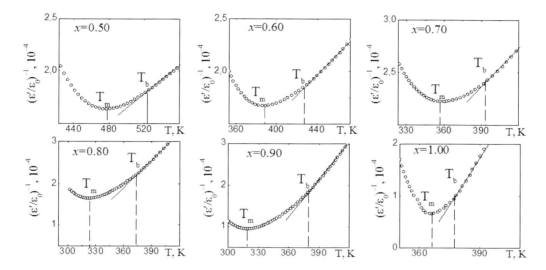

Figure 3. Dependencies $(\varepsilon'/\varepsilon_0)^{-1}(T, f = 10^5$ Hz) for the solid-solution system of $(1-x)$BiFeO$_3$ – xPb(Fe$_{0.5}$Nb$_{0.5}$)O$_3$ at $x = 0.50 – 1.00$ in the temperature range (300 – 580) K. Solid lines are an illustration of validity of Curie – Weiss law.

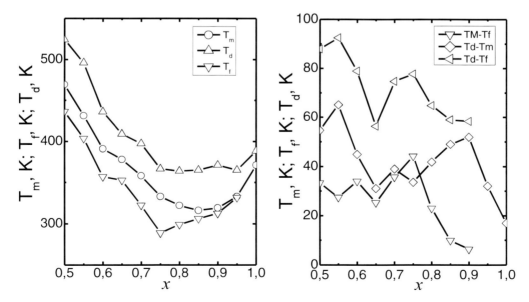

Figure 4. Illustration of validity of Vogel – Fulcher law fulfilment for solid solution system $(1-x)$BiFeO$_3$ – xPb(Fe$_{0.5}$Nb$_{0.5}$)O$_3$ at $x = 0.5 – 1.0$.

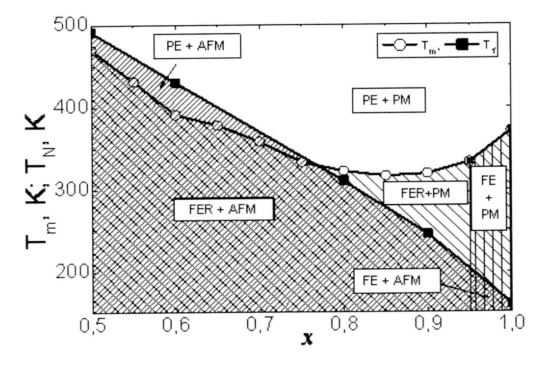

Figure 5. Concentration dependencies of T_m and T_N for the solid-solution system of $(1-x)$ BiFeO$_3$ – xPb(Fe$_{0.5}$Nb$_{0.5}$)O$_3$ in the concentration range $x = 0.5 - 1.0$. Values of T_N at $x = 0.8$ and 0.9 were taken from work [1], values of T_N at $x = 0.5$, 0.6, and 1.0 were obtained from temperature dependencies of the line Mössbauer spectra intensities. PE is paraelectric, FE is ferroelectric, FER is ferroelectric – relaxor, AFM is antiferromagnetic, and PM is paramagnetic.

It is clear that in spectra the investigated solid solutions system can outline at least 5 areas at phase diagram with different phase content.

The solid solutions area at $x = 0.7 - 0.8$ attracts attention where maximum blurring of ferroelectric phase transition takes place on the one hand and maximum coupling the ferroelectric and antiferroelectric phase transitions take place at nearly identical temperatures located near to room temperature.

CONCLUSION

The obtained results can be used for a development of multifunctional materials based on bismuth ferrite and lead iron niobate and applied in different microelectronic devices.

ACKNOWLEDGMENT

The work was supported by Ministry of Education and Science of the Russian Federation (Federal program: contract No. 14.575.21.0007; contract No. 3.1246.2014/K, pr. No. 1927, theme No. 213.01-2014/012-IG).

REFERENCES

[1] Krainik, N. N.; Huchua, N. P.; Berezhnoy, A. A.; Tutov, A. G. *Solid State Physics*, 1965, vol. 7, 132 - 142.

[2] Shilkina, L. A.; Pavlenko, A. V.; Reznichenko, L. A.; Verbenko, I. A. *Proceedings of the Third International Symposium "Physics of Lead-Free Piezoactive and Related Materials. (Analysis of Current State and Prospects of Development), LFPM-2013,"* Rostov-on-Don - Tuapse, 2014, vol. 1, 312-313.

In: Proceedings of the 2015 International Conference … ISBN: 978-1-63484-577-9
Editors: Ivan A. Parinov, Shun-Hsyung Chang et al. © 2016 Nova Science Publishers, Inc.

Chapter 38

THE STRUCTURE, MICROSTRUCTURE, DIELECTRIC SPECTROSCOPY AND THERMAL PROPERTIES OF $BiFeO_3$ MODIFIED WITH DY

S. V. Khasbulatov[1,], A. A. Pavelko[1], L. A. Shilkina[1], L. A. Reznichenko[1], G. G. Hajiyev[2], A. G. Bakmaev[2], M.-R. M. Magomedov[2], Z. M. Omarov[2] and V. A. Aleshin[1]*

[1]Research Institute of Physics, Southern Federal University,
Rostov-on-Don, Russia
[2]H. I. Amirkhanov Institute of Physics of Daghestanian Scientific
Center of the Russian Academy of Sciences, Makhachkala, Russia

ABSTRACT

The chapter presents the results of a comprehensive study of the crystal structure, grain structure, dielectric and thermal properties of high-temperature multiferroics $Bi_{1-x}Dy_xFeO_3$ ($x = 0.05$-0.20). The regularities of formation of the phase and grain structure, electrical and dielectric properties of objects at room temperature were established. In the temperature range 300-970 K detailed X-ray study of $Bi_{0.9}Dy_{0.1}FeO_3$ was conducted, the results obtained were compared with the high-temperature anomalies of dielectric and thermal properties. The assumptions about the nature of the observed phenomena were suggested.

Keywords: multiferroics, rare-earth elements, bismuth ferrite, structure, microstructure, dielectric and thermal properties

---

[*] Corresponding author: S. V. Khasbulatov. Research Institute of Physics, Southern Federal University, 344090 Rostov-on-Don, Russia. E-mail: said_vahaevich@mail.ru.

1. Introduction

Multiferroics represent a broad class of materials that combine the ferroelectric, ferromagnetic and ferroelastic properties and are currently being studied in detail in connection with the potential for their use in new devices, based on mutual monitoring of magnetic and electric fields [1]. Bismuth ferrite $BiFeO_3$ and its solid solutions are convenient objects for the creation of magnetoelectric materials because of high temperatures of electric (Curie temperature, $T_C = 1083$ K) and magnetic (Neel temperature, $T_N \sim 643$ K) orderings [2].

2. Objects and Methods

The research objects were $Bi_{1-x}Dy_xFeO_3$ ceramics (where $x = 0.05$-0.20, $\Delta x = 0.05$). Samples obtained by two-stage solid-phase synthesis followed by sintering by conventional ceramic technology.

X-ray study of the objects held by powder diffraction method using a DRON-3 diffractometer ($CoK\alpha$-radiation, Bragg-Brentano focusing scheme). Perovskite cell parameters were calculated by the standard technique [2]. Change in the phase composition of the samples and the behavior of the structural parameters with temperature controlled by the diffraction reflections $(111)_c$ to $(200)_c$ and $(220)_c$.

Investigation of polycrystalline (grain) structure of the samples performed in reflected light on an optical microscope Neophot 21 and on the inverted high-precision microscope Leica DMI 5000M.

Temperature dependencies of relative dielectric permittivity, $\varepsilon/\varepsilon_0$, at frequencies 25-$2\cdot10^6$ Hz were investigated in the range 300-900 K at the special setup with use of precision LCR meter Agilent E4980A. Thermal conductivity (λ) and thermal diffusivity (χ) measured at the facility LFA-457 "MicroFlash" (NEZSCH), a heat capacity (C_p) defined by means of the differential scanning calorimeter DSC-204 F1 (NEZSCH).

3. Results and Discussion

X-ray diffraction analysis, conducted at room temperature, showed (Table 1) that all investigated solid solutions contain impurity phases $Bi_{25}FeO_{40}$ (cubic symmetry), $Bi_2Fe_4O_9$ (rhombic symmetry), usually associated with formation of $BiFeO_3$ [5] and the phases with a garnet-type structure ($Ln_3Fe_5O_{12}$) (cubic symmetry) [6], the concentration of which varies depending on the concentration of Dy. The relative intensity of the strongest line of impurity phases (I_{imp}/I_{per}) against the Dy concentration increases with increasing x, passing through a minimum at $x = 0.10$.

Table 1 also shows that when $x = 0.10$ objects undergo a transition from the rhombohedral (Rh) phase to morphotropic region (MR) comprising a mixture of Rh and rhombic (R) with the monoclinic (M) subcell phases (the last has $GdFeO_3$-type structure [7, 8]). Figure 1 shows the microstructure pictures of studied objects with $x = 0.10$ and 0.20. With increasing x the main phase average crystallite size increases. Usually it leads to a weakening of the dielectric properties by increasing the internal electromechanical losses.

Table 1. Relative intensity of the strongest line of the impurity phase (I_{imp}/I_{per}), symmetry and density of $Bi_{1-x}Dy_xFeO_3$ solid solutions

x	I_{imp}/I_{per}	Symmetry	P_{exp}, g/cm^3	P_{theor}, g/cm^3	P_{rel}, %
0.05	10 (25-1-40)* 10 (2-4-9)	Rh	7.48	8.36	89.44
0.10	8 (25-1-40) 5 (2-4-9)	Rh + Traces of R	7.64	8.31	91.95
0.15	21 (25-1-40) ~30Dy$_3$Fe$_5$O$_{12}$	Rh + Traces of R	7.41	8.25	89.79
0.20	35 (25-1-40) ~60Dy$_3$Fe$_5$O$_{12}$	Rh + Traces of R	7.10	8.19	86.67

*(25-1-40) corresponds to $Bi_{25}FeO_{40}$; (2-4-9) – $Bi_2Fe_4O_9$.

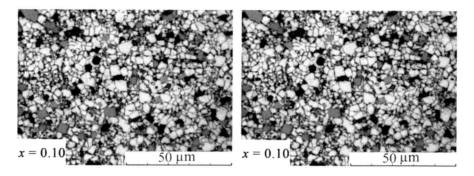

Figure 1. Microstructures of $Bi_{1-x}Dy_xFeO_3$ ceramic samples; concentration of Dy is shown in the lower left corner of the photos.

Results of the dielectric dispersion study at high temperatures (in the heating mode) shown in Figure 2. As can be seen from the figure, the considered dependencies show two types of anomalies. The first type locates in the temperature range 400-500 K and is of the form of one or two, depending on the temperature change direction, strong dispersive relaxing maxima of $\varepsilon/\varepsilon_0$. We shall discuss it below in details. The second type demonstrates the appearance of the broad step forming on mentioned dependencies in the temperature range 600-750 K (alleged antiferromagnetic transition area), with exception of the sample with $x = 0.10$. Composition with $x = 0.10$ was the subject of particular attention on the one hand, due to the absence of anomalies in the dielectric properties in T_N region, on the other hand – by reason of the optimal ratio of ceramic and electrical properties at room temperature: the smallest amount of impurities led to the smallest σ and tgδ and the largest $\varepsilon/\varepsilon_0$.

Figure 3 shows dependencies of the structural parameters and m_P of $Bi_{0.9}Dy_{0.1}FeO_3$ ceramics on temperature, dashed lines highlight areas of anomalous behavior of the cell volume and the boundary between the two-phase, Rh + R, and single-phase, R, states. Below 673 K parameters of R-phase did not calculate, because its structure has exhibited instability (distortion of weak X-ray lines profiles).

Figure 3 shows that the two-phase region exists in the range of 293 K $\leq T <$ 853 K with phase ratio substantially unchanged up to 573 K.

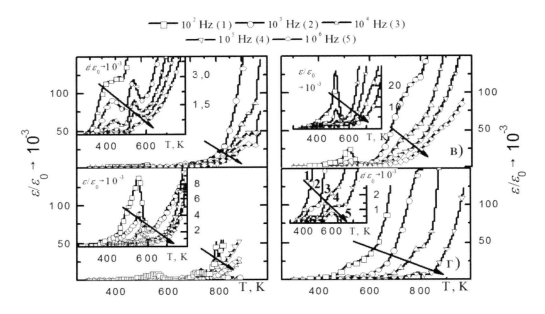

Figure 2. Dependencies of $\varepsilon/\varepsilon_0$ for $Bi_{1-x}Dy_xFeO_3$ ceramic samples on temperature (heating mode), x = 0.05 (a) 0.10 (b) 0.15 (c) and 0.20 (d).

Table 2. Temperature ranges of invar effect

IE area	Temperature range, K
I	$373 \leq T < 433$
II	$453 \leq T < 493$
III	$513 \leq T < 573$
IV	$623 \leq T < 653$
V	$673 \leq T < 713$
VI	$733 \leq T < 773$

Above that temperature, the content of R-phase begins to increase up to T = 643 K gradually, and then rapidly to 773 K with a slight slowdown in the 773-853 K range, near transition to the R-phase (853 K).

The inset shows a temperature region where m_R increase gradually. It is clear that there are some discontinuities in m_R curve.

Figure 3 also shows that in the two-phase region $V_{Rh}(T)$-dependence has a stepped character, since it includes 6 regions of constancy (the invar effect, IE). Furthermore, on the $\alpha_{Rh}(T)$-dependence one can see skips within a IE (I) field, or on the boundaries of the (II, III, IV) regions. The IE temperature ranges listed in Table 2.

Figure 4 presents fragments of X-ray diffraction patterns including lines $(111)_C$ and $(200)_C$ of $Bi_{0.9}Dy_{0.1}FeO_3$, that illustrate the changes in the structure in different temperature ranges. Figure 4 shows that at 293 K at the base of a single 200 line of Rh phase there is a maximum from the side of large θ angles, which with increasing temperature transforms into X-ray line 200 of R-phase. The width of this peak varies according to the temperature changes. So at T = 383 K the peak washout and the line profile is more consistent with single-phase state with a bimodal size distribution of coherent scattering.

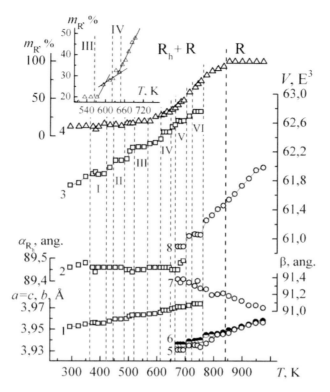

Figure 3. Dependencies of parameters and volume of the perovskite cell of $Bi_{0.9}Dy_{0.1}FeO_3$ and contained rhombic phase, m_R, at temperature: a_{Rh} (1), $α_{Rh}$ (2), V_{Rh} (3), m_R (4), $a_M = c_M$ (5), b_M (6), $β$ (7), V_M (8). The inset shows the change in m_R in the 530-690 K range. The dotted lines highlight areas of anomalous behavior of the cell volume and the boundary between the two-phase, Rh + R, and single-phase, R, states.

This implies that at these temperatures R phase still structurally not fully formed, but corresponds to a package of planes with interplanar spacing $d ≈ 1.95$ Å. The studies show a multiple changes in a slope of $σ(T)$ dependence measured during heating in the 450-640 K temperature range (Figure 5).

A broad hysteresis forms in the same region, shown in the inset, which presents the experimental data, obtained during heating and cooling processes, and provided in Arrhenius plot. It can be seen that the curve measured in the cooling mode has monotonous shape. It experiences a change in the slope in the antiferromagnetic transition area with activation energy of charge carriers decreasing from ≈ 0.75 eV to ≈ 0.35 eV as the material heats above T_N.

By increasing the temperature above 600 K, the $ε/ε_0(T)$-curve undergoes an exponential increase, which corresponds to the beginning of rapid growth of R-phase content (Figure 3). So it leads to decreasing the Rh-phase ratio and loose its stability and thus to decreasing the ferroelectric properties of objects.

Figure 5 shows the temperature dependencies of $σ$, $ε/ε_0$, $λ$, $χ$, C_p and $α$ of $Bi_{0.90}Dy_{0.10}FeO_3$ ceramics.

Dependencies of thermal characteristics on the concentration of the introduced modifiers are as usual with the minima and maxima [3, 4] near the T_N [2].

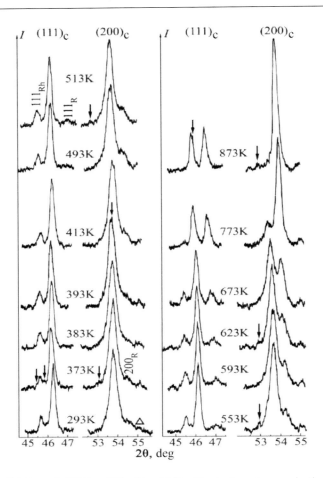

Figure 4. X-ray lines $(111)_C$ and $(200)_C$ of $Bi_{0.9}Dy_{0.1}FeO_3$, illustrating changes in the structure in the two-phase area and in the Rh + R → R transition area. $Bi_2Fe_4O_9$ line is denoted by triangle.

CONCLUSION

In the study of phase and grain structures, electrical and dielectric properties of $Bi_{1-x}Dy_xFeO_3$ type multiferroics ($x = 0.05$-0.20) at room temperature, the regularities of their formation were established.

It is shown that at $x = 0.10$ the studied object reaches the optimal correlation of ceramic and electrical properties: the least amount of impurities leads to the smallest σ and tgδ and the largest $\varepsilon/\varepsilon_0$ values.

Detailed X-ray studies of $Bi_{0.9}Dy_{0.1}FeO_3$ samples were carried out in a wide temperature range, which revealed stepped character of $V_{Rh}(T)$ dependence change, comprising 6 regions of constancy (the invar effect). The results of high-temperature studies of electric and thermal properties of the multiferroics were analyzed and compared with changes in the crystal structure of the object.

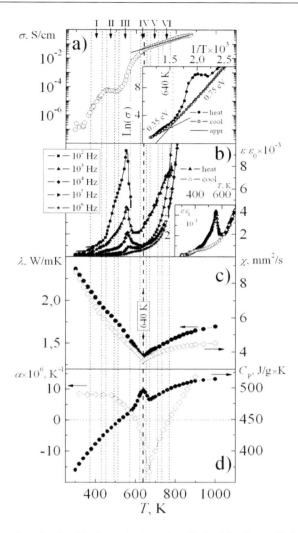

Figure 5. Dependencies of σ (a), $\varepsilon/\varepsilon_0$ (b), λ, χ (c), C_p and α (d) for $Bi_{0.90}Dy_{0.10}FeO_3$ ceramics on temperature.

ACKNOWLEDGMENTS

This work was financially supported by the Ministry of Education and Science of the Russian Federation: Grant of the President of the Russian Federation No. MK-3232.2015.2; Agreement No. 14.575.21.0007 (Federal Target Program); themes Nos. 1927, 213.01-2014/012-ВГ and 3.1246.2014/K (the basic and project parts of the State task).

REFERENCES

[1] Zvezdin, A. K., Pyatakov, A. P. *Sov. Phys. Usp.* 2004, vol. 47, 416–421.
[2] Smolenskii, G. A., Chupis, I. E. *Sov. Phys. Usp.* 1982, vol. 25, 475–493.

[3] Magomedov, Ya. B., Hajiyev, G. G. *TVT*. 1990, vol. 28, 185–186 (in Russian).

[4] Magomedov, M.-R. M., Kamilov, I. K., Omarov, Z. M. et al., *Instruments and Experimental Techniques*. 2007, vol. 4, 165.

[5] Denisov, V. M., Belousov, N. V., Zhereb, V. P. et al., *Journal of Siberian Federal University. Chemistry*. 2012, vol. 5, 146–167 (in Russian).

[6] Smith, J., Wayne, H. *Ferrites, Foreign Literature*, Moscow, 1962 (in Russian).

[7] *Powder Diffraction File*. Data Card. Inorganic Section. Set 47, card 67, JCPDS. Swarthmore, Pennsylvania, US, 1948.

[8] Khomchenko, V. A., Karpinsky, D. V., Kholkin, A. L. et al., *J. Appl. Phys.* 2010, vol. 108, 074109.

In: Proceedings of the 2015 International Conference …   ISBN: 978-1-63484-577-9
Editors: Ivan A. Parinov, Shun-Hsyung Chang et al.   © 2016 Nova Science Publishers, Inc.

Chapter 39

PREPARATION, STRUCTURE, MICROSTRUCTURE, DIELECTRIC AND THERMAL PROPERTIES OF BISMUTH FERRITE WITH HOLMIUM, ERBIUM AND YTTERBIUM

S. V. Khasbulatov[1,], A. A. Pavelko[1], L. A. Shilkina[1], V. A. Aleshin[1], I. A. Verbenko[1], G. G. Hajiyev[2], Z. M. Omarov[2], A. G. Bakmaev[2], O. N. Razumovskaya[1] and L. A. Reznichenko[1]*

[1]Research Institute of Physics, Southern Federal University,
Rostov-on-Don, Russia
[2]H.I. Amirkhanov Institute of Physics of Daghestanian Scientific Center
of the Russian Academy of Sciences, Makhachkala, Russia

ABSTRACT

The phase and polycrystalline (grain) structures as well as dielectric and thermal properties of $Bi_{1-x}REE_xFeO_3$ (where REE – Ho, Er and Yb, $x = 0.05 − 0.20$, $\Delta x = 0.05$) magnetoelectric solid solutions were investigated.

The regularities of the objects' phase and grain structures, and electrical and dielectric properties formation at room temperature were established.

Keywords: multiferroics, bismuth ferrite, rear earth elements, phase composition, thermal properties, permittivity

1. INTRODUCTION

Bismuth ferrite is a representative of multiferroics – a class of materials that have both magnetic and electric ordering. $BiFeO_3$ attracts attention due to its inherent high temperatures

---

[*] Email: said_vahaevich@mail.ru.

of ferroelectric ($T_C \sim 1083$K) and antiferromagnetic ($T_N \sim 643$K) phase transitions [1]. The bismuth ferrite at room temperature has R3c space group. Its crystal structure is characterized by rhombohedrally distorted perovskite cell, very close to cubic one. In the area of temperatures below the T_N bismuth ferrite has a complex spatially-modulated magnetic structure of cycloid type, which suppress ferromagnetic properties [2].

The necessary condition for the occurrence of magnetoelectric effect is the destruction of his space-modulated spin structure, which can be achieved by doping bismuth ferrite with rare earth elements. The represented work is continuation and development of the complex studying of a high-temperature multiferroic bismuth ferrite, modified by rare-earth elements, *REE*, undertaken earlier by the authors [3, 4].

2. OBJECTS AND METHODS

Objects of the present study were $Bi_{1-x}REE_xFeO_3$ type solid solutions with *REE*: Ho, Er, Yb; $0.00 \leq x \leq 0.20$, $\Delta x = 0.05$, obtained by two-stage solid-phase synthesis followed by sintering, by conventional ceramic technology at $T_{1sint} = 1073$ K, $T_{2sint} = 1173$ K, $\tau_{1,2\ sint} = 6$ h, $T_{synt} = 1203$ K, $\tau = 1.5$ h (Ho); at $T_{1sint} = 1073$ K, $T_{2sint} = 1093$ K, $\tau_{1,2sint} = 10$ h, $T_{synt} = 1183$ K, $\tau = 2$ h (Er); and $T_{1sint} = 1073$ K, $T_{2sint} = 1123$ K, $\tau_{1,2sint} = 6$ h, $T_{synt} = 1163$ K, $\tau = 1.5$ h (Yb).

X-ray study of the objects held by powder diffraction method using a DRON-3 diffractometer (CoKα-radiation, Bragg-Brentano focusing scheme). Perovskite cell parameters were calculated by the standard technique [5]. Change in the phase composition of the samples and the behavior of the structural parameters with temperature were controlled by the diffraction reflections $(111)_c$ to $(200)_c$ and $(220)_c$.

Investigation of polycrystalline (grain) structure of the samples performed in reflected light on an optical microscope Neophot 21 and on the inverted high-precision microscope Leica DMI 5000M.

Temperature dependencies of relative dielectric permittivity, $\varepsilon/\varepsilon_0$, at frequencies $25\text{--}2 \cdot 10^6$ Hz were investigated in the range of $300 - 900$ K at the special setup with use of precision LCR meter Agilent E4980A.

Thermal diffusivity (χ) were measured at the facility LFA-457 "MicroFlash" (NEZSCH), a heat capacity (Cp) was defined by means of the differential scanning calorimeter DSC-204 F1 (NEZSCH).

3. RESULTS AND DISCUSSION

Figure 1 demonstrates dependencies of the objects' structural parameters on concentration of the modifiers.

The results showed that all solid solutions contain the impurity phases $Bi_2Fe_4O_9$, $Bi_{25}FeO_{40}$, and $Ho_3Fe_5O_{12}$, $Er_3Fe_5O_{12}$ which quantity increases with growth of x.

The intensity of strong lines of these phases at $x = 0.15$ and 0.20 reaches 50%. Characteristics of impurity phases: $Bi_{25}FeO_{40}$ – cubic symmetry, S.G. I23, $a = 10.181$ Å, $Bi_2Fe_4O_9$ – rhombic symmetry, S.G. Pbam, $a = 7.965$ Å, $b = 8.44$ Å, $c = 5.994$ Å, $Ho_3Fe_5O_{12}$, $Er_3Fe_5O_{12}$ – cubic symmetry, S.G. Ia3d, $a = 12.301$ Å.

Basic perovskite cell of the objects has rhombohedral symmetry for all values of x.

Figure 1 shows that reduction of volume, V_{exp}, with growth of x does not correspond to dependence of $V_{theor}(x)$: it is observed a slight fall in the range $0.00 < x \leq 0.05$, and when $x > 0.05$, V_{exp} practically unchanged. Here V_{theor}, V_{exp} are the experimental and theoretical volumes of a perovskite cell, respectively; V_{theor} calculated using the formula

$$V_{theor} = \left[\frac{\sqrt{2}[(1-x)\, n_{Bi}\, L_{BiO} + x\, n_{Yb}\, L_{XO}] + 2\, n_{Fe}\, L_{FeO}}{(1-x)\cdot n_{Bi} + x\cdot n_X + n_{Fe}} \right]^3 ,$$

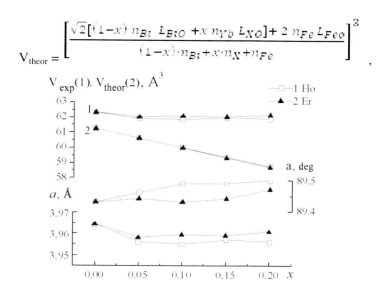

Figure 1. Dependencies of the structural characteristics of the solid solutions on concentration of modifier.

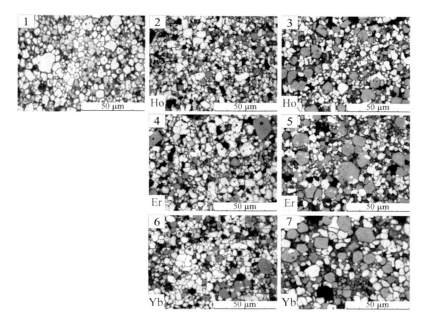

Figure 2. Fragments of the microstructure of $Bi_{1-x}REE_xFeO_3$ ceramic samples; *REE*: Ho, Er, Yb. 1. $BiFeO_3$; 2. $x = 0.10$; 3. $x = 0.20$ (Ho); 4. $x = 0.10$; 5. $x = 0.20$ (Er); 6. $x = 0.10$; 7. $x = 0.20$ (Yb).

where L is the length of relaxed cation-oxygen bond in view of the coordination number of the cation to oxygen, n is the valency of cations, the calculation was performed using the ionic radii according to Belov-Bokiy [6]. Thus, it is possible to draw a conclusion that only a small amount of Ho and Er atoms dissolve in $BiFeO_3$. The rest of those form ferrite with garnet structure, and Bi – other two impurity compounds.

Figure 2 presents fragments of microstructures of pure and modified with Ho, Er, Yb (x = 0.10; 0.20 for each *REE*) $BiFeO_3$ samples.

Black areas rounded and irregular in shape are pores. They are distributed inhomogeneously over the surface. Etching revealed crystallite boundaries: closed dark lines around the lighter areas (ceramic grains). It was found that the polycrystalline structure of the modified bismuth ferrite ceramics are multiphase microstructures, which include the basic coherent "light" phase, and one or more non-basic local "gray" phases. The main phase is represent by crystallites, the size and shape of which vary widely. With growth of x the average size of crystallites both the main phase, and nonbasic "gray" phases increases, also there is an expected increase in a percentage ratio of the last. Usually it leads to weakening of the dielectric properties by increasing the internal electromechanical loss, space charge accumulation at the interfaces, micro and mesoscopic areas having different electrical properties.

The analysis of results of dielectric spectroscopy (Figure 3) showed that the considered dependencies undergo anomalies of two types.

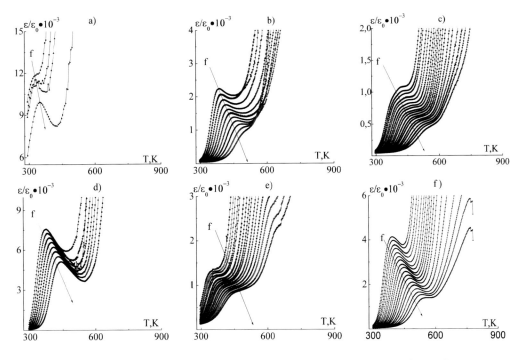

Figure 3. Dependence of $\varepsilon/\varepsilon_0$ of $Bi_{1-x}REE_xFeO_3$ ceramic samples (*REE*: Ho, Er, Yb) on the temperature in the frequency range $25 - 2 \cdot 10^6$ Hz; arrows show the increase in the frequency, f; (a) $BiFeO_3$; (b) Ho – x = 0.10, c) x = 0.20; d) Er – x = 0.10, e) x = 0.20; f) Yb – x = 0.20.

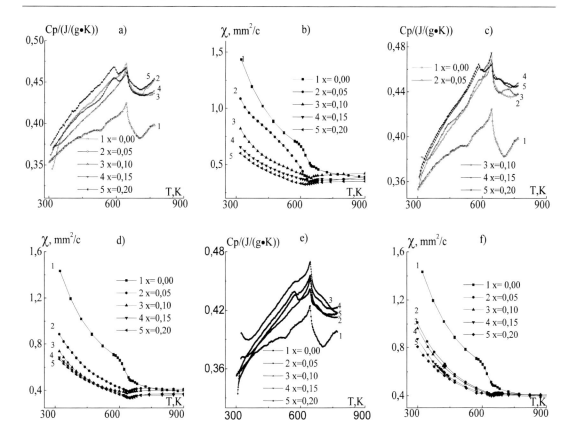

Figure 4. Temperature dependencies of the thermal diffusivity (χ), specific heat (Cp) of studied Bi$_{1-x}$REE$_x$FeO$_3$ ceramic samples; (a, b) – Ho-modified, (c,d) – Er-modified, (e, f) – Yb-modified.

The first one exists in the temperature range 400 – 500 K in the form of one or two, depending on the direction of change of temperature, highly dispersive maxima of $\varepsilon/\varepsilon_0$, having relaxation character. The increase in REE concentrations leads to complication of the registered dielectric spectra (their shift, enhancing the dispersion and formation of additional anomalies – ill-defined peaks in the $\varepsilon/\varepsilon_0(T)$ curves at $T = 600$ K, more noticeable in BiFeO$_3$ with Er (Figure 3, e). Similar phenomena are known as Maxwell-Wagner relaxation, observed earlier in [7] and associated with the accumulation of free charges at the interfaces of components on the background of the interlayer, and interfacial intraphase rearrangements.

The dependencies of thermal characteristics on concentration of the modifiers presented in Figure 4 are conventional: with minima and maxima [8, 9] near Neel temperature ($T_N \sim 643$ K) [9].

ACKNOWLEDGMENTS

This work was financially supported by the Ministry of Education and Science of the Russian Federation: Grant of the President of the Russian Federation No. MK-3232.2015.2;

Agreement No. 14.575.21.0007 (Federal Target Program); themes Nos. 1927, 213.01-2014/012-ВГ and 3.1246.2014/К (the basic and project parts of the State task).

REFERENCES

[1]   Wong, I.; et al. *Science*. 1965, vol. 299, 52-69

[2]   Zvezdin, A.K.; Pyatakov, A.P. *Sov. Phys. Usp.* 2004, vol. 47, 416–421.

[3]   Khasbulatov, S.V.; et al. Bulletin of the Kazan Technological University. 2014, vol. 17, 142-143 (in Russian).

[4]   Verbenko, I.A.; et al. Bulletin of the Russian Academy of Sciences: Physics. 2010, vol. 74, 1141-1143.

[5]   Fesenko, E.G. Perovskite Family and Ferroelectricity, Atomizdat, Moscow, 1972, 248 p. (in Russian)

[6]   Bokiy, G.B. Introduction to Crystal Chemistry, Moscow State University Press, Moscow, 1954 (in Russian).

[7]   Miller, A.I.; et al. *Ecology of Industrial Production*. 2012, *vol. 2*, 65–74 (in Russian).

[8]   Kallaev, S.N.; et al. *JETP Letters*. 2013, vol. 97, 470-472.

[9]   Amirov, A.A.; et al. *Physics of the Solid State*. 2009, vol. 51, 1189-1192.

In: Proceedings of the 2015 International Conference … ISBN: 978-1-63484-577-9
Editors: Ivan A. Parinov, Shun-Hsyung Chang et al. © 2016 Nova Science Publishers, Inc.

Chapter 40

MAGNETODIELECTRIC INTERACTIONS IN $BI_{0.5}LA_{0.5}MNO_3$ CERAMICS IN TEMPERATURE RANGE OF $10 - 120$ K

A. V. Pavlenko[1,2,], A. V. Turik[2], L. A. Reznitchenko[2] and Yu. S. Koshkid'ko[3]*

[1]Southern Scientific Center of the Russian Academy of Sciences,
Rostov-on-Don, Russia
[2]Research Institute of Physics, Southern Federal University,
Rostov-on-Don, Russia
[3]International Laboratory of High Magnetic Fields and Low Temperatures,
Wroclaw, Poland

ABSTRACT

Dielectric and magnetic characteristics of manganite bismuth-lanthanum $Bi_{0.5}La_{0.5}MnO_3$ ceramics are measured in the temperature range of $T = 10 - 120$ K. It is shown that the contribution of magnetostriction effects must be taken into account for establishing mechanisms of interaction between electric and magnetic subsystems in $Bi_{0.5}La_{0.5}MnO_3$ ceramics.

Keywords: magnetodielectric effect, bismuth–lanthanum manganite, solid solutions

1. INTRODUCTION

In $Bi_{0.5}La_{0.5}MnO_3$ (BLM) ceramics near the temperature $T_C = 120$ K there is a ferromagnetic phase transition [1], and at $T = 77$ K there are the well-defined magnetodielectric effect (MDE) [2] and magnetoresistance, the causes of which are not properly understood to date. In this work, to establish laws of interaction of electric and magnetic subsystems in BLM ceramics, we studied the temperature dependencies of the dielectric (relative permittivity $\varepsilon/\varepsilon_0$ and conductivity σ) and magnetic (magnetization M and magnetostriction ω) characteristics in the tmperature range of $T = 10$–120 K.

2. OBJECTS, METHODS OF PREPARATION AND INVESTIGATION OF THE SAMPLES

The objects of the investigation were ceramics with the $Bi_{0.5}La_{0.5}MnO_3$ composition. The solid solutions were synthesized by the method of solid phase reactions from oxides Bi_2O_3, Mn_2O_3, and La_2O_3 of high-pure grade by the two stage calcination with intermediate milling at temperatures $T_1 = 1173$ K and $T_2 = 1273$ K and holding times $\tau_1 = 10$ h and $\tau_2 = 2$ h. Ceramics was sintered at $T_{sy} = 1293$ K for 2 h. Complex permittivity $\varepsilon^* = \varepsilon' - i\varepsilon''$ (ε' and ε'' are the real and imaginary parts of ε^*, respectively) in the temperature $(10 - 200)$ K and frequency $(10^2 - 10^6)$ Hz ranges was measured using a Wayne Kerr 6500 B precision impedance analyzer. An E4339 B High Resistance Meter Agilent was used to measure resistivity ρ.

Magnetization M and volum magnetostriction ω were measured in the temperature range of $(4.2 - 120)$ K at the International Laboratory of Strong Magnetic Fields and Low Temperatures (Wroclaw, Poland).

3. EXPERIMENTAL RESULTS. DISCUSSION

Obtained samples are practically undoped and monophase at room temperature. Also they have the cubic crystal structure [3 – 4]. Grain composition corresponds to settable composition in cation components. Cations Mn^{3+} and Mn^{4+} are present in the crystal structure. Experimental dependencies of $\varepsilon/\varepsilon_0(T)$, $\sigma(T)$ and $M(T)$ at $T = (10 - 120)$ K are shown in Figure 1.

Lowering temperature leads to decreasing $\varepsilon/\varepsilon_0(T)$ with dispersion intensification and forming slightly defined maxima at $T_1 \sim 25$ K, $T_2 \sim 40$ K and $T_3 \sim 55$ K. The peculiarities at the $M(T)$ and $\sigma(T)$ dependencies are observed at T_1 and T_2, but they are almost invisible at T_3. It shows that succession of phase tranformations is the result of temperature decrease. Different character of detected correlations in dielectric and magnetic characteristics testifies that there are various mechanisms of interaction between electric and magnetic subsystems. Moreover, each of these mechanisms may be prevalent in the particular temperature range. Experimental $M(B)$ and $\omega(B)$ dependencies of BLM ceramics at $T = (4.2 - 100)$ K are shown in Figures 2 and 3.

Dependencies $M(B)$ show that magnetization of BLM ceramics is rather high. Moreover, significant change of magnetization after saturation in demagnetizing field was observed at all cases. Dependencies $\omega(B)$ show that hgh volume magnetostriction at $T = 4.2$ K and 50 K ($T < T_C$) detects in the all research range, and volume magnetostriction at $T = 100$ K ($T < T_C$) and $B = (0 - 6$ T) misses. If $B > 6$ T then volume magnetostriction becomes apparent and increases with B increment. It may be due to the fact that applying magnetic field leads to the shift of T_C to the high temperature range.

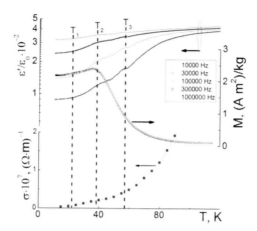

Figure 1. Experimental dependencies of $\varepsilon/\varepsilon_0(T)$, $\sigma(T)$ and $M(T)$ for BLM ceramic at $T = 10 - 120$ K ($\Delta T = 0.4 - 1$ K) at frequencies $f = (10^4 - 10^6)$ Hz. The dependencies were obtained in the cooling mode.

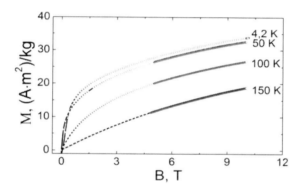

Figure 2. Experimental dependencies of $M(B)$ for BLM ceramic at $T = 4.2$ K, 50 K, 100 K and 150 K.

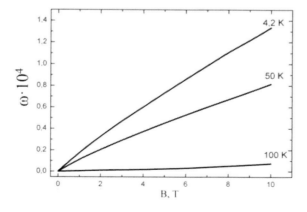

Figure 3. Experimental dependencies of $\omega(B)$ for BLM ceramic at $T = 4.2$ K, 50 K and 100 K.

The obtained results show that contribution of the magnetostriction effects must be taken into account when mechanisms of electric and magnetic subsystems interaction are determined. In particular appearance of colossal magnetoresistance and ferromagnetism in lanthanum manganite relates to phase separation into ferromagnetic amd antiferromagnetic (isolating) zones because of irregularity and heterogeneity of composition. Transition of ceramics BLM to ferromagnetic and dielectric phases at lower temperatures indicates volume prevalence of dielectric antiferromagnetic phase part. Presence in such ceramic materials interphase boundaries of different nature (including inportant for colossal magnetoresistance grain boundaries) can lead to appearance of interlaminar polarization and relaxation effects assotiated with charge accumulation on the boundary surface. However growth and deformation of ferromagnetic areas under magnetic field can lead to change of relaxation time and increment of the role of tunneling processes. It must be taken into account in analysis of temperature, frequency and field dependences of the MDE in such materials.

ACKNOWLEDGMENTS

This work was supported by the Ministry of Education and Science of the RF (Federal program: State contract No. 14.575.21.0007; Projects Nos. 213.01-11/2014-21, 3.1246.2014/K, 213.01-2014/012 and SP-1689.2015.3).

REFERENCES

[1] Pavlenko, A.V.; Turik, A.V.; Reznichenko, L.A.; Koshkid'ko, Yu.S. *Phys. Solid State*, 2014, vol. 56(6), 1137–1143.

[2] Pavlenko, A. V.; Turik, A. V.; Reznichenko, L. A.; Shilkin, L. A.; Konstantinov, G. M. *Tech. Phys. Lett.* 2013, vol. 39(1), 78–84.

[3] Kochur, A.G.; Kozakov, A.T.; Nikol'skii, A.V.; Guglev, K.A.; Pavlenko, A.V.; Verbenko, I.A.; Reznichenko, L.A.; Shevtsova, S.I. *Phys. Solid State.* 2013, vol. 55(4), 743–747.

[4] Kochur, A.G.; Kozakov, A.T.; Nikolskii, A.V.; Googlev, K.A.; Pavlenko, A.V.; Verbenko, I.A.; Reznichenko, L.A.; Krasnenko. *J. of Electron Spectroscopy and Related Phenomena.* 2012, vol. 185, 175.

In: Proceedings of the 2015 International Conference …   ISBN: 978-1-63484-577-9
Editors: Ivan A. Parinov, Shun-Hsyung Chang et al.   © 2016 Nova Science Publishers, Inc.

Chapter 41

MATHEMATICAL SIMULATION OF PROCESSES OF THE ELECTROMASS TRANSFER IN CONTROLLED ELECTROCHEMICAL RESISTANCE

T. P. Skakunova[*]*, Yu. Ya. Gerasimenko, D. D. Fugarov and I. V. Tarasov*
Don State Technical University,
Rostov-on-Don, Russia

ABSTRACT

In the paper, we solved the following problems:

(i) A general case of the problem of electromass transfer in a controlled electrochemical resistance (CER);
(ii) developed a mathematical model of CER as part of an electrical circuit;
(iii) obtained a CER transfer function, based on the assumption that "input" of the object is controlled by current, and "output" – by conductance in the main circuit.

Keywords: mathematical modeling, electrolyte, concentration, cathode, plate, electric current density, mass-transfer

1. THE MAIN PHYSICAL AND CHEMICAL ASSUMPTIONS ACCEPTED IN CASE OF SIMULATION OF CER

In the case of mathematical simulation of physical fields in CER the following assumptions are accepted [1–4]:

---

[*] E-mail: spu-45@donstu.ru.

(i). all processes occurring in the object of research are isothermal;
(ii). electric fields in all fragments of CER are potential and plane-parallel;
(iii). all physical and chemical parameters of CER materials are constants;
(iv). the kinetics of all electrode processes in the object of research are controlled by diffusion stage in electrolyte;
(v). we neglect the migration phenomenon in electric field of the electrolyte.

2. MATHEMATICAL SIMULATION OF CONCENTRATION FIELD OF ELECTROLYTE

We consider, that all physical fields in CER are plane-parallel and do not depend on z-coordinate (Figure 1).

Let us place the metallic layer on a substrate of CER. By this, the metallic layer is much thinner in comparison with the geometrical sizes of the substrate and thickness of an electrolyte layer.

Based on the second Fick's law, a concentration field of the electrolyte $C(x; y; t)$ is subject to the following initial boundary value problem:

$$\frac{\partial C}{\partial t} = D\left(\frac{\partial^2 C}{\partial x^2} + \frac{\partial^2 C}{\partial y^2}\right), \quad x \in [0;l],\ y \in [0;h]; \tag{1}$$

$$C(x; y; 0) = C_0; \tag{2}$$

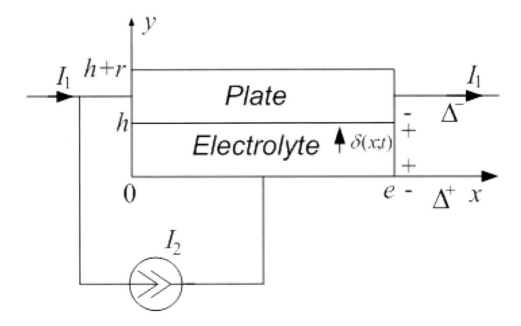

Figure 1. Transverse section of CER.

$$\frac{\partial C}{\partial y}(x;0;t) = \frac{N I_2(t)}{al} ; \tag{3}$$

$$\frac{\partial C}{\partial y}(x;h;t) = N\delta(x;t) ; \tag{4}$$

$$\frac{\partial C}{\partial x}(0;y;t) = 0 ; \tag{5}$$

$$\frac{\partial C}{\partial x}(l;h;t) = 0 ; \tag{6}$$

where D is the diffusion coefficient of electrolyte, C_0 is the initial concentration of electrolyte, a is the CER thickness, $\delta(x;t)$ is the unknown distribution of current density on substrate; N is an electrode, kinetic constant.

It is most convenient to solve this problem applying the Laplace operator method. With this purpose, we introduce the following designations:

$$\overset{oo}{\underset{o}{C(x;y;t)}} = \overset{oo}{C(x;y;p)} ; \overset{oo}{\underset{o}{I_2(t)}} = \overset{oo}{I_2(p)} ; \overset{oo}{\underset{o}{\delta(x;y;t)}} = \overset{oo}{\delta(x;y;p)} .$$

The analytical description of a concentration field of an electrolyte in the Laplace operator form has been present in Refs. [5, 6] as

$$\overset{o}{C(x;y;p)} = \frac{C_0}{p} + \left(\frac{N \beta_0(p)}{\sqrt{\frac{p}{D}} \, sh\sqrt{\frac{p}{D}} h} ch\sqrt{\frac{p}{D}} y - \frac{N \overset{o}{I_2(p)}}{a\sqrt{l} \cdot \sqrt{\frac{p}{D}} \, sh\sqrt{\frac{p}{D}} h} ch\sqrt{\frac{p}{D}} (h-y) \right) \frac{1}{\sqrt{l}} +$$

$$+ \sum_{k=1}^{\infty} \frac{N \beta_k(p)}{\sqrt{\frac{p}{D} + \mu_k^2} \, sh\sqrt{\frac{p}{D} + \mu_k^2} h} ch\sqrt{\frac{p}{D} + \mu_k^2} y \cdot X_k(x) \tag{7}$$

where $\beta_0(p)$ and $\beta_k(p)$ are the coefficients of distribution of the current density.

By using the formula (7), it is possible to calculate at this stage a concentration field only in the case of uniform distribution of the current density on substrate, when

$$\overset{o}{\delta(x;p)} = \frac{\overset{o}{I_2(p)}}{al} \tag{8}$$

By dividing this density in Fourier series, $\dfrac{\overset{o}{I_2}(p)}{al} = \sum\limits_{k=0}^{\infty} \beta_k(p) X_k(x)$, we obtain the following relationships:

$$\beta_0(p) = \frac{\overset{o}{I_2}(p)}{a\sqrt{l}} \tag{9}$$

$$\beta_k(p) = 0, \ k = 1, 2, \dots \tag{10}$$

By substituting relationships (9) and (10) into (7), we can define the concentration field of electrolyte in the case of uniform distribution of current density in the form:

$$\overset{o}{C}(y;p) = \frac{C_0}{p} + \overset{o}{I_2}(p)\frac{N}{al\sqrt{\dfrac{p}{D}}ch\sqrt{\dfrac{p}{D}}\dfrac{h}{2}}sh\sqrt{\frac{p}{D}}(y - \frac{h}{2}) \tag{11}$$

In the case of switching on input circuit of CER on a direct current, we have

$$I_2(p) = \frac{I_0}{p} \tag{12}$$

By substituting relationship (12) into (11), we obtain

$$\overset{o}{C}(y;p) = \frac{C_0}{p} + I_0 \frac{N}{al\,p\sqrt{\dfrac{p}{D}}ch\sqrt{\dfrac{p}{D}}\dfrac{h}{2}}sh\sqrt{\frac{p}{D}}(y - \frac{h}{2}) \tag{13}$$

3. MATHEMATICAL SIMULATION OF ELECTRIC FIELD IN ELECTROLYTE LAYER

Into frameworks of the accepted assumptions on potential character of electric field, unchanged physical and chemical parameters of electrochemical system, and also plane-parallel geometry of the electric field, the Laplace differential equation will model rather scalar electric potential $\varphi_y(x;y;t)$ [7–9]:

$$\frac{\partial^2 \varphi_{\acute{Y}}}{\partial x^2} + \frac{\partial^2 \varphi_{\acute{Y}}}{\partial y^2} = 0, \ x \in [0;l], \ y \in [0;h] \tag{14}$$

To this equation, we add the following boundary conditions:

$$\frac{\partial \varphi_{\dot{y}}}{\partial x}(0; y; t) = 0 ; \tag{15}$$

$$\frac{\partial \varphi_{\dot{y}}}{\partial x}(l; y; t) = 0 ; \tag{16}$$

$$\frac{\partial \varphi_{\dot{y}}}{\partial y}(x; 0; t) = -\frac{I_2(t)}{al\gamma_{\dot{y}}} ; \tag{17}$$

$$\frac{\partial \varphi_{\dot{y}}}{\partial y}(x; h; t) = -\frac{\delta(x; t)}{\gamma_{\dot{y}}} \tag{18}$$

where $\gamma_{\dot{y}}$ is the specific conductivity of electrolyte, $\delta(x;t)$ is the unknown distribution of density of an electric current, used already in the case of calculation of the concentration field.

The solution of the boundary-value problem (14 – 18) by using the operator method reduces to the following relationship:

$$\overset{o}{\varphi_{\dot{y}}}(x; y; p) = \left(-\frac{\overset{o}{I_2(p)}}{a\sqrt{l}\gamma_{\dot{y}}} y + B_0(p) \right) X_0(x) - \frac{1}{\gamma_{\dot{y}}} \sum_{k=1}^{\infty} \frac{\beta_k(p) ch\dfrac{k\pi}{l} y}{\dfrac{k\pi}{l} sh\dfrac{k\pi}{l} h} X_k(x) \tag{19}$$

The analysis of the obtained expression for the potential of electrolyte $\overset{o}{\varphi_{\dot{y}}}(x; y; p)$ shows that it is defined with currency up to additive constant. It follows from that the term $B_0(p)$ remains indefinite.

4. Mathematical Simulation of Electric Field in Substrate of CER

Potential character of electric field in the substrate and its constant physical and chemical parameters allow one to describe the plane-parallel field by using the scalar electric potential subjecting to the Laplace equation:

$$\frac{\partial^2 \varphi_c}{\partial x^2} + \frac{\partial^2 \varphi_c}{\partial y^2} = 0 , \quad x \in [0; l], \quad y \in [h; h + r] \tag{20}$$

To the last equation, we add the boundary conditions describing interaction of the substrate with environment:

$$\frac{\partial \varphi_c}{\partial x}(0; y; t) = -\frac{I_1(t) - I_2(t)}{a r \gamma_c}, \quad y \in [h; h+r] \; ; \tag{21}$$

$$\frac{\partial \varphi_c}{\partial x}(l; y; t) = -\frac{I_1(t)}{a r \gamma_c}, \quad y \in [h; h+r]] \; ; \tag{22}$$

$$\frac{\partial \varphi_c}{\partial y}(x; h+r; t) = 0, \quad x \in [0; l] \; ; \tag{23}$$

$$\frac{\partial \varphi_c}{\partial y}(x; h; t) = -\frac{\delta(x; t)}{\gamma_c}, \tag{24}$$

where $I_1(t)$ is the current in circuit of CER; $I_2(t)$ is the circuit current controlling the CER; γ_c is the specific conductance of the substrate material; $\delta(x; t)$ is the unknown distribution of density of electric current on the interface "substrate – electrolyte."

The solution of the boundary-value problem (20) – (24) reduces to the following relationship:

$$\overset{o}{\varphi}_c(x; y; p) = \left(\frac{I_2(p)}{2 a r \sqrt{l} \gamma_c} y^2 - \frac{\overset{o}{I_2}(p)}{a r \sqrt{l} \gamma_c}(h+r) y + v_0(p) \right) \frac{1}{\sqrt{l}} -$$

$$- \frac{\overset{o}{I_2}(p)}{2 a r l \gamma_c} x^2 - \frac{1}{a r \gamma_c}[\overset{o}{I_1}(p) - \overset{o}{I_2}(p)] + \frac{l}{\gamma_c} \sum_{k=1}^{\infty} \frac{ch \dfrac{k\pi}{l}(h+r-y)}{k \pi sh \dfrac{k\pi}{l} r} \beta_k(p) X_k(x) \tag{25}$$

Analyzing the expression, we see that potential $\overset{o}{\varphi}_c(x; y; p)$ is found by additive constant $v_0(p)$, that is proper for Neumann problem (the second boundary-value problem for Laplace equation).

Moreover, we note that in the expression for $\overset{o}{\varphi}_c(x; y; p)$, the operator constant $\beta_k(p)$ remains indefinite.

We can calculate his constant only after studying the physical and chemical processes occurring on the interface "substrate – electrolyte."

5. Matching the Physical Fields on the Interface "Electrode – Electrolyte" in CER

To obtain the operation characteristics of CER, it is necessary to execute matching a concentration and electric fluid on the boundary "electrode – electrolyte." In engineering calculations is quite admissible to assume that the current density $\delta(x;t)$ is uniformly distributed on a surface of electrodes. It ensures the next simplified expression for the potential of electrolyte:

$$\overset{o}{\varphi_{\dot{y}}}(y;p) = -\frac{\overset{o}{I_2}(p)}{al\,\gamma_{\dot{y}}}y + \frac{B_0(p)}{\sqrt{l}} \tag{26}$$

At the interface "electrode – electrolyte," it is arisen an accumulation of electric potentials:

(i) on the positive electrode:

$$\Delta^+(p) = \frac{g_{10}}{p} + g_{11}\overset{o}{C}(0;p) \,; \tag{27}$$

(ii) on the negative electrode:

$$\Delta^-(p) = \frac{g_{10}}{p} + g_{11}\overset{o}{C}(h;p) \,, \tag{28}$$

where $g_{10} > 0$, $g_{11} > 0$ are constants of a linear approximation of the Nernst equation.

The values of concentration $\overset{o}{C}(0;p)$ and $\overset{o}{C}(h;p)$ are calculated from (11) at $y = 0$ and $y = h$, respectively:

$$\overset{o}{C}(0;p) = \frac{C_0}{p} - \overset{o}{I_2}(p)\frac{N\,sh\sqrt{\dfrac{p}{D}}\dfrac{h}{2}}{al\sqrt{\dfrac{p}{D}}\,ch\sqrt{\dfrac{p}{D}}\dfrac{h}{2}} \,; \tag{29}$$

$$\overset{o}{C}(h;p) == \frac{C_0}{p} + \overset{o}{I_2}(p)\frac{N\,sh\sqrt{\dfrac{p}{D}}\dfrac{h}{2}}{al\sqrt{\dfrac{p}{D}}\,ch\sqrt{\dfrac{p}{D}}\dfrac{h}{2}} \,. \tag{30}$$

By substituting (29) and (30) into (27) and (28), we find:

$$\Delta^+(p) = \frac{g_{10}}{p} + g_{11}\left(\frac{C_0}{p} - \overset{o}{I}_2(p) \frac{N\,sh\sqrt{\frac{p}{D}}\frac{h}{2}}{al\sqrt{\frac{p}{D}}\,ch\sqrt{\frac{p}{D}}\frac{h}{2}} \right); \tag{31}$$

$$\Delta^-(p) = \frac{g_{10}}{p} + g_{11}\left(\frac{C_0}{p} + \overset{o}{I}_2(p) \frac{N\,sh\sqrt{\frac{p}{D}}\frac{h}{2}}{al\sqrt{\frac{p}{D}}\,ch\sqrt{\frac{p}{D}}\frac{h}{2}} \right) \tag{32}$$

Voltage drop in the controlling circuit of CER is defined trivially:

$$\overset{o}{u}_{\acute{y}}(p) = \overset{o}{\varphi}_{\acute{y}}(0;p) - \overset{o}{\varphi}_{\acute{y}}(h;p), \tag{33}$$

where $\overset{o}{\varphi}_{\acute{y}}(0;p)$ and $\overset{o}{\varphi}_{\acute{y}}(h;p)$ are defined by expression (26) at $y = 0$ and $y = h$, respectively.

Then the computation of $\overset{o}{u}_{\acute{y}}(p)$ from (33) leads to the expression:

$$\overset{o}{u}_{\acute{y}}(p) = \overset{o}{I}_2(p)\frac{h}{al\,\gamma_{\acute{y}}} \tag{34}$$

By using the second law of Kirchhoff for the controlling circuit of CER, we can write:

$$\overset{o}{E}_c(p) = -\overset{o}{\Delta}^+(p) + \overset{o}{\Delta}^-(p) + \overset{o}{u}_{\acute{y}}(p) + \overset{o}{I}_2(p)R_0, \tag{35}$$

where $\overset{o}{E}_c(p)$ is the image of electric signal on CER input, $\overset{o}{E}_c(p)$ is the internal resistance of the signal generator.

After substitution of expressions (31), (32) and (34) into (35), we obtain:

$$\overset{o}{E}_c(p) = \overset{o}{I}_2(p)\left(\frac{2g_{11}N\,sh\sqrt{\frac{p}{D}}\frac{h}{2}}{al\sqrt{\frac{p}{D}}\,ch\sqrt{\frac{p}{D}}\frac{h}{2}} + \frac{h}{al\,\gamma_{\acute{y}}} + R_0 \right) \tag{36}$$

By introducing the following designations:

$$\overset{o}{Z}(p) = \frac{2g_{11}N\,sh\sqrt{\dfrac{p}{D}}\dfrac{h}{2}}{al\sqrt{\dfrac{p}{D}}\,ch\sqrt{\dfrac{p}{D}}\dfrac{h}{2}}\,; \tag{37}$$

$$R_{\acute{y}} = \frac{h}{al\,\gamma_{\acute{y}}}\,; \tag{38}$$

$$R = R_{\acute{y}} + R_0\,, \tag{39}$$

and taking into account (37) - (39), we reduce formula (36) to the next relationship:

$$\overset{o}{E}_c(p) = \overset{o}{I}_2(p)\left(\overset{o}{Z}(p) + R\right) \tag{40}$$

6. MATHEMATICAL SIMULATION OF OUTPUT CONDUCTANCE OF CER

The equivalent conductance of the CER is determined by an output circuit in the form:

$$Y_S(p) = Y_c + \overset{o}{Y}(p)\,, \tag{41}$$

where $Y_c = \dfrac{r\,a\,\gamma_c}{l}$ is the substrate conductivity, $\overset{o}{Y}(p)$ is the conductivity (on output of circuit) of a metallic layer deposited on the substrate.

It is shown in [10] that $\overset{o}{Y}(p)$ is defined as

$$\overset{o}{Y}(p) = \frac{k\,\gamma\,\overset{o}{E}_c(p)}{l^2\,\rho\,p\left(Z(p) + R\right)} \tag{42}$$

If $\overset{o}{E}_c(p)$ is the input value of the CER signal, and $\overset{o}{Y}(p)$ is the output value of conductivity, then the CER transfer function is defined by the expression:

$$W(p) = \frac{k\gamma}{l^2 \rho p \left(Z(p) + R\right)}$$

(43)

Thus, by knowing the CER transfer function (43), we can calculate all of its response characteristics for electric circuits and systems, containing in the governing equations.

REFERENCES

[1] Gerasimenko, Yu. Ya. Mathematical Modeling of Electrochemical Systems. SRSTU Press, 2009, Novocherkassk, pp. 1 – 314.

[2] Gerasimenko, Yu. Ya.; Gerasimenko, E. Yu. Matthematical Modeling of Electrochemical System with Diffusive Hyperbolic Control of Electrod Kinetics. IWK Information Technology and Electrical Engineering – Devices and Syistems, Materials and Technology for the Future, Germany, Ilmenau, 7 – 10 September, 2009, pp. 391 - 392.

[3] Gerasimenko, Yu. Ya.; Tsygulev, N. I.; Gerasimenko, A. N.; Gerasimenko, E. Y.; Fugarov, D. D.; Purchina, O. A. *Life Science Journal*, 2014, vol. 11(12), 265 - 269.

[4] Gerasimenko, Yu. Ya.; Gerasimenko, E. Yu. Matthematical Modeling of a Secondary Cell Breakdown Current Curve Through Fixed Resistance. IWK Information Technology and Electrical Engineering – Devices and Syistems, Materials and Technology for the Future, Germany, Ilmenau, 7 – 10 September, 2009, pp. 389 - 390.

[5] Beyer, W. H. CRS Standard Mathematical Tables and Formulae. CRS Press, Boca Raton, 1991, pp. 1 – 609.

[6] Davis, B. Integral Transforms and Their Applications. Springer-Verlag, New York, 1978.

[7] Butkcov, E. Mathematical Physics. Addison-Wesley, Reading, Mass., 1968.

[8] Zwillinger, D. Handbook of Differential Equations. Academic Press, Boston, 1989. pp. 1 – 673.

[9] Farlow, S. J. Partial Differential Equations for Scientists and Engyneers. John Wiley and Sons, New York, 1982.

[10] Newton, G. C.; Gould, L. A.; Kaiser, J. F. Analitical Design of Linear Feedback Controls, John Wiley and Sons, New York, Chapman and Hall, London, 1957.

III. Mechanics

In: Proceedings of the 2015 International Conference …   ISBN: 978-1-63484-577-9
Editors: Ivan A. Parinov, Shun-Hsyung Chang et al.   © 2016 Nova Science Publishers, Inc.

Chapter 42

MECHANICAL PROPERTIES OF THE BULK COMPOSITE NANOMATERIAL CONSISTING OF ALBUMIN AND CARBON NANOTUBES

Alexander Gerasimenko, Levan Ichkitidze[*], *Vitaly Podgaetsky and Sergei Selishchev*

National Research University of Electronic Technology "MIET,"
MIET, Zelenograd, Moscow, Russian Federation

ABSTRACT

The mechanical properties of a bulk composite nanomaterial prepared from the aqueous dispersion of bovine serum albumin and carbon nanotubes under laser radiation are investigated. In the experiment, the varied parameters were radiation power and time, nanotube type and concentration, drying time and temperature. The obtained nanomaterial exhibits high hardness (~350 MPa) and tensile strength (~35 MPa) at the low density ($1200 - 1250$ kg/m^3). The hardness of the prepared nanomaterial exceeds that of dried albumin by a factor of 4–5. The attained maximum hardness and tensile strength are compatible with the corresponding parameters of such well-studied materials as organic glass (hardness ~150 MPa), aluminum (hardness ~200 MPa and tensile strength ~100 MPa), and human porous bone tissue (hardness ~ 500 MPa and tensile strength $15 - 50$ MPa).

Keywords: albumin, carbon nanotubes, laser radiation, bulk composite nanomaterial, hardness, tensile strength

---

[*] Corresponding Author, E-mail: leo852@inbox.ru.

1. INTRODUCTION

Carbon nanotubes (CNTs) exhibit high values of the mechanical characteristics. In particular, the Young's modulus E of single-wall carbon nanotubes (SWCNTs) is ~ 5 TPa [1], which exceeds the value for high-resistance steel by a few orders of magnitude. The situation is similar in multiwall carbon nanotubes (MWCNTs). Embedding of CNTs in different composite materials (polymers, ceramics, etc.) even in minor concentrations (C ~ 0.1 wt.%) significantly improves the mechanical properties of these composites. In addition, using small amounts of CNTs, the thermal, electrical, and other parameters of materials were improved, which stimulated deep interest in application of CNTs in modern engineering, including medical practice [2 – 4]. For the last two decades, bone tissue fillers from artificial hydroxyapatite (HAp) have been used in medicine. The implants fabricated on the base of HAp or by embedding CNTs have much worse mechanical (hardness, strength, etc.), physical (porosity, thermal and electric conductivity, etc.), chemical (inertness and tolerance to other materials), and biological (bioresorption and regeneration of biological tissues) parameters as compared with the natural bone tissue [5 – 8].

The use of CNTs as additives significantly improved the properties of different biological materials. Study of the dose effect, i.e., the dependence of mechanical parameters of a nanocomposite consisting of plasticized amylumand MWCNTs on the concentration of the latter showed the duplication of the σ and E values at $C \approx 3$ wt.%. However, the attained maximum values $\sigma \approx 5$ MPa and $E \approx 40$ MPa were still much lower than the corresponding parameters of the human porous bone tissue (HPB, σ ~ 50 MPa, $E \approx 15$ GPa, and hardness $H_v \approx 500$ MPa [9]). The authors of [10, 11] observed the dose effect in the material consisting of chitosan and MWCNTs and reported the parameters $C = 2$ wt.%, $\sigma \approx 70$ MPa, $H_v \approx 150$ MPa, and $E \approx 2$ GPa, which are also noticeably worse than the values for the human bone tissue.

The analysis of many studies on risk and safety of application of CNTs and nanomaterials and also products fabricated on their base allows drawing the following conclusions:

I.   the toxicity of CNTs depends on the efficiency of their purification from different impurities, including catalytic metals;
II. the toxicity of SWCNTs is higher than that of MWCNTs;
III. the toxicity of SWCNTs is lower than that of asbestos particles subcutaneously introduced in test animals (muridae);
IV. functionalized CNTs are less toxic than non-functionalized ones [12 – 16].

According to the latter conclusion, nanomaterials for biomedical applications should be created on the base of functionalized CNTs. This approach was used in [17 – 20] for fabrication of composite materials from a bovine serum albumin (BSA) matrix filled with CNTs. In this case, the BSA matrix functionalizes CNTs [17 – 22].

The aim of this study was to investigate the mechanical properties of bulk composite nanomaterials (BCNMs) synthesized from albumin and carbon nanotubes (BSA+SWCNT or BSA+MWCNT) under laser radiation.

2. Materials

Sample preparation and experimental conditions were considered in detail in [17 – 20]. Here, we provide just a brief description. As a matrix material, BSA with 98% purity (Heat Shock Isolation, Amresco Inc., USA) was used. The fillers were MWCNT and CWCNT produced by different companies, including Taunit MWCNTs of the MWCNT-MD type (Komsomolets plant, Tambov, Russia) [23], and carboxylated SWCNTs of the Pastek-SWCNT-90 type (OOO Uglerod Chg) [24].

Samples of the BCNM with the composition BSA+SWCNT or BSA+MWCNT were prepared as follows:

I. BSA was dissolved in distilled water in a concentration of 25 wt.%. The obtained solution was dispersed in a magnetic stirrer for the time $t = 1 - 2$ h and exposed in an ultrasonic bath at the temperature $T = 40 - 50°C$ for the time $t = 2$ h.

II. The albumin solution was added with CNTs in the concentration $C = 1 - 4$ g/l ($0.1 - 0.4$ wt.%). The obtained aqueous dispersion with the composition BSA+CNT was dispersed in an ultrasonic bath for $3 - 4$ h.

III. The BSA+CNT dispersion was decanted by removal of the forming precipitates for the time $t = 24$ h.

IV. The obtained dispersion was irradiated by a diode laser with the generation wave length $\lambda_{gen} = 0.97$ μm, the radiation power on the fiber output $N \leq 10$ W, and the specific power of ≤ 5 W/cm^2 for $2 - 10$ min, up to evaporation of water and obtaining a dark nanomaterial.

V. The obtained BNCM samples were dried in air at room temperature or at $T = 30°C$.

Reference samples were materials consisting of BSA and K-354 carbon black (BSA+K-354) prepared using the analogous procedure.

Depending on the experimental parameters (albumin concentration, CNT type and concentration, laser irradiation power and time, and drying temperature and time), the BCNM consistency varied from paste to glass-like. The high quality of the obtained nanoproduct was confirmed by the absence of whitish areas of denaturized albumin. Then a no material retained its form and strength for a year and more, while the analogous products obtained from aqueous solutions of albumin with CNT using other techniques (heating or ultrasonic exposure) without laser irradiation decomposed into separate flakes for a few days after drying. The same occurred upon ordinary drying of the albumin aqueous solutions.

3. Experimental Methods

Below we present typical view of the prepared samples: BSA dried from the albumin aqueous solution (25 wt.% BSA) at room temperature (Figure 1a) and BCNM dried from the aqueous solution 25 wt.% BSA+0.4 wt.% MWCNT at a temperature of 30°C (Figure 1b). Figure 1 shows the BCNM obtained under laser radiation using the technique described above (dispersion 25 wt.% BSA+0.4 wt.% MWCNT). It should be noted that the samples obtained using the laser technique retained their form for over 3 years of storing under normal conditions. The other samples prepared without thermostating, i.e., without laser irradiation, spontaneously decomposed into small pieces upon storing for more than 20 h under normal conditions.

The nanomaterial samples were tablets 16 – 20 mm in diameter with a thickness of 4 – 5 mm and a mass of 1.5 – 2 g. Density ρ of the samples was measured by hydrostatic weighting in NEFRAS-S2_80/120 benzine. Vickers hardness of the BCNM samples was determined on PMT-3M microhardness tester. For this purpose, a BCNM piece with a linear size of 4 – 6 mm was fixed on a substrate using an epoxy adhesive. After drying at room temperature, the sample was grinded and polished. Finishing was made using diamond paste No. 2/1. These polished samples were used also for obtaining microphotographs of the BCNM surface.

Tensile strength σ was measured on samples in the form of plate bridges with a width of 2 mm and a thickness of 1 mm with a narrow central area with the square $S_n \approx 0.5 - 0.6$ mm^2. The plate with fixed ends was extended at a constant loading growth rate of ~10 N/min. At the moment of the sample break, the load force was measured with an AIGUZP-500N digital dynamometer accurate to ~ 0.1N. In all the measurements, the break occurred at the narrow plate place. The strength was calculated as $\sigma = F_l/S_n$, where F_l is the load force at the break.

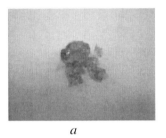 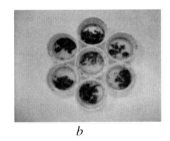 

a   *b*   *c*

Figure 1. BSA and BCNM (25 wt.% BSA+0.4 wt.% MWCNT) prepared after evaporation of the liquid component from the BSA solution and from the BSA+CNT dispersion using different techniques: (*a*) BSA, (*b*) BCNM after thermostating at
$T = 30°C$, and (*c*) BCNM after laser beam irradiation.

The average hardness was calculated using no less than 10 measurements at different sample points distant from one another by 150 – 250 μm. A coordinate table allowed moving the samples along two coordinate axes accurate to 10 μm. The average errors of H_V were determined at a confidence probability of 0.9. The average tensile strength was determined from the σ values obtained at the break of no less than three plates formed from one tablet.

4. RESULTS AND DISSCUSION

Figure 2 presents the hardness and tensile strength values for different materials, including the investigated BSA and BCNM.

It can be seen that H_V and σ values for the BCNM with CNTs are higher than the values for the BSA matrix by a factor $3 - 5$ and are comparable with the corresponding parameters of the human porous bone tissue or aluminum. Of special interest are density ρ, specific hardness H_V/ρ, and specific strength σ/ρ. These parameters are present in Table 1 together with the values for the investigated materials and some other well-studied materials. Since the prepared BCNM has a low density, which exceeds the density of water by only 25%, fairly high hardness (~ 350 MPa) and tensile strength (~ 35 MPa), the specific hardness and tensile strength of this material are comparable with those of the natural human porous bone tissue (see Table 1).

The hardness of the composite material BSA+K-354 carbon black (~ 50 MPa) was lower than that of the dried BSA (~ 70 MPa); i.e., addition of carbon in the form of fine carbon black particles with an average size of ≤ 10 μm did not lead to strengthening of the composite material. This result indicates an importance of introduction of CNTs for strengthening the BCNM.

Note that the obtained values of mechanical parameters are characterized by high repeatability in numerous experiments. In many case, the BCNM showed an insignificant hardness spread. Figure 3 shows typical BCNM surface microphotographs obtained on a PMT-3M microscope. One can see cross-shaped indentor traces and characteristic cracks. The indentor trace size at different 500 μm ×500 μm surface are as was nearly the same. The hardness values determined on the seareas were also similar. Therefore, we can conclude that the nanomaterial is homogeneous, at least on small areas. In addition, we observed the correlation between the crack density (the number of cracks per unit square of the BCNM surface) and the hardness of the samples, similar to that reported in [25]: as the CNT density was increased, the crack density decreased to $\leq 2 - 3$ cracks per 10 mm^2.

The crack density is especially high on the dried BSA surface [26], which apparently strongly degrades the strength of the material. However, CNTs not only strengthen the BSA matrix, but also eliminate small cracks (Figure 4). The crack width in the investigated BCNM samples was ≤ 50 μm. In the cracks with a width of ≤ 5 μm, we observed CNTs connecting the opposite crack edges (see Figure 1). The sample contains SWCNTs $\sim$ 1nm in diameter, but they mainly have a shape of bundles $\sim$ 7 nm in diameter or larger and are aggregated in accordance with the technical data [24]. In view of this, we may assume that here we observed the CNT bundles in the crack gap.

As was shown in [17 – 20], the bulk composite nanomaterials based on the water-protein dispersion of CNTs can acquire high mechanical performances under exposure to laser radiation during fabrication. However, the physical mechanism of this effect is still unclear. We can just make some assumptions.

Strengthening of materials by adding CNTs is the well-known fact. Nevertheless, the physical picture of this phenomenon is not always obvious. Moreover, laser radiation can differently affect the matrix and CNTs comprising a sample.

First, laser radiation of the liquid suspension can affect nanotubes with high conductivity and orient them mainly along the electric field of the laser radiation wave. This mechanism can lead to the strengthening of the matrix.

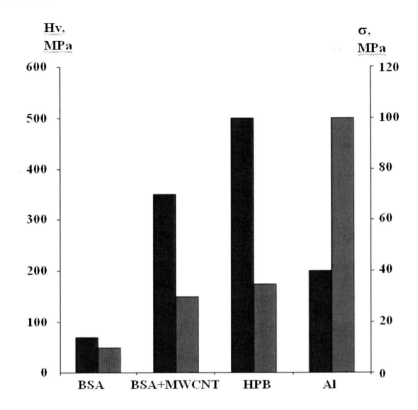

Figure 2. Maximum values of mechanical parameters for the BCNM obtained using the laser technique (25 wt.%BSA+0.4 wt.% MWCNT) and for other materials: BSA – bovine serum albumin; BSA+MWCNT – nanomaterial composed with bovine serum albumin and carbon nanotubes; HPB – human porous bone; Al – aluminum; hardness is shown with blue and tensile strength, with red.

Table 1. Comparable physical and mechanical parameters of different materials*

Material	ρ, kg/m^3	H_V/ρ, MPa/(kg/m^3)	σ/ρ, MPa/(kg/m^3)
BCNM 25 wt.% BSA+0.4 wt.% MWCNT	1260 ± 40	0.278	0.024
BCNM 25 wt.% BSA+0.4wt.% SWCNT	1250 ± 40	0.256	0.026
Albumin (dried)	1050 ± 40	0.067	–
BSA+K-354 carbon black	1100 ± 40	0.045	–
Human bone tissue [9]	1930	0.259	0.026
Aluminum	2700	0.074	0.037
Cast iron	7800	0.064	0.026

*Note: The relative error in determining the values of H_V/ρ and σ/ρ were ≤ 8%

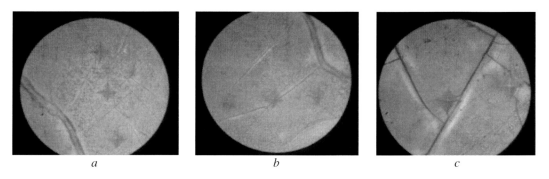

Figure 3. Microphotographs of the BCNM surface: the distance between the cross-shaped indentor traces is (*a, b*) 150 and (*c*) 500 μm.

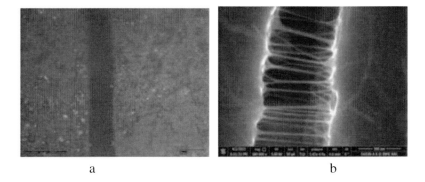

Figure 4. Typical cracks on the surface of BCNM (25 wt.% BSA+0.3wt.%S WCNT): (*a*) scale 20 μm (optical microspe) and (*b*) scale 500 nm (electron microscope).

Second, the electromagnetic field of laser radiation induces the electromotive force in conducting nanotubes. Plasma currents can flow between the close CNTs and weld them at the contact places. In this case, CNTs will be connected with one another and the CNT structure will exhibit the properties of a bulk strengthening frame. The latter will make the matrix strong and conductive. The rough estimates are $\geq 10^9$ A/m^2 for the density of current flowing between the neighboring CNTs and $\sim 10^5$ W/m^2 for the power density of laser radiation at a wave length of 970 nm. In the estimation, we assumed the CNT length to be ≥ 1 μm, the CNT width to be ≤ 10 nm, the square of the contour formed by a nanotube to be ~0.15 μm^2, the width of the dielectric spacer between nanotubes to be ≤ 5 nm, and the specific conductivity for an individual CNT to be $\geq 10^7$ S/m. The estimated current density is rather high; therefore, almost any material can melt under the action of such a current. Possibly, this occurs with CNTs at the contact places; i.e., nanomelting takes place. Our assumption can be related to stimulation of electric conductivity of the layers of a composite nanomaterial consisting of the carboxymethylcellulose matrix filled with CNTs in a concentration of ~ 4 wt.% [27]. Here, we established that the electric conductivity of the material layers increases by a factor of 4 – 6 due to the effect of laser radiation, while thermal annealing led to the electric conductivity growth by a factor of no more than 2 – 3.

Third, nanotubes and their bundles overlap the cracks (Figure 4b) and apparently prevent the crack growth. As a consequence, the number of cracks decreases, they are strengthened, and the mechanical strength of a material increases.

The same aqueous dispersion BSA+CNT demonstrated its high efficiency when used as a biological solder for laser welding of biological tissues [28, 29]. The laser welds of skin of pig and cartilage had strength of a few MPa. The physical mechanism of weld strengthening at the laser welding with the use of the BSA+CNT dispersion (biological solder) is apparently analogous to the mechanism of the investigated formation of the strong bulk composite nanomaterial under laser radiation.

CONCLUSION

Study of the composite nanomaterials based on the bovine serum albumin matrix filled with carbon nanotubes demonstrated the possibility of fabricating a bulk nanomaterial with the high mechanical parameters at the low CNT filler concentration (0.1 − 0.4 wt.% of MWCNTs or SWCNTs). In particular, the hardness exceeds the value for the BSA matrix by a factor of 5 − 7 and attains 350 MPa and the tensile strength attains 35 MPa. In addition, the bulk nanomaterial is light, since its density (1.25 g/cm^3) is higher than the densities of water and albumin by only ~ 20%. We obtained the specific hardness H_v / ρ ~ 0.24 MPa/(kg/m^3) and the specific strength σ / ρ ~ 0.025 MPa/(kg/m^3), while the corresponding parameters of the human porous bone tissue are H_v / ρ ~ 0.26 MPa/(kg/m^3) and σ / ρ ~ 0.026 MPa/(kg/m^3) [9].

Despite the fact that the question of the toxicity of CNTs causing heated discussions, studies on the development of new type of nanotube, or the development of new materials based on them, or use them in new directions, conducted with vigor and obtained important results. For example, it was found that the structure of the CNT nested cups are less toxic than MWCNT. The level of cytotoxicity of nanotubes can be considered such as a biocompatible material suitable for use as a means of delivering drugs to the desired area of the body [30]. Experiments in *vitro* showed stimulation growth of cells of neuroblastoma Neuro-2a and Normal human embryo fibroblast on the substrates coated with a layer BCNM [31]. Similar BCNM, we studied in this paper. In experiments in mice in *vitro* and in *vivo* first obtained evidences that CNTs can overcome the blood-brain barrier [32]. Nanotubes can also penetrate into the nuclei of pancreatic cancer cells [33]. Additional modification (selection of the type of CNT, their size, methods of functionalization, etc.) makes the carbon nanotubes and nanomaterials based on them entirely biocompatible [34]. Consequently, in this perspective, nanomaterials with CNT fit into new biocompatible materials of the 21st century [35, 36], including BCNM which we investigated now.

Thus, the composite nanomaterial consisting of the albumin matrix and carbon nanotube filler in both the bulk and liquid form will be useful for medical practice, including the creation of bone tissue implants and biological solders for laser welding of tissues.

ACKNOWLEDGMENT

We are grateful to Profs. Yu. P. Masloboev for useful discussions and D. I. Ryabkin for help in the experiments.

This work was financially supported by the Ministry of Education and Science of the Russian Federation (agreement No. 14.575.21.0089, RFMEFI57514X0089).

REFERENCES

[1] Yakobson, B.; Avouris, Ph. *Carbon Nanotubes, Topics Appl. Phys*. 2001, *vol. 80*, 287–327.

[2] Grobert, N. (2007). Carbon nanotubes – becoming clean. *Materials Today, vol.10(1-2)*, 28–35.

[3] Sinha, N.; Yeow, J. T.-W. *IEEE Transactions on Nanobioscience*. 2005, *vol. 4(2)*, 180–195.

[4] Zhang, D., Yi, C.; Zhang, J.; et al. *Nanotechnology*. 2007, *vol. 18*, 475102, (9 p.).

[5] Bonfield, W.; Tanner, E. *Source Materials World*. 1997, *vol. 5(1)*, 18–20.

[6] Fritsch, A.; Dormieux, L.; Hellmich, C.; Sanahuja J. *J. Biomed. Mater. Res. A*. 2009, *vol. 88(1)*, 149–161.

[7] Kovaleva, E.S.; Putlayev, V.I.; Filippov, Ya.Yu.; et al. Carbonated hydroxyapatite nanopowders for preparation of bioresorbable materials. *Mat.-wiss. u. Werkstofftech*. 2008, *vol. 39(11-12)*, 1–8.

[8] Kealley, C.; Elcombe, M.; van Riessen, A.; Ben-Nissan, B. *Physica B*, 2006, *vol. 385*, 496–498.

[9] Hench, L.; Jones, R. *Biomaterials, Artificial Organs and Tissue Engineering*. Woodhead Publishing. 2005, pp. 1 – 304.

[10] Wang, S.-F.; Shen, L.; Zhang, W.-D.; Tong, Y.-J. *Biomacromolecules*. 2005, *vol. 6*, 3067–3072.

[11] Abarrategi, A.; Gutierrez, M.C.; Moreno-Vicrnte, C.; et al. *Biomaterials*. 2008, *vol. 29(1)*, 94–102.

[12] Gusev, A.A.; Fedorova, I.A.; Tkachev, A.G.; et al. *Nanotechnologies in Russia*. 2012, *vol. 7(9-10)*, 509-516 (In Russian).

[13] Zhu, Y.; Zhang, X.; Zhu. J.; et al. *Int. J. Mol. Sci*. 2012, *vol. 13*, 12336–12348.

[14] Tolstikova, T.G.; Morozova, E.A.; Khvostov M.V.; et al. *Chemistry for Sustainable Development*. 2010, *vol. 18*, 445–452.

[15] Jackson, P.; Jacobsen, N.R.; Baun A.; et al. *Chemistry Central Journal*. 2013, *vol. 7(154)*, 1–21.

[16] Ichkitidze, L.P.; Komlev, I.V. *Lasers in Science, Engineering, Medicine*. 2010, *vol. 21*, 106–113 (In Russian).

[17] Andreeva, I. V.; Bagratashvili, V. N.; Ichkitidze, L. P.; et al. *Biomedical Engineering*. 2009, *vol. 43(6)*, 241–248.

[18] Ageeva, S.A.; Eliseenko, V.I.; Gerasimenko, A.Yu.; et al. 2011, *Biomedical Engineering, vol. 44(6)*, 233-236.

[19] Gerasimenko, A.Yu.; Dedkova, A.A.; Ichkitidze, L.P.; et al. *Optics and Spectroscopy*. 2013, *vol. 115(2)*, 283–289.

[20] Ichkitidze, L.; Podgaetsky, V.; Prihodko, A.; et al. *Materials Sciences and Applications*. 2012, *vol. 3(10)*, 728–732.

[21] Fu, K.; Huang, W.; Zhang, D.; et al. *J. Nanosci. Nanotech*. 2002, *vol. 2(5)*, 457–461.

[22] Zhao, X.; Liu, R.; Chi, Z.; et al. *J. Chem. Phys., B*. 2010, *vol. 114(16)*, 5625–5631.

[23] www.nanotam@yandex.ru.

[24] Krestinin, A.V.; Kharitonov, A.P.; Shul'ga, Yu.M.; et al. *Nanotechnologies in Russia.* 2009, *vol. 4*, 60–78 (In Russian).

[25] Ichkitidze, L.P.; Podgaetsky, V.M.; Ponomareva, O.V.; Selishchev, S.V. *Izvestia VUZov. Phisica.* 2010, *vol. 53(3/2)*, 125–129.

[26] Chudinov, A.V.; Rumyantseva; Lobanov, A.V.; et al. *Bioorganic Chemistry.* 2004, *vol. 30(1)*, 99–104 (In Russian).

[27] Ichkitidze, L.; Podgaetsky, V.; Selishchev, S.; et al. *Materials Sciences and Applications.* 2013, *vol. 4(5A)*, pp. 1–7.

[28] Gerasimenko, A.Yu.; Gubarkov, O.V.; Ichkitidze, L.P.; et al. *Semiconductors.* 2011, *vol. 45(13)*, 93–98.

[29] Ichkitidze, L.P.; Komlev, I.V.; Podgaetsky, V.M.; et al. *The method of laser welding of biological tissue.* Russian Federation Patent No. 2425700, 2011 (In Russian).

[30] Hanui, H.; Saito, N.; Matsuda Y.; et al. *International Journal of Nanomedicine*, 2014, *vol. 9(1)*, 1979–1990.

[31] Bobrinetskiy, I.; Gerasimenko, A.; Ichkitidze, L.; et al. *American Journal of Tissue Engineering and Stem Cell.* 2014, *vol. 1(1)*, 27–38.

[32] Kafa, H.; Wang, J.; Rubio, N. *Biomaterials.* 2015, *vol. 53*, 437–452.

[33] Andreoli, E.; Suzuki, R.; Orbaek, A.W.; et al. *J. Mater. Chem. B.* 2014, *vol. 2*, 4740–4747.

[34] Geng, J.; Kim, K.; Zhang J.; et al. *Nature.* 2014, *vol. 514(7524)*, 612–615.

[35] Hench, L.; Thompson, I. *J. R. Soc. Interface.* 2010, *vol. 7*, S379–S391.

[36] Gerasimenko, A.Yu.; Ichkitidze, L.P.; Podgaetsky, V.M.; Selishchev, S.V. *Biomedical Engineering.* 2015, *vol. 48(6)*, 310–314.

In: Proceedings of the 2015 International Conference …
Editors: Ivan A. Parinov, Shun-Hsyung Chang et al.

ISBN: 978-1-63484-577-9
© 2016 Nova Science Publishers, Inc.

Chapter 43

DYNAMIC CONTACT PROBLEM FOR A HETEROGENEOUS LAYER WITH A LIQUID SHEET ON A NON-DEFORMABLE FOUNDATION

M. A. Sumbatyan[1,], A. Scalia[2] and E. A. Usoshina[1]*
[1]Southern Federal University, Rostov-on-Don, Russia
[2]University of Catania, Catania, Italy

ABSTRACT

We study oscillations of a heterogeneous layer placed above a liquid sheet under an outer oscillating load. The oscillations of the poroelastic medium under condition of complete or particular water-saturation are described by the Biot consolidation theory. By applying the Fourier transform to the governing boundary value problem, we construct Green's matrix. Here are studied dispersive sets of the boundary value problem, and constructed formulae describing wave fields in the near- and far- zones. We also study the influence of the liquid layer on the surface wave fields. On the base of this problem, we solve a contact problem, by expanding the solution to a series by Chebyshev's polynomials.

Keywords: heterogeneous medium, porous material, wave process, Green's matrix, integral represenation

1. INTRODUCTION

The dynamic influence of massive objects on poroelastic foundations are modeled in frames of contact problems, which are of great interest for many researchers. This is due to various applications in seismology, geophysics, and civil engineering. For a homogeneous saturated porous elastic half-space a contact problem is studied in [1-4]. The less studied problems are now the influence of the heterogeneity of the layered porous medium and its

---

[*] E-mail: sumbat@math.sfedu.ru.

degree of saturation on the distribution of the contact stress. The most popular model here is the Biot theory of consolidation [5]. This type of medium well describes the processes in natural geological materials and rocks. In [6] there are listed mechanical characteristics of eleven different types of rocks, determined experimentally, by taking into account watering and water saturation that permits modeling with high accuracy, in frames of Biot's model, of the dynamic processes in real rock media.

To construct the solution, we propose an integral approach, which has previously been developed for elastic media; this describes correctly the waves irradiated to infinity [7]. We develop Green's matrix for the heterogeneous layer in the form, which permits stable calculations with arbitrary variation of arguments. Here are obtained integral relations describing wave fields, which are analyzed numerically in the near field, at the same time, in the far field here is used an asymptotic approach. On the base of this problem, the solution to the contact problem is constructed as an expansion by Chebyshev's polynomials.

2. FORMULATION OF THE PROBLEM

We study oscillations of a heterogeneous water-saturated and gas-saturated layer located above a layer of liquid. The layer of nonviscous liquid is located on a non-deformable foundation. In the Cartesian coordinate system, the liquid and the solid layers occupy the following domains, respectively:

$$
\begin{aligned}
-\infty < x < \infty, \quad -h_2 \le y \le 0, \\
-\infty < x < \infty, \quad 0 \le y \le h_1.
\end{aligned}
\tag{1}
$$

A harmonic outer load $Pe^{-i\omega t}$ is applied to the free surface of the layer, over the finite interval: $-a < x < a, \ y = h_1$. The displacements of the two-phase heterogeneous medium consisting of the elastic skeleton and the pores, filled of a mixture of viscous liquid and gas, are defined by Biot's equations [5]:

$$
\rho_{11}\frac{\partial^2 u_i}{\partial t^2} + \rho_{12}\frac{\partial^2 v_i}{\partial t^2} + b\left(\frac{\partial u_i}{\partial t} - \frac{\partial v_i}{\partial t}\right) = \sigma^s{}_{ij,j};
\tag{2}
$$

$$
\rho_{12}\frac{\partial^2 u_i}{\partial t^2} + \rho_{22}\frac{\partial^2 v_i}{\partial t^2} - b\left(\frac{\partial u_i}{\partial t} - \frac{\partial v_i}{\partial t}\right) = \sigma^f{}_{,i} \ ; i = 1,2,
$$

where $u_i(x,y,t), \ v_i(x,y,t), \ i = 1,2$ are components of the displacement vector in solid and liquid phases, respectively. Under assumption of linearity and isotropy of the porous solid phase, the Hook law for porous-elastic water saturated medium has the form:

$$\sigma_{ij}^s = Ae\delta_{ij} + 2Ne_{ij} + Q\varepsilon\,\delta_{ij}, \quad \sigma^f = Q\upsilon + R\varepsilon,$$

$$\upsilon = div\overline{u}, \quad \varepsilon = div\overline{v}, \quad \Gamma_{ij} = \sigma_{ij}^s + \delta_{ij}\sigma^f. \tag{3}$$

where Γ_{ij} is the tensor of full stress, e_{ij} and ε_{ij} are tensors of deformation for solid and liquid phases; δ_{ij} is Kronecker's delta: $\delta_{ii} = 1, \delta_{ij} = 0 \; i \neq j$, $b = \eta\,m^2/k_o$. Moreover, σ_{ij}^s is the stress tensor in the elastic skeleton, σ_{ij}^f is the stress acting in the liquid pores, A, N, Q, R are mechanical characteristics of the heterogeneous medium, which are expressed through volumetric compressibility of the elastic, fluid and gas phases [5], [6], $\rho_{12} < 0$ is the coefficient of dynamic coupling between the elastic skeleton and the fluid, ρ_{11}, ρ_{12}, ρ_{22} are the coefficients of dynamic mass density expressed through ρ_s, ρ_f, the mass densities of the solid and liquid media, coefficient b depends on the porosity of the medium m, upon the coefficient of viscosity in the pores and the coefficient of penetrability.

The displacement vector $\overline{w}\{w_i(x, y, t)\}$, $i = 1, 2$ and the pressure $p_0(x, y, t)$ in the non-viscous fluid of the lower layer is expressed in terms of wave potential $\varphi(x, y, t)$:

$$p_0(x, y, t) = -\rho_0 \frac{\partial\varphi}{\partial t}, \quad \overline{w} = grad\varphi. \tag{4}$$

Over the lower boundary between the liquid layer and the rigid foundation, the normal displacement is trivial:

$$w_2(x, y, t) = 0, \quad y = -h_2. \tag{5}$$

On the boundary between the fluid and the poroelastic media at $y = 0$, we assume a free filtration of the fluid through the boundary. For this one should provide equal normal stresses in the porous medium and the pressure in the fluid, the condition of trivial tangential stress in the porous medium, the continuity of the fluid motion inside and outside the porous medium:

$$\sigma_{22}^s + \sigma^f = (m-1)p_0, y = 0; \quad \sigma_{12}^s = 0;$$

$$\sigma^f = -mp_0; \quad (1-m)\frac{\partial u_2}{\partial t} + m\frac{\partial v_2}{\partial t} = \frac{\partial w_2}{\partial t}. \tag{6}$$

On the upper face of the layer $y = h_1$, on a finite-length interval, there is applied a uniform load with

$$\sigma_{22}^{s} + \sigma^{f} = -Pe^{-iwt},$$
$$u_2 = v_2, \ |x| \le a, \qquad (7)$$
$$\sigma_{22} = 0, \ \sigma_{12}^{s} = 0, \ |x| > a$$

To complete the formulation of the problem, one should use the radiation condition at infinity.

3. CONSTRUCTION OF THE SOLUTION

Since the oscillations are assumed harmonic, the time-dependent factor e^{-iwt} is hidden in all functions all over below. Let us represent the displacements as a sum of the three potentials:

$$\overline{u} = grad(\Lambda_1 + \Lambda_2) + rot\overline{\Lambda},$$
$$\overline{v} = grad(m_1\Lambda_1 + m_2\Lambda_2) + m_3 rot\overline{\Lambda}, \qquad (8)$$
$$\overline{\Lambda} = (0,0,\Lambda_3).$$

Let us further apply the Fourier transform to equations (1), (2):

$$\widehat{u}(\alpha, y) = \int_{-\infty}^{\infty} \overline{u}(x, y)e^{i\alpha x}dx .$$

Then set (1), (2) disintegrates to the three wave equations:

$$\Delta\widehat{\Lambda}_i + k_i^2 \widehat{\Lambda}_i = 0, \ i = 1,2,3, \ k_i = \frac{w}{V_i};$$
$$\widehat{L}_k(\alpha, y) = \int_{-\infty}^{\infty} L_k(x, y)e^{-i\alpha x}dx. \qquad (9)$$

In accordance with (7) in the heterogeneous layer, there can propagate waves of three independent types, V_1, V_2 are the wave speeds of "fast" and "slow" longitudinal waves, V_3 is the wave speed of the transverse wave. The fast longitudinal wave corresponds to an in-phase motion of the skeleton and the fluid; this propagates with a slow attenuation. In the slow longitudinal wave, the motions in the skeleton and in the fluid are anti-phased this attenuates more rapidly. If the connection between the elastic and the fluid phases is weak, then the wave speeds of the longitudinal waves approach respective wave speeds in the solid and the fluid media, separately. The displacements in the poroelastic medium are expressed in terms of solutions to wave equations (9), in the following form:

$$\hat{u}(\alpha, y) = \underline{\underline{B}}^s \overline{C}^s + \underline{\underline{B}}^k \overline{C}^k;$$

$$\underline{\underline{B}}^s = \begin{pmatrix} -i\alpha\hat{c}_1(y) & -i\alpha\hat{c}_2(y) & \gamma_3\hat{c}_3(y) \\ \gamma_1^2\hat{s}_1(y) & \gamma_2^2\hat{s}_2(y) & -i\alpha\gamma_3\hat{s}_3(y) \end{pmatrix};$$

$$\gamma_i = \sqrt{\alpha^2 - k_i^2}, \; i = 1,2; \tag{10}$$

$$\hat{s}_i(y) = (e^{\gamma_i(y-h_1)} - e^{-\gamma_i(y+h_1)})/\gamma_i;$$

$$\hat{c}_i(y) = e^{\gamma_i(y-h_1)} + e^{-\gamma_i(y+h_1)}.$$

Matrix $\underline{\underline{B}}^k$ is obtained from matrix $\underline{\underline{B}}^s$ by the change of $\hat{c}_i(y)$ by $\hat{s}_i(y)$ and of $\hat{s}_i(y)$ by $\hat{c}_i(y)$. Here $C^{s,k}$ are three-dimensional vectors of unknown constants. The elements of matrix (10) does not contain exponential functions increasing at infinity, their expressions are given in [8].

In order to satisfy boundary conditions, in the two-dimensional problem, we have 8 unknown constants. As a result, by satisfying boundary conditions (4) – (6), we come to a set of linear algebraic equations in respect to $C^{s,k}$. The coefficients of this set are some functions of the Fourier parameter. The solution of this set by a method of analytical transformation on computer, for example by using computer mathematical software Maple, seems to be inefficient. By such an approach, it is impossible to provide stable computations for large values of the argument. For this reason, the satisfaction to the boundary conditions is arranged in three steps, see [8]. As a result, we come to the representation for Green's matrix, which in the present case is a non-symmetric matrix of the following form:

$$\underline{\underline{G}}(\alpha, y) = \frac{1}{\Delta} \begin{pmatrix} g_{11} & g_{12} \\ i\alpha & i\alpha \\ g_{21} & g_{22} \end{pmatrix} \tag{11}$$

$$\Delta = s_1 s_{21} - s_2 s_{22} + s_3 s_{23}, \quad s_{21} = s_5 s_7 - s_6 s_8 + s_9, \quad s_{22} = s_4 s_7 + \alpha^2(s_6 s_{10} + s_{11}),$$

$$s_{23} = -s_4 s_8 + \alpha^2(s_5 s_{10} - s_{12}), \quad \gamma_{3+i} = q_6 \gamma_i^2 - k_i^2(q_4 + q_5 m_i), \; i = 1,2,$$

$$q_{6+i} = -k_i^2(q_{12} + q_{22}m_i), \quad \gamma_6 = \alpha^2 - 0.5 k_3^2, \quad \gamma_7 = \frac{iwp_0 mk_3^2 cth\gamma_0 h_2(1 - m + mm_2)}{2\gamma_0 q_8 H},$$

$$\gamma_0 = \sqrt{\alpha^2 - k_0^2}, \quad k_0 = \frac{w}{V_0}, \quad \gamma_8 = (q_7\gamma_5 - q_8\gamma_4)/q_8 q_6, \quad \gamma_9 = (q_8 - m\gamma_5)\gamma_7/q_6 m,$$

$$\hat{s}_i = \hat{s}_i(h_1)/\gamma_i, \quad \hat{c}_i = \hat{c}_i(h_1), \; i = 1,2,3, \quad q_{11} = \frac{A + 2N}{H}, \quad q_{12} = \frac{Q}{H},$$

$$q_{22} = \frac{R}{H}, \quad H = A + 2N + Q + R, \quad q_6 = \frac{2N}{H}, q_4 = q_{11} + q_{12} - q_6, \quad q_5 = q_{11} + q_{22}.$$

The explicit representations for m_i are present in [8]. Here V_0 is the velocity of wave propagation in the non-viscous fluid. For the sake of brevity, let us give the analytical expressions for the matrix element g_{22}, the expression for g_{ij} is given in [9]:

$$m_{i1} = 1 - m_i, \quad s_1 = \alpha^2(\gamma_1^2 \hat{s}_1 + \gamma_2^2 \hat{s}_2 q_9) + \gamma_6 \gamma_8 \hat{s}_3, \quad s_2 = \gamma_2^2 \hat{c}_1 + \gamma_9 q_5 \hat{c}_2 + q_6 q_8 \hat{c}_3,$$

$$s_3 = m_{11} \gamma_1^2 \hat{s}_1 + q_9 m_{21} \gamma_2^2 \hat{s}_2 + m_{31} \gamma_8 \hat{s}_3, \quad s_4 = \alpha^2(\gamma_2^2 \gamma_7 \hat{s}_2 / \gamma_6 - \hat{c}_3) + \gamma_9 \hat{s}_3,$$

$$s_5 = (\gamma_5 \gamma_7 \hat{c}_2 + q_6 \gamma_9 \hat{c}_3 - \alpha^2 \gamma_3^2 q_6 \hat{s}_3) / \gamma_6, \quad s_6 = (m_{21} \gamma_2 \gamma_7 \hat{s}_2 + m_{31} \gamma_9 \hat{s}_3 - m_{31} \alpha^2 \hat{c}_3) / \gamma_6,$$

$$s_7 = m_{21} \hat{c}_2 - m_{11} \hat{c}_1, \quad s_8 = \gamma_4 \hat{s}_1 - \gamma_5 \hat{s}_2, \quad s_9 = m_{21} \hat{c}_2 \gamma_4 \hat{s}_1 - m_{11} \gamma_5 \hat{c}_1 \hat{s}_2,$$

$$s_{10} = \hat{c}_1 - \hat{c}_2, \quad s_{11} = (m_1 - m_2) \hat{c}_1 \hat{c}_2, \quad s_{12} = \gamma_4 \hat{c}_2 \hat{s}_1 - \gamma_2 \hat{c}_1 \hat{s}_2, \quad s_{18} = s_1 s_6 - s_3 s_4,$$

$$g_{22} = -\gamma_1^2 s_{22} \hat{s}_1(y) + \gamma_2^2 s_{42} \hat{s}_2(y) + s_{44} \hat{s}_3(y) + (s_{18} + s_{25}) \hat{c}_1(y) +$$

$$+ (s_{24} - s_{18}) \hat{c}_2(y) - \alpha^2 / \gamma_6 s_{32} \hat{c}_3(y),$$

$$s_{23+i} = (s_i \gamma_i m_{i1} - \alpha^2 \gamma_3 s_3) \hat{c}_i, \quad i = 1,2, \quad s_{40+i} = (-1)^{i+1} q_9 s_{2i} + \gamma_7 / \gamma_6 s_{3i},$$

$$s_{42+i} = (-1)^{i+1} \gamma_8 s_{2i} + \gamma_9 / \gamma_6 s_{3i}.$$

The elements of Green's matrix are oscillating functions decreasing at infinity as some power functions, meromorphic in the complex plane. With $\alpha \to \infty$, $\Delta = O(\alpha^4)$.

4. RESULTS OF NUMERICAL ANALYSIS AND CONCLUSION

The denominator of the integrand has complex-valued zeroes, located near the real axis. With frequency increasing, the bahaviour of the neutral curves rapidly changes, the imaginary part of the zeroes increases.

The dislacement field in the poroelastic layer placed above the liquid strip, for arbitrary load $\bar{q}(x, y)$, on the free space can be represent in the form

$$\bar{u}(x, y) = \frac{1}{2\pi} \int_{R_1} \underline{\underline{G}}(\alpha, y) \hat{q}(\alpha, y) \alpha e^{i\alpha x} d\alpha. \tag{12}$$

When formulating the contact problem, the integrand in (12) is unknown, this can be sought as a series by the Chebyshev polynomials, with an explcit extraction of the square-root singularity near the punch ends. The presence of the liquid sheet significantly modifies the wave process on the free surface of the layered medium. The dispersion of the heterogeneous layer is considerably greater than for similar homogeneous layer placed without friction above the rigid foundation. For low frequencies, the character of the surface waves rapidly modifies, due to accumulation of energy. This has been confirmed by experiments.

ACKNOWLEDGMENT

The first author is grateful to the Russian Scientific Foundation (RSCF), for its support by Project 15-19-10008.

REFERENCES

[1]   Trofimchuk, A.N.; Gomilko, A.M.; Savitsky O.A. *Dynamics of Porous Elastic Fluid-Saturated Media*, Kiev: Naukova Dumka, 2003 (in Russian).

[2]   Molotkov. L.A. *Wave Propagation in Porous and Cracked Media on the Base of Effective Biot's Models and Layered Media*, St.-Petersburg: Nauka, 2001 (in Russian).

[3]   Scalia, A.; Sumbatyan, M.A. *J. Elasticity*, 2000, *vol. 60*, 91 – 102.

[4]   Jin, B.; Jin, H. *Soil Dyn. Eathquake Eng.* 1999, *No.18*, 437 – 443.

[5]   Biot, M.A. *J. Acoust. Soc. America*, 1956, *vol. 28*, 168 – 178.

[6]   Chao-Lung, Yeh; Wei-Cheng, Lo; Jan, Chyan-Deng. *American Geophysical Union, Fall Meeting*, 2006, *No. 12*.

[7]   Vorovich, I.I.; Babeshko, V.A. *Dynamic Problems of the Elasticity Theory with Mixed Boundary Conditions for Non-classical Domains*, Nauka, Moscow, 1976 (in Russian).

[8]   Suvorova, T.V.; Usoshina, E.A., Construction of Green's matrix for poroelastic layer oscillating over a layer of fluid, Proc. XII Intern. Conf. "Modern Problems of Continuum Mechanics," Rostov-on-Don, 2008, *vol. 2*, 127 – 132 (in Russian).

[9]   Edelman, I.; Wilmanski, K. *Continuum Mech. Thermodyn.*, 2002, 25 – 44.

In: Proceedings of the 2015 International Conference … ISBN: 978-1-63484-577-9
Editors: Ivan A. Parinov, Shun-Hsyung Chang et al. © 2016 Nova Science Publishers, Inc.

Chapter 44

THE DEVELOPMENT OF METHODS FOR THE DETERMINATION OF THERMAL AND TRIBOLOGICAL CHARACTERISTICS OF THE FRICTION SURFACES

P. G. Ivanochkin[1,], S. I. Builo[2,†], I. V. Kolesnikov[1,‡] and N. A. Myasnikova[1,§]*

[1]Rostov State Transport University, Rostov-on-Don, Russia
[2]Vorovich Mathematics, Mechanics and Computer Sciences
Institute, Southern Federal University, Rostov-on-Don, Russia

ABSTRACT

The results of the experiment confirmed the validity of subsurface maximum temperature in metal tribosystem predicted by the theory. During the experiment there were used the diagnostic method of temperature field based on the use of surface acoustic Rayleigh waves. We established that the cause of the maximum temperature was not on the surface, but in the surface layer caused by the change of boundary conditions.

To diagnose the condition of the tribomating surface "metal-composite material," we conducted spectral analysis of AE signals and made a conclusion that before the destruction of the frictionally transferred film there existed a large number of discrete frequencies on the surface of tribocontact in the spectrum recorded with AE. The essence of AE method lays in the analysis of extremely weak ultrasonic radiation accompanying any change in the structure of the observed material.

While investigating friction pairs we found a strong distortion and overlapping of AE signals. Thus, it is necessary to carry out the recovery procedure not by the pulse intensity, but by the intensity of oscillations flow of signals recorded on the surface of the friction pair.

---

[*] P.G. Ivanochkin, e-mail: ivanochkin_p_g@mail.ru.

[†] S. I. Builo, e-mail: bsi@math.sfedu.ru.

[‡] I.V. Kolesnikov, e-mail: kvi@rgups.ru.

[§] N.A.Myasnikova, e-mail: myasnikova@rgups.ru.

The observed strong correlation between friction factor with the AE flow acts allows one to conduct a remote rapid assessment of the friction factor according to the AE tests data without direct measurement of this factor.

Keywords: surface of tribocontact, friction coefficient, temperature field, surface acoustic wave, thermal characteristics, acoustic emission, friction transfer film, frequency analysis, initial stage of destruction

1. Determination Methods of the Friction Surfaces Thermal Characteristics

The ascertainment of the specific behavior of the surface layers of metal polymer - tribocontact is one of the central problems in tribotechnology. Therefore, for a deeper knowledge of the processes at the contact, it is necessary to develop not only theoretical models, but also diagnostic methods that take into account changes in the scope and boundary layer.

Polymer-based materials in significantly greater extent than metals are susceptible to the impact of many factors. Among these factors is friction, the influence of the environment, and temperature. In the process of friction under the influence of heat, there are significant changes in material properties in the volume and at the surface layers, influencing the mechanical, physical and tribological properties. The specificity of polymer materials is also that their work in tribosystems accompanies with the processes of triboelectrification, diffusion and degradation with the formation of chemically active products, which react with the metal counterbody forming a transfer film on it.

Despite numerous and significant studies of the phenomenon of friction transfer, the question of mechanism of formation of the transferred layer is not completely explored and ascertained. Recent researches found out the variety of friction transfer shapes with solid and focal films formations - the fragments of "third body" in contacts "polymer-polymer," "metal-polymer." The complexity of the problem is that the volumes of the body actively involved in the process of friction are extremely low and there are no experimental methods for studying the processes occurring in the surface layer.

In present, there are theoretical models developed for the analytical study of the processes at the contact that not only take into account the changes in volume, but also in a thin surface layer, as shown in [1]. In tribotechnology the asymptotic methods can be used to solve such problems. However, carrying out such calculations requires enormous amount of time and does not allow conducting analysis of the thermo-elastic characteristics of the frictional tribosystem. In such cases, one of the most effective solutions is the finite element method. We calculated the temperature field of a particular pairing "shaft - partial bearing," as the most typical and widespread tribomating.

The calculations made by the method of regularization of singularly perturbed problems and the finite element method showed the presence of subsurface maximum, which means that in the friction area it was created negative temperature gradient [2]. In addition, we defined that the whence of the negative gradient was the maximum temperature was not on the surface but in the surface layer. It was caused by both changing boundary conditions (there were heating under the bearing and cooling outside it) and unsteady loading of metal-polymer interfacing. Notwithstanding, to confirm theoretical research, it is necessary to have a procedure for measuring the temperature field in a thin surface layer.

The use of existing methods for the determination of the temperature field does not allow conducting its monitoring in a thin surface layer during the friction process. To estimate the temperature field in the surface layer of the tribo interfacing, we worked out the technique of layer precision measurement of velocity and attenuation of ultrasonic Rayleigh waves. For this purpose, a twin-turbo prop with acoustic line performing the role of the friction surface is set on the reciprocating friction machine. On the right and left ends of the acoustic line, there are wedge-mounted piezoelectric transducers with piezoelectric plates. Under an influence of generator, electric signal transducer excites a bulk wave that spreads at the speed υ_{o6} towards the direction of acoustic line surface and in contact with it transforms into a surface acoustic wave. The physical properties of the material determine the speeds of bulk wave υ_{o6} in the piezoelectric transducer wedge and a surface acoustic wave υ_a in the acoustic line. Efficient excitation of a surface wave occurs at a ratio, $\sin \alpha = \upsilon_{o6} / \upsilon_a$, i.e., the angle of wave incidence α is selected in such a way that the projection of the velocity vector of the bulk wave in the wedge of the surface acoustic line is numerically equal to the modulus of the velocity vector of the surface wave. Then, the spatial period of the elastic deformation on the bottom surface of the wedge is equal to the length of a surface acoustic wave in the acoustic line. This is due to the effective transformation of the bulk wave into the surface one, which output transforms into a bulk with the help of the same wedge-shaped piezoelectric transducer. Then the bulk wave with the help of piezoelectric crystal plates transforms into an electric signal recorded by the oscilloscope. Counterbody moves across reciprocating the direction of propagation of the surface wave. When changing the boundary conditions (load, speed, heat transfer) temperature field changes in acoustic line that leads to heterogeneous perturbed surface layer. At the same time, there is a strong correlation between the shape of the curve dispersion of the surface wave and the properties of this layer. In result of changes of relative fringes of resonance peaks, we obtain the frequency dependence of the quality factor and running attenuation, respectively. This frequency dependence gives us information about change of the temperature field in depth. Indeed, the surface energy of the Rayleigh wave is localized in the layer thickness $\lambda - 1.5\lambda$ and hence the depth of the wave penetration depends on its frequency.

Thus, by varying the frequency of the surface wave excitation, we get a picture of the distribution of the temperature field in the surface layer.

Moreover, to quantify the temperature field there were preliminary altimeter graphs establishing the connection between the indices of apparatus and the temperature. Then we conducted the friction measurements. In result of the research, we established, that, depending on the operating conditions (load, speed) and heat removal the maximum temperature zone locates at a different distance from the friction surface.

2. THE USE OF THE ACOUSTIC EMISSION FOR TRIBOLOGICAL CHARACTERISTICS ANALYSIS

The research of the processes occurring in the friction zone in real scale of time, involves considerable difficulties because of the lack of direct access to the contact zone. A very effective way of obtaining information on the wear processes is to record acoustic signals in friction zone, i.e., the method of acoustic emission (AE). The method presents analysis of AE parameters of weak ultrasonic radiation accompanying any changes in the structure of the observed material. The mechanism of the elastic impulse relates to the asperity summits deformation of the interacting surfaces and breaking adhesive bonds arising from the external friction. Thus, the processes of formation and destruction of the frictionally transferred film at the metal-polymer tribo interfacing will accompany by AE radiation.

Currently there are various options of statistical treatment of AE signals: emission method, the method of amplitude distribution analysis, the method of frequency analysis, correlation analysis method, the method of spectral density, estimation of the standard deviation. Several researches [3-4] relate the value of standard deviation of AE signal to the parameters of wear. The study [5] represents the relationship between wear and a spectral width of random fluctuations of AE signals amplitudes.

In addition, to evaluate the characteristics of frictional interaction, the pairs "roller – bush" (from composite material) continued development of the method of acoustic emission (AE) diagnostics of friction. The method of AE uses to analyze the extremely weak ultrasonic radiation accompanying any change in the structure of the material. Tests were carried out on the machine SMT-1 friction in sliding on a "partial-shaft liner." The authors studied the relationship between the AE parameters in a wide (30–500 kHz) frequency range and the features of the friction interaction in the roller - bush pair. The AE transducer was compressed to the end of the brake shoe by a "magnetic chuck" through a coupling medium layer. Obtained by the transducer the AE pulses were amplified by a preamplifier. Then they go to the input of an A_Line 32D (Interyunis Co.) digital measuring device.

The radiation mechanism of the AE pulses themselves in the friction processes has an interdisciplinary complicated physical–chemical–mechanical nature. In our opinion, it can be present in the following way. It is well known that in friction processes adhesive bonds form between directly osculating irregularities of contacting surfaces and broken. The collective breaking of these bonds is an event of a physicochemical interaction, that is accompanies by radiation of an elastic pulse and can be considered as a certain AE event [6]. We established that in a wide range of physicochemical processes, AE pulses of a detectable level emitted actually. Therefore, there are good prospects for developing qualitative methods for estimation of the kinetics of the processes investigating the parameters of the accompanying acoustic emission [7, 8].

In the absence of a lubricant, AE of a much higher level should be expected because of the predominance of the physical mechanical mechanism of the AE in the course of deformation of the material of irregularities itself. It is well known that real contacts "metal-tail" occur only on the tops of irregularities, whose contact area is a small fraction of the nominal area. Therefore, even under small loads, the tops of the irregularities deform, which must also inevitably lead to radiation of the AE events of a sufficiently high level owing to alterations in the internal structure of the deformed material.

From analysis of the obtained signals, it follows that the friction interaction processes accompany by such a large number of radiated AE events that it leads to an almost complete superimposition of the recorded AE pulses. Investigations show that practically all AE diagnostic devices of domestic and foreign manufacturers, including such well-known systems as Spartan and A-Line 32D, miss more than 90% of the AE events with a high intensity of the AE sources inside the body owing to superimposition [9, 10].

To eliminate this drawback, the authors developed and proposed an interdisciplinary approach that permits restoration of the true, i.e., radiated inside a body, AE event stream using their detectable signals. We obtained the following relationship for the case of restoration by the oscillations [9-10]:

$$\dot{N}_a = \dot{N}/(f - \dot{N})\tau. \tag{1}$$

Here, $\dot{N}_a$ is the intensity, i.e., the number per time unit of the AE event stream radiated (restored) inside the material;

$\dot{N}$ is the count rate, i.e., the number per time unit of the oscillations of the recorded AE pulses;

f is the basic frequency of the AE pulses, which is close to the transducer's resonance;

τ is the time constant of the AE pulse reverberation in the specimen and the transducer.

The reverberation time constant τ can be easily estimated by the propagation of a short calibration signal.

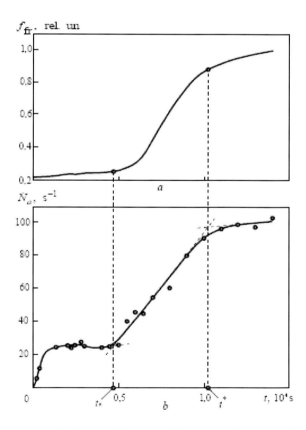

Figure 1. (a) Dependence of the friction coefficient and (b) restored intensity of the accompanying AE stream on the time during the testing of the roller-shoe pair with a multilayer nanostructured antifriction coating.

Figure 1 shows simultaneous graphs of changes in the friction coefficient $f_\text{fr}(t)$ and of the intensity of the AE event stream $\dot{N}_a(t)$, which was restored with the use of (1), during the test of the multilayer antifriction coating. It can be seen that these dependencies are similar, S-shaped, and have critical points.

At the steady-state friction stage, the intensity of the AE event stream maintains an approximately constant value and the appearance of the lower break in the friction coefficient at time moment t_* accompanies by a corresponding lower break in the restored intensity of the AE event stream. Interestingly, this break occurs before the appearance of discrete frequencies in the AE spectrum. The occurrence of the upper break in the intensity of the AE event stream at time moment t^* practically coincides with a similar break in the friction coefficient.

Investigation of the structure of the friction surfaces of the specimens showed that the lower break in the AE event flow at $t = t_*$ corresponded to the transition point from the steady-state to the initial stage of destruction of the friction layer (transfer film on the metal surface), or for example, by increasing the temperature or load. The upper break at $t = t^*$ correlated with the transition point to the beginning of the friction stage without the protective layer.

Thus, it becomes possible to diagnose the initial stage of a precollapse condition of the antifriction transfer film by revealing the lower break in the restored stream of the concurrent AE events and to diagnose the beginning of the friction stage itself without the protective layer by recording the upper break of the AE event stream. The strong correlation of the friction factor with the AE event stream must allow rapid evaluation of the friction coefficient according to the data of AE tests without directly measuring this coefficient in the future.

CONCLUSION

To estimate the temperature field in the surface layer of tribo interfacing we worked out the technique of layer precision measurement of velocity and attenuation of ultrasonic Rayleigh waves. By changing the frequency of the surface waves' excitation and having built a special dependence of quality factor and running attenuation, respectively, we get a picture of the temperature field distribution in the surface layer. It is shown that the position of this zone is displaced from the surface with decreasing intensity, and increasing the heat elimination from the friction surface.

We developed a method of diagnostics of the stage of coating catastrophic destruction with aim to increase the amplitude and emergence of a large number of discrete frequencies in the spectrum of recorded AE. For the first time two specific points of the AE acts flow, corresponding to the change of friction stages, were found. A method of diagnostics of the early stage of pre-demolishing state of frictionally transferred film because of the emergence of the lower turning point of the reduced flow acts related AE and diagnostics of the initial friction stage without the protective layer on the upper registration fracture of AE acts flow.

The strong correlation of coefficient of friction with AE acts flow in the future will allow conducting its remote rapid assessment without direct measurement. Preliminary experiments confirm the prospects of this approach to the AE diagnosis of the materials for friction units, which should provide a significant increase in safety and reliability of products by monitoring parameters accompanying acoustic radiation.

ACKNOWLEDGMENTS

The Russian Science Foundation (grant number 14-29-00116) supported this study.

REFERENCES

[1] Kolesnikov, V. I. *Thermo-Physical Processes in the Metal-Tribosystems*. Science, Moscow, 2003, pp. 1 – 279 (in Russian).

[2] Kolesnikov, V. I.; Chebakov, M. I.; Kolesnikov, I. V.; Lyapin, A. A. *Transport. Science, Technology, Management. Scientific Information Proceedings*, 2015, *No. 1*, 6-11 (in Russian).

[3] Kannatcy-Asibu, E.; Dornfeld, D. A. *J. Eng. Ind.*, 1981, *vol. 103(3)*, 330-340.

[4] Jiaa, C.L.; Dornfeld, D.A. *Wear*, 1990, *vol. 139(2)*, 403-424.

[5] Kozyrev, Yu. P.; Sedakov, E. B. Methods of Non-Destructive Testing of Wear of Polymeric Materials (for example Polyimides), Based on the Correlation and Spectral Analysis of the Amplitude Fluctuations of Flow of Acoustic Emission. In: *Proceedings of the International Congress "The Mechanics and Tribology of Transport Systems 2003,"* RSTU Press, Rostov-on-Don, 2003, *vol. 1*, 416-420.

[6] Builo, S. I.; Ivanochkin, P. G.; Myasnikova, N. A. *Russian Journal of Nondestructive Testing,* 2013, *vol. 49(6)*, 318-322.

[7] Builo, S. I.; Kuznetsov, D. M.; Gaponov, V. L. Chapter 13. Acoustic Emission Diagnostics of the Kinetics of Physicochemical Processes in Liquid and Solid Media. pp. 193–208, In: *Advanced Materials. Studies and Applications. Ivan A. Parinov, Shun-Hsyung Chang and Somnik Theerakulpisut* (Eds.). New York: Nova Science Publishers, 2015, pp. 1 – 527.

[8] Builo, S. I.; Kuznetsov, D. M. *Russian Journal of Nondestructive Testing*, 2010, *vol. 46(9)*, 686–691.

[9] Builo S.I. Chapter 15. Physical, Mechanical and Statistical Aspects of Acoustic Emission Diagnostics. pp. 171–184. In: *Physics and Mechanics of New Materials and Their Applications.* Ivan A. Parinov, Shun-Hsyung Chang (Eds.). New York: Nova Science Publishers, 2013, pp. 1 – 444.

[10] Builo, S. I. *Physicomechanical and Statistical Aspects of Increasing the Reliability of Results of Acoustic-Emission Testing and Diagnostics*, Southern Federal University Press, Rostov-on-Don, 2008, pp. 1-192 (in Russian).

[11] Builo, S. I. *Russian Journal of Nondestructive Testing*, 2008, *vol. 44(8)*, 517–526.

In: Proceedings of the 2015 International Conference …   ISBN: 978-1-63484-577-9
Editors: Ivan A. Parinov, Shun-Hsyung Chang et al.   © 2016 Nova Science Publishers, Inc.

Chapter 45

INVESTIGATION OF THE INDENTER TEMPERATURE AND SPEED EFFECT DURING INSTRUMENTED INDENTATION ON THE MECHANICAL PROPERTIES OF CARBON STEELS

E. V. Sadyrin[1,], L. I. Krenev[1], B. I. Mitrin[1], I. Yu. Zabiyaka[1], S. M. Aizikovich[1] and S. O. Abetkovskaya[2]*

[1]Research and Educational Center "Materials,"
Don State Technical University, Rostov-on-Don, Russia
[2]A. V. Luikov Heat and Mass Transfer Institute, National
Academy of Sciences of Belarus, Minsk, Belarus

ABSTRACT

In the present paper we have studied the heating effect of an indenter and a sample, as well as the speed of indentation, on the hardness and the Young's modulus of carbon steels. The carbon steel used in the study as a sample was steel 40Cr. We have fulfilled the experiments on NanoTest 600 Platform 3 indentation test machine and used Oliver and Pharr method to analyze the results. Heating was carried out in two modes. Experiments were divided by series consisting of eight indents, each. The parameters of each indentation in a series, we kept constant. The duration of one indentation varied from 30 to 150 s in different series to investigate the dependence of calculated mechanical properties from the indentation speed. The results indicate that for a correct determination of material properties at elevated temperatures using the standard software of the device and Oliver – Pharr method should be maintained the same temperature for a sample and an indenter.

---

[*] Corresponding author: E. V. Sadyrin. Research and Educational Center "Materials," Don State Technical University, Rostov-on-Don, Russia. E-mail: evgeniy.sadyrin@gmail.com.

Keywords: instrumented indentation, hardness, Young's modulus, heating, indentation speed

1. INTRODUCTION

Parts of machines, structures and microelectronics are subject to significant heating in operation. Raising temperature causes change in physical and mechanical properties of materials from which the components are made that leads to deterioration in both strength and functional characteristics. The necessary condition for the industrial development of new materials and coatings is to consider the dependence on temperature of mechanical properties of materials. These high-tech industries include biotechnology, nanoelectronics, energetics and mechanical engineering. The modern instrumented indentation test machines [1] allow carrying out experiments under controlled temperature and have several advantages [2] over traditional methods of materials testing, such as tension/compression tests and dynamic methods [3], which used before to study temperature dependencies of mechanical properties. Instrumented indentation methods allow one to carry out repetitive tests on one sample (without its destruction), impose less stringent requirements for the preparation of the samples, make it possible to research films and layers with a thickness from few atoms to several micrometers, recording changes in their properties due to the influence of the substrate and the environment.

The aim of the present study is to investigate the heating effect of the indenter and the sample, as well as the speed of indentation, on the hardness H and the Young's modulus E of carbon steels, determined from the results of instrumented indentation. Oliver and Pharr method was used during analysis of the results. The carbon steel used in the study as a sample was steel 40Cr.

2. EXPERIMENTAL STUDY

We have fulfilled the experiments on NanoTest 600 Platform 3 indentation test machine [4], equipped with the cabinet, in which the temperature kept constant and functional block "Nanotest," allowing one to apply load in the range from 0.5 mN to 500 mN. In all the experiments, we used a Berkovich indenter with the diamond tip with the radius of curvature at the top ~ 100 nm. The indenter is adapted for experiments at temperatures up to 500°C.

"Hot stage" construction used to perform heating, namely thermally insulating ceramic block with a sample holder. Between it and a pendulum, the thermal protection screen mounted, preventing overheating of the platform. The indenter heated with special heating element. The sample mounted on the sample holder, using the high-temperature silicone gasket maker Red RTV Silicone Gasket Maker (Doctor Chemical Corp., US). It took 24 hours from gluing the sample to the first experiment that corresponded to the requirements of the gasket maker. During the experiment, the temperature was controlled by the temperature sensors Eurotherm 2216e (Eurotherm, UK).

In order to protect the platform from mechanical vibrations, it was placed on the antivibration air-damped table. The optical microscopes, positionally synchronized with the indenter, can be used not only to monitor the residual indentation imprint after unloading, but also to seek the place for indents free from all sorts of surface defects (4 lens are used, the maximum zooming up is 400-fold) [5]. It is important that the measurement results were not affected by the presence of overlaps and sags in the contact area, caused by the previous series of indentations. For this purpose, the minimum distance between imprints was kept at least five times greater than the largest diameter of the imprint [6]. Also, the indents were positioned so that to space from the boundaries of the sample by a distance of at least three of their diameters.

First, we have carried out a test experiment at room temperature. Calculated Young's modulus E of the sample was 3.17 ± 0.23 GPa, calculated hardness H was 212.19 ± 4.9 GPa.

After that, a heating was carried out in two modes:

1. the heating of the indenter and the sample to the same temperature with further indentation (isothermal contact); the results represent monotonically decreasing dependence of the calculated hardness from 3.19 GPa at 85°C to 2.63 GPa at 200°C; the calculated value of the Young's modulus shows a non-monotonic behavior, previously mentioned by [3] and others;
2. the heating of the indenter and the sample to the different temperatures (T_{ind} is 85°C, whereas T_{sam} increased from 85°C to 200°C from one experiment to other) with further indentation (non-isothermal contact); the results presented an increasing dependence of the calculated hardness from 3.19 GPa at the sample temperature 85°C to 5.07 GPa at the sample temperature 200°C; they presented also a monotonically decreasing dependence of the calculated Young's modulus from 192.16 GPa at the sample temperature 85°C to 136.75 GPa at the sample temperature 200°C (Figures 1 and 2).

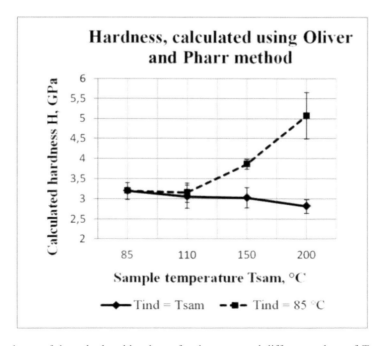

Figure 1. Dependence of the calculated hardness for the same and different values of T_{ind} and T_{sam}.

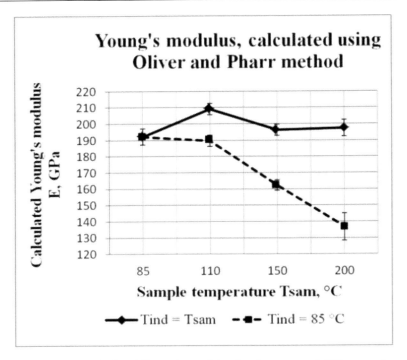

Figure 2. Dependence of the calculated Young's modulus for the same and different values of T_{ind} and T_{sam}.

Experiments divided by series consisting of eight indents each. The parameters of each indentation in a series were kept constant. The duration of one indentation varied from 30 to 150 s that corresponded to different indentation speeds. This duration consisted of the following three time periods:

1. the time of applying a test load, s;
2. the holding time under load, s;
3. the time of removing of the test load, s.

Further, referring to the indentation time, we shall use the following notation: $t_1/t_2/t_3$, where t_1 is the time of applying test load, t_2 is the holding time under load, t_3 is the time of removing of the test load. After removing 90% of maximum load, the remaining load was held for 60 seconds for thermal drift rate measurements [5].

We found out that in the selected range, the indentation speed had no effect on calculated values of H and E at the same temperature of the indenter and the sample ($T_{ind} = T_{sam}$), whereas at $T_{ind} \neq T_{sam}$ the difference amounted 19% for E and 36% for H. This may be due to the heat flux between the indenter and the sample (Figures 3 and 4).

CONCLUSION

Dependence of Young's modulus of the material on temperature has been determined from the results of instrumented indentation during isothermal contact of indenter and sample.

The dependence we obtained is similar to that presented in the literature [3]. It made possible to suggest the instrumented indentation as an alternative method to determine the Young's modulus dependence on temperature.

We also carried out a number of experiments on non-isothermal instrumented indentation. Standard software of the test device was used for analysis of experimental data. Calculated mechanical characteristics are found to be significantly affected by temperature difference between the sample and the indenter, leading to physically unrealistic results.

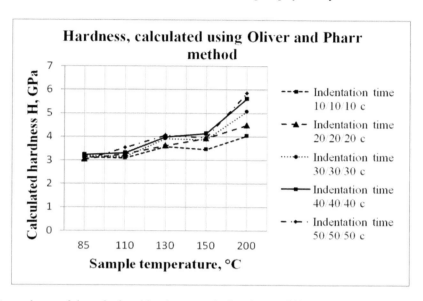

Figure 3. Dependence of the calculated hardness on the heating at different indenter speeds (indenter temperature is 85°C, sample temperature increases from 85°C to 200°C).

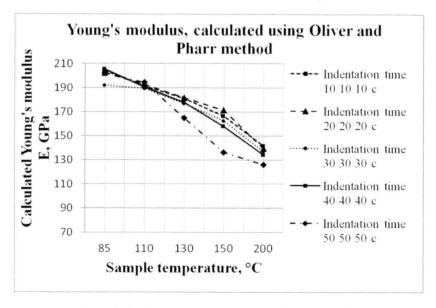

Figure 4. Dependence of the calculated Young's modulus on the heating at different indenter speeds (indenter temperature is 85°C, sample temperature increases from 85°C to 200°C).

Therefore, when using the standard methods for determining the mechanical properties of materials, it is required to perform simultaneous and uniform heating of both the indenter and the sample during the test. Today, rather small amount of devices for instrumented indentation have ability to satisfy these requirements.

Interpretation of the indentation results in the case of temperature difference between the indenter and sample requires the development of appropriate mathematical methods.

ACKNOWLEDGMENTS

This work was supported by RFBR grants Nos. 13-08-01435-a, 14-07-00343-a, 15-07-05208-a, 15-07-05820-a, 15-57-04084-Bel_mol_a. Aizikovich S.M. acknowledges financial support from Ministry of Education and Sciences of Russia in the framework of "Organization of Scientific Research" Government Assignment. The authors express gratitude to Yu. L. Goncharova for her help with English editing.

REFERENCES

[1] *Nanotest Platform* [Internet], 2001 [updated 2015 March; cited 2015 May 18]. http://www.micromaterials.co.uk.

[2] Golovin, Yu. I. *Nanoindentation and its Capabilities*. Mashinostroenie, Moscow, 2009 (in Russian).

[3] Rabotnov, Yu. N. *Strength of Materials*. Nauka, Leningrad, 1962 (in Russian).

[4] Nanocenter of Don State Technical University [Internet], 2008 [updated 2014 December 15; cited 2015 May 18]. http://nano.donstu.ru.

[5] *Technical Committee ISO/TC 164. ISO 14577-1:2002. Metallic Materials – Instrumented Indentation Test for Hardness and Material Parameters – Part 1: Test Method*. ISO, Geneva, 2002.

[6] Isaenkova, M. G., Perlovich, Yu. A., Golovinm, Yu. I. *Using the Nanoindentation Platform for Evaluating the Mechanical Properties of Materials: Laboratory Workshop*. MIFI, Moscow, 2008 (in Russian).

In: Proceedings of the 2015 International Conference …   ISBN: 978-1-63484-577-9
Editors: Ivan A. Parinov, Shun-Hsyung Chang et al.   © 2016 Nova Science Publishers, Inc.

Chapter 46

NUMERICAL-ANALYTICAL SOLUTION FOR DEFLECTIONS AND FORCE FACTORS DUE TO CONCENTRATED LOAD ON CANTILEVER PLATE

Yu. E. Drobotov[*] *and G. A. Zhuravlev*
Vorovich Institute of Mathematics, Mechanics and Computer Sciences,
Southern Federal University, Rostov-on-Don, Russia

ABSTRACT

The present chapter contains a numerical-analytical solution for deflections and force factors acting on a cantilever plate of infinite length due to a concentrated load, which are evaluated in the form of optimal quadrature formulae. It is demonstrated, that the obtained formulae are suitable for thin plates of different materials, for example, for duraluminium, steel and ceramal plates. By comparison with other numerical-analytical solutions it is shown that the new solution is qualitative improved.

Keywords: theory of elasticity, plane problem, method of local approximations, plate, Kirchhoff plate theory, numerical-analytical solution, deflections, moments, shearing force

1. INTRODUCTION

The present paper contains the basics of the method of local approximations (MLA), which was proposed by G. A. Zhuravlev [1, 2]. The purpose of the method is the exploration of the stressed state of elastic complex-shaped bodies with loaded ledges. The idea of the MLA is to combine two plane problems of theory of elasticity in order to form the solution for different force factors (shearing force, bending and twisting moments) and the respective elements of the stressed state for the three-dimensional problem of theory of elasticity. The

---

[*] E-mail: yu.e.drobotov@yandex.ru.

first one is the problem of the force factors due to a distributed load on a cantilever plate. This problem is described, for example, in Ref. [3] on the base of [4]. The calculated force factors are supposed to be used in the second plane problem, namely the problem of the stressed state of an elastic rod with two deep symmetric hyperbolic recesses. The considered rod is loaded by a pure moment, and the solution for this case is given in Ref. [5]. Using the force factors from the first plane problem to obtain the stresses in the second plane problem, we get an opportunity to distinguish the components of the full stress state of a three-dimensional ledge, approximated by these two plane models, by their provenance. The feature of separate analysis of the effect of each of the force factors on the stressed state in the area of a geometric concentrator is the main advantage of the MLA. The first plane problem is set for an infinitely long cantilever plate under the influence of a transverse concentrated load within the limits of Kirchhoff plate theory. The authors obtained a numerical-analytical solution, based on high-accuracy quadrature formulae. This solution has been applied to obtain the values of respective improper integrals in an exact solution of the problem. A similar solution can find in Ref. [6], where Gaussian quadrature is applied. The advantage of the solution, provided in the present paper, is that the new quadrature formulae are considered as the optimal ones in numerical integration. In order to estimate their accuracy, we compare our results with the results of the finite elements method (the FEM) and also the solution [4] for deflections and the solution [7] for force factors.

2. STATEMENT OF PROBLEM

On the analogy with [1] the plate is considered in the form of an infinitely long strip of width A and thickness h, fixed along one of its long edges. The transverse concentrated load F is applied at an arbitrary point $P(c,0)$ of the plate, where the Cartesian coordinate system $Oxyz$ is located thus xy is the middle plane of the plate, Oy belongs to the fixed side and the line of F belongs to the xz plane.

Let the plane $x = c$ divides the plate and the functions, which are determined in the areas $0 \le x \le c$ and $c \le x \le A$ have indices 1 and 2, respectively. According to [1], if weight of the plate does not consider, the displacements $W_j(x,y)$, $j = 1,2$, satisfy the biharmonic equation:

$$\nabla^4 W_j(x,y) = 0, \ j = 1,2,\tag{1}$$

which is valid everywhere except the point P. Hereinafter $\nabla^4 = \dfrac{\partial^4}{\partial x^4} + 2\dfrac{\partial^4}{\partial x^2 \partial y^2} + \dfrac{\partial^4}{\partial y^4}$ is

the biharmonic operator.

The boundary conditions for the force factors (shear force, bending and twisting moments), written in displacements, form the following equations set:

$$
\left.\begin{aligned}
&W(0,y) = W_{1,x}(0,y) = 0; \\
&\left[W_{2,xx} + \mu W_{2,yy}\right]_{(A,y)} = 0; \\
&\left[W_{2,xxx} + (2-\mu)W_{2,xyy}\right]_{(A,y)} = 0; \\
&W_1(c,y) = W_2(c,y); \\
&W_{1,x}(c,y) = W_{2,x}(c,y); \\
&\left[W_{1,xx} + \mu W_{1,yy}\right]_{(c,y)} = \left[W_{2,xx} + \mu W_{2,yy}\right]_{(c,y)},
\end{aligned}\right\}
\tag{2}
$$

where μ is the Poisson's ratio and the lower letter indexes specify the respective partial derivatives.

Besides the conditions (2), shear force, which is continuous along $x = c$ except the point P, satisfies the transitional condition. The latter was formulated in [1] as

$$
-D\left[W_{1,xxx} + W_{1,yyx}\right]_{(c,y)} + D\left[W_{2,xxx} + W_{2,yyx}\right]_{(c,y)} = \int_0^\infty \frac{2f\sin(\alpha\delta)}{\pi\alpha}\cos(\alpha y)d\alpha
\tag{3}
$$

where the force function $F_{(y)}$ is considered as the concentrated load F and, considering the condition $2f\delta \underset{\delta\to 0}{\to} F$, the following equations set is obtained:

$$
\left.\begin{aligned}
&F_{(y)} = f, \, |y| \le \delta; \\
&F_{(y)} = 0, \, \delta < |y| < \infty.
\end{aligned}\right\}
\tag{4}
$$

The functions $W_j(x,y)$, $j = 1,2$, are defined in the form:

$$
W_j(x,y) = \int_0^\infty f_j(x,\alpha)\cos(\alpha y)d\alpha,
\tag{5}
$$

where

$$
f_j(x,\alpha) = (A_j + B_j\alpha x)\cosh(\alpha x) + (C_j + D_j\alpha x)\sinh(\alpha x),
\tag{6}
$$

A_j, B_j, C_j, D_j are assumed functions of α-variable.

Then, the equations (2) and (3) can be written as

$$\left.\begin{aligned}
&f_1(0,\alpha)=f_{1,x}(0,\alpha)=0;\\
&f_{2,xx}(A,\alpha)-\alpha^2\mu f_2(A,\alpha)=0;\\
&f_{2,xxx}(A,\alpha)-\alpha^2(2-\mu)f_{2,x}(A,\alpha)=0;\\
&f_1(c,\alpha)-f_2(c,\alpha)=0;\\
&f_{1,x}(c,\alpha)-f_{2,x}(c,\alpha)=0;\\
&f_{1,xx}(c,\alpha)-f_{2,xx}(c,\alpha)=0;\\
&f_{2,xxx}(c,\alpha)-f_{1,xxx}(c,\alpha)=\frac{F}{\pi D},
\end{aligned}\right\} \tag{7}$$

where $D=\dfrac{Eh^3}{12(1-\mu^2)}$ is the flexural rigidity of the plate (E is the Young's modulus, μ is the Poisson's ratio).

This equation set, solved for A_j, B_j, C_j, D_j, $j=1,2$, provides the following functions:

$$f_1(x,c,\alpha)=\frac{F}{2\pi ADγ}\left[-x\alpha\cosh(x\alpha)\sum_{i=1}^{4}a_ik_i+\sinh(x\alpha)\sum_{i=4}^{8}a_ik_i\right],$$

$$f_2(x,c,\alpha)=f_1(c,x,\alpha),$$

$$γ=\alpha^3\left[5+2(A\alpha)^2(\mu-1)^2+(2+\mu)\mu-(\mu^2+2\mu-3)\cosh(2A\alpha)\right]$$

Thus,

$$W_j(x,y,c)=\int_0^{\infty}f_j(x,c,\alpha)\cos(y\alpha)d\alpha,\ j=1,2. \tag{8}$$

i	k_i	a_i
1	$5+2A(A-c)\alpha^2(\mu-1)^2+(2+\mu)\mu$	$\cosh(c\alpha)$
2	$-(\mu^2+2\mu-3)$	$\cosh\left[(c-2A)\alpha\right]$
3	$\alpha(\mu-1)(2A-c)(\mu-1)$	$\sinh(c\alpha)$
4	$\alpha c(\mu-1)(\mu+3)$	$\sinh\left[(c-2A)\alpha\right]$
5	$5+(2A^2-2Ac+cx)\alpha^2(\mu-1)^2+(2+\mu)\mu$	$\cosh(c\alpha)$
6	$-(1+cx\alpha^2)(\mu^2+2\mu-3)$	$\cos\left[(c-2A)\alpha\right]$
7	$\alpha\left[-c(\mu-1)^2+2A^2x\alpha^2(\mu-1)^2-2A(cx\alpha^2-1)(\mu-1)^2+4x(\mu+1)\right]$	$\sinh(c\alpha)$
8	$c\alpha(\mu^2+2\mu-3)$	$\sinh\left[(c-2A)\alpha\right]$

3. Numerical-Analytical Solution for Deflections

We successfully attempted to use quadrature formulae to obtain approximate values of the improper integrals (8) in [6]. By using the Gaussian quadrature we obtained high-accuracy formulae for $j=1$ in (8). However, taking into account the function behavior (growing oscillation when the argument aspires to infinity) for $j=2$, the formula applicable for $j=1$ does not provide any acceptable accuracy of the results in this case. In the present paper, we provide $W_j(x,y,c) = \int_0^\infty \frac{F_j(x,y,c,\alpha)}{1+\alpha^2} d\alpha = \frac{\pi}{2n} \sum_{k=1}^{n} F_j\left[x,y,c,\tan\frac{(2k-1)\pi}{4n}\right]$ quadrature formulae for the improper integrals (8), which are considered as the best ones.

It is accepted (for example, [8, 9]) in the theory of quadratures to call a quadrature formula the best one, if it has the minimum estimation of the remainder for fixed nodes or arbitrary nodes and weights. To put it otherwise, the best quadrature formula is optimal from the viewpoint of Sard or from the viewpoint of Nikolsky. In the present work we base our result on the results of Ref. [8], thus consider the following functional class.

Definition 1. Let $[a;b]$ be a finite or infinite interval. We denote as $W^{(r)}L_p[a;b]$, $1 \le p \le \infty$ ($r=0,1,2,...$; $W^{(0)}L_p[a;b] \equiv L_p[a;b]$) the class of functions $f(x)$, which have absolutely continuous derivatives $f^{(r-1)}(x)$ of order $r-1$ and derivatives $f^{(r)}(x)$ of order r, which belong to $L_p[a;b]$ and satisfy the condition

$$\left\| f^{(r)} \right\|_p := \left\| f^{(r)} \right\|_{L_p} = \left(\int_a^b \left| f^{(r)}(x) \right|^p \right)^{\frac{1}{p}} \le 1. \tag{9}$$

Since we are interested in absolute continuity on the infinite interval $[0;\infty)$, we note, that the class of absolute continuous on $[0;\infty)$ functions can be defined through the notion of functions, absolute continuous on a finite interval $[a;b]$. Indeed, let $x = x(y)$ be continuously differentiable mapping of unique correspondence, which acts from $[a;b]$ to $[0;\infty)$, and $\tilde{f}(y) = f[x(y)]$. Then we note $AC[0;\infty] = \{f(x): \tilde{f}(y) \in AC[a;b]\}$ as the class of absolute continuous on $[0;\infty)$ functions.

Note also that every absolute continuous function has a bounded variation on an interval of finite length. Monotone functions are functions of a bounded variation.

In this paper, we use the following result from the paper [8] to build quadrature approximations:

Theorem 1. Among the quadrature formulae in the form of $\int_0^\infty \frac{f(t)}{1+t^2}\,dt = \sum_{k=1}^n p_k\, f(t_k) + R_n(f)$ the unique formula

$$\int_0^\infty \frac{f(t)}{1+t^2}\,dt = \frac{\pi}{2n}\sum_{k=1}^n f\left[\tan\frac{(2k-1)\pi}{4n}\right] + R_n(f) \tag{10}$$

is the best one for the functional class $W^{(1)}L_1[0;\infty)$.

The error of the quadrature formula (10) can be calculated by the formula:

$$\varepsilon_n\left(W^{(1)}L_1[0;\infty);\left(1+t^2\right)^{-1}\right) = \frac{\pi}{4n}. \tag{11}$$

We denote $F_j(x,y,c,\alpha) := \left(1-\alpha^2\right)f_j(x,c,\alpha)\cos(\alpha y)$, $j=1,2$, and show further, that $F_j \in W^{(1)}L_1[0;\infty)$, $j=1,2$. The change of variable

$$\alpha = 1 - \frac{1}{1-\tau}, \tag{12}$$

transforms $[0;\infty)$ into $[0;1)$. We use also dimensionless coordinates:

$$\xi = \frac{x}{A},\ \eta = \frac{y}{A},\ \zeta = \frac{c}{A} \tag{13}$$

as a matter of convenience. Then common shapes of the plots for the functions $F_j(\xi,\eta,\zeta,\tau)$, $j=1,2$, are present in Figure 1 for three chosen materials (duraluminium, steel and ceramal). The values of material constants are:

(i) duraluminium: $E = 74 \cdot 10^9$ Pa, $\mu = 0.34$ ($D = 60.5243$ N·m);
(ii) steel: $E = 210 \cdot 10^9$ Pa, $\mu = 0.3$ ($D = 153.846$ N·m);
(iii) ceramal: $E = 623 \cdot 10^9$ Pa, $\mu = 0.21$ ($D = 434.495$ N·m).

These forms of plots repeat themselves for different values of η and ζ from domains of the functions. The plots in Figure 1 shows, that the functions are monotonous in $(0;1)$, so they are absolutely continuous.

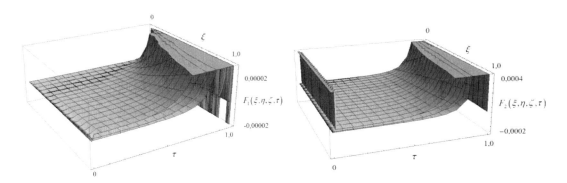

Figure 1. Common shapes of the plots of $F_1(\xi,\eta,\zeta,\tau)$ (the left plot) and $F_2(\xi,\eta,\zeta,\tau)$ (the right plot), where η and ζ can have different values from the domains of the functions, $F = 2$ kN and other parameters corresponds to duraluminium, steel and ceramal.

To obtain the estimation (11), we study maxima of the functions $|F_{j,\alpha}(\xi,\eta,\zeta,\tau)|$, $j = 1,2$, where τ is characterized by expression (12) in $(0;1)$, found by numerical methods. The plots in Figure 2 provide guidance on behavior of the respective maxima for $\eta = 0$ and three chosen demonstrative materials. Thus, we can conclude (on the base of properties of the improper integrals), that the estimation of type (11) is true for $|F_{j,\alpha}(\xi,\eta,\zeta,\tau)|$, $j = 1,2$.

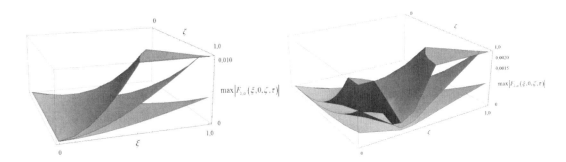

Figure 2. Plots of the maxima of $|F_{1,\alpha}(\xi,0,\zeta,\tau)|$ (the left picture) and $|F_{2,\alpha}(\xi,0,\zeta,\tau)|$ (the right picture) for $F = 2$ kN and different materials.

So, $F_j \in W^{(1)}L_1[0;\infty)$, $j = 1,2$, for all of three demonstrative materials and Theorem 1 can be applied. As a result, we obtain the following formula for calculation of approximate values of the deflections (8):

$$W_j(x,y,c) = \int_0^\infty \frac{F_j(x,y,c,\alpha)}{1+\alpha^2} d\alpha = \frac{\pi}{2n}\sum_{k=1}^n F_j\left[x,y,c,\tan\frac{(2k-1)\pi}{4n}\right] \qquad (14)$$

with the error (11). Based on expression (11), an optimal value of n in Equation (14) can be found. Thus, $n = 785$ provides the value of the error as 0.1%. At the same time, Equation (14) realizes efficiently in computer algebra systems and n can increase. However, it is necessary to take into account the growing oscillation of $F_2(x,y,c,\alpha)$ when $\alpha \to \infty$. In an attempt to decrease the error (11) it seems to be a rational increasing the value of n, but the points of the interval of high and rapid oscillation of $F_2(x,y,c,\alpha)$ would be used. An investigation of this question is beyond the present paper and here we only choose n in (14) on the base of validity of the results and smallness of their deviation from the results of the FEM.

Tables 1 – 3 compare the results of [4] and the values of (14) with the results of FEM modeling for deflections. We built a model of the plate in ANSYS with the following characteristics:

Length of the plate l, m	Thickness of the plate h, м	Width of the plate A, м	Value of the concentrated load F, N	Finite elements type	Multiplicity of splitting of the sides: the side of the size l, the side of the size h, the side of the size A
0.2	0.002	0.01	2000	SHELL63	400
					4
					40

Table 1. Deflections of the plate

ξ \ ζ	0.25	0.50	0.75	1.00
0.25	0.70022	1.1266	1.4482	1.8059
	0.69519	1.1255	1.6056	2.0649
	0.69947	1.1256	1.4478	1.8058
0.50	1.1266	3.2915	4.8826	6.4369
	1.1255	3.2856	4.8829	6.6209
	1.1189	3.2906	4.8817	6.4380
0.75	1.4482	4.8826	9.1631	13.131
	1.6056	4.8829	9.1037	13.159
	1.4466	4.8767	9.1714	13.135
1.00	1.8059	6.4369	13.131	21.623
	2.0649	6.6209	13.159	21.725
	1.8070	6.4383	13.125	21.660

Commentary: First row values in every cell are calculated by FEM, second row values are based on [4] and third row values are defined by using (14). The values for $\xi \le \zeta$ are obtained with the error 0.04% ($n = 2000$) and for $\xi > \zeta$, the error is equal to 0.09% ($n = 840$). All of the values corresponds to $\eta = 0$ and must be multiplied by 10^{-5} that be written in meters.

Table 2. Deflections of the plate (multiplied by 10^{-5} m) for $\zeta = 0.75$

ξ \ η	0	0.25	0.50	1.00	1.50	2.00
0	0	0	0	0	0	0
	0	0	0	0	0	0
	0	0	0	0	0	0
0.25	1.4482	1.3252	1.0521	0.53413	0.2379	0.098841
	1.6056	0.9642	0.8566	0.4800	0.2247	0.05876
	1.4478	1.3244	1.0526	0.52909	0.24493	0.093086
0.75	9.1631	8.3127	6.7672	3.7634	1.8267	0.81635
	9.1078	4.6718	2.4249	0.6704	0.1912	0.0550
	9.1636	8.3513	6.7857	3.7203	1.9674	0.77383
1.0	13.131	12.171	10.073	5.7402	2.8429	1.2921
	13.151	12.174	10.076	5.7063	2.8428	1.2911
	13.120	12.159	10.222	5.4953	3.0802	1.2884

Commentary: First row values in every cell are calculated by FEM, second row values are based on [4] and third row values are defined by using (14). The values for $\xi \leq \zeta$ are obtained with the error 0.04% ($n = 2000$) and for $\xi > \zeta$, the error is equal to 0.09% ($n = 840$).

Table 3. Deflections of the plate (multiplied by 10^{-5} m) for $\zeta = 1.0$

ξ \ η	0	0.25	0.50	1.00	1.50	2.00
0	0	0	0	0	0	0
	0	0	0	0	0	0
	0	0	0	0	0	0
0.25	1.8059	1.6927	1.4137	0.78290	0.36648	0.15685
	2.06488	1.69245	1.41107	0.782087	0.366629	0.156831
	1.8058	1.6926	1.4136	0.78172	0.36716	0.15614
0.75	13.131	12.171	10.073	5.7402	2.8429	1.2921
	13.1507	12.1741	10.0761	5.70634	2.84283	1.29106
	13.135	12.171	10.073	5.6882	2.9175	1.2411
1.0	21.623	19.426	15.755	8.9119	4.4508	2.0473
	21.704	19.4611	15.7411	8.91331	4.45252	2.04832
	21.6599	19.562	15.815	9.08227	4.9149	1.9182

Commentary: First row values in every cell are calculated by FEM, second row values are based on [4] and third row values are defined by using (14). The values for $\xi \leq \zeta$ are obtained with the error 0.04% ($n = 2000$) and for $\xi > \zeta$, the error is equal to 0.09% ($n = 840$).

4. NUMERICAL-ANALYTICAL SOLUTION FOR FORCE FACTORS

As the functions $W_j(x,y,c)$, $j=1,2$, and their partial derivatives of the respective orders in (2) are continuous, Leibniz integral rule can be applied. Based on the definition and basic properties of improper integrals and an analysis of the respective integrands, which is similar to one that has place in the second paragraph of the paper, Theorem 1 can be applied again. Thus, we obtain the following formulae for the force factors:

$$Q_{jx}(x,y,c) = -\frac{\pi D}{2n} \sum_{k=1}^{n} F_j^{Q_x}\left[x,y,c,\tan\frac{(2k-1)\pi}{4n}\right], \tag{15}$$

$$M_{jx}(x,y,c) = -\frac{\pi D}{2n} \sum_{k=1}^{n} F_j^{M_x}\left[x,y,c,\tan\frac{(2k-1)\pi}{4n}\right], \tag{16}$$

$$M_{jxy}(x,y,c) = -\frac{\pi D(1-\mu)}{2n} \sum_{k=1}^{n} F_j^{M_{xy}}\left[x,y,c,\tan\frac{(2k-1)\pi}{4n}\right], \tag{17}$$

$$F_j^{Q_x}(x,y,c,\alpha) = \left\{\left[f_j(x,c,\alpha)\cos(\alpha y)\right]_{xxx} + \left[f_j(x,c,\alpha)\cos(\alpha y)\right]_{xyy}\right\}(1-\alpha^2) \tag{18}$$

$$F_j^{M_x}(x,y,c,\alpha) = \left\{\left[f_j(x,c,\alpha)\cos(\alpha y)\right]_{xx} + \mu\left[f_j(x,c,\alpha)\cos(\alpha y)\right]_{yy}\right\}(1-\alpha^2) \tag{19}$$

$$F_j^{Q_x}(x,y,c,\alpha) = \left[f_j(x,c,\alpha)\cos(\alpha y)\right]_{xy}(1-\alpha^2). \tag{20}$$

A solution for the force factors is present, for example, in Refs. [4, 7]. The paper [7] contains approximate formulae for M_x and Q_x, which can be used for comparison with the values of (15) and (16). Thus, for a steel plate ($\mu = 0.3$) of infinite length there is a formula for bending moment:

$$M_x = F\left\{\left[\frac{a_1}{\left(|\eta|+a_2\right)^2+\zeta^2} + a_3\zeta + a_4\right]\xi^2 + \left[\frac{a_5\zeta+a_6}{\eta^2+\zeta^2} + a_7\zeta + a_8\right]\xi - \frac{\zeta^2\left(a_9+a_{10}\zeta^2\right)}{\eta^2+\zeta^2}\right\} \tag{21}$$

where the values of the coefficients are

a_1	a_2	a_3	a_4	a_5
-0.097647	-0.373938	-0.162076	0.153598	0.679365
a_6	a_7	a_8	a_9	a_{10}
-0.0255795	0.212308	-0.255501	0.335756	0.154592

Table 4. The values of moments, calculated with (21) (the first line of each cell) and with (15) (the second line) for $\eta = 0$

ζ \ ξ	0	0.1	0.2	0.3	0.4	0.5	0.6	0.7	0.8
0.1	-674.604	-	-	-	-	-	-	-	-
	-709.801	569.188	51.5469	22.2932	11.6929	6.1052	3.22759	1.63527	0.686237
0.2	-683.879	-183.456	-	-	-	-	-	-	-
	-646.513	-171.01	696.628	110.713	53.1683	28.5958	15.4763	8.13293	3.72584
0.3	-699.339	-348.031	-9.52	-	-	-	-	-	-
	-670.647	-326.611	-34.4004	767.641	150.339	76.0161	41.5454	22.0783	10.4942
0.4	-720.981	-452.127	-192.749	57.1529	-	-	-	-	-
	-703.624	-426.332	-199.887	37.1503	809.841	175.318	89.9498	47.8839	23.0467
0.5	-748.808	-530.954	-320.216	-116.596	79.9076	-	-	-	-
	-739.572	-505.212	-313.365	-132.237	78.3343	834.126	187.918	93.947	45.4029
0.6	-782.818	-598.977	-420.697	-247.977	-80.8171	80.7823	-	-	-
	-781.107	-575.158	-405.627	-251.508	-94.3395	99.7079	843.203	187.246	85.3411
0.7	-823.012	-663.023	-507.63	-356.833	-210.631	-69.0254	67.9849	-	-
	-829.375	-642.403	-487.663	-350.489	-218.555	-77.874	102.906	834.932	168.651
0.8	-869.39	-726.699	-588.059	-453.47	-322.932	-196.445	-74.0099	44.3744	-
	-884.741	-710.365	-565.761	-439.867	-323.975	-209.136	-83.4835	82.872	800.092
0.9	-921.951	-792.083	-666.018	-543.757	-425.298	-310.642	-199.789	-92.7395	10.5073
	-947.395	-781.14	-643.654	-525.969	-421.455	-324.262	-227.639	-121.831	20.35

Table 5. The values of shearing forces, calculated with (22) (the first line of each cell) and with (16) (the second line) for $\eta = 0$

ζ \ ξ	0	0.1	0.2	0.3	0.4	0.5	0.6	0.7	0.8
0.5	-43307.4	-47741.5	-43607.3	-30831.5	-9819.55	-	-	-	-
	269.825	233.959	226.723	256.342	436.122	2072.3	-288.38	-98.554	-49.887
0.6	-57004.5	-55473.1	-48894.4	-37272.7	-20848.6	-49.0527	-	-	-
	233.43	204.788	194.052	201.029	239.287	424.766	2065.14	-291.929	-98.235
0.7	-61779.6	-57368.6	-50217.2	-40337.9	-27864.7	-13031.8	3861.22	-	-
	209.56	185.495	173.705	173.141	186.715	229.653	418.906	2062.96	-289.54
0.8	-59096.6	-54496.0	-48347.3	-40650.5	-31471.8	-20936.4	-9213.83	3501.0	-
	194.138	172.737	160.465	156.228	160.866	178.395	224.833	417.953	2067.57
0.9	-49905.4	-47528.4	-43767.6	-38605.3	-32065.0	-24209.1	-15133.2	-4957.75	6183.06
	184.603	164.496	151.744	145.287	145.242	153.235	173.949	224.292	424.248

and the formula for the shear force is

$$Q_x = \frac{F}{A}\left(\frac{t_1}{\eta^2 + |\eta| t_2 + t_3} + t_4\right),$$ (22)

where

$$t_i = b_{3i-2}\,\xi^2 + b_{3i-1}\,\xi + b_{3i}\,,\ i = 1,...,4\,,$$ (23)

$$b_j = C_{3j-2}\,\zeta^2 + C_{3j-1}\,\zeta + C_{3j}\,,\ j = 1,...,12\,,$$ (24)

k	C_k	k	C_k	k	C_k	k	C_k
1	−5.70039	10	14.3017	19	−1.44539	28	9.89275
2	10.3014	11	−28.4647	20	4.06981	29	−16.5577
3	−5.28244	12	14.1892	21	−3.66913	30	7.00416
4	4.93001	13	2.60439	22	1.50211	31	−5.551411
5	−7.7914	14	−2.02051	23	−2.31709	32	8.46253
6	3.17001	15	−0.116368	24	1.44754	33	−2.97045
7	−1.65402	16	−0.937562	25	−1.06376	34	1.16429
8	2.72311	17	1.11118	26	−1.99425	35	−1.68405
9	−0.778444	18	−0.351077	27	−0.615883	36	0.424682

The formula (21) can be applied for $0.1 \le \zeta \le 0.9$; $0 \le \xi \le \zeta - 0.1$; $|\eta| \le 1$, as the formula (22) is suitable for $0.5 \le \zeta \le 0.9$; $0 \le \xi \le \zeta - 0.1$; $|\eta| \le 1$.

Tables 4, 5 provide comparison of (16) with (21) and (15) with (22).

As we can see, the formula (21) gives a good correlation of the results compared with the results of (15), and Table 5 demonstrates qualitative improvement of the results. Thus, the values decrease while ξ moves off ζ to the sealing line and the value of shearing force at the point of loading is approximately the value of the concentrated load F.

CONCLUSION

The present chapter contains a numerical-analytical solution for deflections and force factors on a cantilever plate of infinite length due to a concentrate load, which are evaluated in the form of optimal quadrature formulae. It is demonstrated, that the obtained formulae are suitable for thin plates of different materials, for example, for duraluminium, steel and ceramal plates. By comparison with other numerical-analytical solutions it is shown, that the new one is qualitative improved. However, the question about optimal number of summands in the respective series is open, but empirically found values are 2000 for $j = 1$ and 750 for $j = 2$ in (14), while for (22), for example, n can be increased to 2000 for $j = 2$.

ACKNOWLEDGMENT

This research was partially supported by the Southern Federal University grant (No. 213.01.-2014/03VG).

REFERENCES

[1]   Zhuravlev, G.A.; Iofis, R.B. *Hypoid Gears. Problems and Development*, Rostov State University Press, Rostov-on-Don, 1978 (in Russian)

[2]   Zhuravlev, G. A.; Prokopiev, P.S. In: *Proceedings of the International Conference on Motion and Power Transmissions*, The Japan Society of Mechanical Engineers, Tokyo. 23–26 November 1991, 866–870.

[3]   Zhuravlev, G.A.; Onishkova, V.M. *VINITI*, Moscow. No 46-B, 02.01.1986 (in Russian).

[4]   Jaramillo, T.J. *J. Appl. Mech.* 1950, *vol. 17(1)*, 67.

[5]   Neueber, G. *Stress Concentration*. State Publishers of Technical and Theoretical Literature, Moscow, 1947 (in Russian).

[6]   Drobotov, Y. E.; Zhuravlev, G. A. In: *Modern Problems of the Theory of Machines*. Novokuznetsk. 2015, 224 – 231 (in Russian).

[7]   Zhuravlev, G.A.; Onishkova, V.M. *VINITI, Moscow*, No 6266-B87, 17.07.1987 (in Russian).

[8]   Parvonaeva, Z. A. Best Quadrature and Cubeture Formulae for Several Classes of Functions, Dushanbe. 2011 (in Russian).

[9]   Korneichuk, N. P.; Lushpai, N. E. *Izv. Akad. Nauk SSSR Ser. Mat.*, 1969, *vol. 33(6)*, 1416–1437.

In: Proceedings of the 2015 International Conference ...   ISBN: 978-1-63484-577-9
Editors: Ivan A. Parinov, Shun-Hsyung Chang et al.   © 2016 Nova Science Publishers, Inc.

Chapter 47

THEORY AND EXPERIMENT IN THE ULTRASONIC NONDESTRUCTIVE TESTING FOR ARRAYS OF SPATIAL DEFECTS OF THE ELASTIC MATERIALS

*N. V. Boyev**

Vorovich Mathematics, Mechanics and Computer Sciences Institute,
Southern Federal University, Rostov-on-Don, Russia

ABSTRACT

In the present work the authors propose a deterministic model of ultrasonic wave diffraction with multiple reflections by the defects located in an elastic solid of finite size. The model is tested in the case of three cylindrical defects. In this particular example there is performed a comparison between the theoretical results and the results of a certain natural experiment.

Keywords: scattering of ultrasonic waves, spatial defects, high-frequency diffraction

1. INTRODUCTION

In the scanning of arrays of defects in the elastic solids by combined ultrasonic sensors [1] the information about their position and shape can be obtained from the characteristics of single and multiple reflections. The complete study of such a problem is possible only on the basis of a three-dimensional problem of the high-frequency diffraction of ultrasonic waves. Taking into account the frequency range in the ultrasonic nondestructive testing (US NDT) of materials, the study of the three-dimensional multiple diffraction of elastic waves can be performed within the framework of the geometrical diffraction theory [2, 3].

It is also very interesting to study the high-frequency multiple diffraction by stochastic methods. However, as a first step, it is quite natural to consider such problems within the

---

*E-mail: boyev@math.rsu.ru.

framework of a deterministic model for defects of a certain canonical shape. The principal aim of the present work is to perform a comparative analysis of the longitudinal wave's amplitude, in the process of single and multiple reflections, in the elastic solid of finite size.

2. THE DETERMINISTIC MODEL AND EXPERIMENTAL MEASUREMENTS

The model is realized in a laboratory sample. As an elastic finite-size solid there is used a steel specimen of a rectangular parallelepiped shape, the thickness is 40 mm, the height is 78 mm, the length is 300 mm.

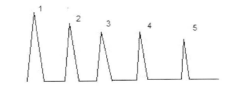

1 - probing impulse
2 - impulse, reflected from hole 1 (coincides by the time of flight with the one reflected from hole 3)
3 - impulse, reflected from hole 2
4 - impulse, reflected from hole 1 and then re-reflected from hole 3
5 - impulse, reflected from hole 1 and firther re-reflected by holes 2 and 3

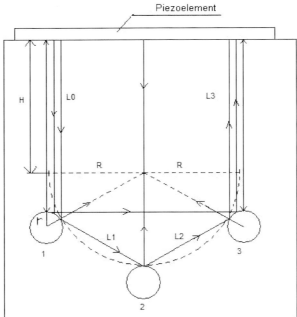

Figure 1. Sketch of the sample with a sensor and a diagram from the flaw detector's display.

On the distance H from the upper face, there is placed a sensor of longitudinal ultrasonic (US) waves of 2.5 MHz frequency and 30 mm diameter. There are arranged, along a circular arc of radius R, three equal co-axial cylindrical holes of radius r whose generatrix is perpendicular to the side walls. The draft of the sample with the holes and the sensor is presented in Figure 1. As an electronic device for the experimental investigations we use a standard ultrasonic flaw detector, which works jointly with a PC.

3. THE RAY THEORY SOLUTION

The sensor of the longitudinal US waves allows us to evaluate the location of the cylindrical holes in the sample. The same sensor is used to record the amplitude of the longitudinal waves back-scattered from the surface of a defect. It can register the signal in the double reflection from defects 1 and 3, as well as in the triple reflection from all three obstacles. The trajectories of respective US rays are shown in Figure 1. Let us note that all trajectories are certain lines located in a unique plane. The general theory of single and multiple reflections of the longitudinal waves from curved surfaces in the elastic medium have been proposed in [3, 4]. Explicit expressions for the wave amplitude in the reflected wave have been obtained in frames of the geometrical diffraction theory in [5], and on the basis of the modified Kirchhoff diffraction theory in [6]. Similar results are obtained with the use of the asymptotic multidimensional stationary phase method, applied to certain diffraction integrals, see [7, 8]. Let us write out these formulas for the longitudinal displacement u_p in the case of single, double, and triple reflections. The amplitude of a single reflection is given by the following expression:

$$u_p^{(1)} = Z^{-1/2} B V_{PP}(y^*) \exp\left[ik_p(L_0 + L_1) + \frac{\pi}{4}(\delta_2^{(pp)} + 2) \right];$$

$$Z = (L_0 + L_1)^2 + 2L_0 L_1 (L_0 + L_1)(2H_c \cos^2 \gamma + \widetilde{k} \sin^2 \gamma)\cos^{-1} \gamma + 4L_0^2 L_1^2 K \tag{1}$$

$$B = Q_q k_p^2 (4\pi\mu k_s^2)^{-1}; \ L_0 = |x_0 - y^*|, \ L_1 = |y^* - x|.$$

Formula (1) gives the leading asymptotic term for the longitudinal displacement with $k_p L_0 \gg 1$, $k_p L_1 \gg 1$, $k_p R_1 \gg 1$, $k_p R_2 \gg 1$, where R_1 and R_2 are the principal curvature radii of the surface at the point of mirror reflection y^*. Angle γ is an angle between the direction of incidence $q^0 = \{-\cos\alpha, -\cos\beta, -\cos\gamma\}$ and the unit normal to the surface at point y^*, in the local Cartesian coordinate system. The same angle is formed between the direction of reflection and the unit normal vector.

The case of double ($N = 2$) and triple ($N = 3$) reflection is given as follows:

$$u_p^{(N)} = \frac{Q_q}{4\pi\mu} \frac{k_p^2}{k_s^2} \left(\prod_{n=1}^{N} V_{pp}(y_n^*) \cos\gamma_n \right) \frac{\exp\left[i\left(k_p \sum_{n=0}^{N} L_n + \frac{\pi}{4}\left(\delta_{2N}^{(p)} + 2N\right) \right) \right]}{\left(\prod_{n=0}^{N} L_n \right) \sqrt{\left| \det\left(D_{2N}^{(p)}\right) \right|}}. \tag{2}$$

Here $\delta_{2N}^{(p)} = \text{sign} D_{2N}^{(p)}$ is the difference between the number of positive and the number of negative eigenvalues of the Hessian matrix $D_{2N}^{(p)} = (d_{ij})$, $i,j = 1,2,...,2N$ ($N = 2$, $N = 3$), which is symmetric and banded, with the width equal to 7. The elements of this matrix are defined by the characteristic distances of the US ray, in its flight between the source and the receiver, as well as by principal curvatures at the points of mirror reflection [4].

4. COMPARATIVE ANALYSIS

The measurements are performed for a steel sample. The wave speed for the longitudinal and the transverse wave is $c_p = 5850\,\text{m/s}$, $c_s = 3260\,\text{m/s}$, respectively. For the used frequency 2.5 MHz, the longitudinal wave-length is 2.34 mm. The diameter of the holes is $d = 2r = 3\,\text{mm}$. As an example, let us perform a comparison between the theoretical calculations and the results of experimental measurements, in the case of single (echo) reflections from voids 1 and 3, and in the doubly reflected wave, from defects 1 and 3 sequentially.

From the experiments, there have been obtained the following values of the displacement amplitude in the reflected waves: the amplitude of displacement $u_{pe}^{(1)}$ in the echo singly reflected wave, from defect 1 (see Figure 1) is 60.5 dB; the amplitude of the displacement $u_{pe}^{(1,3)}$ in the doubly reflected wave, sequentially from defects 1 and 3, is 29.5 dB. The difference between these values is 31 dB. For these data the ratio $u_{pe}^{(1)} / u_{pe}^{(1,3)} = 35.5$.

The theoretical calculation of amplitudes $u_p^{(1)}$ and $u_p^{(1,3)}$ is founded upon analytical expressions. For the ratio $u_p^{(1)} / u_p^{(1,3)}$ there is obtained the expression

$$\frac{u_p^{(1)}}{u_p^{(1,3)}} = \frac{L_0^2 \left(r\left(a^2 - 1\right)\left[(a+b)^2 - 1\right] \right)^{1/2}}{2\left(H + \dfrac{R}{2} - \dfrac{r}{2} \right)\sqrt{\left(H + \dfrac{R}{2} - \dfrac{r}{2} \right)} L_1 V_{pp}^2\left(y_1^*\right)} = 51.2, \tag{3}$$

where

$$L_0 = 48.94\,\text{mm}, \qquad L_1 = 12.88\,\text{mm}, \qquad H + 0.5(R + r) = 50\,\text{mm}, \qquad V_{pp} = 0.36,$$

$$a = 1 + L_1/L_0, \; b = 2\sqrt{2}L_1/r.$$

The difference of the values for the ratio $u_p^{(1)}/u_p^{(1,3)}$, between the experimental and the theoretical results, is $20\log(51.2/35.5) = 3\,\text{dB}$. The analogous difference between the theory and the experiment for the ratio $u_p^{(1,3)}/u_p^{(1,2,3)}$, i.e., the ratio of double and triple reflections, is $20\log(u_p^{(1,3)}/u_p^{(1,2,3)}) = 5\,\text{dB}$. With increasing of holes' diameter from 3 mm to 5 mm, the difference between the theoretical and the experimental results decreases, respectively, to 1 dB and 2.5 dB.

When estimating the precision of the theoretical results, it should be taken into account that (i) the steel specimen is made with instrumental errors; (ii) the US wave, emitted by the probe, is not perfectly plane; (iii) the ratio of the wave length (2.34 mm) to the diameter of the holes (3 mm and 5 mm) is not small. With these inaccuracies, it may be accepted that the precision of the Ray diffraction theory applied is quite satisfactory.

CONCLUSION

The basic conclusion, which is extracted from the theoretical and experimental results obtained above, can be formulated as follows. The standard US NDT evaluation technique, operating in the frequency range 2.5 – 10 MHz, permits application of the geometrical high-frequency diffraction theory of elastic waves in solids with acceptable precision, if the characteristic size of the defects to be detected is expressed in mms. This conclusion is valid for both single and multiple reflections.

ACKNOWLEDGMENTS

The author is grateful to the Russian Scientific Foundation (RSCF), for its support by Project No. 15-19-10008.

REFERENCES

[1] Ermolov, I.N. *Theory and Practice of the Ultrasonic Testing*. Mashinostroenie, Moscow, 1981 (in Russian).

[2] Boyev, N.V.; Sumbatyan, M.A. *Doklady Physics*, 2003, vol. 48(10), 540-544.

[3] Boyev, N.V.; Vorovich, I.I.; Sumbatyan, M.A. *Mechanics of Solids*, 1992, No. 5.

[4] Boyev, N.V. *Acoustical Physics*, 2004, vol. 50(6).

[5] Hönl, H.; Maue, A.W.; Westpfahl, K. *Theorie der Beungung*. Springer-Verlag: Berlin, 1961 (in German).

[6] Borovikov, V.A.; Kinber, B.Ye. *Geometrical Theory of Diffraction*, IEE Publ.: London, 1994.

[7] Fedoryuk, M.V. *Steepest Descent Method*. Nauka, Moscow, 1977 (in Russian).

[8] Scarpetta, E.; Sumbatyan, M.A. *Acta Acustica united with Acustica*, 2011, vol. 97(1), 115-127.

In: Proceedings of the 2015 International Conference …   ISBN: 978-1-63484-577-9
Editors: Ivan A. Parinov, Shun-Hsyung Chang et al.   © 2016 Nova Science Publishers, Inc.

Chapter 48

BOUNDARY ELEMENT METHOD IN SOLVING PROBLEMS OF POROVISCOELASTICITY

L. A. Igumnov, A. A. Belov and A. A. Ipatov[*]

Research Institute for Mechanics, Lobachevsky State University of Nizhni Novgorod,
Nizhny Novgorod, Russia

ABSTRACT

Boundary-value problem of three-dimensional poroviscoelasticity is considered. The basic equations for fluid-saturated porous media proposed by Biot are modified by applying elastic-viscoelastic correspondence principle to classical linear elastic model of the solid skeleton. To describe viscoelastic properties of the solid skeleton we use model with weakly singular kernel. Boundary Integral Equations (BIE) method and Boundary-Element Method (BEM) with mixed discretization are applied to obtain numerical results. Solutions are obtained in Laplace domain. Modified Durbin's algorithm of numerical inversion of Laplace transform is used to perform solutions in time domain. One-dimensional analytical solutions are compared with numerical ones in order to verify our numerical approach. An influence of viscoelastic parameter on dynamic responses is studied on the example of prismatic poroviscoelastic solid under Heaviside-type load.

1. INTRODUCTION

Various types of interactions in advanced dispersed media, such as, porous of viscous media, are of a great interest of many disciplines. Wave propagation in porous/viscous media is an important issue of geophysics, geomechanics, geotechnical engineering etc. Study of wave propagation processes in saturated porous continua began from the works of Y. I. Frenkel and M. Biot [1, 2]. The viscoelastic effects of porous media were first introduced by Biot [3]. Although many significant achievements have been made on wave motion in porous media [4], because the complexity of the inertial viscosity and mechanical coupling in porous

---

[*] E-mail: SanSan.com@inbox.ru.

media, most transient response problems can only be solved via numerical methods. There are two major approaches to dynamic processes modeled by means of BEM: solving BIE system directly in time domain [5] or in Laplace or Fourier domain followed by the respective transform inversion [6]. The Laplace transform is the main method in dealing with transient response of porous media. Shanz and Cheng [7] presented an analytical solution in the Laplace domain and analytical time-domain solution without considering viscous coupling effect, and then developed the governing equation for saturated poroviscoelastic media by introducing the Kelvin–Voigt model and obtained an analytical solution in the Laplace domain for the 1D problem [8].

In present paper, we use special material constants (case where Poison's ratio equals 0) and boundary conditions for 3D formulation to make solution similar to 1D solution. It is performed to verify numerical approach. Moreover, a classical three-dimensional formulation is employed for study of viscoelastic parameters influence to displacement responses in poroviscoelastic prismatic solid.

2. PROBLEM FORMULATION AND SOLUTION METHOD

Homogeneous body Ω in three-dimensional Euclidean space R^3 is considered, with the boundary $\Gamma = \partial\Omega$. It is assumed, that body Ω is isotropic poroviscoelastic.

The set of differential equations of poroelasticity for displacements $\hat{u}_i$ and pore pressure $\hat{p}$ in Laplace domain (s is the transform parameter) take the following form [4]:

$$G\hat{u}_{i,jj} + \left(K + \frac{1}{3}G\right)\hat{u}_{j,ij} - (\alpha - \beta)\hat{p}_{,i} - s^2(\rho - \beta\rho_f)\hat{u}_i = -\hat{F}_i ; \tag{1}$$

$$\frac{\beta}{s\rho_f}\hat{p}_{,ii} - \frac{\phi^2 s}{R}\hat{p} - (\alpha - \beta)s\hat{u}_{i,i} = -\hat{a} ; \tag{2}$$

$$\beta = \frac{k\rho_f\phi^2 s^2}{\phi^2 s + s^2 k(\rho_a + \phi\rho_f)}, R = \frac{\phi^2 K_f K_s^2}{K_f(K_s - K) + \phi K_s(K_s - K_f)},$$

where G, K are material constants from elasticity, ϕ is the porosity, k is the permeability, α is the Biot's effective stress coefficient, ρ, ρ_a, ρ_f are the bulk, apparent mass and fluid densities, respectively, $\hat{F}_i, \hat{a}$ are the bulk body forces per unit volume.

Following types of boundary conditions for Ω are considered:

$$u_l(x,s) = f_l(x,s), \, u_4(x,s) = p(x,s) = f_4(x,s), x \in \Gamma^u, \, l = \overline{1,3} ;$$

$$t_l(x,s) = g_l(x,s), \, t_4(x,s) = q(x,s) = g_4(x,s), \, x \in \Gamma^\sigma, \, l = \overline{1,3} .$$

Γ^u and Γ^σ are the parts of boundary Γ, where corresponding generalized displacements and generalized tractions are prescribed.

Direct approach of boundary integral equations method is given [4, 9, 10]:

$$\begin{bmatrix} \hat{u}_j \\ \hat{p} \end{bmatrix} = \int_\Gamma \begin{bmatrix} \hat{U}^s_{ij} & -\hat{P}^s_j \\ \hat{U}^f_i & -\hat{P}^f \end{bmatrix} \begin{bmatrix} \hat{t}_i \\ \hat{q} \end{bmatrix} d\Gamma - \int_\Gamma \begin{bmatrix} \hat{T}^s_{ij} & -\hat{Q}^s_j \\ \hat{T}^f_i & -\hat{Q}^f \end{bmatrix} \begin{bmatrix} \hat{u}_i \\ \hat{p} \end{bmatrix} d\Gamma , \tag{3}$$

where $\begin{bmatrix} \hat{U}^s_{ij} & -\hat{P}^s_j \\ \hat{U}^f_i & -\hat{P}^f \end{bmatrix}, \begin{bmatrix} \hat{T}^s_{ij} & -\hat{Q}^s_j \\ \hat{T}^f_i & -\hat{Q}^f \end{bmatrix}$ are the matrices of fundamental and singular solutions.

Displacement vector at internal points is connected with boundary displacements and tractions as follows:

$$u_l(x,s) = \int_{\Gamma_k} U^s_{lj}(x,y,s) t_j(y,s) d_y S - \int_{\Gamma_k} T^s_{lj}(x,y,s) u_j(y,s) d_y S , l = 1,2,3, \quad x \in \Omega, \tag{4}$$

where U_{lj} and T_{lj} are the components of fundamental and singular solution tensors.

Analytical solutions of one-dimensional problem of axial force $t_2 = N /\mathrm{m}^2$ for displacements and pore pressure can be obtained as follows [4]:

$$\tilde{u}_y = \frac{S_0}{E(d_1\lambda_2 - d_2\lambda_1)} \left[\frac{d_2(e^{-\lambda_1 s(l-y)} - e^{-\lambda_1 s(l+y)})}{s(1 + e^{-2\lambda_1 sl})} - \frac{d_1(e^{-\lambda_2 s(l-y)} - e^{-\lambda_2 s(l+y)})}{s(1 + e^{-2\lambda_2 sl})} \right];$$

$$\tilde{p} = \frac{S_0 d_1 d_2}{E(d_1\lambda_2 - d_2\lambda_1)} \left[\frac{e^{-\lambda_1 s(l-y)} - e^{-\lambda_1 s(l+y)}}{1 + e^{-2\lambda_1 sl}} - \frac{e^{-\lambda_2 s(l-y)} - e^{-\lambda_2 s(l+y)}}{1 + e^{-2\lambda_2 sl}} \right].$$

Poroviscoelastic solution is obtained from poroelastic solution by means of the elastic-viscoelastic correspondence principle, applied to skeleton's elastic constants K and G in Laplace domain. Forms of $\hat{K}(s)$ and $\hat{G}(s)$ in case of model with weakly singular kernel of Abel type are following: $\hat{G} = \dfrac{G}{1 + hs^{\alpha-1}}, \hat{K} = \dfrac{K}{1 + hs^{\alpha-1}}$, h and α are model parameters [10, 11].

Thus, we get poroviscoelastic solution in time domain with the help of the Laplace transform inversion. Durbin's algorithm [12] with variable integrating step (5) is used for numerical inversion of Laplace transform.

$$f(0) = \sum_{k=1}^{\infty} \frac{(f(\alpha + i\omega_k) + f(\alpha + i\omega_{k+1}))}{2\pi} \Delta_k ,$$

$$f(t) \approx \frac{e^{\alpha t}}{\pi} \sum_{k=1}^{\infty} \frac{(f(\alpha + i\omega_k)e^{it\omega_k} + f(\alpha + i\omega_{k+1})e^{it\omega_{k+1}})}{2} \Delta_k \tag{5}$$

Integral representation and BIE with integral Laplace transform are used. Fundamental and singular solutions are considered in term of singularity isolation. Numerical realization of boundary element approach for solving problems of poroelasticity/poroviscoelasticity needs a special calculating method [4].

Numerical scheme is based on the Green-Betti-Somigliana formula. To perform numerical results double accuracy calculations are used. Regularized BIE (6) are considered in order to introduce BE discretization [9].

$$\alpha_\Omega u_k(x,s) + \int_\Gamma \left(T_{ik}(x,y,s)u_i(y,s) - T_{ik}^0(x,y,s)u_i(x,s) - U_{ik}(x,y,s)t_i(y,s) \right) d_y\Gamma = 0,$$

$$\alpha_\Omega u_k(x,t) + \int_\Gamma \left(F_k^1(x,y,t) - F_k^2(x,y,t) \right) d_y\Gamma = 0,$$

$$(x \in \Gamma), \quad t = [t_1,t_2,t_3,q]^T, \quad u(u_1,u_2,u_3,p),$$

$$(6)$$

BE discretization process is the process of surface fragmentation into boundary elements. Shape functions are quadratic polynomials of interpolation. Unknown boundary fields are integrated by using interpolation node values. Boundary-element schemes are based on the mixed approximation of boundary functions: with biquadratic boundary approximation, bilinear displacement approximation and constant elements for tractions.

2. METHOD VERIFICATION

The problem of poroviscoelastic rod under Heaviside type load $t_2 = 1\,\text{N/m}^2$ is considered. Material constants are: $K = 4.8 \cdot 10^9\,\text{N/m}^2$, $G = 7.2 \cdot 10^9\,\text{N/m}^2$, $\rho = 2458\,\text{kg/m}^3$, $\phi = 0.19$, $K_s = 3.6 \cdot 10^{10}\,\text{N/m}^2$, $\rho_f = 1000\,\text{kg/m}^3$, $K_f = 3.3 \cdot 10^9\,\text{N/m}^2$ $k = 1.9 \cdot 10^{-10}$ $\text{m}^4/(\text{N} \cdot \text{s})$, $\nu = 0$ (it is important, because we simulate 1D solution through 3D solution).

The problem has analytical solution in Laplace domain [4], which just transforms back to time domain with the help of the Durbin's method for Laplace transform numerical inversion.

In Figure 1, the comparison of numerical and analytical solutions is shown. Maximum relative error is less than 2% at considered time interval. Our numerical approach demonstrates a satisfactory accuracy, so we can apply it to 3D problems.

3. NUMERICAL RESULTS

Problem of a prismatic poroviscoelastic column clamped at one end and subject to a Heaviside type load at another end is considered (see Figure 2). Poroelastic material is Berea sandstone with the following constants: $K = 8 \cdot 10^9\,\text{N/m}^2$, $G = 6 \cdot 10^9\,\text{N/m}^2$, $\rho = 2458\,\text{kg/m}^3$, $\phi = 0.66$, $K_s = 3.6 \cdot 10^{10}\,\text{N/m}^2$, $\rho_f = 1000\,\text{kg/m}^3$, $K_f = 3.3 \cdot 10^9\,\text{N/m}^2$, $k = 1.9 \cdot 10^{-10}\,\text{m}^4/(\text{N} \cdot \text{s})$.

Solid divided into two subdomains. To obtain numerical solution for displacements at reference point B, a boundary element mesh, which consisted of 520 elements in each subdomain, was employed (see Figure 1).

Figure 3 gives the comparison of poroelastic and poroviscoelastic (case of weakly singular kernel model) solutions with different viscoelastic parameters.

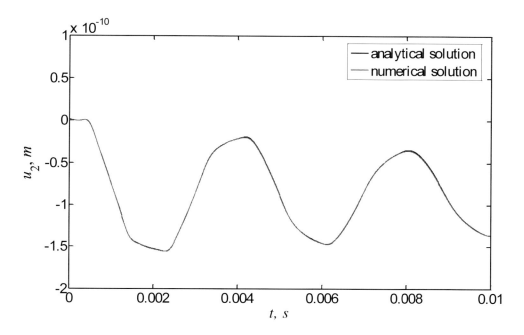

Figure 1. Numerical and analytical poroviscoelastic solutions comparison in case of weakly singular kernel model, $h = 1$ and $\alpha = 0.7$.

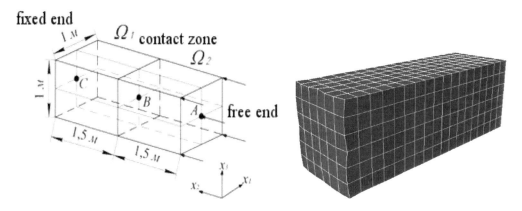

Figure 2. Problem formulation (left) and boundary element mesh visualization (right).

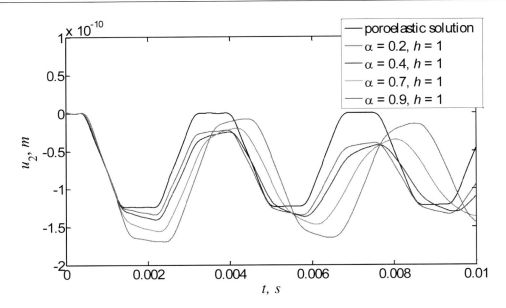

Figure 3. Displacement u_2 at points B in case of considered model for different values of viscoelastic parameters.

CONCLUSION

An influence of viscoelastic parameters on displacement responses is demonstrated on the example of a problem of the prismatic poroviscoelastic solid under Heaviside-type load. Boundary integral equations method and boundary element method are applied in order to solve three dimensional boundary-value problems. We verified numerical approach based on BEM by comparison with analitics. The poroviscoelastic media modelling is based on Biot's theory of porous material in combination with the elastic-viscoelastic corresponding principle. Viscous properties are described by model with weakly singular kernel.

ACKNOWLEDGMENT

The work was supported by RFBR under grants Nos. 14-08-31415, 14-08-31410, 14-08-00811, 15-08-02814.

REFERENCES

[1] Biot, M. A., *J. Acoust. Soc. Am.*, 1941, vol. 12, 155.
[2] Biot, M. A., *J. Acoust. Soc. Am.*, 1956, vol. 28(2), 168.
[3] Biot, M. A., *J. Acoust. Soc. Am.*, 1956, vol. 27, 459.
[4] Schanz, M., Wave Propagation in Viscoelastic and Poroelastic Continua, *Springer*, Berlin, 2001.

[5]   Nardini, D.; Brebbia, C. A., Boundary Element Methods in Engineering, C. A. Brebbia (Ed.), *Springer,* Berlin, 1982.

[6]   Babeshko, V. A., *Proceedings of the USSR Academy of Sciences,* 1985, vol. 284, 73.

[7]   Schanz, M.; Cheng, *AHD Acta Mech.,* 2000, vol. 145, 1.

[8]   Schanz, M.; Cheng, *AHD J Appl Mech.*, 2001, vol. 68, 192.

[9]   Bazhenov, V.G.; Igumnov, L.A. Boundary Integral Equations and Boundary Element Methods in Treating the Problems of 3D Elastodynamics with Coupled Fields PhysMathLit, Moscow, 2008 (In Russia).

[10] Ugodchikov, A.G.; Hutoryanskii, N.M. Boundary Element Method in Deformable Solid Mechanics, Kazan State University Press, Kazan, 1986 (In Russia).

[11] Cristensen R. Introduction to the Theory of Viscoelasticity, *Mir,* Moscow, 1974 (In Russia).

[12] Durbin, F. *The Computer Journal.* 1974, vol. 17, 371.

In: Proceedings of the 2015 International Conference … ISBN: 978-1-63484-577-9
Editors: Ivan A. Parinov, Shun-Hsyung Chang et al. © 2016 Nova Science Publishers, Inc.

Chapter 49

DESCRIPTION OF NON-LINEAR VISCOELASTIC DEFORMATIONS BY THE 3D MECHANICAL MODEL

A. D. Azarov[1,] and D. A. Azarov[2,†]*

[1] Vorovich Mathematics, Mechanics and Computer Sciences Institute,
Southern Federal University, Rostov-on-Don, Russia
[2] Don State Technical University, Rostov-on-Don, Russia

ABSTRACT

In order to describe large deformations, many different types of constitutive relations (such as Signorini's, Murnaghan's, Mooney-Rivlin's etc.) exist within the framework of the non-linear theory of elasticity. The potential energy in these relations is represented as the functions of the strain tensor's invariants. This chapter highlights a new approach to the issue. A "force-elongation" relation of the model and, consequently, "stress-strain" constitutive law of the continuum are modelled through the calculations of geometric and force characteristics of the model. This relation is demonstrated for a case of uniaxial strain.

Keywords: mechanical model, constitutive law, non-linear viscoelastisity

1. MODEL EQUATIONS FOR A TRIAXIAL STRETCHING

Let us consider a certain elementary cubic volume of a contiuum with a mechanical structure integrated. This structure is a system of viscoelastic deformable bonds (rods), describing interactions between opposite and adjacent sides of the cube. We call the structure a "3d mechanical model." The model's parameters can be divided into two groups: geometrical and mechanical. The geometrical parameters are: the bonds' lengths and the angles between them. The mechanical parameters are: elastic rigidities and viscosities of the

---

[*] E-mal: polyani49@mail.ru.
[†] E-mal: danila_az@mail.ru.

bonds. The bonds' lengths after deformation determine the shape of the elementary volume of the continuum. The bonds between the adjacent sides of the cube are very important, as they specify the relations between the longitudinal and transverse deformations. In the most general case, if the mechanical parameters (such as elasticity coefficients of the bonds) are unequal, the structure defines the anisotropic continuum. For an isotropic material there are two types of elasticity coefficients, one of them being responsible for the properties of the bonds between adjacent sides and the other for the properties of the bonds between opposite sides. A "force-elongation" relation of the model corresponds to "stress-strain" constitutive law of the continua in main axes. The general view of the model is shown in Figures 1 and 2. The rods (bonds) are hinged to the sides' centers A_i of the cube. The model equations for a triaxial stretching of orthotropic continuum derive from the simple geometric dependencies as well as from the equlibrium equations of the external forces and internal force reactions of the bonds.

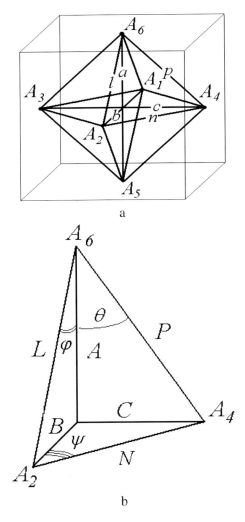

Figure 1. Geometry of the model bonds: (a) pre-deformation elementary volume and the model (where $a = b = c = 1$, $l = p = n = \sqrt{2}$); (b) angles of the deformed model.

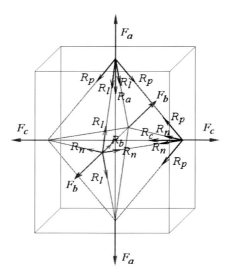

Figure 2. Triaxial stretching: external forces and internal reaction forces of the model.

Simple geometrical dependencies for a pre-deformed state are:

$$l^2 = a^2 + b^2;\ p^2 = a^2 + c^2;\ n^2 = b^2 + c^2 \qquad (1)$$

The corresponding dependencies for the deformed state are:

$$L^2 = A^2 + B^2;\ P^2 = A^2 + C^2;\ N^2 = B^2 + C^2, \qquad (2)$$

where

$$A = a + \delta_a;\quad B = b + \delta_b;\quad C = c + \delta_c;$$
$$L = l + \delta_l,\quad P = p + \delta_p,\quad N = n + \delta_n. \qquad (3)$$

Here δ_i is the change of length (either positive or negative) of the i-th ($i = a, b, c, n, p, l$) bond. All bonds extend under triaxial stretching by external forces, i.e., $\delta_i > 0$, and angles between them change as follows:

$$\cos\varphi = A/L,\quad \cos\psi = B/N,\quad \cos\theta = A/P. \qquad (4)$$

The external forces and model reaction forces (see Figure 2) satisfy the conditions of equilibrium in the joints:

$$F_a = R_a + 2R_l \cos\varphi + 2R_p \cos\theta,\quad F_b = R_b + 2R_l \sin\varphi + 2R_n \cos\psi,$$
$$F_c = R_c + 2R_p \sin\theta + 2R_n \sin\psi. \qquad (5)$$

The rigidities of the bonds are the generalized integral estimations of internal force interactions in the material, but no real intermolecular or interatomic interactions, studied by physical-chemical sciences. At the same time, the positive and negative rigidities are possible, that is supposed similar to the mutual attraction and repulsion forces in the real material.

In the simplified case, the model reaction force of each bond determines by the correspondent bond elongation as follows:

$$R_i = k_i \delta_i, \quad i = a,b,c,l,p,n. \tag{6}$$

In the paper, we use differential relations instead of the algebraic dependencies (6). These relations are Standard Linear Solid model (SLS model or Zener model, see Figure 3):

$$\begin{cases} R_i = k_{i0} \delta_i(t) - k_{i1} \eta_i(t) \\ \dot{\eta}_i = -\tau_i^{-1}(\eta_i(t) - \delta_i(t)) \end{cases} \tag{7}$$

where $i = a,b,c,l,p,n$, and $k_{i\infty} = k_{i0} - k_{i1}$ are the equilibrium values of the elasticity coefficients.

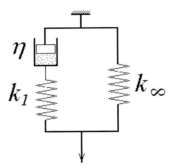

Figure 3. SLS model.

The inner variables $\eta_i(t)$ are the calculated deformations of the viscous fluid dashpots of the SLS model. These variables define the rheological properties of the bonds' reactions.

The system has zero initial conditions:

$$\eta_a(0) = \eta_b(0) = \eta_l(0) = \eta_n(0) = 0.$$

2. UNIAXIAL STRETCHING

The problem of uniaxial stretching of isotropic material by F_a force ($F_b = F_c = 0$) is analized below.

The notations in an isotropic case are left unchanged for the sake of illustration, even for parameters with similar values. Here the instant elasticities of the bonds are

Description of Non-linear Viscoelastic Deformations ... 371

$k_{a0} = k_{b0} = k_{c0}$, $k_{l0} = k_{p0} = k_{n0}$. The coefficients k_{i1} and the times of relaxation τ_i may take different values even in isotropic case (due to the fact, that initially isotropic material gains anysotropic properties under deformation).

We derive the following formulae from the relations (2) – (4):

$$\delta_b = \delta_c, \ \delta_l = \delta_p, \ \text{angles} \ \psi = 45°, \ \theta = \varphi, \ \delta_n = \sqrt{2}\delta_b = l\delta_b,$$

$$\delta_l = \widetilde{\delta}_l(\delta_a, \delta_b) = -l \pm \sqrt{l^2 + q_l}, \ q_l = 2\delta_a + \delta_a^2 + 2\delta_b + \delta_b^2.$$

Then we obtain the system of five differential equations of the model with five unknowns:

$$z_1(t) = \eta_b(t), \ z_2(t) = \eta_l(t), \ z_3(t) = \eta_n(t),$$
$$z_4(t) = \delta_b(t), \ z_5(t) = \delta_l(t).$$

The notations for the system are:

$$D = (Q + \sin\varphi \cdot P)^{-1},$$
$$Q = k_{b0} + 2[k_{l0}\delta_l(t) - k_{l1}\eta_l(t)](l + \delta_l(t))^{-1} + l^2 k_{n0},$$
$$P = 2k_{l0}\sin\varphi - 2[k_{l0}\delta_l(t) - k_{l1}\eta_l(t)](1 + \delta_b(t)(l + \delta_l(t))^{-2}.$$

Thus the system can be written down as follows:

$$\begin{cases} \dot{z}_1(t) = -\tau_b^{-1}(z_1(t) - z_4(t)) \\ \dot{z}_2(t) = -\tau_l^{-1}(z_2(t) - z_5(t)) \\ \dot{z}_3(t) = -\tau_n^{-1}(z_3(t) - lz_4(t)) \\ \dot{z}_4(t) = D \ [-\cos\varphi \cdot P \cdot \dot{\delta}_a(t) - k_{b1}\tau_b^{-1}(z_1(t) - z_4(t)) - \\ \quad - 2k_{l1}\sin\varphi \cdot \tau_l^{-1}(z_2(t) - z_5(t)) - lk_{n1}\tau_n^{-1}(z_3(t) - lz_4(t))] \\ \dot{z}_5(t) = \cos\varphi \cdot \dot{\delta}_a + \sin\varphi \cdot D \cdot [-\cos\varphi \cdot P \cdot \dot{\delta}_a(t) - k_{b1}\tau_b^{-1}(z_1(t) - z_4(t)) - \\ \quad - 2k_{l1}\sin\varphi \cdot \tau_l^{-1}(z_2(t) - z_5(t)) - lk_{n1}\tau_n^{-1}(z_3(t) - lz_4(t))] \end{cases}$$

Solving the system by numerical methods, we get the unknown functions $z_i(t)$ (which characterize the bonds' elongations) and then the force reactions of the bonds. The reaction force R_a and the inner variable $\eta_a(t)$ may be found independently out of Equations (7) at $i = a$. The stretching force may be calculated as:

$$F_a(t) = \left[k_{a0}\delta_a(t) - k_{a1}\eta_a(t)\right] + 4\left[k_{l0}z_5(t) - k_{l1}z_2(t)\right]\frac{1 + \delta_a(t)}{l + z_2(t)}.$$

The geometrical restriction on the value of unknown transverse deformation $|\delta_b| < 1$ is an important condition, providing the effectiveness of numerical solution of the system, in spite of a possible large value of δ_a.

In the elastic case, for infinitesimal deformations and after linearization of equations, we obtain:

$$\delta_{0l} = l^{-1}(\delta_{0a} + \delta_{0b}),$$
$$\delta_{0b} = -k_l\delta_{0a}/(k_b + k_l + 2k_n), \quad F_a = k_a\delta_{0a} + 2k_l(\delta_{0a} + \delta_{0b}).$$

Since the material is isotropic, i.e., $k_a = k_b$, $k_l = k_n$, we can evaluate the Young's modulus E, the Poisson's ratio v, shear and bulk moduli G and K:

$$v = -\delta_{0b}/\delta_{0a} = k_l/(k_a + 3k_l),$$
$$E = F_a/4\delta_{0a} = (k_a + k_l)\cdot(k_a + 4k_l)/(4\cdot(k_a + 3k_l)),$$
$$G = E/(2(1+v)) = (k_a + k_l)/8, \quad K = E/(3(1-2v)) = (k_a + 4k_l)/12$$

The restrictions on the model parameters: $(k_a + k_l) > 0$, $(k_a + 4k_l) > 0$, $(k_a + 3k_l) > 0$ follow from the natural conditions $G > 0$, $K > 0$, $E > 0$, $-1 < v < 1/2$.

3. NUMERICAL EXAMPLE

The results of numerical evaluations are shown in Figures 4 – 7. The considered material has the following parameters:

$$k_{a0} = k_{b0} = -500; \quad k_{l0} = k_{n0} = 1000; \quad k_{a1} = -150;$$
$$k_{b1} = -200; \quad k_{l1} = 250; \quad k_{n1} = 450;$$
$$\tau_a = 15; \quad \tau_b = 25; \quad \tau_l = 10; \quad \tau_n = 20.$$

The asymmetric behavior of the plots during loading-unloading demonstrates rheological delay of reactions.

The energies of the bonds calculated numerically. It should be noted, that the general energy of the model is positive, though the energies of the certain bonds can become negative.

Description of Non-linear Viscoelastic Deformations ... 373

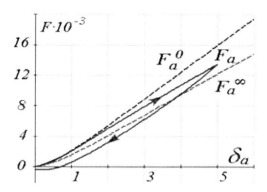

Figure 4. "Force-elongation" dependence under loading-unloading; $F_a^o(t)$ is the "instant-elastic" behavior of the model with elastic bonds' properties k_0, $F_a^\infty(t)$ is the elastic relation with equilibrium bonds' properties $k_\infty = k_0 - k_1$.

In the simplest case the linear theory of visco-elasticity uses two three-parametric differential models with two various relaxation times. For example: visco-elastic operator $\overline{E}$ uses τ_E time, and the $\overline{V}$ operator - τ_V time. Similarly, the proposed model uses two relaxation times: τ_a and τ_l. However, the transverse strain dependent on longitudinal strain has two relaxation time parameters τ_1, τ_2, which determine from the quadratic equation. As a result, the relaxation function $E(t)$ is characterized by four relaxation time parameters: τ_a, τ_l, τ_1 and τ_2. Thus the existence of many relaxation times typical for polymeric materials has its reflection in the model.

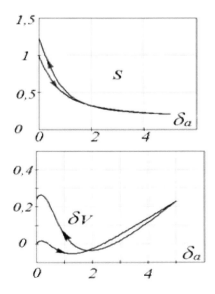

Figure 5. Cross-section area S and relative volume δV under loading-unloading.

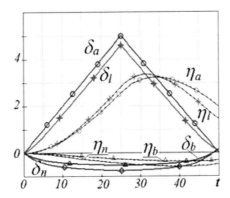

Figure 6. Bonds' elongations of the model and the inner variables as the time functions.

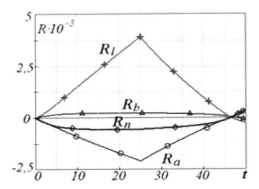

Figure 7. Internal force reactions of the model's bonds as the time functions.

The proposed model is not only used to approximate experimental data, but claims to become a new method for representing the non-linear mechanical properties of materials on the base of physically non-contradictory statements.

The necessity of unique measurement of the transverse deformation becomes clear by comparing the results for two different examples of the model. The calculations carried out show that despite the similarity of the "force-elongation" plots, the model demonstrates different properties of changing the cross-section area and relative volume, as well as the internal bonds' reactions. It means that the correct development of the constitutive relations of the materials is impossible based only on the "force-elongation" dependencies.

The constant parameters for viscous and elastic bonds' properties (7) are used in the current chapter, which is enough for describing below-medium visco-elastic deformations. Note that large deformations result in structural changes within the material, and thus the mechanical characteristics of the bonds must be the functions, not the constants. Further development of the model is required for describing the complex S-type "force-elongation" curves of the rubbers. Three typical zones can be seen on these curves for a wide class of rubber-like materials. Each zone corresponds with a particular type of inter-molecular bonds, determined by the essential reconstruction of the material structure. Thus, for each zone we should use its own elasticity coefficients and viscosity functions depending on the elongations.

The proposed model has been partially discussed in [1, 2, 3].

REFERENCES

[1] Azarov, A. D.; Azarov, D. A. *Vestnik DSTU*, 2011, *vol. 11(2)*, 147–156 (in Russian).

[2] Azarov, A. D.; Azarov, D. A. Comparison of 3d mechanical model with Murnaghan's constitutive law. *Proc. XVI Int. Conf. on Modern Problems of Mechanics of Solid Medium, Southern Federal University Press, Rostov-on-Don*, 2012, *vol. 1*, 5–9 (in Russian).

[3] Azarov, D. A., Nonlinearly deformable 3d mechanical model of incompressible elastic materials. *Proc. XV Int. Conf. on Modern Problems of Mechanics of Solid Medium, Southern Federal University Press, Rostov-on-Don*, 2011, *vol. 1*, 11-15 (in Russian).

In: Proceedings of the 2015 International Conference …   ISBN: 978-1-63484-577-9
Editors: Ivan A. Parinov, Shun-Hsyung Chang et al.   © 2016 Nova Science Publishers, Inc.

Chapter 50

DETERMINATION OF STRAIN CHARACTERISTICS

A. I. Kozinkina[1,] and Y. A. Kozinkina[2]*

[1]Don State Technical University, Rostov-on-Don, Russia
[2]Southen Federal University, Rostov-on-Don, Russia

ABSTRACT

It is present the experimental procedure based on using the acoustic emission (AE) data to study the stepwise character of deformation and fracture of polycrystalline materials. The AE-parameters analyzed in order to determine the characteristic points of the transition of the defect accumulation process from one stage to other. A quantitative estimation of discontinuities concentration is given. The proposed method, based on the concepts of plastic-destruction deformation, can serve as a base for monitoring and diagnostics of the material strength in real time scale for determining the conditions of formation of microdefects and macrodefects.

Keywords: stress, strain, imperfection, strain characteristics, elastic limit, yield point, destruction, strength, stages of deformation, transition points, polycrystalline materials, acoustic emission

1. INTRODUCTION

In order to understand phenomena of strength and fracture of solids, it is first necessary to identify the real degree of imperfection and take into account its effect on the strain characteristics of materials. Moreover, the present models of mechanics should be based on experimental data and physical mechanisms of the fracture process.

It has been known that $\sigma - \varepsilon$ plots give a visual characterization of the main mechanical properties of polycrystalline materials. Such plots reflect the functional stress–strain relationships. The most widely used plots are those obtained by using uniaxial tension as

---

[*] E-mail: akozinkinz@mail.ru.

shown in Figure 1 a, b. They are rather easy to analyze and provide such important characteristics as elastic limit σ_e (point E), yield point σ_y (point Y), and point of stability loss or "neck" formation σ_n (point N).

The vast experimental data presented in the book of Bell [1] showed that the stress–strain relationship, in the case of uniaxial tension of crystalline solids, is described most accurately by the parabolic function. Therefore, to reveal separate stages of deformation, it is most convenient to represent the $\sigma - \varepsilon$ relationship by a linear function in the $\sigma^2 - \varepsilon$ coordinates. This dependence is in good agreement with the relationship between yield stress and dislocation density known from the physical theory of deformation hardening [2]:

$$\sigma = \alpha G b \rho^{1/2},$$

where G is the shear modulus, b is the Burgers vector, ρ is the dislocation density, and α is a constant.

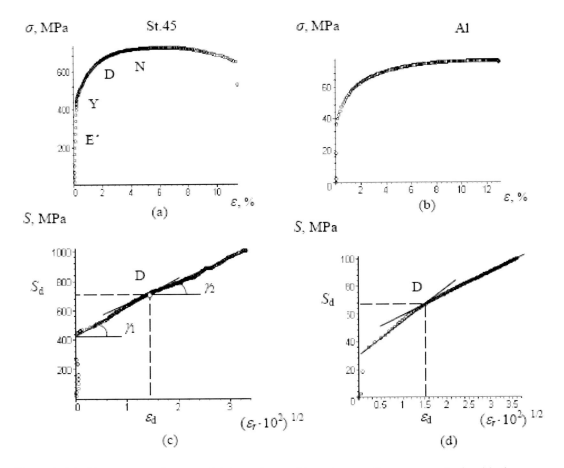

Figure 1. (a), (b) stress – strain curve; (c) dependence of true stress on the square root of residual longitudinal strain and (d) dependence of true stress on the square root of residual transverse strain during tension of specimen.

The stepwise character of deformation, which associates with a particular mechanism and extent of irreversible damage, can be revealed most clearly in the plots $S - \varepsilon_r^{1/2}$, where S is the true strain and ε_r is the residual deformation [3]. The true strain in a specimen determined as

$$S = \sigma /(1-\varepsilon_\perp)^2,$$

where $\varepsilon_\perp$ is the transverse strain of a specimen. The residual deformation calculated as

$$\varepsilon_r = \varepsilon - \sigma / E,$$

where E is the Young modulus and ε is the longitudinal stran of the specimen.

The structural analysis of various metals and their alloys, showed that, in tension in the deformation hardening region before "neck" formation (Figure 1a), there is some point D between points Y and N, after which, strictly speaking, a material cannot be regarded as continuous. Unlike crack formation in the "neck," which results to material fracture into pieces, this process called destruction [2]. In the mechanics of deformed solids, the destruction phenomenon studied by L. M. Kachanov [4] and Yu. N. Rabotnov [5]. They introduced damage parameters describing the changes in the effective cross-section area due to the growth of discontinuities. This new approach of plastic-destruction deformation provided the base for development of methods for the detection and quantitative assessment of fracture [6, 7].

From the viewpoint of non-equilibrium dynamics, such processes are characterized by non-equilibrium phase transitions at special points (bifurcation points) at which material properties change. The parameters that determine the transition points are universal and invariant [8]. They are most informative at prediction of physic and mechanical features of materials. For most of materials, a selection of three stages, namely initial, basic and final ones, we assume to be worthwhile. The initial stage is due to appearance of local stresses and accumulation of defects at micro-level in size of order 1 nanometers. The basic stage consists in formation of incipient voids and its coalescence up to the critical size of order $1 - 10^2$ microns and larger, which are commensurable with the sizes of structural heterogeneity of materials. At the given stage, thus, it forms a local critical loosening which leads to occurrence of the neck or the main crack. The final stage consists in sharp localization of plastic deformation, growth of macro-defect, and avalanche destruction.

2. METHODS

The acoustic emission method may be considered as a direct method for studying the kinetics of damage accumulation in the sense that to each collective event of damaging material structure there corresponds a primary elastic pulse. The process of its radiation represents an event of acoustic emission (AE). Hence, the measurement of intensity of AE events stream and its total number allows investigating the kinetics of deformation and destruction [9].

We investigated the integral characteristic of the stream of AE events defined by

$$N_{an} = \int_0^t \dot{N}_a(t)dt \bigg/ \int_0^\tau \dot{N}_a(t)dt,$$

where N_{an} is the normalized total number of AE acts, $\dot{N}_a$ is the current intensity of the stream of AE acts, t is the current time, and τ is the loading time to failure.

The test block consisted of Instron loading machine and hardware complex of detection and analysis of signals. To ensure that the stream of AE events matches the stream of damages, we used a procedure for recovering the true parameters of the stream of AE events from the signals recorded [10]. To reduce the non-uniformity of the amplitude-frequency characteristic and the reverberation time, we used transducers consisting of a large number of thin unidirectional piezocrystals. A feature of such transducers is selectivity toward longitudinal oscillations, which greatly increases their noise immunity [11]. Since fracture characteristics are structure-sensitive and depend on the composition, machining, and thermal treatment of a material, the specimens were preliminary annealed.

3. EXPERIMENTAL RESULTS AND DISCUSSION

During uniaxial tension of round annealed samples of various polycrystalline materials, experimental data have shown that relationship between the integrated characteristic of AE events stream and total strain is close to parabolic function. In logarithmic coordinates as shown in Figure 2 it is approximated by a polygonal line with three characteristic bend points separating the stages of deformation and damage accumulation.

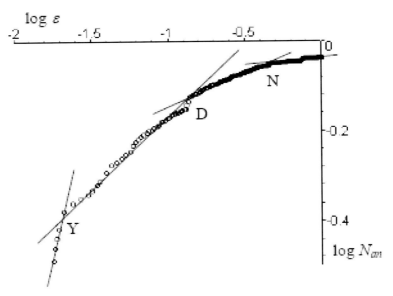

Figure 2. Logarithmic plot of the total number of AE events on longitudinal strain during tension of steel 45 specimen.

Determination of Strain Characteristics 381

A comparison between mechanical destruction diagram and AE diagram has shown that the first point corresponds to the yield point (point Y), the second point corresponds to the point of material destruction (point D) and the third point (N) corresponds to the moment of neck formation. Furthermore, the singularities expose on corresponding dependencies at loading high-strength tool materials with very low AE activity.

Regression analysis and statistical data processing showed that, with a probability of 0.95, the confidence intervals for the slopes of the approximating lines do not overlap.

Moreover, at consideration of body plastic deformation in view of formation and growth of voids, it is possible to show, that their concentration c, defined as ratio between volume of voids and volume of a material, is determined by a destruction component of residual strain [12]. Hence, the geometry of AE diagrams leads to the formula:

$$c = \left(\varepsilon_r^{1/2} - \varepsilon_d\right)^2 \left(1 - \frac{\gamma_2}{\gamma_1}\right)^2$$

According to experimental results, we obtained ε_r-dependencies of the defect concentration for various materials:

(i) for steel 45: $c = 0.35(\varepsilon_r - 0.118)^2$;
(ii) for steel U9: $c = 0.65(\varepsilon_r - 0.178)^2$;
(iii) for Al: $c = 0.53(\varepsilon_r - 0.168)^2$;
(iv) for Cu: $c = 0.68(\varepsilon_r - 0.224)^2$.

In Table are listed the numerical values of c, calculated by the AE procedure considered above. The magnitudes of c_0 were calculated using the microstructural data obtained with the help of electronic microscopy and low-angle X-ray diffraction.

Material	c_0	c		
		in vicinity of point D	between point D and point N	at point N
Steel 45	$5.2 \cdot 10^{-7}$	$3.5 \cdot 10^{-7}$	$8.8 \cdot 10^{-4}$	$3.5 \cdot 10^{-3}$
Steel U9	$5.2 \cdot 10^{-7}$	$1.7 \cdot 10^{-7}$	$4.2 \cdot 10^{-4}$	$1.7 \cdot 10^{-3}$
AO commercial aluminum	$1.4 \cdot 10^{-4}$	$3.5 \cdot 10^{-7}$	$8.7 \cdot 10^{-4}$	$3.5 \cdot 10^{-3}$
MO commercial copper	$4.1 \cdot 10^{-3}$	$5.1 \cdot 10^{-6}$	$1.3 \cdot 10^{-2}$	$5.1 \cdot 10^{-2}$

As it follows from Table 1, accuracy of determination of the defect concentration by using AE data and the proposed method is comparable with the accuracy of its determination by X-ray diffraction and electron microscopy methods. The obtained data also correspond to various estimates of critical defect concentration at the point N, which were obtained in [13]. Thus, in accordance with the statistical theory of failure, the "neck" formation occurs upon reaching a local concentration of micro-discontinuities $c \approx 10^{-3} - 10^{-2}$.

CONCLUSION

Thus, an experimental procedure for constructing the strain diagrams using AE data was developed. It allows one to investigate the kinetics of deformation and failure of various materials, to identify stages of deformation and damage accumulation, to determine the parameters of transition points and the conditions of formation of microdefects and macrodefects.

The results obtained can be used to directly assess the extent of fracture and determine the preliminary failure condition of structural materials. One can see that the acoustic emission reflects the changes in the deformation mechanisms earlier than the mechanical behavior of a specimen. Furthermore, the change in the deformation stage expresses more sharply.

Upon reaching the point of fracture in a polycrystalline material, a quantity of defects forms. The values given in Table 1 can be considered as the lower-bound estimate of c. It is apparent that accuracy of defect concentration determination on the base of AE data correlates with accuracy of X-ray diffraction method and electronic microscopy method.

The proposed method, based on the concepts of plastic-destruction deformation, is a new approach not only to determining the conditions of formation of micro- and macro-defects and strength criteria, but also to diagnostics of preliminary failure state. It can serve as a base for monitoring and diagnostics of the strength of materials in real time scale.

REFERENCES

[1] Bell, G.F. *Experimental Fundamentals of Mechanics of Solids under Strain*, Filin, A.P. (Ed.), Nauka, Moscow, 1984 (In Russian).

[2] Ivanova, V.S.; Gorodienko, L.K.; Geminov, V.N.; et al., *Function of Dislocations under Hardening and Fracture of Metals*, Ivanova, V.S. (Ed.), Nauka, Moscow, 1965.

[3] Rybakova, L.M. *Vestnik Mashinostroeniya*. 1993, *vol. 8*, 32–37 (In Russian).

[4] Kachanov, L.M. *Izv. AN SSSR, OTN*. 1958, *vol. 8*, 26–31.

[5] Rabotnov, Yu. N. *Issues of Strength of Materials and Constructions*. AN SSSR Press, Moscow, 1959.

[6] Krajcinovic, D. *Int. J. Solids and Structures*, 2000, *vol. 37*, 267–277.

[7] Berezin, A. V.; Kozinkina, A. I. *Journal of Machinery Manufacture and Reliability*, 2002, *vol. 3*, 102–108.

[8] Kozinkina, A. I., *Russian Journal of Non-destructive Testing*, 1999, *vol. 9*, 95–99.

[9] James, D. R.; Carpenber, S. H. *J. Applied Physics*, 1971, *vol. 12*, 4685–4698.

[10] Builo, S. I.; Kozinkina, A. I. *Physics Solid State*. 1996, *vol. 11*, 1844-1845.

[11] Kozinkina, A. I., Berezin, A. V. *Engineering and Automation Problems*, 2008, *vol. 1*, 75–78 (In Russian).

[12] Kozinkina, A. I.; Kozinkina, Y. A. *Journal of Applied Mechanics and Technical Physics*. 2010, *vol. 6*, 913–917.

[13] Cheremskoi, P. G.; Slezov. V. V.; Betekhtin, V. I, *Pores in Solids*, Energoatomizdat, Moscow, 1990 (In Russian).

In: Proceedings of the 2015 International Conference … ISBN: 978-1-63484-577-9
Editors: Ivan A. Parinov, Shun-Hsyung Chang et al. © 2016 Nova Science Publishers, Inc.

Chapter 51

FREQUENCY DEPENDENCES OF THE COMPLEX MATERIAL CONSTANTS FOR POROUS PZT PIEZOCERAMICS

A. A. Naumenko, M. A. Lugovaya, E. I. Petrova, I. A. Shvetsov and A. N. Rybyanets*

Institute of Physics, Southern Federal University, Rostov-on-Don, Russia

ABSTRACT

Many advanced piezoelectric materials are lossy and show significant dispersion in their material constants. In this paper, frequency dependencies of complex dielectric, piezoelectric and elastic constants of porous PZT piezoceramics with the composition of $Pb_{0.45}Ti_{0.45}Zr_{0.53}(W_{1/2}Cd_{1/2})_{0.02}O_3$ and relative porosity 15-25% were studied using piezoelectric resonance analysis. This produced a set of material constants at each resonance frequency which, when plotted as a function of resonance frequency, gave a good representation of the dispersion in the 'constants'. Experimental samples were fabricated in the shape of disks and were polarized on thickness. The frequency dependencies of the complex constants were determined by analyzing the fundamental and higher resonances (third and fifth harmonics) for the thickness extensional mode of piezoelectric disk vibrations. Strong spatial dispersion of the complex elastic moduli for porous ceramics samples was found. Microstructural and physical mechanisms of elastic, piezoelectric and dielectric losses and spatial dispersion for porous piezoceramics were discussed.

1. INTRODUCTION

Large-scale production of composite materials and their intensive use in ultrasonic transducers requires not only overcoming technological problems, but also a psychological revolution in fundamental views of technologists and designers of piezoceramic material.

---

* e-mail: arybyanets.jr@gmail.com.

The change from prevailing industry inclination to manufacture high-density, chemically perfect and structurally flawless ceramic materials to the fabrication and use of "imperfect," low density and "defective" materials (lead-metaniobate ceramics, porous ceramics and composites) has taken a long time [1-4]. Commercialization of composite materials has lead also to the development of new concepts of material and ultrasonic transducers designing [5-8]. Demands on medical and NDT ultrasonic transducer performance have increased in recent years. Low-Q piezoceramics and piezocomposite materials are widely used for wide-band NDT ultrasonic transducers with high sensitivity and resolution. Some of these advanced materials are lossy, and direct use of IEEE Standards for material constant determination leads to significant errors. The modeling and design of piezoelectric devices by finite element methods, among others, rely on the accuracy of the dielectric, piezoelectric and elastic coefficients of the active material used which is, commonly, an anisotropic ferroelectric polycrystal. The accurate description of piezoceramics must include the evaluation of the dielectric, piezoelectric and mechanical losses, accounting for the out-of-phase material response to the input signal. Porous piezoceramics based on different piezoceramic compositions were proposed as a promising candidate for lead metaniobate replacement in wide-band ultrasonic transducers [9, 10]. Intensive research and technological work, as well as improvements in fabrication methods, have allowed large-scale manufacture of porous piezoelectric ceramics with reproducible and controllable porosity and properties [11, 12].

A comprehensive study including аштшеу finite difference modeling, impedance spectroscopy characterization and pulse-echo ultrasonic measurements of different porous ceramics are described in this Chapter.

2. EXPERIMENTAL SCENARIO

Porous PZT piezoceramics $Pb_{0.45}Ti_{0.45}Zr_{0.53}(W_{½}Cd_{½})_{0.02}O_3$ (PCR-1) with different porosity and pore sizes were chosen as model samples for comparison with piezoelectric resonance analysis (PRAP) and ultrasonic measurements (Figure 1):

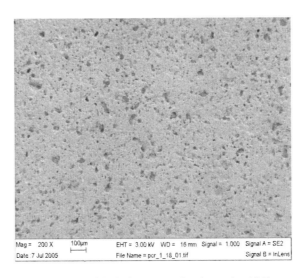

Figure 1. SEM micrographs of porous PCR-1 piezoceramics (porosity 18%, pore size 30 μm).

Experimental samples of porous piezoceramics in porosity range 15-25% were prepared by burning out of organic powders using specially developed technology [3, 4]. Mixtures of organic crystal powders with different particle shapes, decomposition temperatures and grain-size distribution (from 30 to 50 µm) were used for producing the pore volumes. The porous ceramics elements with vacuum deposited Cr/Ni electrodes were poled on air at cooling from 380 °C to 90 °C for 1 min by applying a dc electric field of 1 – 1.5 kV/mm.

Specimens for optical microscopy were ground and polished using a specially designed procedure for composites comprising components with different hardness.

Complex elastic, dielectric and piezoelectric coefficients of porous ceramics and ceramic matrix piezocomposite elements were determined by impedance spectroscopy method using Piezoelectric Resonance Analysis (PRAP) software [8, 12]. Measurements were made on Solartron Impedance/Gain-Phase Analyzer SL 1260 and Agilent 4291 Impedance Analyzer.

Sound velocity and attenuation of longitudinal and shear waves in different directions for polarized and non-polarized samples were measured by pulse–echo and through-transmit ultrasonic methods in the frequency range 1 – 10 MHz using digital oscilloscope LeCroy Wave Surfer 422, Olympus 5800, 5077 pulser/receivers using standard Olympus transducers.

The microstructure of polished, chemically etched, and chipped surfaces of composite samples was observed with optical (NeoPphot-21) and scanning electron microscopes (SEM, Karl Zeiss) [4]. Density and shrinkage coefficients of the sintered porous ceramics and composite bodies were measured by weighing and measuring the volume, as well as by the hydrostatic weighing method.

3. RESULTS AND DISCUSSION

3.1. Ultrasonic Measurements

Figures 2 shows echo-pulse characteristics for PCR-1 porous ceramics measured at transmitted pulse frequencies 2.25 MHz and 5 MHz.

It is readily seen from Figure 2 that because of ultrasonic wave scattering and spatial dispersion, the received echo-pulses are strongly distorted and shifted down in frequency. Thus, these pulse-echo measurements of elastic properties of ceramic composites as function of frequency are inaccurate and ambiguous.

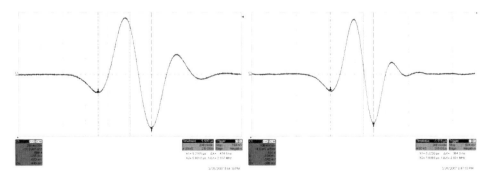

Figure 2. Echo-pulse characteristics for porous ceramics element: transmitted pulse frequency 2.25 MHz and 5 MHz; received pulse frequency 2.1 MHz and 2.6 MHz, respectively.

3.2. Resonance Measurements and Wave Pro Simulations

Figure 3 shows an example of impedance spectra and PRAP approximations for the thickness extensional (TE) mode of PCR-1 porous ceramic disks (Ø10 × 0.5 mm).

Table 1 summarizes complex constants for a PCR-1 disk (porosity 18%) obtained using PRAP analysis for thickness and radial resonance modes, along with corresponding IEEE Standard results. Additional physical parameters of PCR-1 porous ceramics are as follows: $Z_A = 25.5$ Mrayl, $\rho = 6.6$ g/cm^3, $V_t = 3875$ m/c, $Q^t_M = 25$, $T_C = 340°C$ [2].

It is readily observed in Tables 1 that for moderately low-Q porous ceramics, the results obtained by PRAP and the IEEE Standard methods are practically similar. However, even here, PRAP gives sets of complex constants, thereby allowing an accounting for and an analysis of the losses. PRAP can be used also to determine material properties of ceramics from higher order harmonics [8].

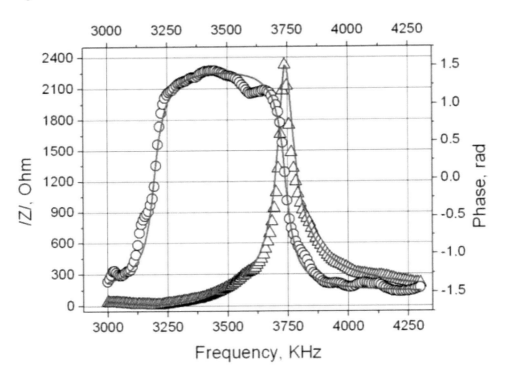

Figure 3. Impedance spectrum and PRAP approximation for TE mode of PCR-1 disk.

Figure 4 shows results of PRAP analysis of material properties along with the simulation results of Wave 3000 Pro program for complex C^D_{33} and C^E_{33} obtained from the higher harmonics of the TE mode of a PCR-1 porous ceramic disk.

It is readily observed that c'^E_{33} (real), as well as c''^E_{33} (imaginary), increase with the frequency because of the Rayleigh scattering (dispersion) of high frequency ultrasonic waves on pores. In its turn, c'^D_{33} (real) decreases and c''^D_{33} increases significantly with the frequency as a result of the electromechanical contribution to c^D_{33} (note that k_t value measured on higher harmonics drops drastically with the harmonics number).

Table 1. PRAP analysis results for different vibrational modes of PCR-1 disk

Parameter	Real	Imaginary	IEEE Standard	Error, %
Radial Extensional (RE) Mode				
f_p (Hz)	222222	754.256	221340	0.4
s_{11}^E (m²/N)	$1.75 \cdot 10^{-11}$	$-8.91 \cdot 10^{-14}$	$1.74 \cdot 10^{-11}$	0.6
s_{12}^E (m²/N)	$-4.99 \cdot 10^{-12}$	$1.48 \cdot 10^{-13}$	$-4.87 \cdot 10^{-12}$	2.4
$-d_{31}$ (C/N)	$7.92 \cdot 10^{-11}$	$-4.36 \cdot 10^{-12}$	$7.76 \cdot 10^{-11}$	2.0
ε_{33}^T (F/m)	$4.35 \cdot 10^{-9}$	$-1.97 \cdot 10^{-10}$	$4.27 \cdot 10^{-9}$	1.8
k_p	0.479629	-0.01101	0.47302	1.4
σ^P	0.284483	0.009907	0.278866	2.0
e_{31} (C/m²)	6.32055	-0.22792	6.16393	2.5
s_{66}^E (m²/N)	$4.5 \cdot 10^{-11}$	$1.18 \cdot 10^{-13}$	$4.46 \cdot 10^{-11}$	0.9
c_{66}^E (N/m²)	$2.22 \cdot 10^{10}$	$5.841 \cdot 10^7$	$2.24 \cdot 10^{10}$	0.9
Thickness Extensional (TE) Mode				
f_p (Hz)	$3.73 \cdot 10^6$	37771.9	$3.73 \cdot 10^6$	0.0
k_t	0.566103	-0.00555	0.56646	0.1
c_{33}^D (N/m²)	$9.91 \cdot 10^{10}$	$2.01 \cdot 10^9$	$9.93 \cdot 10^{10}$	0.2
c_{33}^E (N/m²)	$6.74 \cdot 10^{10}$	$1.99 \cdot 10^9$	$6.75 \cdot 10^{10}$	0.2
e_{33} (C/m²)	8.3419	-0.25107	-	-
h_{33} (V/m)	$3.81 \cdot 10^9$	$-1.17 \cdot 10^8$	-	-
ε_{33}^S (F/m)	$2.19 \cdot 10^{-9}$	$-1.33 \cdot 10^{-10}$	-	-

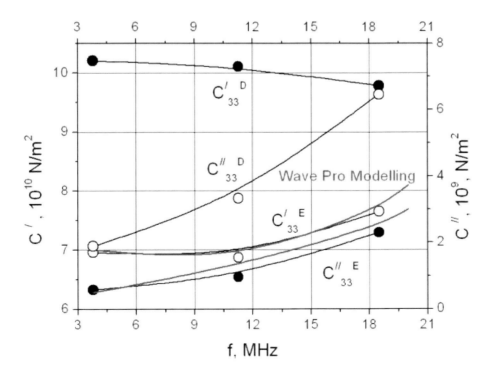

Figure 4. Frequency dependence of $c'_{33}{}^D$, $c'_{33}{}^E$ (real) and $c''_{33}{}^D$, $c''_{33}{}^E$ (imaginary) obtained by PRAP and Wave 3000 Pro program for higher orders peaks of TE mode of a PC-1 porous ceramic disk.

Figure 5 shows the frequency dependencies of ultrasonic velocities $V_t^D = \sqrt{\frac{c_{33}^{/D}}{\rho}}$ and $V_t^E = \sqrt{\frac{c_{33}^{/E}}{\rho}}$ and corresponding attenuation coefficients $\alpha(V_t^D) = \frac{c_{33}^{//D}\omega_0}{2c_{33}^{/D}}$ and $\alpha(V_t^E) = \frac{c_{33}^{//E}\omega_0}{2c_{33}^{/E}}$ obtained from PRAP results the higher harmonics of the TE mode of PCR-1 porous ceramic disks with different fundamental resonant frequencies. The following standard notations were used above: $c_{33}^{/D}$, $c_{33}^{/E}$ are the real and $c_{33}^{//D}$, $c_{33}^{//E}$ are the imaginary parts of elastic stiffness, ρ is the density, ω_0 is the cyclic frequency.

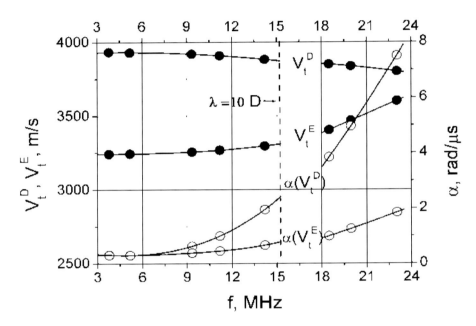

Figure 5. Frequency dependencies of ultrasonic velocities V_t^D and V_t^E and corresponding attenuation coefficients $\alpha(V_t^D)$ and $\alpha(V_t^E)$ obtained from PRAP results for higher harmonics of the TE mode of PCR-1 porous ceramic disks with different fundamental frequencies. (Vertical dashed line depict boundary between Rayleigh and stochastic scattering regions).

The least squares approximations of the frequency dependencies of Figure 5 have shown that in the frequency range corresponding to Rayleigh scattering of high frequency ultrasonic waves on pores ($\lambda \gg D$, where D is the average pores diameter, λ is the wavelength), attenuation coefficients $\alpha(V_t^D)$ and $\alpha(V_t^E)$ grow with the frequency approximately as f^4. At further frequency growth scattering mechanism changes from Rayleigh type to stochastic one ($4 \leq \lambda/D \leq 10$) and frequency dependencies of $\alpha(V_t^D)$ and $\alpha(V_t^E)$ are approximated by f^2 function.

The corresponding frequency dependence of ultrasonic velocity V_t^E defined by the elastic modulus C_{33}^E is approximated by f^3 function (normal dispersion type) in Rayleigh range and by linear f function that agreed with the theoretical predictions of Section I for the dispersion (Equations 3 and 4). In its turn, ultrasonic velocity V_t^D defined by the elastic modulus C_{33}^D, decreases with the frequency (anomalous dispersion) as result of the electromechanical contribution to C_{33}^D according the equation $C_{33}^D = C_{33}^E/(1-k_t^2)$ (note that effective value of k_t

measured on higher harmonics drops drastically with the harmonics number as $k_{eff.n}^2 = \frac{8k_t^2}{((2n+1)\pi)^2)}$ [3]).

CONCLUSION

A comprehensive study including finite difference 3D-simulation, impedance spectroscopy, and pulse-echo ultrasonic measurements of porous ceramics with strong spatial dispersion and high losses was carried out. Piezoelectric resonance analysis methods for automatic, iterative evaluation of complex material parameters were present, along with the full sets of complex constants for porous ceramics and ceramic matrix piezocomposites.

It was shown that pulse-echo measurements of frequency dependencies of elastic properties for dispersive and lossy ceramic composites are inaccurate and ambiguous.

In its turn, piezoelectric resonance measurements give accurate and reproducible results that well agree with the results of 3D finite-difference simulations.

Microstructural and physical mechanisms of losses and dispersion in porous ceramics were also considered. In conclusion, the advantages of low-Q porous ceramics were defined with reference to wide-band ultrasonic transducers applications.

ACKNOWLEDGMENTS

Work supported by the Ministry of Education and Science of RF 3.1246.2014/K (theme No. 213.01-11/2014-66).

REFERENCES

[1]   Wersing, W. In: *Piezoelectric Materials in Devices*. N. Setter (Ed.); Swiss Federal Institute of Technology, Lausanne, Switzerland. 2002, 29-66.
[2]   Rybyanets, A. N. *IEEE Trans. UFFC*, 2011, *vol. 58*, 1492-1507.
[3]   Rybyanets, A. N. *Ferroelectrics*, 2011, *vol. 419*, 90-96.
[4]   Rybyanets, A. N., Rybyanets, A.A. *IEEE Trans. UFFC*. 2011, *vol. 58*, 1757-1774.
[5]   Rybyanets, A. N. In: *Piezoelectrics and Related Materials: Investigations and Applications*. Ivan A. Parinov (Ed.). Nova Science Publishers Inc. NY. 2012, Chapter 5, 143-187.
[6]   Rybyanets, A. N., Naumenko, A. A. In: *Physics and Mechanics of New Materials and Their Applications*. Ivan A. Parinov, Shun Hsiung-Chang (Eds.). Nova Science Publishers Inc., NY. 2013, Chapter 1, 3-18.
[7]   Rybyanets, A. N., Zakharov, Y. N., Raevskii, I. P., Akopjan, V. A., Rozhkov E. V., Parinov, I. A. In: *Physics and Mechanics of New Materials and Their Applications*, Ivan A. Parinov, Shun Hsiung-Chang (Eds.). Nova Science Publishers Inc. NY. 2013, Chapter 22, 275-308.
[8]   Rybianets, A. N.; Tasker, R. *Ferroelectrics*. 2007, *vol. 360*, 90-95.

[9] Rybyanets, A. N.; Naumenko, A. A.; Shvetsova, N. A. In: *Nano- and Piezoelectric Technologies, Materials and Devices*. Ivan A. Parinov (Ed.). Nova Science Publishers Inc. NY. 2013, Chapter 1, 275-308.

[10] Rybjanets, A. N.; Nasedkin, A. V.; Turik, A.V. *Integrated Ferroelectrics*, 2004, *vol. 63*, 179-182.

[11] *IEEE Standard on Piezoelectricity*. ANSI/IEEE Std. 1987, 176.

[12] *PRAP (Piezoelectric Resonance Analysis Program)*. TASI Technical Software Inc. www.tasitechnical.com.

In: Proceedings of the 2015 International Conference …   ISBN: 978-1-63484-577-9
Editors: Ivan A. Parinov, Shun-Hsyung Chang et al.   © 2016 Nova Science Publishers, Inc.

Chapter 52

FINITE ELEMENT MODELING OF LOSSY PIEZOELECTRIC ELEMENTS

A. A. Naumenko, S. A. Shcherbinin, A. V. Nasedkin and A. N. Rybyanets*

Institute of Physics, Southern Federal University,
Rostov-on-Don, Russia

ABSTRACT

Theoretical aspects of the effective moduli method for an inhomogeneous piezoelectric media were examined. Different models of representative volume were considered. Based on these models and using finite element method (FEM) the full set of effective moduli for PZT porous ceramics having wide porosity range was calculated. A novel approach for optimization of finite element modeling (FEM) of lossy piezoceramic elements has been proposed. The procedure of optimization has consisted in sequential and iterative application of FEM and piezoelectric resonance analysis (PRAP) to complex electric impedance spectra of piezoceramic elements. For validation of proposed optimization procedure, FEM calculations of standard shape piezoelements (disks, bars and rods) made from dense and porous PZT-type piezoceramics were fulfilled using FEM ANSYS package. Comparison shows a good agreement between initial and calculated complex sets of material constants including losses ($Q = C'/C''$, where C' and C'' are the real and imaginary parts of related material constants).

1. INTRODUCTION

In recent years low-Q piezoceramics and piezocomposite materials are widely used for wide-band medical and NDT ultrasonic transducers with high sensitivity and resolution [1–7].

---

* E-mail: arybyanets.jr@gmail.com.

Porous piezoceramics based on different piezoceramic compositions were proposed as a promising candidate for lead metaniobate ceramics replacement in wide-band ultrasonic transducers and sensors [2, 3].

Porous piezoceramics have great advantage in low acoustic impedance and higher efficiency compared to conventional dense piezoceramics. Intensive research and technological work, as well as improvements in fabrication methods, have allowed large-scale manufacture of porous piezoelectric ceramics with reproducible and controllable porosity and properties [3, 8].

The material properties of porous piezoceramics with mixed connectivities (3–0, 3–3 and 0–3) have been evaluated using different theoretical models [9, 10]. The majority of new advanced ceramics and composite materials are lossy and direct use of IEEE Standards [11] for material constant determination can lead to significant errors.

Iterative methods [8, 9, 12] provide a means to accurately determine the complex coefficients in the linear range of poled piezoceramics from complex impedance resonance measurements.

The piezoelectric resonance analysis method and program (PRAP) has been proposed for the full set of standard geometries and resonance modes needed to complete complex characterization in a wide range of materials with very high and moderate losses [13–16].

Standard FEM packages that are widely used for modeling of piezoelements and devices do not take into account losses (mechanical losses can be included in calculations by implicit manner). Sets of material constants used for FEM calculations also do not contain losses data, except of Q_M for the radial mode of vibrations. As a result, FEM calculations of real piezoelements and devices can give inadequate results for lossy materials (composites, porous ceramics etc.).

This chapter presents a novel approach for optimization of finite element modeling of lossy piezoceramic elements.

2. OPTIMIZATION OF FINITE ELEMENT MODELS FOR POROUS CERAMIC PIEZOELEMENTS

Procedure of optimization has consisted in sequential and iterative application of FEM and Piezoelectric Resonance Analysis (PRAP) to complex electric impedance spectra of piezoceramic elements. For validation of proposed optimization procedures, FEM calculations of standard shape piezoelements (disks, shear plates, bars, and rods) made from porous PZT-type piezoceramics were fulfilled using FEM ANSYS software package [15].

Finite element modeling of the effective constants of porous piezoelectric ceramics based on the effective moduli method was performed using ANSYS software package [15]. Different models of representative volume were considered: piezoelectric cubes with one cubic and one spherical pore inside, cubic volume evenly divided in partial cubic volumes, with a portion randomly declared as pores, and the like ones. For modeling porous piezoceramics with 3-0/3-3 connectivity type, we used the representative volume with rigid skeleton structure (Figure 1) [15].

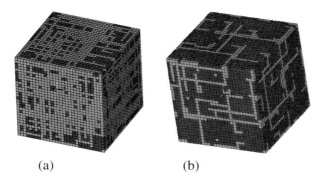

Figure 1. Examples of representative volumes used for FEM calculations of porous piezoceramics: (a) 3-3 connectivity, porosity of 20%; (b) 3-0 connectivity, porosity of 80%.

Optimization procedures were following:

1. Direct FEM calculations of complex electric impedance spectra by using sets of material constants for standard piezoceramic elements.
2. Analysis of the calculated impedance spectra by using the piezoelectric resonance analysis program (PRAP) and by deriving complex sets of material constants related to specific piezoceramic element.
3. Correction of initial material constants, including losses ($Q = c'/c''$, where c' and c'' are the real and imaginary parts of relative material constant).

Validation of the optimization procedures was performed using the FEM ANSYS package. FEM-generated complex impedance spectra for standard-shaped piezoelements (disks, bars, shear plates and rods) made from "hard" porous PZT piezoceramics were processed by PRAP-software to derive a complex set of material constants.

For the experiments, we applied the PRAP automatic iterative method [8, 12] to the full set of standard geometries and resonance modes needed to complete a characterization of hard PZT porous piezoceramics with moderate loss factors.

This software uses a generalized form of Smits's method [12] to determine material properties for any common resonance mode, and a generalized ratio method for the radial mode [8] valid for all material Q's. The software routinely generates an impedance spectrum from the determined properties to indicate the validity of the results. Measurements of electric parameters were conducted by using the Solartron Impedance/Gain-Phase Analyzer SL 1260 and Agilent 4291A Impedance Analyzer.

Figure 2 shows measured complex impedance spectra and PRAP approximations for the thickness (TE), radial (RE), modes of "hard" porous piezoceramics with relative porosity of 21% and average pore size of 50 μm.

Table 1 summarizes the complex constants of the porous ceramics obtained using PRAP analysis for a full set of standard geometries and resonance modes. Additional physical parameters of porous ceramics measured by ultrasonic and hydrostatic weighting methods are by following: $Z_A = 18.5$ MRayl, $\rho = 6.25$ g/cm^3, $V_t = 2960$ m/s, $Q_M^t = 150$, ($T_C = 340$ °C) [2, 5]. Experimental aspects of porous ceramics application in ultrasonic transducers and devices considered in [15, 16].

Table 1. Complex constants of "hard" porous ceramics

Parameter	Real	Imaginary	Parameter	Real	Imaginary
Shear Thickness (ST) Mode			Length Extensional (LE) Mode		
f_p (Hz)	$1.60 \cdot 10^6$	–	f_p (Hz)	$2.30 \cdot 10^5$	–
k_{15}	0.56	– 0.013	k_{33}	0.55	$-8.8 \cdot 10^{-4}$
C^E_{55} (N/m²)	$1.60 \cdot 10^{10}$	$5.80 \cdot 10^8$	S^D_{33} (m²/N)	$2.12 \cdot 10^{-11}$	$-5.92 \cdot 10^{-14}$
C^D_{55} (N/m²)	$2.32 \cdot 10^{10}$	$3.50 \cdot 10^8$	S^E_{33} (m²/N)	$3.03 \cdot 10^{-11}$	$-1.26 \cdot 10^{-14}$
S^D_{55} (m²/N)	$4.31 \cdot 10^{-11}$	$-6.49 \cdot 10^{-13}$	d_{33} (C/N)	$2.54 \cdot 10^{-10}$	$-1.35 \cdot 10^{-12}$
S^E_{55} (m²/N)	$6.25 \cdot 10^{-11}$	$-2.27 \cdot 10^{-12}$	g_{33} (V·m/N)	0.04	$-7.1 \cdot 10^{-5}$
e_{15} (C/m²)	7.51	– 0.14	ε^T_{33} (F/m)	$6.85 \cdot 10^{-9}$	$-1.5 \cdot 10^{-11}$
h_{15} (V/m)	$9.65 \cdot 10^8$	$-1.32 \cdot 10^7$	Radial Extensional (RE) Mode		
d_{15} (C/N)	$4.69 \cdot 10^{-10}$	$-2.56 \cdot 10^{-11}$	f_{p1} (Hz)	$1.03 \cdot 10^5$	–
g_{15} (Vm/N)	0.04	– 0.001	S^E_{11} (m²/N)	$1.96 \cdot 10^{-11}$	$-3.10 \cdot 10^{-14}$
ε^T_{11} (F/m)	$1.13 \cdot 10^{-8}$	$-2.92 \cdot 10^{-10}$	S^E_{12} (m²/N)	$-6.37 \cdot 10^{-12}$	$2.30 \cdot 10^{-14}$
ε^S_{11} (F/m)	$7.77 \cdot 10^{-9}$	$-3.60 \cdot 10^{-11}$	d_{31} (C/N)	$5.92 \cdot 10^{-11}$	$-5.19 \cdot 10^{-14}$
Thickness Extensional (TE) Mode			ε^T_{33} (F/m)	$6.44 \cdot 10^{-9}$	$-3.09 \cdot 10^{-12}$
f_p (Hz)	$1.48 \cdot 10^6$		k_p	0.29	$-8.84 \cdot 10^{-5}$
k_t	0.52	– 0.04	σ^P	0.33	$-6.3 \cdot 10^{-4}$
C^D_{33} (N/m²)	$5.49 \cdot 10^{10}$	$3.70 \cdot 10^8$	S^E_{66} (m²/N)	$5.19 \cdot 10^{-11}$	$-1.08 \cdot 10^{-13}$
C^E_{33} (N/m²)	$4.02 \cdot 10^{10}$	$2.66 \cdot 10^8$	C^E_{66} (N/m²)	$1.93 \cdot 10^{10}$	$4.0 \cdot 10^7$
e_{33} (C/m²)	10.08	– 1.07	Physical Parameters		
h_{33} (V/m)	$1.47 \cdot 10^9$	$-7.10 \cdot 10^7$	ρ (kg/m³)	$6.25 \cdot 10^3$	–
ε^S_{33} (F/m)	$6.87 \cdot 10^{-9}$	$-4.00 \cdot 10^{-10}$	Z_A (MRayl)	18.5	–

Figures 3 shows complex impedance spectra generated by FEM using the constants shown in Table 1 and PRAP approximations for TE and RET modes of porous piezoceramics. The complex constants of "hard" porous piezoceramics, obtained using FEM generated spectra are summarized in Table 2. It is readily seen from Tables 1 and 2 that the complex material constants resulting from FEM-generated impedance spectra are very close to measured ones, including imaginary parts (losses). For other low-Q materials, additional corrections of material constants (Q effective in ANSYS) with recurring optimization procedure should be performed.

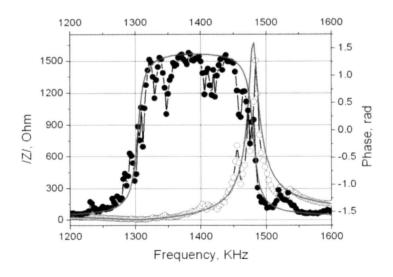

Figure 2. (Continued).

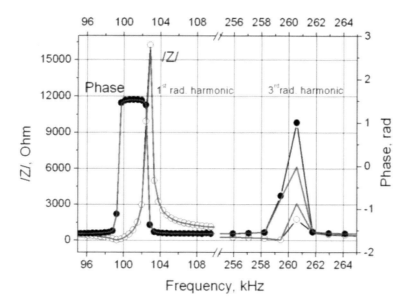

Figure 2. Measured impedance spectra and PRAP approximations for TE- and RE-modes of porous piezoceramic disk Ø20 × 1 mm^2.

Table 2. Complex constants obtained using PRAP analysis of FEM-generated spectra for hard porous piezoceramics

Parameter	Real	Imaginary	Parameter	Real	Imaginary
Shear Thickness (ST) Mode			Length Extensional (LE) Mode		
f_p (Hz)	$1.66 \cdot 10^6$	–	f_p (Hz)	$2.8 \cdot 10^5$	–
k_{15}	0.57	– 0.048	k_{33}	0.57	$- 7.02 \cdot 10^{-4}$
C^E_{55} (N/m²)	$1.67 \cdot 10^{10}$	$1.48 \cdot 10^9$	S^D_{33} (m²/N)	$2.04 \cdot 10^{-11}$	$-1.038 \cdot 10^{-13}$
C^D_{55} (N/m²)	$2.47 \cdot 10^{10}$	$1.77 \cdot 10^8$	S^E_{33} (m²/N)	$3.04 \cdot 10^{-11}$	$- 1.9 \cdot 10^{-13}$
S^D_{55} (m²/N)	$4.03 \cdot 10^{-11}$	$- 2.89 \cdot 10^{-13}$	d_{33} (C/N)	$2.55 \cdot 10^{-10}$	$- 1.27 \cdot 10^{-12}$
S^E_{55} (m²/N)	$5.93 \cdot 10^{-11}$	$- 5.25 \cdot 10^{-12}$	g_{33} (V·m/N)	0.04	$- 1.48 \cdot 10^{-4}$
e_{15} (C/m²)	9.49	– 1.26	ε^T_{33} (F/m)	$6.48 \cdot 10^{-9}$	$- 7.9 \cdot 10^{-12}$
h_{15} (V/m)	$8.5 \cdot 10^8$	$- 2.4 \cdot 10^7$	Radial Extensional (RE) Mode		
d_{15} (C/N)	$5.56 \cdot 10^{-10}$	$- 1.25 \cdot 10^{-11}$	f_{p1} (Hz)	$1.03 \cdot 10^5$	–
g_{15} (Vm/N)	0.035	– 0.0012	S^E_{11} (m²/N)	$1.99 \cdot 10^{-11}$	$- 8.83 \cdot 10^{-14}$
ε^T_{11} (F/m)	$1.63 \cdot 10^{-8}$	$- 3.063 \cdot 10^{-9}$	S^E_{12} (m²/N)	$- 6.62 \cdot 10^{-12}$	$2.53 \cdot 10^{-14}$
ε^S_{11} (F/m)	$1.12 \cdot 10^{-8}$	$- 1.17 \cdot 10^{-9}$	d_{31} (C/N)	$6.37 \cdot 10^{-11}$	$- 8.42 \cdot 10^{-13}$
Thickness Extensional (TE) Mode			ε^T_{33} (F/m)	$6.64 \cdot 10^{-9}$	$- 2.19 \cdot 10^{-11}$
f_p (Hz)	$1.48 \cdot 10^6$	–	k_p	0.3	– 0.002
k_t	0.52	– 0.006	σ^P	0.33	0.0028
C^D_{33} (N/m²)	$5.49 \cdot 10^{10}$	$3.145 \cdot 10^8$	S^E_{66} (m²/N)	$5.29 \cdot 10^{-11}$	$- 1.26 \cdot 10^{-13}$
C^E_{33} (N/m²)	$4.03 \cdot 10^{10}$	$5.92 \cdot 10^8$	C^E_{66} (N/m²)	$1.89 \cdot 10^{10}$	$4.49 \cdot 10^7$
e_{33} (C/m²)	9.26	– 0.35	Physical Parameters		
h_{33} (V/m)	$1.57 \cdot 10^9$	$- 2.9 \cdot 10^7$	ρ (kg/m³)	$6.25 \cdot 10^3$	–
ε^S_{33} (F/m)	$6.87 \cdot 10^{-9}$	$- 3.29 \cdot 10^{-10}$	Z_A (MRayl)	18.5	–

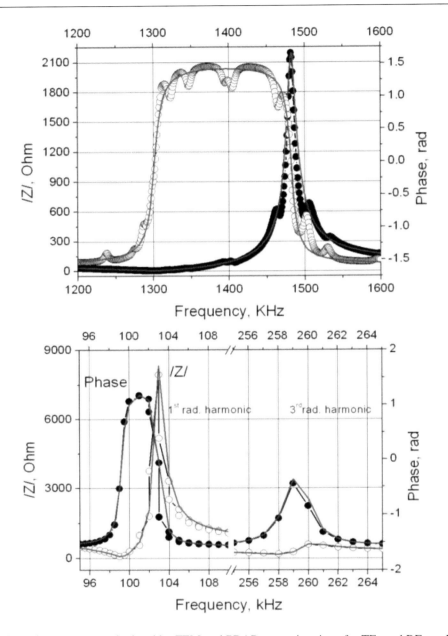

Figure 3. Impedance spectra calculated by FEM and PRAP approximations for TE- and RE-modes of porous piezoceramic disk Ø20 × 1 mm^2.

CONCLUSION

A novel approach for optimization of finite element modeling (FEM) of lossy piezoceramic elements was considered. Different FEM models of porous piezoceramics based on effective moduli method were developed and considered.

For the models confirmation, the porous piezoceramics with well controllable porosity were manufactured and measured. The result of FEM modeling shows that the representative volume model for 3-3 connectivity type with rigid skeleton gives better results compared to other examined models and describes experimental results for porous ceramics in the relative porosity range of $0 - 70\%$ very well.

Piezoelectric resonance analysis methods for automatic, iterative evaluation of complex material parameters were described, along with full sets of complex constants for various ceramic piezocomposites. Complex sets of material constants for "hard" porous piezoceramics were obtained by PRAP analysis of electric impedance spectra measured by using standard porous piezoceramic elements.

Electric impedance spectra were calculated with help of measured sets of material constants using FEM ANSYS software package. Complex sets of material constants were derived from FEM generated complex impedance spectra using PRAP analysis. Comparison shows a good agreement between initial and calculated complex sets of material constants including losses ($Q = C''/C''$, where c' and c'' are the real and imaginary parts of related material constants).

ACKNOWLEDGMENTS

Work supported by the Ministry of Education and Science of RF 3.1246.2014/K (theme № 213.01-11/2014-66).

REFERENCES

[1]   Wersing, W. In: *Piezoelectric Materials in Devices*. N. Setter (Ed.); Swiss Federal Institute of Technology, Lausanne, Switzerland. 2002, 29-66.

[2]   Rybyanets, A. N. *IEEE Trans. UFFC*, 2011, *vol. 58*, 1492-1507.

[3]   Rybyanets, A. N. *Ferroelectrics*, 2011, *vol. 419*, 90-96.

[4]   Rybyanets, A. N., Rybyanets, A. A. *IEEE Trans. UFFC*. 2011, *vol. 58*, 1757-1774.

[5]   Rybyanets, A. N. In: *Piezoelectrics and Related Materials: Investigations and Applications*. Ivan A. Parinov (Ed.). Nova Science Publishers Inc. NY. 2012, Chapter 5, 143-187.

[6]   Rybyanets, A. N.; Naumenko, A. A. In: *Physics and Mechanics of New Materials and Their Applications. Ivan A. Parinov, Shun Hsiung-Chang (Eds.). Nova Science Publishers Inc., NY.* 2013, Chapter 1, 3-18.

[7]   Rybyanets, A. N.; Zakharov, Y. N.; Raevskii, I. P.; Akopjan, V. A.; Rozhkov E. V.; Parinov, I.A. In: *Physics and Mechanics of New Materials and Their Applications*, Ivan A. Parinov, Shun Hsiung-Chang (Eds.). Nova Science Publishers Inc. NY. 2013, Chapter 22, 275-308.

[8]   Rybianets, A. N.; Tasker, R. *Ferroelectrics*. 2007, *vol. 360*, 90-95.

[9]   Rybyanets, A. N.; Naumenko, A. A.; Shvetsova, N. A. In: *Nano- and Piezoelectric Technologies, Materials and Devices*. Ivan A. Parinov (Ed.). Nova Science Publishers Inc. NY. 2013, Chapter 1, 275-308.

[10] Rybjanets, A. N.; Nasedkin, A. V.; Turik, A. V. *Integrated Ferroelectrics*, 2004, *vol. 63*, 179-182.

[11] *IEEE Standard on Piezoelectricity*. ANSI/IEEE Std. 1987, 176.

[12] *PRAP (Piezoelectric Resonance Analysis Program)*. TASI Technical Software Inc. www.tasitechnical.com.

[13] Rybjanets, A. N.; Razumovskaja, O. N.; Reznitchenko, L. A.; Komarov, V.D.; Turik, A. V. *Integrated Ferroelectrics*, 2004, *vol. 63*, 197-200.

[14] Rybianets, A. N. *Ferroelectric*, 2007, *vol. 360*, 84-89.

[15] Nasedkin, A.; Rybjanets, A.; Kushkuley, L.; Eshel, Y. *Proc. 2005 IEEE Ultrason. Symp.* 2005, 1648-1651.

[16] Rybianets, A. N.; Nasedkin, A. V. *Ferroelectrics,* 2007, *vol. 360*, 57-62.

In: Proceedings of the 2015 International Conference … ISBN: 978-1-63484-577-9
Editors: Ivan A. Parinov, Shun-Hsyung Chang et al. © 2016 Nova Science Publishers, Inc.

Chapter 53

ELECTROMECHANICAL RESPONSE CHARACTERIZATION OF PIEZOELECTRIC MATERIALS

S. A. Shcherbinin[1,], I. A. Shvetsov[1], M. A. Lugovaya[1], A. A. Naumenko[1] and A. N. Rybyanets[1]*

[1]Institute of Physics, Southern Federal University,
Rostov-on-Don, Russia

ABSTRACT

The chapter presents a basic theory for the electromechanical response characterization method (STEP), modern apparatus setups and software along with the experimental results for wide range of piezo- and ferroelectric materials and devices. The STEP method combines large signal modeling of the mechanical and electrical behavior of ferroelectric, piezoelectric, electrostrictive, and antiferroelectric materials with a comprehensive interface for acquisition of strain **(S),** stress (T), electric field **(E),** and electric displacement or electric polarization (P) curves as a function of time. The basic STEP package includes modules for piezoelectric, electrostrictive, ferroelectric, and antiferroelectric materials. The STEP method was applied for a wide range of ferroelectric, piezoelectric, electrostrictive, and antiferroelectric materials. The experimental examples of hysteresis and strain loops along with the analysis results for materials were present and discussed.

1. INTRODUCTION

Demands on medical and NDT ultrasonic transducer performance have increased in recent years. Low-Q piezoceramics and piezocomposite materials are widely used for wideband NDT ultrasonic transducers with high sensitivity and resolution.

---

[*] E-mail: step_scherbinin@list.ru.

Some of these advanced materials are lossy and direct use of IEEE Standards for material constant determination leads to significant errors. The modeling and design of piezoelectric devices by finite element methods, among others, rely on the accuracy of the dielectric, piezoelectric and elastic coefficients of the active material used which is, commonly, an anisotropic ferroelectric polycrystal. The accurate description of piezoceramics must include the evaluation of the dielectric, piezoelectric and mechanical losses, accounting for the out-of-phase material response to the input signal.

The basic techniques for finding material constants of piezoelectric materials outlined in the IEEE Standard on Piezoelectricity (1987) [1]. These methods work for many of the most widely used commercial piezoceramics based on lead-titanate-zirconate (PZT) compositions that are high-Q_M and high-coupling coefficient piezoelectric materials. However, there is a general agreement that their use in many new piezoelectric materials such as porous ceramics, piezoelectric polymers or piezoelectric composites may lead to significant errors.

Furthermore, the current IEEE Standard does not comprehensively account for the complex nature of material coefficients, as it uses only the dielectric loss factor (tan δ) and the mechanical quality factor (Q_M) to account for loss. Numerous techniques using complex material constants have been proposed to take into account losses in low-Q_M materials and to overcome limitations in the IEEE Standard [2-4]. Iterative methods [4, 5] provide a means to determine accurately the complex coefficients in the linear range of poled piezoceramics from measurements of complex impedance resonance.

The piezoelectric resonance analysis method and program (PRAP) has been proposed for the full set of standard geometries and resonance modes needed to complete complex characterization in a wide range of materials with very high and moderate losses [6-9]. The electromechanical response characterization program (STEP) combining large signal modeling of the mechanical and electrical behavior of ferroelectric, piezoelectric, electrostrictive, and antiferroelectric materials with a comprehensive interface for acquisition of strain (S), stress (T), electric field (E), and electric displacement or electric polarization (P) curves as a function of time was proposed in [10].

The chapter presents basic theory for the methods, modern apparatus setups and software along with broad experimental results for wide range of piezoelectric materials, ultrasonic transducers and devices.

2. PIEZOELECTRIC MATERIALS MODELING

A phenomenological model derived from thermodynamic potentials mathematically describes the property of piezoelectricity. The derivations are not unique and the set of equations describing the piezoelectric effect depends on the choice of potential and the independent variables used. In the case of a sample under isothermal and adiabatic conditions, and ignoring higher order effects, the elastic Gibbs function may be described by

$$G_1 = -\frac{1}{2}\left(s_{ijkl}^D T_{ij} T_{kl} + 2g_{nij} D_n T_{ij}\right) + \frac{1}{2}??_{mn}^T D_m D_n, \tag{1}$$

where g is the piezoelectric voltage coefficient, s is the elastic compliance, and β is the inverse permittivity. The independent variables for this equation are stress T and dielectric displacement D. The superscripts of the coefficients designate the independent variable that is held constant when defining the coefficients, and the subscripts define tensors that take into account the anisotropic nature of the material.

The linear equations of piezoelectricity for this potential are determined from the derivative of Gl and are:

$$S_{ij} = -\frac{\partial G_1}{\partial T_{ij}} = s^D_{ijkl} T_{kl} + g_{nij} D_n \tag{2}$$

$$E_m = \frac{\partial G_1}{\partial D_m} = \beta^T_{mn} D_n - g_{nij} T_{ij} \tag{3}$$

where S is the strain and E is the electric field. The above equations are usually simplified to a reduced form by noting that there is a redundancy in the strain and stress variables.

The elements of the tensor are contracted to a 6 × 6 matrix with 1, 2, 3 designating the normal stress and strain and 4, 5 and 6 designating the shear stress and strain elements. Other representations of the linear equations of piezoelectricity derived from the other possible thermodynamic potentials are shown below. This set of equations includes the above equation in contracted notation.

$$S_p = s^D_{pg} T_g + g_{pm} D_m;$$

$$E_m = \beta^T_{mn} D_m - g_{pm} T_p;$$

$$S_p = s^E_{pg} T_g + d_{pm} E_m;$$

$$D_m = \varepsilon^T_{mn} E_n + d_{pm} T_p;$$

$$T_p = c^E_{pg} S_g - e_{pm} E_m;$$

$$D_m = \varepsilon^S_{mn} E_n + e_{pn} S_p;$$

$$T_p = c^D_{pg} S_g - h_{pm} D_m;$$

$$E_m = \beta^S_{mn} D_n - h_{pm} S_p. \tag{4}$$

where d, e, g, h are piezoelectric constants, s and c are the elastic compliance and stiffness, and ε and β the permittivity and the inverse permittivity. Fortunately, due to symmetry, many of the elements in the reduced matrix are either zero or dependent. For PZT (C∞/C6v symmetry for polycrystalline poled materials), the number of constants can be reduced to 5 independent elastic constants, 3 independent piezoelectric constants and 2 independent dielectric constants. The reduced matrix in relation to the polarization axis (3 directions) for PZT is then

$$\begin{bmatrix} S_1 \\ S_2 \\ S_3 \\ S_4 \\ S_5 \\ S_6 \\ D_1 \\ D_2 \\ D_3 \end{bmatrix} = \begin{bmatrix} s_{11}^E & s_{12}^E & s_{13}^E & 0 & 0 & 0 & 0 & 0 & d_{13} \\ s_{12}^E & s_{11}^E & s_{13}^E & 0 & 0 & 0 & 0 & 0 & d_{13} \\ s_{13}^E & s_{13}^E & s_{33}^E & 0 & 0 & 0 & 0 & 0 & d_{33} \\ 0 & 0 & 0 & s_{55}^E & 0 & 0 & 0 & d_{15} & 0 \\ 0 & 0 & 0 & 0 & s_{55}^E & 0 & d_{15} & 0 & 0 \\ 0 & 0 & 0 & 0 & 0 & 2(s_{11}^E . s_{12}^E) & 0 & 0 & 0 \\ 0 & 0 & 0 & 0 & d_{15} & 0 & \varepsilon_{11}^T & 0 & 0 \\ 0 & 0 & 0 & d_{15} & 0 & 0 & 0 & \varepsilon_{11}^T & 0 \\ d_{13} & d_{13} & d_{33} & 0 & 0 & 0 & 0 & 0 & \varepsilon_{33}^T \end{bmatrix} \begin{bmatrix} T_1 \\ T_2 \\ T_3 \\ T_4 \\ T_5 \\ T_6 \\ E_1 \\ E_2 \\ E_3 \end{bmatrix} \tag{5}$$

The other linear equations discussed above can also be expressed in terms of matrix notation. There are only 10 independent constants in all representations. The IEEE Standard on Piezoelectricity (1987) contains the appropriate equations that allow one to convert from one set of equations/matrix to another. The matrix above expresses the relationships between the 10 independent material constants and variables S, T, E, and D. Ideally, under small fields and stresses, and for materials with low losses within a limited frequency range, these 10 constants contain all information that is required to predict the behavior of the material under application of stress, strain, electric field, or charge to the sample surface. In practice, most materials display dispersion, non-linearity, and have measurable losses.

The piezoelectric analysis module of STEP [10] can be used to measure and analyze the high signal quasistatic piezoelectric, dielectric, and elastic behavior of materials.

In the case of a ferroelectric piezoelectric material, the models assume that the maximum applied field is below the coercive field of the material at the temperature of measurement.

In many materials, a hysteresis is present and the general equations do not account for hysteresis. Hysteresis can identify by an open curve (cigar shape) that has an average linear slope over the range of the measurement. In order to account for hysteresis we have used Rayleigh's Law as discussed in [11]. Using this approach, the linear relationships are generalized to include a first order correction to account for the change in slope and hysteresis as a function of the field.

The linear relationships may be generalized and written as:

$$Y_i = (\alpha_{ij} + 2a_2 X_m) X_j \pm a_2 (X_m^2 - X_j^2) \tag{6}$$

where Y_i and X_j are the dependent and independent variables (for example, Strain S versus electric field E), α_{ij} is the small signal coefficient (for example, d_{ij}) and X_m is the maximum value of the independent variable attained during the measurement (for example, E_m). a_2 is a coefficient that describes the linear increase in slope and hysteresis as a function of X_m. The signs + or − are used to designate the top curve (+) and the bottom curve (−). The equation shown above describes a family of curves for Y versus X, where the slope and hysteresis of the curve depend on the maximum extent of the independent variable and the coefficient a_2. This equation can be further generalized to include displacements from the origin coordinates of the curve Y_0 and X_0 (see Figure 1) as

$$Y_i - Y_0 = (\alpha_{ij} + 2a_2 X_m)(X_j - X_0) \pm a_2 \left(X_m^2 - (X_j - X_0)^2 \right), \tag{7}$$

where Y for the top curve is

$$Y_T = Y_i - Y_0 = (\alpha_{ij} + 2a_2 X_m)(X_j - X_0) \pm a_2 \left(X_m^2 - (X_j - X_0)^2\right) \tag{8}$$

and Y for the bottom curve is

$$Y_B = Y_i - Y_0 = (\alpha_{ij} + 2a_2 X_m)(X_j - X_0) \pm a_2 \left(X_m^2 - (X_j - X_0)^2\right). \tag{9}$$

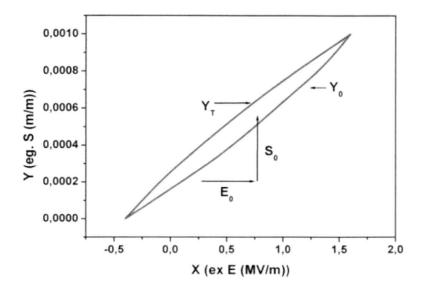

Figure 1. Top and bottom curves for a material displaying hysteresis in the strain vs field curves. The curves are offset from the origin by Y0 = S0 and X0 = E0.

An alternative representation of the model can be obtained by taking the average and difference curves of the top and bottom curves.

The average curve is:

$$Y_{AVG} = \frac{(Y_T + Y_B)}{2} = (\alpha_{ij} + 2a_2 X_m)(X_j - X_0). \tag{10}$$

This is linear in X and slope is seen to increase with the extent of the maximum value of X and the coefficient alpha a_2. When considering the average curve, we then have an effective coefficient α_{ij}^{eff}:

$$\alpha_{ij}^{eff} = \alpha_{ij} + 2a_2 X_m. \tag{11}$$

The difference or hysteresis curve is described by

$$Y_{HYS} = \frac{(Y_T + Y_B)}{2} = a_2 \left(X_m^2 - (X_j - X_0)^2\right). \tag{12}$$

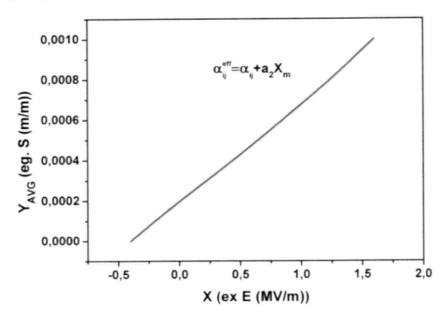

Figure 2. Average curve for the hysteresis loop shown in Figure 1.

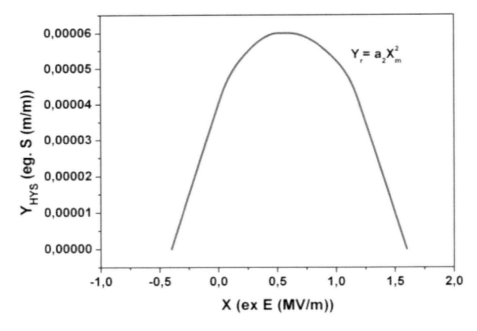

Figure 3. Difference curve for the hysteresis loop shown in Figure 1.

At X = X0, the value of YHYS is Yr, the remnant value of Y. The average and difference curves for the above example are shown in Figures 2 and 3. The coefficient a_2 can be determined from the remnant hysteresis:

$$a_2 = \frac{Y_R}{X_m^2} \tag{13}$$

Electromechanical Response Characterization of Piezoelectric Materials   405

and the initial or small signal value of the linear relationship can be determined from the slope of the average curve:

$$\alpha_{ij} = \alpha_{ij}^{eff} - 2a_2 X_m. \tag{14}$$

CONCLUSION

The chapter presents a basic theory for the electromechanical response characterization method (STEP), modern apparatus setups and software along with the experimental results for wide range of piezo- and ferroelectric materials and devices. The electromechanical response characterization method combines large signal modeling of the mechanical and electrical behavior of ferroelectric, piezoelectric, electarostrictive, and antiferroelectric materials with a comprehensive interface for acquisition of strain (**S**), stress (T), electric field (**E**), and electric displacement or electric polarization (P) curves as a function of time.

The basic STEP package includes modules for piezoelectric, electrostrictive, ferroelectric, and antiferroelectric materials. At the core of the STEP analytical engine is a modified form of the Levenberg-Marquardt algorithm that simultaneously fits models to related sets of data. This permits the determination of the best fit of a model to multiple data curves. Note that this is quite different from conventional approaches where the bottom and top parts of a hysteresis curve are alternately fitted. A significant feature of STEP is the optional plug-in Basic Data Acquisition Module. This module in principle allows one to measure any electromechanical response and acquired through a large variety of instrument drivers.

These drivers allow both the direct and indirect control of any one of the variables strain, stress, electric field or electric displacement, and the measurement of any two of these variables. The measurements can be as quasi-static as dynamic, depending on the instrumentation employed.

ACKNOWLEDGMENT

Work supported by the RSF grant No. 15-12-00023.

REFERENCES

[1]   IEEE Standard on Piezoelectricity. ANSI/IEEE Std. 1987, 176.
[2]   Holland, R. IEEE Trans. Sonics Ultrason. 1967, SU-14, 18.
[3]   Sherrit, S.; Wiederick, H. D.; Mukherjee, B. K. *SPIE Proceedings*, 1997, vol. 3241, 327.
[4]   Alemany, C.; Pardo, L.; Jimenez, B.; Carmona, F.; Mendiola, J.; Gonzalez, A. M. *J. Phys. D: Appl. Phys.*, 1994, vol. 27, 148.

[5] Pardo, L.; Alemany, C.; Ricote, J.; Moure, A.; Poyato, R.; Alguero, M. Proc. Intern. Conf. Material Technology and Design of Integrated Piezoelectric Devices. Couurmayeur, Italy. 2004, 145.

[6] PRAP (Piezoelectric Resonance Analysis Program). TASI Technical Software Inc. www.tasitechnical.com.

[7] Rybyanets, A. N.; Tasker, R. *Ferroelectrics,* 2007, vol. 360, 90-95.

[8] Rybyanets, A. N.; Rybyanets, A. A. *IEEE Trans. UFFC*, 2011, vol. 58, 1757-1774.

[9] Rybyanets, A. N. *IEEE Trans. UFFC*, 2011, vol. 58, 1492-1507.

[10] STEP (Electromechanical Response Characterisation Program). TASI Technical Software Inc. www.tasitechnical.com.

[11] Damjanovic, D.; Demartin M. *J. Phys. D. Appl. Phys.*, 1996, vol. 29, 2057-2060.

In: Proceedings of the 2015 International Conference … 　ISBN: 978-1-63484-577-9
Editors: Ivan A. Parinov, Shun-Hsyung Chang et al. 　© 2016 Nova Science Publishers, Inc.

Chapter 54

DIELECTRIC, PIEZOELECTRIC AND ELASTIC PROPERTIES OF PZT/PZT CERAMIC PIEZOCOMPOSITES

N. A. Shvetsova[1,], M. A. Lugovaya[1], I. A. Shvetsov[1], D. I. Makariev[1] and A. N. Rybyanets[1]*

[1]Institute of Physics, Southern Federal University,
Rostov-on-Don, Russia

ABSTRACT

A new method of PZT/PZT ceramic piezocomposites fabrication is proposed. Samples of piezocomposites with the volume fraction of components 0-100% for different PZT compositions were fabricated and measured. Complex sets of elastic, dielectric, and piezoelectric parameters of piezocomposites were determined by ultrasonic and impedance spectroscopy method using piezoelectric resonance analysis (PRAP) software. Microstructure of polished, chemically etched, and chipped surfaces of piezocomposite samples was observed with optical and scanning electron microscopes. New PZT/PZT ceramic piezocomposites are characterized by a unique spectrum of the electrophysical properties unachievable for standard PZT ceramic compositions and fabrication methods and can be useful for wide-band ultrasonic transducer applications.

1. INTRODUCTION

Limited transducer materials are currently available for use in ultrasonic transducer design in Non Destructive Testing (NDT) inspection and medical diagnostics systems, considering combinations of high sensitivity (efficiency), resolution (bandwidth) and operational requirements (working temperature). There are currently no commercially available piezoceramic materials that offer high piezoelectric properties together with low Q,

---

* E-mail: yfnfif_71@bk.ru.

low acoustic impedance, electromechanical anisotropy and high operating temperatures (>250°C) [1].

Those that are available offer either high piezoelectric properties (lead zirconate titanate – PZT, lead magnesium and zinc niobates – PMN, PZN) or high electromechanical anisotropy (lead titanate – PT) and low Q, low acoustic impedance (lead metaniobate – PN) [2].

To date, practically all broadband transducers are based on strongly externally damped PZT or lead metaniobate (PN) ceramics. PN ceramics show the better properties ($Q_M^t \sim 8$, $Z_A \sim 18$) and are widely used at high temperatures, but the piezoelectric properties of PN are weak. In addition, PN is very difficult for mass production that is reflected in its relatively high cost. Therefore, efforts have been made to replace lead metaniobate by 1-3 piezocomposite materials. However, to date the production costs of these components well exceeds the common price of a complete standard transducer. Moreover, there are considerable technical limitations with this type of transducer, in particular, with respect to the maximum allowable ambient temperature and pressure [2].

Porous piezoceramics based on different piezoceramic compositions were proposed as a promising candidate for PN replacement in wide-band ultrasonic transducers [3, 4]. Despite their long history and unique characteristics, so far porous piezoelectric ceramics have not been commercialized and their application in transducers and devices is limit [2-4]. Intensive research and technological work, as well as improvements in fabrication methods, have allowed large-scale manufacture of porous piezoelectric ceramics with reproducible and controllable porosity and properties [5-7].

Over the past years, considerable advances have been made to improve the mechanical properties of ceramics using ceramic composite approaches. Numerous technologies based on incorporation of functional ceramics into structural ones and *vice versa* have been developed, and novel design ideas have applied in the field of functional ferroelectric ceramics [8, 9]. However, the problem of property trade-off, i.e., the deterioration of electromechanical properties, remains unsolved. This chapter presents a new method of low-Q PZT/PZT ceramic piezocomposites fabrication as well as complex sets of elastic, dielectric, and piezoelectric parameters of the piezocomposites.

2. EXPERIMENTAL PROCEDURES

At first, a line of chemically, thermally, and technologically compatible ceramic matrix and scattering phase materials were chosen. Different types of raw PZT powders and milled PZT piezoceramic particles as well as pre-sintered PZT granules were used as initial components for ceramic composite preparations. Special pressing and firing regimes and porosifiers were used for the formation of microporous piezoceramic matrices. Sintering of the green bodies was carried out at special thermal profiles to prevent cracking caused by difference in shrinkage and thermal expansion coefficients of the composite components. Figure 1 shows two examples of ceramic composite microstructures.

Complex electrical coefficients of piezocomposite elements were determined by impedance spectroscopy using Piezoelectric Resonance Analysis (PRAP) software [10, 11]. Measurements were made on Solartron Impedance/Gain-Phase Analyzer SL 1260.

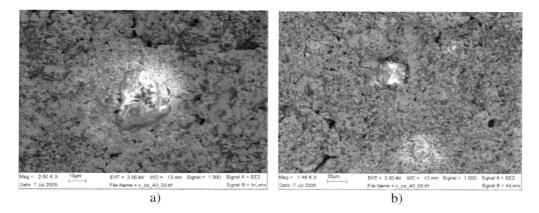

Figure 1. SEM micrographs of ceramic composite structures: (a) pre-sintered PZT granules in porous PZT matrix; (b) milled dense PZT particles in porous PZT matrix.

Sound velocity and attenuation of longitudinal and shear waves in different directions of polarized and non-polarized samples were measured by pulse–echo and through-transmit ultrasonic methods in the frequency range 1 – 5 MHz using digital LeCroy oscilloscope, Olympus pulser/receivers and various Olympus transducers. Microstructures of polished, chemically etched, and shipped surfaces of composite samples were observed with scanning electron microscopes (SEM, Karl Zeiss).

3. RESULTS AND DISCUSSION

PZT/PZT ceramic piezocomposites composed by the hard PZT matrix with randomly distributed pre-sintered PZT granules with a mean particle diameter ~ 30 µm and volume fraction from 0 up to 100 mass.% were chosen as model samples for illustration of the "damping by scattering" approach.

Figure 2 shows shrinkage coefficient K_{sh}^{diam}, theoretical ρ^{theor} and measured ρ^{exper} density, as well as relative porosity $P\%$ of ceramic composite on concentration of pre-sintered ceramic granules mass.%. It is readily seen in Figure 2 that K_{sh}^{diam} decreases drastically with mass.% caused by the increase of the non-shrinking phase concentration (pre-sintered ceramic granules), which prevents shrinkage of ceramic matrix and leads to microporosity appearance.

It is readily observed in Figure 2 that the density of the ceramic composite drops and relative porosity grows rapidly with concentration mass.%, which corresponds well to shrinkage coefficient behavior.

Figure 3 shows piezoelectric moduli d_{33}, d_{31}, relative dielectric constant $\varepsilon_{33}^{T}/\varepsilon_0$, as well as electromechanical coupling factors for thickness k_t, longitudinal k_{31} and radial k_p vibration modes of ceramic composites as function of concentration of pre-sintered ceramic granules mass.%.

The dielectric constant of the composite $\varepsilon_{33}^{T}/\varepsilon_0$ decreases drastically with mass.% because porosity grows (i.e., the dielectric constant of air is much less than for PZT ceramics). The piezoelectric modulus d_{33} for ceramic composite has minor changes in all mass.% range caused by continuity of rigid "quasi-rod" ceramic skeleton in the polarization direction (sample thickness).

Reduction in relative area of the piezoceramic phase in this case is compensated by an increase in relative pressure applied to the ceramic skeleton. Reduction in $|d_{31}|$ with mass.% is obvious and is caused by alteration of quasi-rod ceramic skeleton continuity in a lateral direction (i.e., the lateral size of elements usually twenty times more than the thickness).

The behaviour of electromechanical coupling factor k_p and k_{31} of ceramic composites with concentration of pre-sintered ceramic granules mass.% is determined by a decrease of piezomodulus d_{31} and dielectric permittivity ε_{33}^T and a competing increase of elastic compliances S_{11}^E and S_{12}^E according to following relation: $k_p^2 = 2d_{31}^2/[\varepsilon_{33}^T(S_{11}^E + S_{12}^E)]$ and $k_{31}^2 = d_{31}^2/(\varepsilon_{33}^T S_{11}^E)$. The main reason for the decrease of k_p and k_{31} is mentioned above alteration of piezoceramics skeleton continuity in a lateral direction and, as the consequence, an increase in corresponding elastic compliances of porous ceramics. The electromechanical coupling factor k_t slightly increases with mass.% due to partial removal of mechanical clamping of a porous piezoceramic structure in lateral direction (electromechanical coupling factor for piezoceramic rods equal to k_{33}, and mechanical clamping of porous piezoceramic skeleton by air is negligible).

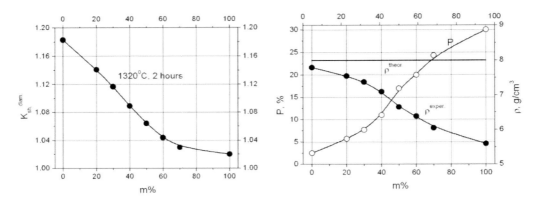

Figure 2. Dependencies of shrinkage coefficient $K_{sh}^{diam.}$, theoretical ρ^{theor} and measured ρ^{exper} density, as well as relative porosity $P\%$ of ceramic composites as function of concentration of pre-sintered ceramic granules mass.%.

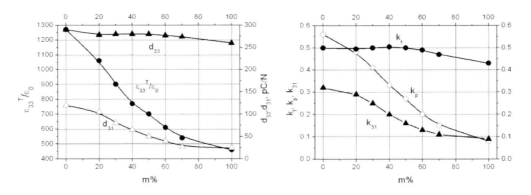

Figure 3. Dependencies of piezoelectric moduli d_{33}, d_{31}, relative dielectric constant $\varepsilon_{33}^T/\varepsilon_0$, electromechanical coupling factors k_t, k_{31} and k_p on concentration of pre-sintered ceramic granules mass.% for ceramic composites PZT/PZT.

At a further increase in mass.%, the electromechanical coupling factor k_t decreases slightly because of a reduction in e_{33} piezoconstant, $k_t^2 = e_{33}^2/(\varepsilon_{33}^S C_{33}^D)$.

The interrelationship of resulting values of considered electromechanical coupling factors at any mass.% is satisfactorily described by the following approximated formula: $k_t^2 \approx (k_{33}^2 + k_p^2)/(1 - k_p^2)$.

The dependencies of elastic moduli C_{33}^D, C_{33}^E and elastic compliances S_{11}^E, S_{11}^E on the concentration of pre-sintered ceramic granules mass.% for ceramic composites PZT/PZT are shown in Figure 4.

Elastic compliances S_{11}^E and S_{11}^D of ceramic composite increase with concentration mass.% because of porosity growth (stiffness decreasing). S_{11}^E and S_{11}^D increase approximately 4 – 5 times in a concentration range of 0 – 100%. At the same time, elastic moduli C_{33}^D and C_{33}^E decrease very fast with concentration mass.% and practically linearly up to $m = 60\%$. With further mass.% growth, C_{33}^D and C_{33}^E decrease slowly, as a result of a rearrangement of porous composite structure and formation of quasi-rod structure in thickness direction. The average decrease in C_{33}^D and C_{33}^E in concentration range 0 – 100% is to 7 – 9 times.

The difference in C and S behavior is caused by an inequality in porous composite structure in thickness and lateral direction (the lateral size of standard piezoceramic elements usually twenty times more than the thickness).

A complete characterization also must include frequency dependencies of loss tan (i.e., the ratio of imaginary and real parts of corresponding elastic constants) for each concentration or porosity value. The full complex set of measured parameters for ceramic composites PZT/PZT is present in Table 1.

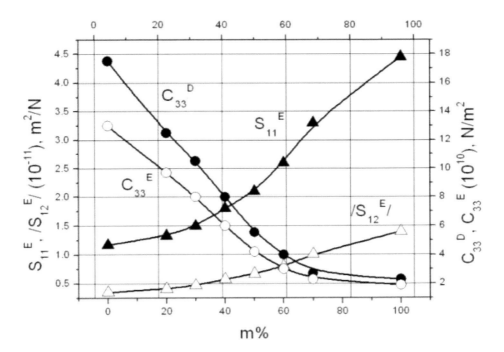

Figure 4. Elastic moduli C_{33}^D, C_{33}^E and elastic compliances S_{11}^E, S_{11}^D as a function of concentration of pre-sintered ceramic granules mass.%.

Table 1. Complex set of parameters of PZT/PZT composite disks for different concentrations of pre-sintered ceramic granules $m\%$

Parameter/mass.%	0	20	60	100
Porosity, vol. %	2.5	5.75	20	30
d_{33} quasi static (pC/N)	290	280	280	260
f_s rad (kHz)	113.4	109	84.4	70.3
S'^E_{11}, 10^{-11} (m²/N)	1.17	1.33	2.60	4.48
S''^E_{11}, 10^{-14} (m²/N)	-1.25	-2.33	-19.44	-56.73
S'^E_{12}, 10^{-12} (m²/N)	-3.49	-4.09	-7.77	-13.22
S''^E_{12}, 10^{-14} (m²/N)	0.485	9.61	32.06	43.21
d'_{31} (pC/N)	120	98	45	20
d''_{31} (pC/N)	-0.0974	-1.4713	-1.7495	-2.0596
ε'^T_{33}, 10^{-9} (F/m)	11.20	9.35	5.86	4.05
ε''^T_{33} 10^{-12} (F/m)	-3.93	-56.32	-32.52	-23.81
k'_p	0.561	0.473	0.201	0.109
k''_p	-0.0002	-0.0056	-0.0015	-0.0073
e'_{31} (C/m²)	14.7	10.8	2.6	0.8
e''_{31} (C/m²)	-0.0038	-0.1480	-0.0088	-0.0430
f_s^{thick} (kHz)	2010	1789	1076	941
k'_t	0.498	0.498	0.516	0.440
k''_t	-0.00899	-0.01842	-0.02829	-0.03118
C'^D_{33}, 10^{10} (N/m²)	17.2	12.48	4.04	2.33
C''^D_{33}, 10^8 (N/m²)	7.71	12.24	15.99	19.89
C'^E_{33}, 10^{10} (N/m²)	13.01	10.19	2.99	1.91
C''^E_{33}, 10^8 (N/m²)	21.2	21.9208	23.552	25.3558
e'_{33} (C/m²)	19.6	14.2	6.6	3.5
e''_{33} (C/m²)	-0.1	-0.15	-0.30	-0.50
ε'^S_{33}, 10^{-9} (F/m)	7.62	6.59	4.41	3.22
ε''^S_{33}, 10^{-9} (F/m)	-0.81	-1.75	-0.23	-0.43

CONCLUSION

New ceramic piezocomposites composed by pre-sintered piezoceramic granules embedded in porous piezoceramic matrix were developed and investigated. PZT/PZT ceramic piezocomposites are characterized by a unique spectrum of the electrophysical properties unachievable for standard PZT ceramic compositions fabricated by standard methods and can be useful for wide-band ultrasonic transducer applications.

ACKNOWLEDGMENT

Work supported by the RSF grant No. 15-12-00023.

REFERENCES

[1] Wersing, W. In: Piezoelectric Materials in Devices. Setter, N. (Ed.); Swiss Federal Institute of Technology, Lausanne, Switzerland. 2002, 29 - 66.

[2] Rybyanets, A. N. *IEEE Trans. UFFC*, 2011, vol. 58, 1492 - 1507.

[3] Rybyanets, A. N., Rybyanets, A. A. *IEEE Trans. UFFC*, 2011, vol. 58, 1757 - 1774.

[4] Rybyanets, A. N. *Ferroelectrics*, 2011, vol. 419, 90 - 96.

[5] Rybyanets, A. N. Ceramic Piezocomposites: Modeling, Technology, and Characterization. In: Piezoceramic Materials and Devices. Parinov I. A. (Ed.), Nova Science Publishers Inc., New York, 2010, Chapter 3, 113 - 174.

[6] Rybyanets, A. *Proc. 2007 IEEE Ultrason. Symp.*, 2007, 1909 - 1912.

[7] Rybianets, A. N.; Nasedkin, A. V. *Ferroelectrics*, 2007, vol. 360, 90 - 95.

[8] Rybianets, A. N.; Razumovskaya, O. N.; Reznitchenko, L. A.; Komarov V. D.; Turik, A. V. *Integrated Ferroelectrics*, 2004, vol. 63, 197-200.

[9] Rybjanets, A. N.; Nasedkin, A. V.; Turik A. V. *Integrated Ferroelectrics*, 2004, vol. 63, 179 - 82.

[10] Rybianets, A. N.; Tasker, R. *Ferroelectrics*, 2007, vol. 360, 90 - 95.

[11] Rybyanets, A. N.; Naumenko, A. A.; Shvetsova, N. A. Characterization Techniques for Piezoelectric Materials and Devices. In: Nano- and Piezoelectric Technologies, Materials and Devices. Parinov I. A. (Ed.) Nova Science Publishers Inc., New York, 2013, Chapter 1, 275 - 308.

In: Proceedings of the 2015 International Conference …   ISBN: 978-1-63484-577-9
Editors: Ivan A. Parinov, Shun-Hsyung Chang et al.   © 2016 Nova Science Publishers, Inc.

Chapter 55

SURFACE ACOUSTIC WAVES METHOD FOR PIEZOELECTRIC MATERIAL CHARACTERIZATION

N. A. Shvetsova, A. N. Reznitchenko, I. A. Shvetsov, E. I. Petrova and A. N. Rybyanets*

Institute of Physics, Southern Federal University,
Rostov-on-Don, Russia

ABSTRACT

The paper presents new surface acoustic waves (SAW) method for material parameters assessment and studying the relaxation process in ferroelectrics. Detailed experimental study of a dc electric field influence on the propagation of SAW in PZT and lead titanate piezoceramics with different ferroelectric "hardness" allowed developing a new method for domain orientation and space charge relaxation process studying in piezoelectrics. The qualitative analysis of the influence of domain orientation and space charge relaxation processes on SAW propagation have shown that a character of SAW parameters changes under the influence of the dc electric field depends on a ferroelectric "hardness." Automatic (auto-generator) measurements of $\Delta f/f(E)$ dependencies at different SAW frequencies (different penetration depths of SAW to piezoelectric substrate) provide a new method for measuring of coercive field E_c and relaxation times of space charge and domain orientations on surface and in subelectrode layers of piezoceramics.

1. INTRODUCTION

Surface acoustic waves (SAW) are widely used in acoustoelectronic devices, as well as for material parameters assessment [1]. The effects of external influences on the propagation of surface acoustic waves (SAW) in piezoelectric materials were an object of wide

---

* e-mail: yfnfif_71@bk.ru.

experimental studying [2, 3] mainly because of the possibility of designing controllable signal-processing devices, or compensation of temperature dependences of the SAW parameters. From the other hand, SAW is a powerful ultrasonic method for material parameters measurement and studying a physical process in piezoelectric materials [4, 5]. The influence of a dc electric field E on the SAW propagation has been studied in details for lithium niobate crystals [3]. For the YZ-cut of lithium niobate a weak nonlinear $\Delta\tau/\tau(E)$ was found [3]. It has been shown [6] that lead zirconate titanate (PZT) and lead titanate (PT) piezoceramics can demonstrate significantly bigger $\Delta\tau/\tau$ changes under dc bias than crystals. The dependence $\Delta\tau/\tau(E)$ in this case was nonlinear and $\Delta\tau/\tau$ decreased in time at fixed E. However, the cited studies were not of a systematic nature, and no physical interpretation of the observed effects was offered.

In this Chapter, we report detailed experimental study of the influence of a dc electric field on the SAW propagation in PZT and PT piezoceramics with different ferroelectric "hardness." New SAW method for material parameters assessment and studying of the relaxation process in ferroelectrics is also present.

2. Experimental Results and Discussion

The material chosen for the studies were ferroelectrically "soft" PZT compositions from the morphotropic region with a low coercive field and a high lability of the domain structure, rombohedral and tetragonal PZT compositions with intermediate "hardness" as well as ferroelectrically "hard" tetragonal PZT and PT piezoceramics with high coercive field and strongly clamped domain structure [7, 8, 9]. The samples were rectangular substrates of hot-pressed piezoceramics with dimensions polarized in the thickness direction. Interdigital transducers (IDT) for transmitting and receiving SAW were deposited on the working surface of the samples. Control electrodes were placed on the front and back surfaces of the substrates on the path of SAW propagation between IDT. The delay time and SAW attenuation were measured by the standard rf-pulses superposition techniques [1, 4].

2.1. Ferroelectrically "Soft" Piezoceramics

Figure 1 shows the relative changes in the SAW delay time $\Delta\tau/\tau$ and the additional SAW attenuation α as a function of the applied dc electric field E for ferroelectrically "soft" PZT piezoceramics. Positive values of E correspond to opposite directions of applied dc field and polarization P_R of the piezoceramics. The samples were held for 10 min at each dc field value. Concurrently with the delay time, the changes in the amplitude of the output SAW signal were measured. The curves $\Delta\tau/\tau$ (E) and $\alpha(E)$ shown in Figure 1 have several characteristic regions. The partial depolarization of the piezoceramic SAW delay line in the field interval $0 < E < E_c$ ($E_c \approx 8$ kV/cm is the coercive field at 50 Hz) leads to an increase in the delay time (segment a-b in Figure 1) in the result of decreasing in the contribution of the SAW coupling factor k_s to the effective elastic constants and an enhancement of the piezo-electric interaction of the domains [9-12].

The increase in $\Delta\tau/\tau$ is accompanied by an increase in α (segment a-b in Figure 1) caused by mechanical stresses at the boundaries of the depolarized region. In further increase in the field ($E > E_c$), a polarization reversal of the part of the piezoceramics beneath the control electrodes (the vertical dashed lines b-c-d in Figure 1 was observed. For a detailed study of the behavior of $\Delta\tau/\tau$ and α during piezoceramics polarization reversal, we chose a dc field $E = 7.25$ kV/cm, which ensured complete polarization reversal during ~ 30 min (Figure 2).

The changes in $\Delta\tau/\tau$ during reversal polarization are governed by a number of competing factors, mainly the increase in $\Delta\tau/\tau$ as a result of the further depolarization of the piezoceramics (k_s decrease) and the decrease in $\Delta\tau/\tau$ as a result of the enhancement of the domain clamping (increase in the concentration of 180°-domains and 90°-domain rotations). In the initial stage of reversal polarization, the effects, which lead to an increase in delay time $\Delta\tau/\tau$, dominate. The maximum of $\Delta\tau/\tau(t)$ corresponds to the maximum degree of depolarization of the piezoceramic. As the process of reversal polarization continues, the piezoceramics are become polarized in the opposite direction and $\Delta\tau/\tau$ decreases. Steady values of $\Delta\tau/\tau$ correspond to completion of the polarization reversal (for the chosen value of E).

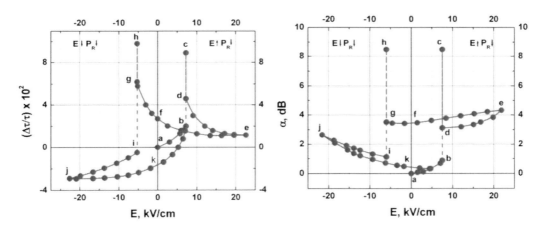

Figure 1. Dependencies of the relative change in the SAW delay time $\Delta\tau/\tau$ (left) and additional attenuation α (right) on the dc electric field E for ferroelectrically "soft" PZT piezoceramics; $\Delta\tau = \tau' - \tau$, where $\tau = 6410$ ns is the delay time at $E = 0$ and τ' is the current value of the delay time; $\alpha = 20 log (u_1/u_2)$, where u_1 and u_2 are the amplitudes of the output SAW signal at $E = 0$ at the current value of E, respectively.

The changes in $\Delta\tau/\tau$ during reversal polarization is accompanied by the changes in α (Figure 1), which is determined by the following factors: (i) reflection and scattering of SAW on the boundaries of the depolarized region; (ii) scattering of SAW by moving domain walls; (iii) scattering of SAW by mechanical stresses due to 90°-domain rotations. The decrease in a (segment c-d) in the final stage of the reversal polarization is due to a decrease in the number of scattering domain walls and a partial relaxation of the mechanical stresses.

The attenuation fluctuations observed in the initial stage of the reversal polarization (Figure 2) are caused by individual switching events in different regions of the ceramics. After the reversal polarization is complete, the polarization of the ceramics coincides with the direction of E. A further increase in E ($E_c < E < 22$ kV/cm) gives a gradual decrease in $\Delta\tau/\tau$ and an insignificant increase in α (segment d-e in Figure 1). A repeated depolarization of the ceramic ($0 < E < 7.25$ kV/cm) leads to a decrease in $\Delta\tau/\tau$ and to an insignificant change in α (segments a-b in Figure 1). The complete loops described by $\Delta\tau/\tau\,(E)$ and $\alpha(E)$ are reproduced in subsequent measurement cycles.

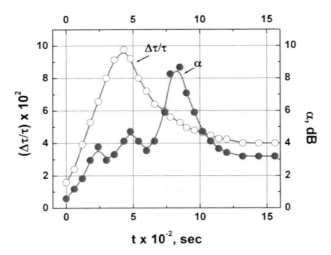

Figure 2. Relative changes in the SAW delay time $\Delta\tau/\tau$ and attenuation α as functions of time t during reversal polarization of a segment of ferroelectrically "soft" PZT piezoceramics delay line at $E = 7.25$ kV/cm.

2.2. Ferroelectrically "Hard" Piezoceramics

Figure 3 shows dependencies of $\Delta\tau/\tau$ and α SAW as functions of the dc electric field E for ferroelectrically "hard" PT piezoceramics.

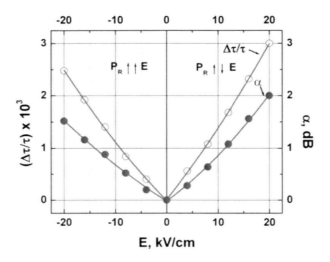

Figure 3. Dependencies of the relative change in the SAW delay time $\Delta\tau/\tau$ and additional attenuation α (maximum final values) on the dc electric field E for ferroelectrically "hard" PT piezoceramics.

Experimental points are maximal values of $\Delta\tau/\tau$ and α for each E value peaking during 3 s after applying E. Values of $\Delta\tau/\tau$ and α at each E are significantly decrease in time (on 80-100% during 60-120 sec). It can be seen from Figure 3, that $\Delta\tau/\tau$ and α increase practically linearly with $|E|$. These dependencies unlike results for ferroelectrically "soft" piezoceramics (Figure 1) cannot be explained by domain reorientation processes (at coinciding directions of P_R and E, a decrease in $\Delta\tau/\tau$ connected with additional polarization of piezoceramics as a result of 90°-domain rotations must be observed [6, 9, 12]. Moreover, the character and time of relaxation demonstrated by $\Delta\tau/\tau$ and α are not typical for domain reorientation processes.

For finding out the physical mechanisms responsible for these changes, the time dependencies of $\Delta\tau/\tau$ and α at the influence of a dc voltage pulses with different polarity have been measured (Figure 4).

At applying a negative voltage pulse (directions E and P_R coincide), $\Delta\tau/\tau$ quickly increases up to $\sim 1.4 \cdot 10^{-3}$ for 2-3 s, and then smoothly falls down up to zero approximately for 90 s (Figure 4). At the end of the voltage pulse $\Delta\tau/\tau$ again increases up to $\sim 1.3 \cdot 10^{-3}$ and falls down up to zero for the same time. Splashes of $\Delta\tau/\tau$ are accompanied by the changes in α repeating trend of the $\Delta\tau/\tau$ (decay time for α reaches 60 s). At applying a positive voltage pulse (directions E and P_R are opposite) behavior of $\Delta\tau/\tau(t)$ and a maximal value $\Delta\tau/\tau \sim 1.8 \cdot 10^{-3}$ is kept, however, decay time is ~ 120 s and $\Delta\tau/\tau$ attain saturation at the level $\sim 0.5 \cdot 10^{-3}$ (Figure 4). At the end of voltage pulse (Off) $\Delta\tau/\tau$ attain maximal value $\sim 1.6 \cdot 10^{-3}$ for 2-3 s and falls down up to zero approximately for 120 s. Character of $\Delta\tau/\tau$ and α changes in all cases is close to exponential decay. Similar dependences of $\Delta\tau/\tau$ and α on E and time t have been obtained for ferroelectrically "hard" PZT piezoceramics. Distinction of the dependencies $\Delta\tau/\tau(E)$ and $\alpha(E)$ for ferroelectrically "soft" (Figure 1) and "hard' (Figure 3) ceramics, as well as the character of $\Delta\tau/\tau(t)$ and $\alpha(t)$-dependencies at applying of dc voltage pulses (Figure 4) allow us to offer a following explanation.

In piezodielectrics at the applying an electric voltage besides a normal leakage current a dielectric absorption current decaying with a time according the law $i_a \sim e^{-\beta t}$ or $i_a \sim t^{-\gamma}$ (β and γ are constants, t is the time) is flows [1, 4, 11]. It is well known, that the transverse dc electric field E applied to the piezosemiconductor causes changes in SAW attenuation and propagation velocity as a result of the interaction of moving charge carriers with $E_-^{\perp}$ SAW (transverse acoustoelectronic effect) [1]. So, flowing through the piezoceramics decaying in time absorption current i_a caused by the external dc electric field E should lead to the changes in $\Delta\tau/\tau$ and α as a result of interaction of migrating space charge carriers with $E_-^{\perp}$ SAW [4]. Apparent increase in $\Delta\tau/\tau$ and α with E (Figure 2) can be caused by the increase in concentration of a charge carriers released from the capture levels and contributed in i_a. The direction of the absorption current in this case is unimportant because of the space charges with low mobility and limited displacement does not leave the SAW localization area at both directions of E.

Time dependencies of $\Delta\tau/\tau$ and α at applying the voltage pulses (Figure 4) are caused by the decrease of i_a in time as a result of space charge migration. The character of the dependencies in all cases is close to the exponential decay and determined by the diffusion law, and captures kinetics of charge carriers on impurity and defect levels [1]. Characteristic times of $\Delta\tau/\tau$ and α relaxation were $60 - 120$ s and are typical for migration process in dielectrics [1, 12]. Nonzero residual level of $\Delta\tau/\tau$ at opposite directions of E and P_R at positive pulse of a dc voltage (Figure 1) can be caused by the intensifying of $E_-^{\perp}$ SAW shielding by the negative carriers of a space charge. Excessive concentration of negative carriers near a working surface is connected with orientation P_R in experimental samples.

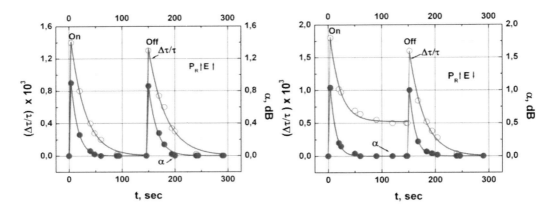

Figure 4. Changes in the SAW delay time $\Delta\tau/\tau$ and attenuation α under the influence of the negative (left) and positive (right) dc voltage pulse with the amplitude $V = 720$ V ($E = 12$ kV/cm) and duration $t = 150$ s for ferroelctrically "hard" PT piezoceramics.

Thus, the changes in the SAW propagation parameters under the influence of dc electric field in ferroelctrically "hard" piezoceramics with high concentration of charge carriers can be interpreted as a result of the interaction of transverse component of SAW electric field $E_{\sim}^{\perp}$ with the migrating carriers of a space charge. Measurements made for PZT piezoceramics with an intermediate ferroelectric "hardness" have revealed, that in the electric fields $E < E_C$ the changes in $\Delta\tau/\tau$ and α, SAW were similar to shown in Figures 3 and 4 for ferroelctrically "hard" piezoceramics. At an increase of electric field $E > E_C$ the dependencies similar shown in Figres 1 and 2 for ferroelctrically "soft" piezoceramics were observed.

CONCLUSION

Detailed experimental study of a dc electric field influence on the propagation of SAW in PZT and lead titanate piezoceramics with different ferroelectric "hardness" allowed one to develop a new method for domain orientation and space charge relaxation process studying in piezoelectrics.

The qualitative analysis of the influence of domain orientation and space charge relaxation processes on SAW propagation have shown that a character of SAW parameters changes under the influence of the dc electric field depends on a ferroelectric "hardness" and is controlled by the following mechanisms:

(i) for ferroelectrically "hard" piezoceramics - by the interaction of SAW with moving carriers of the space charge shielding the applied electric field and preventing polarization switching;

(ii) for ferroelectrically "soft" piezoceramics - by the interaction of SAW with moving domain walls (domain orientation processes);

(iii) for materials with intermediate ferroelectric "hardness," at low electric fields ($E < E_c$) space charge relaxation process are dominate, whereas at high electric fields ($E > E_c$) domain orientation processes are determinative.

Automatic (auto-generator) measurements of $\Delta f/f(E)$ dependencies at different SAW frequencies (different penetration depths of SAW) provide a new method for measuring of coercive field E_c and relaxation times of space charge and domain orientations in surface and subelectrode layers of piezoceramics.

ACKNOWLEDGMENTS

Work supported by the RSF grant No. 15-12-00023.

REFERENCES

[1]   Rybyanets, A.N.; Naumenko, A.A.; Shvetsova, N.A. Characterization Techniques for Piezoelectric Materials and Devices. In: *Nano- and Piezoelectric Technologies, Materials and Devices*, Parinov I. A. (Ed.), Nova Science Publishers Inc., New York, 2013, Chapter 1, 2013, 275-308.

[2]   White, R. *Proc. IEEE,* 1970, *vol. 58(8)*, 68-110.

[3]   Joshi, S.G. *J. Acoust. oc. Amer.* 1982, *vol. 72(12)*, 1872-1877.

[4]   Rybyanets, A. *Proc. 2007 IEEE Ultrason. Symp.* 2007, 1909-1912.

[5]   Rybianets, A.N.; Turik, A.V.; Dorohova, N.V.; Miroshnichenko, E.S. *Sov. Tech Phys. Journ.* 1986, *vol. 31*, 1418–1421.

[6]   Gulyaev, Y.V.; Karinskii, S.S.; Mondikov, V.D. *Sov. Tech. Phys. Lett.* 1975, *vol. 1*, 346.

[7]   Rybianets, A.N.; Turik, A.V. *Izvestia Acad. Nauk USSR. Ser. Phys.* 1987, *vol. 51*, 2244-2248.

[8]   Rybyanets, A.N. *IEEE Trans. UFFC.* 2011, *vol. 58*, 1492-1507.

[9]   Rybyanets, A.N.; Rybyanets, A.A. *IEEE Trans. UFFC.* 2011, *vol. 58,* 1757-1774.

[10] Rybyanets, A.N. Ceramic Piezocomposites: Modeling, Technology and Characterization. In: *Piezoceramic Materials and Devices.* Parinov I.A. (Ed.), Nova Science Publishers Inc., New York. 2010, Chapter 3, 113-174.

[11] Rybianets, A.N.; Tasker, R. *Ferroelectrics.* 2007, *vol. 360,* 90-95.

[12] Rybianets, A.N.; Nasedkin, A.V. *Ferroelectrics,* 2007, *vol. 360*, 90-95.

[13] Rybyanets, A.N. *Ferroelectrics,* 2011, *vol. 419*, 90-96.

In: Proceedings of the 2015 International Conference ... ISBN: 978-1-63484-577-9
Editors: Ivan A. Parinov, Shun-Hsyung Chang et al. © 2016 Nova Science Publishers, Inc.

Chapter 56

DISPERSION RELATION AND RESONANCE FREQUENCIES OF SURFACE ACOUSTIC WAVES EXCITED IN BARIUM TITANATE FILMS

P. E. Timoshenko[1,2,], V. V. Kalinchuk[2], V. B. Shirokov[1,2], M. O. Levi[2] and A. V. Pan'kin[1]*

[1]Southern Federal University, Rostov-on-Don, Russia
[2]South Scientific Center of Russian Academy of Sciences,
Rostov-on-Don, Russia

ABSTRACT

In this article we have studied the propagation characteristics of surface acoustic waves in ferroelectric epitaxial film of barium titanate. The mechanical wave is excited by the interdigital transducer which has a comb-like structure, where the distance between the fingers in the transducer decides the frequency of the wave propagation over the substrate. We have computed a Rayleigh wave type device in COMSOL Multiphysics[TM] platform of finite-element physics-based modeling and simulation, choosing magnesium oxide as the substrate and aluminum metal electrodes as interdigital transducer. The numerical results and the analytical ones obtained by us from the solution based on Green function method are in good agreement. The effect of the lattice misfit strain on heterostructure of barium titanate epitaxial film at room temperature for two low-frequency surface acoustic modes are discussed in present article.

Keywords: MEMS, SAW devices, resonators, FEM, IDT, Lattice misfit strai

1. INTRODUCTION

The devices on surface acoustic waves (SAW) have wide class of applications for analog signals processing in real time includes delay lines, resonators and filters operating at selected

---

* E-mail: P.E.Timoshenko@gmail.com.

frequencies in the range from about 10 MHz to 15 GHz. Due to the growing demand of the high rate of communication data, the need of the increasingly higher operating frequency regime of the electromagnetic spectrum is being experienced very attractive aim of modern acoustoelectronics. There are two ways [1] to achieve desirable parameters: using sound-conducting substrate with a higher sound velocity or decreasing geometrical sizes of transmitting and receiving interdigital transducers (IDT). Both of these methods have limitations. For example, the sound velocity in the substrate is fixed and it is very difficult to make a gap width between contacts of IDT less than $0.5\mu m$ by lithography method.

Ferroelectric thin films with a higher electromechanical coupling coefficient may become an alternative possibility to design SAW devices operating at significantly higher frequencies compared to modern high-speed rate devices.

Recently, thin ferroelectric films of barium titanate $BaTiO_3$ (BT) have been studied extensively due to their potential applications in wide range of microelectronic devices such as nonvolatile ferroelectric random access memories (FeRAMs), sensors and actuators. The high dielectric permittivity of BT in conjunction with the strong dependence of the dielectric constant on the electric field allow one to design relatively well frequency tunable microwave devices for room temperature applications. However, there is a number of fundamental problems, such as compositional and microstructural inhomogeneities, defects and internal stresses, that limit the possible applications of thin BT films compared to bulk ceramics or single crystals.

Internal stresses in epitaxial films are due to the lattice mismatch between film and the substrate. Epitaxy-induced stresses have a strongly pronounced effect on the dielectric response of BT films. The dependence of lattice misfit strain in BT films on the dielectric and material constants using Landau phenomenological theory was present in Refs [2, 3].

The strict numerical modeling is required to take an account influence of thin BT film, complex distributions of electric field, displacements and stresses. Therefore, we have chosen COMSOL Multiphysics^{TM} platform performing the numerical calculations by finite-element method (FEM) [4]. Verifying the accuracy of dispersion curve calculations carried out by comparison of the numerical results and analytical ones obtained from solution based on Green function method [5-6].

In the present paper, we discuss some aspects of simulation of SAW propagation by FEM and estimate effect of the lattice misfit strain on heterostructure of barium titanate epitaxial film grown on magnesium oxide substrate for two low-frequency surface acoustic modes at room temperature.

2. STATEMENT OF THE PROBLEM

The problem of simulating a SAW device is mathematically equivalent to the solution of the differential equations of the piezoelasticity with a given excitation. There are various numerical methods to solve these kinds of equations. For example, FEM is a very popular method. There is a relatively wide range of commercial software's implementing FEM including equations of elasticity and piezoelectricity. It may seem more efficient to use them instead of spending a lot of time and resources to develop and debug specialized software models. Therefore, in this article, we consider propagation characteristics of typical one-port SAW resonator [7-9] using periodic boundary condition in COMSOL Multiphysics™.

2.1. Geometry

Let us consider a BT film (Figure 1, area Ω_1). It occupies the size h_{BT}. Below the film is placed magnesium oxide substrate (Ω_2) having size h_{MgO}. The bottom surface of the substrate is fixed in its position. Upper ($x_3 > 0$) and lower ($x_3 < -h_{BT} - h_{MgO}$) half-spaces are vacuum. Interdigital transducer is a comb-shaped structure of aluminum electrodes (Ω_3, Ω_4) fabricated over the BT film. A voltage applied to the periodic IDTs produces dynamic strains in the substrate and initiates elastic waves that travel along the surface.

The displacements of the surface u_1, u_2, u_3 along x_1, x_2, x_3 directions for Rayleigh wave as shown in Figure 1 are given by the following equations:

$$u_1 = [A_1\exp(-b_1 x_3) + A_2\exp(-b_2 x_3)] \exp(jk(x_1 - ct)),$$

$$u_2 = 0, \; u_3 = [(-b_1/jk)A_1\exp(-b_1 x_3) + (jk/b_2)A_2\exp(-b_2 x_3)] \exp(jk(x_1 - ct)), \quad (1)$$

where A_1, A_2 are the amplitudes, $b_1 = k(1 - c^2/v_l^2)^{1/2}$, $b_2 = k(1 - c^2/v_t^2)^{1/2}$, v_l and v_t are the longitudinal and transverse wave velocities, k is the wave number, c is the velocity of SAW.

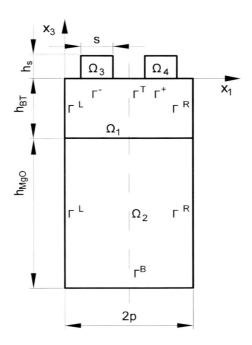

Figure 1. Geometry used in the simulation.

Surface waves propagate in both directions along the coordinate x_1. The coordinate x_3 is the direction of wave attenuation. We assume that the wave parameters do not depend on the coordinate x_2 directed from the observer in Figure 1. In general, there are three components of mechanical displacement u_i, $i = 1, 2, 3$ due to the anisotropy properties of the ferroelectric material. Let us assume that the electric potential is denoted by V and is used to describe the electric field E. The vector $u = (u_1, u_2, u_3, V)^T$ completely characterizes a piezoelectric system. We can define all the mechanical and electrical parameters in the quasi-static approximation using this vector. The fundamental difference between current solution of the problem and general one implemented in many FEM software products is available bias $u_2 \neq 0$, because oscillations are not considered sagittal polarized compared to the classical Rayleigh waves ($u_2 = 0$).

2.2. Constitutive Equations and Boundary Conditions

The linear constitutive equations are governed by the continuum equation of motion, Maxwell's equations under the quasi-static assumption, the strain-mechanical displacement relations and proper boundary conditions. In a homogeneous piezoelectric BT film the stress component at each point depends on applied electric field. The piezoelectric equations [7-9] can be expressed as

$$
\begin{cases}
C_{ijkl}^E \dfrac{\partial^2 u_l}{\partial x_j \partial x_k} + e_{kij} \dfrac{\partial^2 V}{\partial x_j \partial x_k} = \rho \dfrac{\partial^2 u_i}{\partial t^2} \\
e_{jkl} \dfrac{\partial^2 u_l}{\partial x_j \partial x_k} + \varepsilon_{jk}^s \dfrac{\partial^2 V}{\partial x_i \partial x_k} = 0 \qquad , \quad i,j,k,l = 1,2,3 , \\
T_{ij} = C_{ijkl}^E S_{kl} - e_{kij} E_k \\
D_i = e_{ikl} S_{kl} + \varepsilon_{ik}^s E_k
\end{cases} \qquad \square(2)
$$

where, C_{ijkl}^E represents the stiffness tensor for constant electric field, S_{kl} represents strain tensor, e_{kij} is the elastic constant or piezoelectric tensor, T_{ij} is the stress tensor, ρ is the density of the medium, D_i is the electric displacement, E_k is the electric field ($E_i = -\partial V / \partial x_i$), and ε_{jk}^s is the permittivity tensor for constant strain.

The second order system of partial differential equations (2) can be transformed to equation (3) to solve the eigen-frequency problem in COMSOL [7].

$$
ea \cdot \omega^2 \cdot u - \nabla \cdot (c\nabla u) = 0 \tag{3}
$$

where, ea, c are the matrices depending on the material constants. This approach to the solution allows us to take into account three spatial components of mechanical displacements.

Traction of free boundary is given on the top surface of the film. The bottom surface of the substrate is fixed in its position and the displacement in the lateral direction is constrained to be zero ($u = 0$). As the IDTs are periodic in nature, appropriate periodic boundary conditions are applied by using the following equations:

$$
u_i(x+np) = u_i(x) \exp(-j2\pi \gamma n); \; V_i(x+np) = V_i(x) \exp(-j2\pi \gamma n); \; \Gamma_R(u,V) = (-1)^n \Gamma_L(u,V), \tag{4}
$$

where γ is the complex propogation constant. A potential 1V is applied to the electrodes in the simulation. Later, we can determine the necessary parameters of SAW devices using this approach.

3. RESULTS AND DISCUSSION

The geometry of the SAW structure used in the simulation is given in Table 1. Material constants of the film are taken from the paper [3].

Table 1. Dimensions of IDT, film and substrate

Parameter	Value
Aluminum electrode thickness (h_s)	0
Young's modulus of the electrode	70 GPa
Poisson ratio of the electrode	0.35
Density of the electrode	2700 kg/m^3
Periodicity of the electrode (p)	4 µm
Width of the electrode (s)	2 µm
Film thickness (h_{BT})	200 nm, 2 µm
Substrate thickness (h_{MgO})	20 µm (5λ)
Dielectric constant of the substrate (ε^s)	9.8

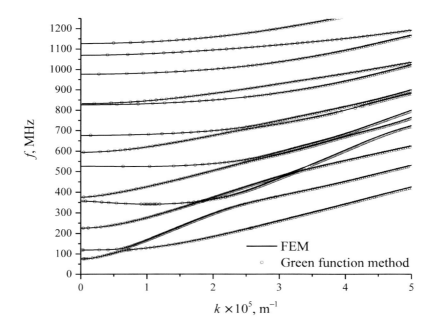

Figure 2. Dispersion curves of the structure with $h_{BT} = 200$ nm obtained by FEM simulation are compared to the analytical ones obtained from the solution based on Green function method [5].

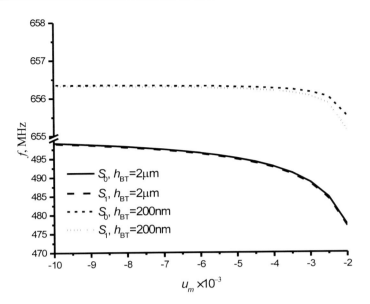

Figure 3. Frequency dependence of the surface modes S_0 and S_1 on the misfit strain u_m for *aa*-phase, when the wavelength is $2p$.

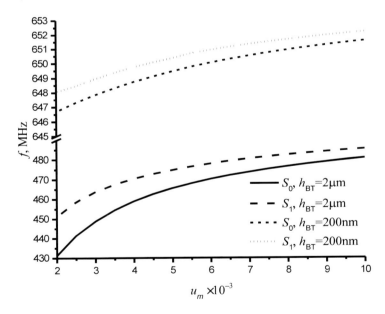

Figure 4. Frequency dependence of the surface modes S_0 and S_1 on the misfit strain u_m for *c*-phase when the wavelength is $2p$.

Dispersion curves obtained by eigen-frequency analysis using COMSOL Multiphysics™ and Green function method [5-6] are present in Figure 2. Curves calculated by FEM have minor differences in the large wave numbers region. The main advantages of FEM are the conservative method, absolute stability, and possibility of creating finite-element approximations on unstructured grids.

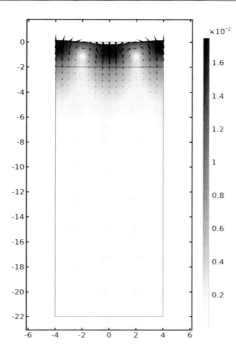

Figure 5. Surface plot of total displacement for surface mode S_0 and misfit strain $u_m = -4 \cdot 10^{-3}$ at the frequency 492.78 MHz.

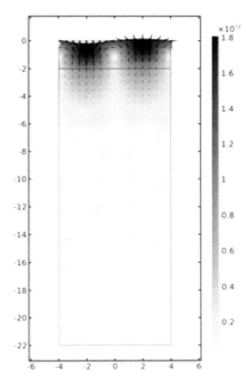

Figure 6. Surface plot of total displacement for surface mode S_1 and misfit strain $u_m = -4 \cdot 10^{-3}$ at the frequency 492.51 MHz.

Figures 3 and 4 show the frequency dependence of the surface S_1 and S_0 modes on misfit strain u_m for aa- and c-phase at a wavelength equal to $2p$. Surface plots of total displacement for surface modes S_0 and S_1 are shown in Figures 5 and 6 for $u_m = -4 \cdot 10^{-3}$ at the frequencies 492.78 MHz and 492.51 MHz, respectively. The frequency changes sharply in phase transition regions. Outside the region of phase transition, the frequency changes slowly. This fact is especially noticeable in case of thin films. Also it should be noted that the order of surface modes is different for aa- and c-phases and in aa-phase the first surface modes are very close placed to each other.

CONCLUSION

We simulated a one-port SAW layered structure consists of ferroelectric barium titanate epitaxial film grown on magnesium oxide substrate using COMSOL Multiphysics. The numerical results for dispersion curves are compared by us with the analytical ones obtained from the solution based on Green function method. The effect of the lattice misfit strain on heterostructure of barium titanate epitaxial film at room temperature for two low-frequency surface acoustic modes are discussed in this article.

Moreover, the results of this study clearly illustrate the effectiveness of using the FEM to model of the acoustic wave propagation problems and demonstrate the potential of the FEM for problems where analytical solution is impossible because of complicated component geometry.

ACKNOWLEDGMENTS

The authors acknowledge the support of the Russian Science Foundation, provided by grant 14-19-01676, for carrying out this research study.

REFERENCES

[1] Mukhortov, Vl. M.; Biryukov, S. V.; Golovko, Yu.; Karapetyan, G. Y.; Masychev, S. I.; Mukhortov, Vas. M. *JTP Letters*. 2011, *vol. 37(5)*, 31–37.

[2] Shirokov, V. B.; Kalinchuk, V.V.; Shakhovoy, R.A.; Yuzyuk, Y. I. *EPL*. 2014, *vol. 108(4)*, 47008 (4p.).

[3] Shirokov, V.B.; Kalinchuk, V.B.; Shakhovoy, R.A.; Yuzyuk, Y.I. *Physics of Solid State*. 2015, in press.

[4] Roger, W.P. "Multiphysics Modeling Using COMSOL 5 and MATLAB (Engineering)," *Mercury Learning and Information*. 2015. p. 700.

[5] Babeshko, V.A.; Kalinchuk, V.V. *Journal of Applied Mathematics and Mechanics*. 2002, *vol. 66(2)*, 275–281.

[6] Belyankova, T.I.; Kalinchuk, V.V.; Lyzhov, V.A. *Journal of Applied Mathematics and Mechanics, vol. 74(6)*, 637-647.

[7] Osetrov, A.V.; Nguen, S.V. *Proceedings of the IEEE Russia. North West Section.* 2011, *No. 1*, 75–78.

[8] Wu, T.T.; Wang, S.M.; Chen, Y.Y.; Wu, T.Y.; Chang, P.Z.; Huang, L.S.; Wang, C.L.; Wu, C.W.; Lee, C.K. *Jpn. J. Appl. Phys.* 2002, *vol. 41*, 6610–6615.

[9] Timoshenko, P.E.; Kalinchuk, V.V.; Shirokov, V.B.; Levi, M.O.; Pan'kin, A.V.; Donets, A.V. Finite element simulation of surface acoustic waves excited in barium-strontium titanate films fabricated on magnesium oxide substrate by interdigital transducers. *Proceedings of 2014 International Conference on Actual Problems of Electron Devices Engineering*, APEDE 2014, *vol. 1*, 283–289.

In: Proceedings of the 2015 International Conference ... ISBN: 978-1-63484-577-9
Editors: Ivan A. Parinov, Shun-Hsyung Chang et al. © 2016 Nova Science Publishers, Inc.

Chapter 57

INVESTIGATION OF FEATURES OF SURFACE WAVE FIELDS IN A MEDIA WITH INHOMOGENEITIES

O. V. Bocharova[1,], V. V. Kalinchuk[1], A. V. Sedov[1] and I. E. Andjikovich[2]*

[1]South Scientific Center of Russian Academy
of Sciences, Rostov-on-Don, Russia
[2]Southern Federal University, Rostov-on-Don, Russia

ABSTRACT

We created a multifunctional measuring system allowing conducting studies, comparing signals, and constructing spectral characteristics by using sensors of various types. We examined features of wave fields at surfaces of structurally inhomogeneous bodies. Numerically and experimentally, we investigated a possibility of determination of defect presence through the parameters of the surface wave field. When solving problems of defects diagnostics, we used the method of image recognition. We carried out the series of computational experiments. The experimental results demonstrated precise spatial distribution of images in the space of recognition depending on sample defect.

Keywords: surface wave field, defect, nondestructive testing

1. INTRODUCTION

A problem of development of techniques for non-destructive testing of state and strength of engineering structures' units and parts is a key to improve the reliability of their operation and to prevent the emergencies. Due to the present-day development of technology of new

---

[*] Corresponding author: O. V. Bocharova. South Scientific Center of Russian Academy of Sciences, Rostov-on-Don, Russia. E-mail: olga_rostov1983@mail.ru.

materials production and increasing requirements for details and assemblies performance, it is necessary to create a simple and effective method of permanent monitoring of the object state not causing its damage [1-6].

2. METHODS AND RESULTS

We created a multifunctional measuring system allowing conducting studies, comparing signals, and constructing spectral characteristics by using sensors of various types. We constructed the experimental setup allowing evaluating a variation of surface wave field in samples of various functional materials in the laboratory conditions. We conducted the tests of the ferroelectric film deformation showing the diversity of its application in various methods of the acoustic testing of surface state.

A block diagram of the experiment is presented in the Figure 1. On the edge of the sample under the action of electromagnetic impact, a surface wave arose. The surface wave's propagation was registered by the accelerometer B&K and was amplified by the charge amplifier B&K2626. The digitizer pad ADC L-Card (E14-140) digitized the obtained signals, and then the computer program PowerGraph processed the signal.

At the same time, we carried out computational experiments based on using the finite element method. We investigated peculiarities of the dynamic process at the surface of a rectangular parallelepiped weakened by a through cylindrical cavity (Figure 2). The diameter of the cavity (D), as well as its depth (h) varied. Surface oscillations were exited in the sample by means of impulse action.

To do the calculations of the wave field at the surface of the sample weakened by the presence of the cavity, we used the package ANSYS and its command language APDL.

In the calculations, we used the following parameters: Young's modulus $E = 22$ MPa, density $\rho = 50$ kg/m^3, Poisson's ratio $\nu = 0.3$.

The wave field was excited by a pulse in a step form with duration of 50 microseconds.

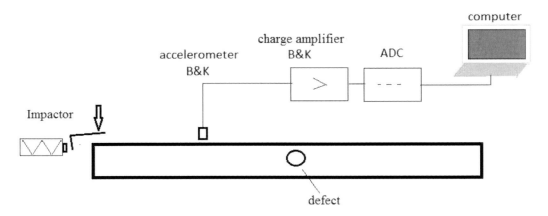

Figure 1. Block diagram of the experiment.

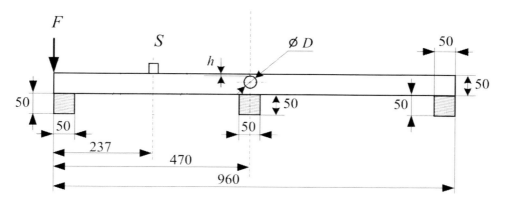

Figure 2. General view of the tested rectangular construction.

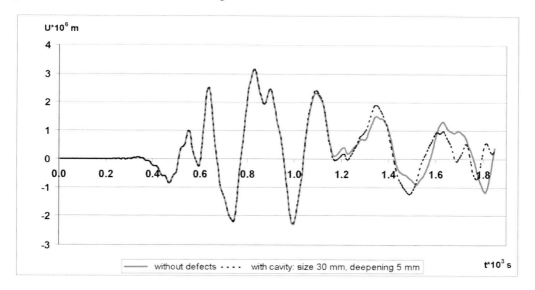

Figure 3. Surface wave field for the sample without defects and the sample with a cavity.

The Figure 3 shows the plots of amplitude of the surface wave for a sample without defects and a samle with a cavity (size equals to 30 mm, depth is 5 mm). The experimental results demonstrate that in the intial time, the amplitudes of the both samples are the same. The difference arises as the wave reflected from a defect reaches the sensor.

When solving problems on defect diagnostics, we have used the method of image recognition. Arising oscillations $f(t), t \in [0, T_n]$ were measured by displacement sensor S, for a time interval T_n sufficient for the arrival of the reflected waves.

By using the method of image recognition, we defined that the type of a defect corresponded to the type of the response function $f(t)$. In this case, there was a steady and unique dependence between the response function type and the defect parameters.

We defined the type of the function $f(t)$ by using the constructive features of spatial diagnostics of ξ. For mapping the generalized topological space C of the function $f(t)$ in the space ξ, we used orthogonal homeomorphic mapping. This mapping can be considered as generalized spectral expansion functions $f_i(t)$ in the orthonormal basis of functions $\{\xi_1, \xi_2, \xi_3, \ldots\}$ of the space ξ:

$$f_i(t) = \sum_{j=1}^{\infty} a_{ij} \xi_j(t), \ (f_i(t) \in \mathbf{C}) \rightarrow ((a_{i1}, a_{i2}, \ldots) \in \xi) \tag{1}$$

Let us consider that functions $f_i(t)$ are represented by a finite number of measurements N, i.e., $f_i(t) \in \mathbf{R}^N$; the last orthogonal transformation can be represented as a finite-dimensional one:

$$f_i(t) = a_0 + \sum_{j=1}^{m} a_{ij} \xi_j(t), \ (f_i(t) \in \mathbf{R}^N) \rightarrow ((a_{i1}, a_{i2}, \ldots, a_{im}) \in \mathbf{R}^m) \tag{2}$$

Plots of functions $f_i(t)$ were converted in accordance with the latter representation to an m-dimensional image $A_i = (a_{i1}, a_{i2}, \ldots, a_{im}) \in \mathbf{R}^m$.

The basis $\{\xi_1, \xi_2, \xi_3, \ldots\}$ is determined as a result of solving a number of following optimization problems:

1. minimization of the mean-square error of the function representation $f(t)$ at a fixed number m of the orthonormal functions:

$$\Delta = \frac{1}{M} \sum_{i=1}^{M} \left\| f_i(t) - \sum_{j=1}^{m} a_{ij} \xi_j(t) - a_0 \right\|_2 = \frac{1}{M} \sum_{i=1}^{M} \sqrt{\sum_{k=1}^{N} \left(f_i(k\Delta t) - \sum_{j=1}^{m} a_{ij} \xi_j(k\Delta t) - a_0 \right)^2} ,$$

where M is the amount of tested samples;

2. fulfillment of the orthonormality condition for the base functions:

$$(\xi_i, \xi_j) = \begin{cases} 1, \ i = j; \\ 0, \ i \neq j, \end{cases} \ i, j = \overline{1, m}.$$

3. fulfillment of the conditions of the greatest spacing images in the specific space, corresponding to the different types of defects:

$$d^2(A_i, A_l) = \frac{1}{M} \sum_{\substack{i,l=1 \\ i \neq j}}^{M} \|A_i - A_l\|^2$$

For simplicity of the physical interpretation, we have realized the defectoscopy of samples in images $A_i = (a_{i1}, a_{i2}) \in \mathbf{R}^2$ in two-dimensional specific space.

We carried out the series of computational experiments. The Figure 4 and Figure 5 show the results of construction of the specific space and location of the images for the samples with a cavity of varied size and depth.

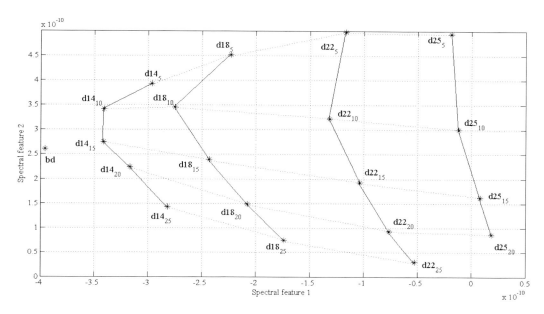

Figure 4. Location of images in the space of recognition for different sizes and depth of defects.

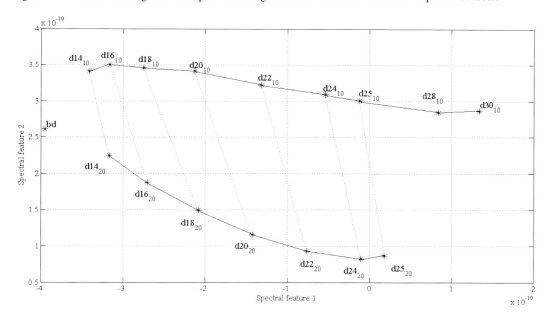

Figure 5. Location of images in the space of recognition for different sizes and depth of defects.

In the figures, bd is the image for the sample without defects, dD_h is the image for the sample with a cavity (of the diameter D and the depth h).

CONCLUSION

The results of the conducted experiments demonstrates that there is a precise spatial distribution of images in the space of recognition depending on the defect of a sample.

ACKNOWLEDGMENTS

The present research was supported by the Russian Science Foundation (grant No. 14-19-01676) and the Russian Foundation for the Basic Research (grant No. 14-08-01213).

REFERENCES

[1] Bocharova, O. V., Vatulyan, A. O. *Acoustical Physics*. 2009, vol. 55(3), 281-288.
[2] Bocharova, O. V. *IOP Publishing Journal of Physics: Conference Series*. 2014, vol. 490(1), 012057.
[3] Denina O. V., Vatulyan A. O. *Journal of Applied Mechanics and Technical Physics*. 2012, vol. 53(2), 266-274.
[4] Alireza, Farhidzadeh; Salvatore, Salamone. *Ultrasonics*, 2015, vol. 57, 198-208.
[5] Dervilis, N., Choi, M., Taylor, S. G., Barthorpe, R. J., Park, G., Farrar C. R. *Journal of Sound and Vibration*. 2014, vol. 333(6), 1833-1850.
[6] Addisson, Salazar; Luis, Vergara; Raúl, Llinares. *Mechanical Systems and Signal Processing*. 2010, vol. 24(6), 1870-1886.

In: Proceedings of the 2015 International Conference … ISBN: 978-1-63484-577-9
Editors: Ivan A. Parinov, Shun-Hsyung Chang et al. © 2016 Nova Science Publishers, Inc.

Chapter 58

VIBROACOUSTICS OF COMPOSITE POLYMERIC SHELLS OF ROTATION REINFORCED BY DISCRETE CIRCULAR RIBS

V. G. Safronenko[1,] and O. I. Safronenko[2]*

[1]Vorovich Institute for Mathematics, Mechanics and Computer Science,
Southern Federal University, Rostov-on-Don, Russia
[2]Department of English for Sciences, Southern Federal University,
Rostov-on-Don, Russia

ABSTRACT

Forced harmonic vibrations of composite polymeric shells of rotation with the polymer binding in the acoustic media are considered. Mathematical and computer models of vibroacoustics for multilayered composite polymeric shells of rotation, reinforced by discrete circular ribs under the influence of vibration loadings have been developed. The Timoshenko model, which takes into account the deformation of transverse displacement and rotation inertia, is used. Physical and mechanical characteristics of a polymer binding are considered in the frameworks of the orthotropic thermoviscoelasticity. The shell is being oscillating in linear acoustic media. It is described by the Helmholtz equation and the equation of continuity for the surface of the shell. To solve this problem, the method based on modeling response of acoustic medium with the help of the local impedance is used. The local impedance is derived from the solution of the model problems. Numerical approach that includes the method of expansion in the Fourier series with the subsequent application of the sweep method is implemented. The amplitude-frequency characteristics and the far field of the sound pressure are investigated.

Keywords: vibroacoustics, shell, composite, polymer, acoustic medium, discrete circular ribs, field of sound pressure

---

*E-mail: safron@math.sfedu.ru.

1. INTRODUCTION

Vibrations, arising in the shell structures in the process of their exploitation result in the increase of the sound radiation level, and they are regarded as hazardous factors. The problem of reducing vibration levels and improving vibroacoustic and dissipative characteristics of the shell structures is crucial for a wide range of applications in advanced technologies. This problem is successfully solved when using composite materials with high coefficients of internal dissipation of vibration energy. Such an approach requires further improvement of mathematical and computational models, which take into account specific properties of composite materials and complexity of the physical and mechanical characteristics of their components. Review of the methods for the solution of the vibroacoustic problems in mechanics of the thin-walled constructions is present in [1]. Let us note that in the solution of vibroacoustic problems it is often necessary to consider heterogeneity of constructions, caused by such factors as discrete circular ribs, irregularly distributed masses and the layered structure of the shell. For this kind of the problems numerical and analytical solutions appear to be more efficient [1].

In further research internal losses of mechnical energy in the shell and ribs are present by the complex moduli. The external reaction is modelled by the local impedance, and the acoustic pressure is calculated with the Helmholtz integral. The developed method also allows us to use the modal analysis, typical for axially symmetric constructions.

When analyzing the sound characteristics the levels of the sound pressure module for the external field $|p|$, remoted from the construction, are taken as the criteria of their acoustic performance. It is important to note that in this case the amplitude-frequency characteristics are plotted in the defined frequency domain. The problem is solved for every oscillation frequency. The solution consists of the two main stages: (i) finding the distribution of the sound pressure and the displacement amplitudes at the boundary of the shell and the acoustic medium; (ii) calculating the pressure in the acoustic medium. It is a well-known fact that such characteristics as the anisotropy of physical and mechanical properties, increased pliability to transverse shear and high nonlinear dependence of thermoviscoelastic properties of the polymer binding on the frequency, the temperature and the load are typical for the polymer composites.

The method of mathematical modeling and the experiment are the main methods of investigation, efficiently and economically feasible in the design of advanced thin-walled structures.

2. METHODS

To study the propagation process of the stationary oscillations in the composite and polymer shell of rotation with the multilayered structure, we used the Timoshenko type theory. The kinematic and deformation equations take the form:

$$U = u + z\varphi_1, \; V = v + z\varphi_2, \; W = w; \; E_{11} = u' + k_1 w, \; E_{22} = v^{\bullet} + \psi u + k_2 w,$$

$$E_{12} = u' + u^{\bullet} - \psi v; \; K_{11} = \varepsilon_1 \varphi_1', \; K_{22} = \varepsilon_1 (\varphi_2^{\bullet} + \psi \varphi_1), \; K_{12} = \varepsilon_1 (\varphi_2' - \psi \varphi_2 + \varphi_1^{\bullet});$$

$$E_{13} = \varphi_1 - \vartheta_1, \ E_{23} = \varphi_2 - \vartheta_2; \ \vartheta_1 = -w' + k_1 u, \ \vartheta_2 = -w' + k_2 u, \tag{1}$$

where

$$\varepsilon_1 = h_*/R_*, \ \psi = A_2'/A_2. \ (\ldots)' = (\ldots)_{,\alpha 1}/A_1 \ (\ldots)' = (\ldots)_{,\alpha 2}/A_2.$$

To represent the dependence between the amplitudes of the generalized forces and the moments with the corresponding amplitudes of deformation, the stiffness matrix of a layered composite is used.

The basic functions describing the natural boundary conditions are as follows:

$$y_1 = S, \ y_2 = M_{11}, \ y_3 = T_{11}, \ y_4 = Q_{11}, \ y_5 = H, \ y_6 = v, \ y_7 = \varphi_1, \ y_8 = u, \ y_9 = w, \ y_{10} = \varphi_2.$$

For the harmonic oscillation in the dimensionless form, the basic system of equations takes the form:

$$y_1' = -2\psi y_1 - T_{22}^{\bullet} - k_2 Q_{22} - \Omega^2 (by_6 + \varepsilon_1 cy_{10}) - q_2,$$

$$y_2' = -\psi(y_2 - M_{22}) - y_5^{\bullet} + y_4/\varepsilon_1 - \Omega^2 (cy_8 + \varepsilon_1 dy_7),$$

$$y_3' = -\psi(y_3 - T_{22}) - y_1^{\bullet} - k_1 y_4 - \Omega^2 (by_8 + \varepsilon_1 cy_7) - q_1,$$

$$y_4' = -\psi y_4 - Q_{22}^{\bullet} + k_1 y_3 + k_2 T_{22} - \Omega^2 by_9 - q_3,$$

$$y_5' = -2\psi y_5 - M_{22}^{\bullet} + Q_{22}/\varepsilon_1 - \Omega^2 (cv + \varepsilon_1 dy_{10}), \tag{2}$$

$$y_6' = E_{12} - y_8^{\bullet} + \psi y_6, \ y_7' = K_{11}/\varepsilon_1,$$

$$y_8' = E_{11} - k_1 w, \ y_9' = E_{13} - y_7 + k_1 y_8,$$

$$y_{10}' = K_{12} \varepsilon_1 + \psi y_{10} - y_7^{\bullet},$$

where

$$\vartheta_2 = -y_9^{\bullet} + k_2 y_6; \ a_{33} = B_{33} D_{33} - A_{33}^2,$$

$$E_{12} = (D_{33} y_1 - A_{33} y_5)/a_{33}, \ K_{12} = (B_{33} y_5 - A_{33} y_1)/(2a_{33}),$$

$$E_{22} = y_5^{\bullet} + \psi y_8 + k_2 y_9, \ K_{22} = y_{10}^{\bullet} + \psi y_7; \ E_{13} = y_4/G_{13},$$

$$Q_{22} = G_{23}(y_{10} + y_9^{\bullet} - k_2 y_6); \ a_{11} = B_{11} D_{11} - A_{11}^2,$$

$$a_1 = y_3 - B_{12} E_{22} - A_{12} K_{22}, \ a_2 = y_2 - A_{12} E_{22} - D_{12} K_{22}, \tag{3}$$

$$E_{11} = (D_{11} a_1 - A_{11} a_2)/a_{11}, \ K_{11} = (B_{11} a_2 - A_{11} a_1)/a_{11};$$

$$T_{22} = B_{12}E_{11} + B_{22}E_{22} + A_{12}K_{11} + A_{22}K_{22},$$

$$M_{22} = A_{12}E_{11} + A_{22}E_{22} + D_{12}K_{11} + D_{22}K_{22}.$$

The stress and strain components are present by the Fourier series along the circumferential coordinate α_2. The obtained quasi-one-dimensional system of the normal type is solved with the orthogonal sweep method. When it is necessary to take into account the sound radiation in the surrounding acoustic medium, one can use the approach developed for the axial symmetric constructions.

To complete relations (1) – (3) it is necessary to determine the dynamic response of the liquid in which the shell is immersed. This can be done with the help of the local impedance modeling method [2, 3]. Having defined the displacement and the pressure fields on the surface of the shell, one can define the field of the dynamic pressure in the acoustic medium with the Helmholtz integral:

$$p(\vec{r}) = \frac{1}{4\pi} \int_S [p(\vec{r}_1) \frac{\partial}{\partial n} (e^{ikR_1} / R_1) - \frac{\partial p(\vec{r}_1)}{\partial n} (e^{ikR_1} / R_1)] dS \tag{4}$$

where S is the middle surface of the shell, k_1 is the wave number, $\vec{r}, \vec{r}_1$ are the radius vectors of points in the liquid and on the shell, $R_1 = |\vec{r}_1 - \vec{r}|$.

Let us consider a circular rib made of a polymeric composite material. The kinematic equations of the Timoshenko type model take into account the deformations of transverse displacement and rotation inertia in the terms of [4] and can be represented as

$$U_k = u_k(\alpha_2) + z\varphi(\alpha_2), \ V_k = v_k(\alpha_2) + z\psi_2(\alpha_2) + \alpha_1\psi_1(\alpha_2), \ W_k = w_k - \beta_1\varphi(\alpha_2). \tag{5}$$

As the typical dimensions of the rib cross-section are much less than its radius, we obtain the following equation for the rib oscillation:

$$Q_{1k}^{\circ} + \omega^2(b_k u_k + c\varphi) + t_k = 0,$$

$$T_k^{\circ} + kQ_{3k} + \omega^2(b_k v_k + c_3\psi_1 + c_1\psi_2) + s_k = 0,$$

$$Q_{3k}^{\circ} - kT_k + \omega^2(b_k w - c_3\varphi) + q_k = 0,$$

$$H_k^{\circ} + kM_{3k} + \omega^2(d_k\varphi + c_1 u - c_3 w) + m = 0, \tag{6}$$

$$M_{3k}^{\circ} - kH_k - Q_{1k} + \omega^2(d_3\psi_1 + d_{13}\psi_2 + c_3 v_k) + m_3 = 0,$$

$$M_{1k}^{\circ} - Q_{3k} + \omega^2(d_1\psi_2 + d_{13}\psi_1 + c_1 v_k) + m_1 = 0,$$

$(\dots)^\circ = (\dots)_{,\alpha 2}/r_k$.

Based on the equations obtained, we formulate conditions for the conjugated segments of the shell, located between the adjacent ribs. External loads on the rib include the influence from the adjacent segments of the shell. The models and methods presented above allow us to calculate the mode and the summary amplitude, i.e., frequency characteristics, and to perform a convergence analysis of solutions.

3. NUMERICAL EXPERIMENT

As an example, let us consider the problem of defining vibroacoustic characteristics of a polymeric composite cylindrical shell, reinforced by a regular set of the circular ribs. The constitutive equations of polymeric filler correspond to the model of a thermoviscoelastic body [5]. The shell is immersed in the acoustic medium (compressed ideal fluid). It is assumed that the wall of the shell consists of 25 layers. Thirteen of them are made of steel $E = 2.1 \cdot 10^{11}$ Pa, $v = 0.3$ and the other twelve are made of the binding polymer with the properties described in [6]. In the notation used in [4], the basic geometric characteristics of the construction in the dimensionless form are as follows:

$r_0 = 0.8$, $r_1 = r_c = 1.375$, $H_k = 2.85$, $L_c = 3.5$, $h_{arm} = 0.08$, $h_{cr} = 0.015$.

Geometric and mechanical characteristics of the ribs and stiffness correspond to those in [4]. Thirty circular ribs have been installed on the shell. The dimensionless parameters of the acoustic medium are $\rho_l = 0.128$, $c_l = 0.28$. Oscillations are induced by the local normal harmonic load, applied to the middle of the shell. The load amplitude is equivalent to the unit dimensionless load $Q = 1$.

Calculations are made in a cyclic way for the dimensionless frequency parameter Ω. Amplitude-frequency characteristics of the normal displacement are plotted and maximum levels of acoustic pressure $|p|$ are identified. Amplitude-frequency characteristics of the field are also plotted based on the sum of the circular modes, as well as on the single mode, taken separately. To transfer levels $|p|$ in the decibels, normalization with respect to the audibility threshold level $p_1 = 2 \cdot 10^{-5}$ Pa has been carried out.

The components of the complex compliance under shear I', I'' in the notation of [6], take the form:

$$I'(\omega) = \int_{r_1}^{r_2} \frac{C(r)}{G(r,v,T)} \left[1 - H(r) \frac{\omega^2}{[\varphi(r,v,T)]^2 + \omega^2} \right] dr , \qquad (7)$$

$$I''(\omega) = \int_{r_1}^{r_2} \frac{C(r)H(r)}{G(r,u,T)} \frac{\varphi(r,v,T)\omega}{[\varphi(r,v,T)]^2 + \omega^2} dr .$$

Such a model excludes bulk relaxation. The boundary conditions correspond to the clamped edges.

To determine the response of a liquid to the oscillations of the shell the impedance of an infinitely long cylindrical shell is used. Figure 1 shows the results of calculations for the mode and summary values of the amplitude-frequency characteristics for the normal displacement in the center of the loading panel.

Figure 2 represents the results for the field of acoustic pressure in the liquid at the distance, equal to 50 radiuses of the shell from the axis of its rotation, passing through the center of the loading panel.

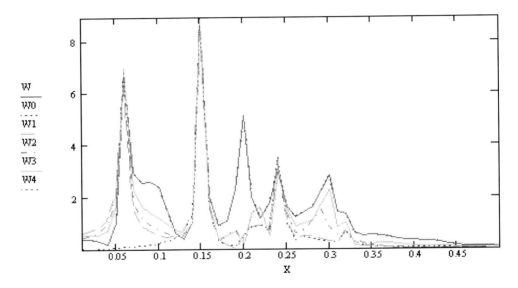

Figure 1. Amplitude-frequency characteristics for the normal displacement.

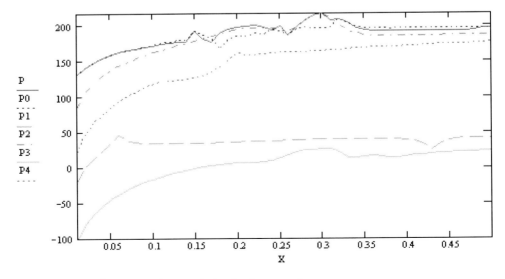

Figure 2. Amplitude-frequency characteristics for the acoustic pressure.

CONCLUSION

Mathematical and computer models of vibroacoustics for multilayered composite polymeric shells of rotation, reinforced by discrete circular ribs under the influence of vibration loadings have been developed. In particular, numerical and analytical approach based on the Fourier-series method, as well as the sweep method, was used. To define the reaction of the surrounding acoustic medium the method of local impedance modeling has been applied. The mode and the summary amplitude-frequency characteristics of the shell, as well as the field of pressure in acoustic medium, have been determined.

REFERENCES

[1]   Vorovich, I. I.; Yudin, A. S.; Safronenko, V. G. *Composite Material Constructions*, 2000, No. 2, 7 – 18 (in Russian).

[2]   Chertock, G. *J. Acoust. Soc. Amer.*, 1964. vol. 36(7), 1305.

[3]   Yudin, A. S.; Ambalova, N. M. *J. Appl. Mech.*, 1989, vol. 25(12), 63 - 71.

[4]   Safronenko, V. G.; Yudin, A. S. Modern Problems of Continuum Mechanics. Proceedings of VI International Conference, Rostov-on-Don, 1998. vol. 2, 154 - 159 (in Russian).

[5]   Ilyushin, A. A.; Pobedrya, B. E. Fundamentals of Thermoviscoelasticity Mathematical Theory. *Nauka,* Moscow, 1970 (in Russian).

[6]   Stepanenko, Yu. P.; Isaev, K. V.; Azarov, A. D. Modern Problems of Continuum Mechanics. Proceedings of II International Conference, Rostov-on-Don, 1996, vol. 1. 118-123 (in Russian).

In: Proceedings of the 2015 International Conference ... ISBN: 978-1-63484-577-9
Editors: Ivan A. Parinov, Shun-Hsyung Chang et al. © 2016 Nova Science Publishers, Inc.

Chapter 59

FORCED OSCILLATIONS OF SHELLS WITH AUXETIC PROPERTIES

A. S. Yudin[*] and S. A. Yudin

Vorovich Institute of Mathematics, Mechanics and Computer Science,
Southern Federal University, Rostov-on-Don, Russia

ABSTRACT

Auxetics are materials which have negative Poisson's ratio; i.e., when stretched, they expand in the direction perpendicular to the applied force. Several studies have revealed an unusual structural behavior of auxetics and their advantages over conventional materials for many applications [1, 2]. This paper deals with forced damped harmonic oscillations of shells made of auxetic materials.

Keywords: auxetics, force oscillations, amplitude-frequency characteristics, vibration damping, influence of the Poisson's ratio

1. INTRODUCTION

The solution to forced harmonic oscillations of cylindrical shells constructed using complex amplitudes and semi-analytic method. For the case of simply supported edges two-dimensional Fourier series can be employed to solve the problem by finding the displacement amplitudes. This ensures the necessary separation of the equations needed to determine the coefficients of Fourier series for different harmonics. The proposed method allows us to build quickly amplitude-frequency characteristics taking into account losses in the material.

---

[*] yudin@math.sfedu.ru.

2. EQUATIONS AND METHOD

Let us consider a circular cylindrical shell reinforced with a grid of longitudinal and circumferential ribs. R denotes the middle surface radius of the shell, while x and φ represent, respectively, the longitudinal and circumferential cylindrical coordinates. Fourier series can be used to decompose the circumferential coordinate φ. The Lamé parameters are given by

$$A_1 = A_2 = R.$$

In order to solve the problem of forced damped harmonic oscillations for this cylindrical shell, we first need to non-dimensionalize all variables using dimensional quantities E_*, v_*, R_*, h_*, ρ_*. E_* is the Young's modulus, v_* is the Poisson's ratio, R_* is the radius of curvature or the linear size, h_* and ρ_* are the thickness and material density, respectively.

Let us introduce the following notations

$$\varepsilon_* = h_*/R_* \, , \ \overline{v}_* = 1 - v_*^2 \, , \ B_* = E_* h_*/\overline{v}_* \, , \ A_* = B_* h_* \, , \ D_* = A_* h_* \, ,$$

$$c_* = [\, E_*/(\rho_* \overline{v}_*)\,]^{1/2} \, , \ F_* = R_* h_*/\overline{v}_* \, , \ J_* = F_* h_*^2 \, , \ S_* = F_* h_* \, . \tag{1}$$

Non-dimensional frequency associated with the circular frequency in Hertz is defined as

$$\Omega = \omega R_*/c_* \, , \tag{2}$$

where c_* is the characteristic sound velocity introduced in (1).

Relations between non-dimensional variables of the problem (brackets with index "N") and their dimensional analogues (brackets with index "D") are given by

$$\{ u, v, w, h_{(\lambda)}, z_{(\lambda)}, \delta, z_{os}, z_{or}, z_s, z_r \}_N = \{...\}_D / h_* \, , \ \lambda = 1, ..., \Lambda;$$

$$\{ \delta_x, \delta_y \}_N = \{...\}_D \overline{v}_* / h_* \, , \ \{ A_1, A_2, l_s, l_r, L, R \}_N = \{...\}_D / R_* \, ,$$

$$\{ k_1, k_2, K_{11}, K_{22}, K_{12} \}_N = \{...\}_D R_* \, ,$$

$$\{ \vartheta_1, \vartheta_2, \varepsilon_{11}, \varepsilon_{22}, \varepsilon_{12}, E_{11}, E_{22}, E_{12} \}_N = \{...\}_D / \varepsilon_* \, ,$$

$$\{ M_{11}, M_{22}, H \}_N = \{...\}_D R_* / D_* \, ,$$

$$\{ T_{11}, T_{22}, S, Q_{11}, Q_{22} \}_N = \{...\}_D R_* / A_* \, ,$$

$$\{ q_1, q_2, q_3 \}_N = \{...\}_D R_*^2 / A_* \, ,$$

$$\{ B_{jl}, B_{j(\lambda)}, B_{pj} \cdot \overline{G}_{(\lambda)} \}_N = \{...\}_D / B_* \, ,$$

$$\{ A_j \}_N = \{...\}_D / A_* \, , \ \{ D_j \}_N = \{...\}_D / D_* \, , \ j = 1, 2, 3;$$

$$\{ E_{k(\lambda)}, G_{k(\lambda)}, E_{pk}, G_{pk} \}_N = \{...\}_D / E_* \, ,$$

$$\{ J_k, I_k, I_k^o \}_N = \{...\}_D / J_* \, ,$$

$$\{ F_k \}_N = \{...\}_D / F_* \, , \ k = s, r \, . \tag{3}$$

Forced Oscillations of Shells with Auxetic Properties 449

Here, $E_* = E$, $v_* = v$, $R_* = R$, $h_* = h$ are the main scaling parameter.

The Fourier series decomposition of the components describing the stress-strain state (SSS) of the shell is performed using cosine waveforms $\cos(n)\varphi$ as basis functions for u, v, θ_1, E_{11}, E_{22}, K_{11}, K_{22}, T_{11}, T_{22}, M_{11}, M_{22}, Q_{11} and sine basis functions $\sin(n)\varphi$ for v, θ_2, E_{12}, K_{12}, S, H, Q_{22}. This decomposition corresponds to the load that is symmetrically applied with respect to the origin of the circumferential coordinate.

Using the complex amplitudes for the n-th circumferential harmonic, we get the equations of the shell oscillation with frequency Ω

$$E_{11n} = u_n', \; E_{22n} = nv_n + w, \; E_{12n} = v_n' - nu_n, \; \theta_{1n} = -w_n', \; \theta_{2n} = nw_n + v_n,$$
$$K_{11n} = \varepsilon_* \theta_{1n}', \; K_{22n} = \varepsilon_* n\theta_{2n}, \; K_{12n} = \varepsilon_*(\theta_{2n}' - n\theta_{1n})/2; \tag{4}$$

$$T_{11n}' + nS_n + \rho_1 \Omega^2 u_n + q_{1n} = 0, \; S_n' + Q_{22n} - nT_{22n} + \rho_1 \Omega^2 v_n + q_{2n} = 0,$$
$$Q_{11n}' + nQ_{22n} - T_{22n} + \rho_1 \Omega^2 w_n + q_{3n} = 0,$$
$$M_{11n}' + nH_n - Q_{11n}/\varepsilon_* = 0, \; H_n' - nM_{22n} - Q_{22n}/\varepsilon_* = 0; \tag{5}$$

$$T_{11n} = B_{11}E_{11n} + B_{12}E_{22n} + A_{11}K_{11n} + A_{12}K_{22n},$$
$$T_{22n} = B_{12}E_{11n} + B_{22}E_{22n} + A_{12}K_{11n} + A_{22}K_{22n},$$
$$M_{11n} = A_{11}E_{11n} + A_{12}E_{22n} + D_{11}K_{11n} + D_{12}K_{22n},$$
$$M_{22n} = A_{12}E_{11n} + A_{22}E_{22n} + D_{12}K_{11n} + D_{22}K_{22n},$$
$$S_n = B_{33}E_{12n} + A_{33} \cdot 2K_{12n}, \; H_n = A_{33}E_{12n} + D_{33} \cdot 2K_{12n}. \tag{6}$$

Let us construct a simple analytical solution to this problem for the case when the shell has stress-free bearing edges (the Navier slip boundary conditions): $v = 0$, $M_{11} = 0$, $T_{11} = 0$, $w = 0$. The following forms of natural vibrations satisfy the homogeneous equations (4) – (6) ($q_{in} = 0$)

$$u_n(x) = u_{nk}\cos(mx), \; v_n(x) = v_{nk}\sin(mx), \; w_n(x) = w_{nk}\sin(mx), \tag{7}$$

where $m/(\pi L)$ and k is the number of longitudinal half waves.

To solve the problem of forced vibrations we need to represent displacements in the form of natural frequencies (7), i.e., expand them as Fourier series in the longitudinal coordinate x. This ensures the necessary separation of the equations (4) – (6) needed to determine the coefficients of Fourier series for different harmonics.

Let us assume that the load is symmetric in the diametrical plane. Then functions of the solution can be decomposed to both sine and cosine Fourier series in the circumferential and longitudinal directions over the ranges $(0, 2\pi)$ and $(0, L)$, respectively. There are no restrictions imposed on the load coordinate along the length of the shell. Here the decomposition of even and odd functions is performed over the range $(-L, L)$ which is formally regarded as a continuation of the solution to the range $(-L, 0)$. We are interested only in the solution over the range $(0, L)$. That is why n denotes the number of waves in the circumferential direction, while k represents the number of half waves along the generatrix of the shell.

Let us consider the load acting normal to the shell and equivalent to the concentrated force Q_3. Let l to be the length of the loading platform along the longitudinal coordinate and

δ_1 to be the segment length of its middle line (as for a trapezoid), so that $S = l\delta_1$ is the area of load application region. Next, 2δ is used to denote the spanning angle in the circumferential direction and (x_1, φ_1) are the coordinates of the platform's center. Non-dimensional values of introduced parameters can be derived as follows

$$\{l, \delta_1\}_N = \{l, \delta_1\}_D / R_*; \{Q_3\}_N = \{Q_3\}_D (1 - v_*^2)/(E_* h_*^2);$$
$$\{q_i\}_N = \{q_i\}_D (1 - v_*^2) R_*^2 /(E_* h_*^2). \tag{8}$$

Therefore, the concentrated force (in its non-dimensional form) "spreads" on this platform and transforms to the distributed load of intensity $q_3 = Q_3/(l\delta_1)$. Taking into account that the shell surface is determined with curvilinear coordinates $\{x \in (0, L); \varphi \in (0, 2\pi)\}$, this distributed load can be expressed in terms of the Heaviside step function H(z), which is equal to 0 when $z < 0$ and to 1 if $z \geq 0$.

Hence, in the region of interest $\{x(\in 0, L) \in \varphi; (0, 2\pi)\}$ the load intensity $q_3(x, \varphi) = q_{3n}\Phi_1(x)\Phi_2(\varphi)$, where

$$\Phi_1(x) = [H(x - (x_1 - l/2)) - H(x - (x_1 + l/2))],$$

$$\Phi_2(\varphi) = [H(\varphi - (\varphi_1 - \delta)) - H(\varphi - (\varphi_1 + \delta))]. \tag{9}$$

$\Phi_2(\varphi)$ is an even function with the following Fourier series decomposition

$$\Phi_2(\varphi) = \delta/\pi + \sum_{k=1}^{N}[2/(k\pi)] \sin(k\delta)\cos(k\varphi_1)\cos(k\varphi). \tag{10}$$

By extending the scope of the problem from $[0, L]$ to $[-L, L]$, $\Phi_1(x)$ can be also expressed as periodic function

$$\Phi_1(x) = \sum_{k=1}^{N}[4/(k\pi)] \sin(ml/2)\sin(mx_1)\sin(mx). \tag{11}$$

Based on these results, it can be concluded that the normal load $q_3(x, \varphi)$ is approximated by double Fourier series $\sin(mx)\cos(k\varphi)$ with coefficients q_{3nk}, the form of which follows from (10), (11).

Now we are in the position to derive the displacements as functions of the cylindrical coordinates and frequency

$$u(x, \varphi, \Omega) = \sum_{k=1}^{M} u_{01}(\Omega)\cos(mx) + \sum_{k=1}^{M}\sum_{n=1}^{N} u_{nk}(\Omega)\cos(mx)\cos(n\varphi),$$

$$v(x, \varphi, \Omega) = \sum_{k=1}^{M}\sum_{n=1}^{N} v_{nk}(\Omega)\sin(mx)\sin(n\varphi),$$

$$w(x, \varphi, \Omega) = \sum_{k=1}^{M} w_{01}(\Omega)\sin(mx) + \sum_{k=1}^{M}\sum_{n=1}^{N} w_{nk}(\Omega)\sin(mx)\cos(n\varphi). \tag{12}$$

The coefficients of (9) can be found from the solution of the third order linear algebraic system

$$B(\Omega) \cdot u(\Omega) = q, \tag{13}$$

where

$$u(\Omega) = \begin{bmatrix} u_{nk}(\Omega) \\ v_{nk}(\Omega) \\ w_{nk}(\Omega) \end{bmatrix}; \ q = \begin{bmatrix} 0 \\ 0 \\ q_{3nk} \end{bmatrix}; \ B(\Omega) = A + \rho_l \Omega^2 E. \tag{14}$$

Here E is the identity matrix, A is the square matrix with components depending on the shell stiffness coefficients and wave parameters. Now, the solution of system (13) can be obtained in the analytical form. The displacements (12) expressed as functions of the cylindrical coordinates and forcing frequency allows us to construct effectively the amplitude frequency characteristic (AFC) of the dynamic compliance at points of interest.

3. CALCULATIONS

As an example, let us consider forced oscillations of the structural anisotropic shells with non-dimensional length $L = \pi$. The loss ratio is $\eta = 0.03$ and the non-dimensional harmonic force applied in the middle of the lateral surface is equal to 0.001.

Poisson's ratio ν is varied from 0.3 for the initial shell down to -0.3. When $\nu = 0.2$, the absolute value amplitude of the input compliance (displacement under the force) has decreased by 8%. When $\nu = 0.1$ this drop was of about 16%, and, finally, when $\nu = 0$ we have observed 19% reduction in the amplitude of the input compliance. Thus, for materials with a positive Poisson's ratio ν, the vibration levels are lower for smaller values of ν. However, if $\nu < 0$ these vibration levels are higher so that the absolute value amplitude is increased by 23-24% demonstrating attractiveness of materials with negative Poisson's ratio for applications in vibration damping.

Figure 1 shows AFC of the input compliance amplitudes plotted versus frequency Ω at fixed longitudinal half-wave for the initial shell with $\nu = 0.3$ (solid line) and the shell with $\nu = -0.3$ (dash-dotted line). The resonant peaks of oscillation mode shapes correspond to the values of circumferential harmonics $3, 4, 2$ and 5, respectively.

Summing up the observations, it can be concluded that Poisson's ratio has the pronounced effect on the shear resonance frequencies as well as the wave amplitude.

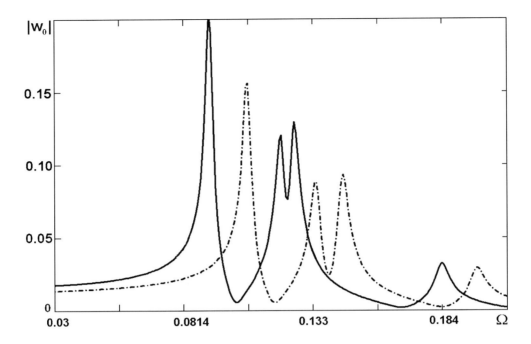

Figure 1. AFC of the input compliance amplitudes for the initial shell with $\nu = 0.3$ (solid line) and shell with $\nu = -0.3$ (dash-dotted line).

ACKNOWLEDGMENTS

This research was supported by the Russian Government article under theme No.213.01-11/2014-29.

REFERENCES

[1] Ken, Evans. *Chem. and Ind. (London)*. 1990. *No. 20*, 654 – 657.
[2] Phan-Thien, N.; Karihaloo, B. L. *Trans. ASME. J. Appl. Mech.* 1994, *vol. 61(4)*, 1001 – 1004.

In: Proceedings of the 2015 International Conference ...   ISBN: 978-1-63484-577-9
Editors: Ivan A. Parinov, Shun-Hsyung Chang et al.   © 2016 Nova Science Publishers, Inc.

Chapter 60

OSCILLATIONS OF COMPRESSIBLE LIQUID FREE SURFACE GENERATED BY PLATE

G. N. Trepacheva[*]

Vorovich Institute of Mathematics, Mechanics and Computer Sciences,
Southern Federal University, Rostov-on-Don, Russia

ABSTRACT

We have solved the mixed boundary-value problem of spatial harmonic surface waves excitation in compressible liquid, placed in a wave tank of finite sizes. The movement of compressible liquid is caused by oscillations of vertical rectangular plate with a given shape. In absence of surface tension, the obtained solution is enclosed. The influence of small energy dissipation is studied.

Keywords: compressible liquid, wave, wave tank, oscillations, ideal liquid, vertical rectangular plate

1. INTRODUCTION

The flat small oscillations of ideal compressible liquid, caused on a free liquid surface of finite depth by oscillations of vertical plate, were studied in [1]. Waves generated by the bending oscillations of infinite strip in the incompressible liquid having finite depth are examined in [2]. The establishment of flat progressive waves under the harmonic oscillations of vertical wall in a deep ideal incompressible liquid was analyzed in [3].

In [1] it is pointed out that assumptions chosen by the authors make it impossible to carry out the limiting transition to the case, when the reflecting wall is infinitely far away from the wave generating wall. Theory proposed in this work allows one to consider similar limiting transition. It corresponds to experimental data concerning progressive waves [4]. Here we

---

[*] E-mail: gal.trep@yan`dex.ru.

454 G. N. Trepacheva

summarize the theory of progressive flat waves, caused by the plate of a given shape [5], in incompressible liquid for the case of spatial waves in compressible environment.

2. METHOD

2.1. The Problem Statement

The problem statement of wave motion, caused by the vertical plate in a heavy compressible liquid, is the following:

$$\Phi_{tt} + \mu\Phi_t = c_0^2 \Delta\Phi, \ \Phi_{tt} + \mu\Phi_t + g\Phi_z + \frac{\alpha}{\rho_0}\Phi_{zzz} = 0, \ z = h;$$

$$\Phi_z = 0, \ z = 0; \ \Phi_x = \frac{\partial u}{\partial t}, \ x = 0, \ |y| < b, \ 0 < z < h;$$

$$\Phi_y = 0, \ y = \pm b, \ 0 < z < h;$$

$$\Phi_x = 0, \ x = a, \ |y| < b, \ 0 < z < h. \tag{1}$$

In (1), c_0 is the speed of sound in the liquid, Φ is the potential of speed, t is the time, x, y are the horizontal coordinates, z is the vertical coordinate, μ is a Rayleigh coefficient of energy dissipation, α is a coefficient of surface tension, ρ_0 is the density of liquid; $u = u(y, z, t)$ is the horizontal velocity of movement of the plate, placed on a vertical wall of the squared pool ($2b$, h and a are the pool width, depth and length, respectively. On sidewalls $(y \pm b)$, on the back wall $(x = a)$ and at the bottom $(z = 0)$ of the pool, impermeability conditions satisfy. On the forward wall of the pool $(x = 0)$ there is a vertical plate. The origin of a Cartesian coordinate system places on the intersection line of pool bottom and its forward wall. Values of pressure P in the liquid and surface elevation ξ are given as

$$P = -\rho_0\left(\Phi_t + \mu\Phi + gz\right), \ 0 < z < h;$$

$$\frac{\partial \xi}{\partial t} = \Phi_z, \ z = h, \tag{2}$$

where g is the acceleration of gravity. The plate makes harmonic oscillations with circular frequency ω:

$$\Phi = \phi e^{i\omega t}, \ \xi = \eta e^{i\omega t}, \ u = u_0 e^{i\omega t}. \tag{3}$$

Substitution of (3) into (1) gives the problem formulation relatively to complex amplitudes of oscillations ϕ, η, u_0, looking as follows:

$$\phi_{xx} + \phi_{yy} + \phi_{zz} + \frac{\omega^2 - i\mu\omega}{c_0^2}\phi = 0;$$

$$\left(\omega^2 - i\mu\omega\right)\phi - \left(g\phi_z + \frac{\alpha}{\rho}\phi_{zzz}\right) = 0, \ z = h;$$

$$\phi_z = 0, \ z = 0;$$

$$\phi_x = 0, \ x = a, \ |y| < b, \ 0 < z < h;$$

$$\phi_x = i\omega u_0, \ x = 0, \ |y| < b, \ 0 < z < h;$$

$$\phi_y = 0, \ y = \pm b. \tag{4}$$

The boundary-value problem (4) is non-uniform because there is a condition $x = 0$ on the plate.

2.2. Speed Potential

We express the speed potential of compressible liquid movement as a superposition of particular solutions of the problem, obtained from (4) for its uniform conditions. This problem can be described by the following double series:

$$\phi(x, y, z) = \sum_{m=0}^{\infty}\sum_{n=0}^{\infty} A_{mn} \cos\left(k_3(m)z\right)\cos\left(k_2(n)(b - y)\right)\times$$

$$\times\left[e^{-m_1(m,n)(2a-x)} + e^{-m_1(m,n)x}\right] \tag{5}$$

The system of functions $\cos k_2(n)(b - y) = \cos(n\pi/2b)(b - y)$ is orthogonal on the segment $-b \leq y \leq b$.

The dispersive equation for the wave number in the vertical direction $k_3(m)$ is obtained in the form:

$$\sin(k_3 h)\left(1 - \beta k_3^2\right) = -\frac{k_p}{k_3}\cos(k_3 h), \ \beta = \frac{\alpha}{\rho g}, \ k_p = \frac{\omega^2 - i\mu\omega}{g} \tag{6}$$

The wave number relating to x-axis, i.e., $m_1(m,n)$ equals to the following relationship:

$$m_1(m,n) = \sqrt{k_2^2(n) + k_3^2(m) - k_0^2}, \ k_0^2 = \frac{\omega^2 - i\mu\omega}{c_0^2}, \ \text{Re}(m_1) > 0, \mu > 0, \tag{7}$$

where k_0 is the wave number of a longitudinal elastic wave in the unbounded compressible liquid; $k_2(n)$ is the wave number for surface wave in the direction of y-axis.

Coefficients A_{mn} are unknown and according to (4) satisfy the condition on the plate:

$$\phi_x = i\omega u_0(z,y), \ x = 0, \ |y| < b \tag{8}$$

Let us multiply both sides of (8) by the product of functions $(2/bh)\cos(k_3(p)z)\cos(k_2(n)(b-y))$ and integrate (8) over the domain $0 < z \leq h$, $-b \leq y \leq b$. Then, let us use the notation:

$$\alpha_{pn} = \frac{2i\omega}{bh} \int_0^h \int_{-b}^b u_0(z,y)\cos(k_3(p)z)\cos(k_2(n)(y-b))dz\,dy,$$

$$p = 0, 1, 2, \dots, \ n = 0, 1, 2, \dots;$$

$$\varepsilon_0 = 2, \ \varepsilon_n = 1, \ n = 1, 2, \dots; \ \beta_{pm} = \frac{2}{h}\int_0^h \cos(k_3(m)z)\cos(k_3(n)z)dz;$$

$$\beta_{mm} = \left[1 + \frac{\sin 2k_3(m)h}{2k_3(m)h}\right] \tag{9}$$

After integration of the transformed formula (9), we obtain the following set of infinite algebraic equations systems, which are necessary to determine the unknown coefficients A_{mn} for (6):

$$\varepsilon_n \sum_{m=0}^{\infty} m_1(m,n)\left[e^{-2am_1(m,n)} - 1\right]\beta_{pm}A_{mn} = \alpha_{pn}; \ n = 0, 1, 2, \dots; \ p = 0, 1, 2, \dots, \tag{10}$$

where n is the number of infinite algebraic equations sets, α is the pool length; $\varepsilon_n, \beta_{pm}$ are the known coefficients computed by (9); α_{pn} are the right sides of infinite algebraic equations defined by (9); A_{mn} are the unknown coefficients of decomposition (5); m is the number of variable in n-th infinite algebraic equations set; p is the number of equation in n-th infinite algebraic equations set.

Now let us consider the construction of asymptotic solutions for each equations set (10) in the field of low frequencies $\omega^2/g \to 0$ and in the field of high frequencies $\omega^2/g \to \infty$ assuming that other parameters of the problem β, h, μ, a take some fixed values. It is necessary to suggest in asymptotical sense that $\beta_{pm} = 0$, if $p \neq m$. As a result, equations sets (10) reduce to diagonal form:

$$\varepsilon_n m_1(p, n)\left[e^{-2am_1(p,n)} - 1\right]\beta_{pp}A_{pn} = \alpha_{pn}, \ p = 0, 1, 2, \ldots; \ n = 0, 1, 2, \ldots \tag{11}$$

From (11) we find unknown coefficients of decomposition (5) looking as follows:

$$A_{mn} = -\frac{\alpha_{mn}}{m_1(m, n)\beta_{mm}\varepsilon_n\left[1 - e^{-2am_1(m,n)}\right]} \tag{12}$$

Coefficients $\beta_{mm} \neq 0$ in (12), and inequality $1 < \beta_{mm} < 2$ is true, if $\mu = 0, m \geq 1$.

Formulae (12) are asymptotical for the following three cases: (i) in the field of low frequencies, (ii) in the field of high frequencies, (iii) in the field of small values of surface tension coefficient (for small β). The regularity of infinite equations sets (10) is proved similarly to [5].

Formulae (12) are precise if there is no surface tension, i.e., for $\beta = 0$, since $\beta_{pm} = 0$, $p \neq m$, $\mu \geq 0$.

In the field of high frequencies, the first modes of free surface oscillations behave almost as well as in the case of impact loading of liquid. On the free surface, $\phi \approx 0$ holds. Physically it means that we neglect the influence of gravity.

The term $e^{-m_1(0,0)x}$, occurring in the solution (5), describes the surface wave with the flat front propagating from the plate and fading along the x-axis. The results obtained here imply as a special case the known results for the problem about a vertical wave maker on the plane [1–5]. The presence of energy dissipation in this work supplements known investigations, since it allows one to optimize the tank length at working off the problem of fading the waves that are reflected from a back wall of the tank.

REFERENCES

[1] Buffalo, V. N.; Buyvol, V. N. Fluctuations and Stability of Deformable Systems in Liquids Buffalo. *Naukova Dumka*, Kiev, 1975, pp. 38 – 58 (In Russian).

[2] Cherkesov, L. V. Unsteady Waves. *Naukova Dumka*, Kiev, 1970, pp. 26 – 32 (In Russian).

[3] Sretensky, L. N. Theory of Wave Movements of Liquid. *Science,* Moscow, 1977, pp. 320 – 335 (In Russian).

[4] Madsen, O. S. *J. Geophysical Rearch*, vol.76, No 36, 1971, pp. 8672 – 8683 (In Russian).

[5] Trepachev, V. V.; Trepacheva, G. N. In: Proceedings of 8th International Conference on Modern Problems of Mechanics of Solid Medium, CVVR, Rostov-on-Don, 2002, pp. 188–192 (In Russian).

V. Applications

In: Proceedings of the 2015 International Conference … ISBN: 978-1-63484-577-9
Editors: Ivan A. Parinov, Shun-Hsyung Chang et al. © 2016 Nova Science Publishers, Inc.

Chapter 61

DEVELOPMENT OF THE MODEL AND STRUCTURE OF THE GaAs P-I-N PHOTODETECTOR FOR INTEGRATED OPTICAL COMMUTATION SYSTEMS

I. Pisarenko and E. Ryndin[*]

Institute of Nanotechnologies,
Electronics and Electronic Equipment Engineering,
Southern Federal University, Taganrog, Russia

ABSTRACT

This chapter is devoted to the problem of the development of a photodetector that we can use in an integrated optical inter-core commutation system of a multi-core ultra-large-scale integrated circuit together with a high-speed A^3B^5 laser-modulator. The aim of this chapter is to present the development of a physical and topological model, modeling methods and a structure of a *GaAs p-i-n* photodetector designed for the operation as part of the on-chip optical commutation system together with the laser-modulator. The model, modeling methods and the software in Octave/MATLAB program proposed in this chapter have allowed us to research photocurrent relaxation processes in *GaAs p-i-n* photodetector and optimize reasonably the constructive and technological parameters of *p-i-n* structures for integrated optical commutation systems.

Keywords: multi-core ultra-large-scale integrated circuit, integrated optical commutation system, high-speed integrated photodetector, *p-i-n* photodiode, physical and topological model, fundamental system of equations of a semiconductor

1. INTRODUCTION

The improvement of integrated circuits (ICs) characteristics by the increase in the number of cores in multi-core systems is considered to be one of the priority directions of the ultra-

---

[*] E-mail: rynenator@gmail.com.

large-scale integrated circuits (ULSIs) development. Traditional metal inter-core connections do not meet modern requirements for the performance and noise immunity of ULSIs. The conception of the design of combined electronic and optical ICs with silicon digital cores and A^3B^5 optical commutation systems is one of possible solutions of this problem [1]. The construction method, models and structures of high-speed integrated injection lasers with double A^3B^5 heterostructures and functionally integrated radiation modulators have been developed and described in references [2, 3]. The modulation frequency of the laser radiation in similar structures can reach the value of 1 THz and above according to the results of the physical and topological modeling. The development of integrated photodetectors is considered to be one of the urgent problems of the implementation of optical interconnections in ULSIs. The basic requirements for the photodetector of an on-chip optical commutation system are technological compatibility with a laser-modulator and performance being sufficient for the adequate detection of modulated laser pulses. An integrated *p-i-n* photodiode is one of possible variations of photodetectors for optical commutation systems from the point of view of the compliance with the specified requirements. The aim of this chapter is to present the development of a model, modeling methods and a structure of a *GaAs p-i-n* photodetector designed for the operation as part of an on-chip optical commutation system together with a high-speed laser-modulator.

2. MODELS AND MODELING METHODS

In this chapter, we have implemented the approach to the research of transient processes in semiconductor photodetectors of integrated optical commutation systems with the use of physical and topological models based on the fundamental system of equations of a semiconductor in the diffusion-drift approximation. In general, this system of equations can be written as follows [4]:

$$\frac{\partial n}{\partial t} = \nabla\big[\mu_n\big(-n\nabla\varphi + \varphi_T\nabla n\big)\big] + R; \tag{1}$$

$$\frac{\partial p}{\partial t} = \nabla\big[\mu_p\big(p\nabla\varphi + \varphi_T\nabla p\big)\big] + R; \tag{2}$$

$$\nabla(\varepsilon \cdot \nabla\varphi) = \frac{q}{\varepsilon_0}(n - p - N), \tag{3}$$

where n, p are the concentrations of electrons and holes; μ_n, μ_p are the mobilities of electrons and holes; t is the time; q is the elementary charge; φ is the electrostatical potential; φ_T is the temperature potential; R is the change in the number of electrons and holes per volume unit and per time unit due to the difference of generation and recombination rates; ε is the dielectric constant of semiconductor; ε_0 is the dielectric constant of vacuum; N is the effective impurity concentration. We have used boundary conditions for a circuit with the current output in the following form [4]:

$$n = \frac{N}{2} + \sqrt{\left(\frac{N}{2}\right)^2 + n_i^2} \; ; \tag{4}$$

$$p = -\frac{N}{2} + \sqrt{\left(\frac{N}{2}\right)^2 + n_i^2} \; ; \tag{5}$$

$$\varphi = \varphi_T \ln \frac{n}{n_i} + U(t) \, , \tag{6}$$

where n_i is the intrinsic concentration of charge carriers; $U(t)$ is the bias voltage applied to the structure at time t.

We have calculated the initial conditions by solving the fundamental system of equations in the stationary form for photodetectors in the state of the thermodynamic equilibrium. For this purpose, we have used the Hummel numerical iterative method of the finite differences [4].

We have solved the non-stationary problem by means of the explicit difference scheme [5]. At first, we calculated the charge carriers' concentrations by solving Equations (1), (2). We transformed the right sides of these equations to the base of the exponents of quasi-Fermi levels and electrostatical potential. We used counter flow scheme for the discretization of continuity equations [6]. We inserted the computed distributions of concentrations in Equation (3) and used them for the electrostatical potential calculation.

We have used the Schockley-Read-Hall model and the model of the Auger recombination for the accounting of the charge carriers' generation and recombination [4]. Our model also takes into account the processes of the charge carriers' optical generation under the influence of laser radiation on photodetectors' structures. We have included the additional summand R_l calculated by the following expression in the value R in Equations (1, 2):

$$R_l = \frac{QP}{E_{ph}} \, , \tag{7}$$

where Q is a coefficient of the photons' absorption in the photodetector's cavity; P is the volume density of the radiation power; E_{ph} is the energy of photons.

We have calculated the charge carriers' mobilities with the use of the analytical model from reference [7]. This model takes into account the dependencies of mobility on concentrations of dopants and the temperature of semiconductor.

We have used the following expressions proposed by Slotbum for the computation of electron and hole current densities in our model [4]:

$$\vec{j}_n = q \varphi_T \mu_n \nabla F_n \exp(\varphi) \; ; \tag{8}$$

$$\vec{j}_p = -q \varphi_T \mu_p \nabla F_p \exp(-\varphi) \, , \tag{9}$$

where $\vec{j}_n, \vec{j}_p$ are the electron and hole current densities; F_n, F_p are the exponents of quasi-Fermi levels for electrons and holes.

Furthermore, we have taken into account the Maxwell displacement current $\vec{j}_d$ when calculating the total current in structure. The corresponding expression can be written as follows [8]:

$$\vec{j}_d = \varepsilon \varepsilon_0 \frac{\partial \vec{E}}{\partial t}, \qquad (10)$$

where $\vec{E}$ is a vector of electrostatical field intensity.

Thus, we have developed the original methodology of the modeling of charge carriers' transfer and accumulation processes in semiconductor photodetectors. This methodology combines the known methods of the solution of mathematical physics' tasks and the modeling of ICs' elements.

We have developed specialized applied software in the Octave/MATLAB program for the implementation of the proposed methodology. This software allows simulating semiconductor *p-i-n* structures with arbitrary electrophysical, constructive and technological parameters.

3. MODELING RESULTS

We have developed the structure of *GaAs p-i-n* photodiode for the approbation of the model, modeling methods and the software described above. This structure is shown in Figure 1, *a*. Mirrors *1* and *3* form optical resonator designed for the gain of photodetector's quantum efficiency, P^+ and n^+ regions ensure ohmic contacts to the device.

We have implemented the modeling for one-dimensional coordinate grid in the assumption that laser pulses with fixed volume power density have illuminated *p-i-n* structures evenly. Total incident radiation power was 1 mW.

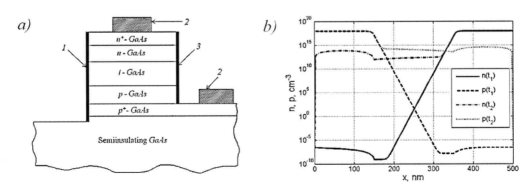

Figure 1. Structure of *GaAs p-i-n* photodiode (*a*): *1* – semi-reflective mirror; *2* – ohmic contacts; *3* – fully reflective mirror; the electron (*n*) and hole's (*p*) concentrations (*b*) before ($t_1 = 1.4$ ps) and after ($t_2 = 3.9$ ps) the leading edge of the optical pulse in one of researched structures.

Processes taking place during the passage of laser pulses through the optical waveguide of integrated optical commutation system have not been taken into account in this chapter. We have used non-stationary physical and topological model of *GaAs p-i-n* photodiode for the research of the performance dependence on constructive and technological parameters of device structure. In this chapter, we have estimated and compared the performance of different variations of *p-i-n* photodetectors in the way presented below.

We have set of the following operating mode for all researched structures:

(i)   the time interval of 7 ps has been considered;
(ii)   the supply voltage has been applied at the time moment of 0.1 ps;
(iii)   structures have been illuminated by single rectangle laser pulses with the duration of 2.5 ps after the end of the supply transition processes;
(iv)   the time moment of the light pulse leading edge was 1.5 ps.

We have compared the following characteristics of *p-i-n* structures with different constructive and technological parameters:

(i)   the maximal absolute value of photocurrent density j_{max} ;

(ii)   the maximal absolute values of current density derivatives over time $\left|\dfrac{dj}{dt}\right|_{max}$ for leading and back edges;

(iii)   the maximal absolute value of the integral for the leading edge

$$Int_f = \frac{1}{j_{max}\,\tau} \int_{t_1}^{t_2} jdt \tag{11}$$

where τ is the duration of laser pulse; t_1, t_2 are the time moments corresponding to leading and back edges of the laser pulse;

(iv)   the maximal absolute value of the integral for the back edge

$$Int_r = 1 - \frac{1}{j_{max}\,\tau} \int_{t_2}^{t_3} jdt \tag{12}$$

where t_2, t_3 are the time moments corresponding to the back edge of the laser pulse and the end of the researched time interval.

We have researched the dependencies of *GaAs p-i-n* photodiode performance on the following constructive and technological parameters:

(i)   the length of *i* region W_i;

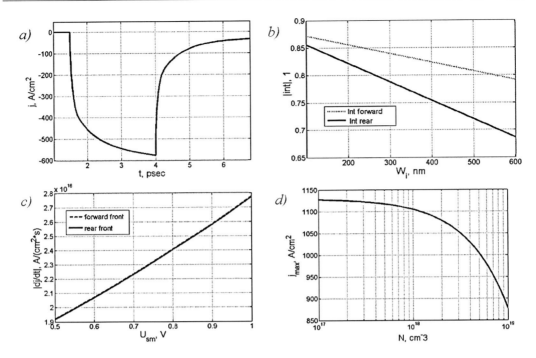

Figure 2. Modeling results: (a) general form of the photocurrent pulse; (b) dependence of integral characteristics on the i region's length; (c) dependence of $\left|\frac{dj}{dt}\right|_{max}$ on the bias voltage; (d) dependence of j_{max} on the concentration of dopants in p and n regions of symmetrical structures.

(ii)   the bias voltage U_{sm};
(iii) the dopants concentration N in p and n regions of symmetrical structures.

Figure 1, *b* and Figure 2 show some of the modeling results obtained by us in this chapter.

The analysis of the modeling results has allowed to evaluate the adequacy of the developed model, modeling methods and the software, to determine the *GaAs p-i-n* structure with the best performance and to compute its prospective characteristics.

CONCLUSION

Thus, the diffusion-drift physical and topological models can be used for the detailed research of photocurrent relaxation processes in semiconductor photodetectors. The model, modeling methods and the software proposed in this chapter have allowed us to optimize reasonably the constructive and technological parameters of *p-i-n* photodetectors' structures for integrated optical commutation systems. The analysis of the simulation results has enabled us to hypothesize that *GaAs p-i-n* photodiodes can be used in on-chip optical commutation systems together with high-speed lasers-modulators for the detection of picosecond laser pulses.

ACKNOWLEDGMENT

The results have been obtained with use of equipment of the Shared Equipment Center and the Research and Educational Center "Nanotechnology" of the Institute of Nanotechnologies, Electronics and Electronic Equipment Engineering, Southern Federal University (Taganrog).

This research is supported by the Russian Foundation for Basic Research (Grants 13-07-00274, 14-07-31234) and the Ministry of Education and Science of the Russian Federation (Project 8.797.2014K).

REFERENCES

[1] Konoplev, B. G.; Ryndin, E. A.; Denisenko, M. A. *News of SFedU. Engineering Sciences*, 2011, vol. 117, 21-27 (in Russian).

[2] Ryndin, E. A.; Denisenko, M. A. *Russian Microelectronics*, 2013, vol. 42, 360-362 (in Russian).

[3] Konoplev, B. G.; Ryndin, E. A.; Denisenko, M. A. *Technical Physics Letters*, 2013, vol. 39, 986-989.

[4] Abramov, I. I.; Haritonov, V. V.; Numerical Modeling of Integrated Circuits' Elements; Vyisheyshaya Shkola, Minsk, 1990 (in Russian).

[5] Samarskiy, A. A.; Gulin, A. V. Numerical Methods; *Nauka,* Moscow, 1989 (in Russian).

[6] Kulikova, I. V.; Lysenko, I. E.; Pristupchik, N. K.; Lysenko, A. S. *News of SFedU. Engineering Sciences*, 2014, vol. 158, 106-111 (in Russian).

[7] Mnacakanov, T. T.; Levinshtein, M. E.; Pomortseva, L. I.; Yurkov, S. N. *Physics and Technics of Semiconductors*, 2004, vol. 38, 56-60 (in Russian).

[8] Savelev, I. V. Course of General Physics, vol. 2: Electricity and Magnetism; Astrel, Moscow, 2005 (in Russian).

In: Proceedings of the 2015 International Conference …   ISBN: 978-1-63484-577-9
Editors: Ivan A. Parinov, Shun-Hsyung Chang et al.   © 2016 Nova Science Publishers, Inc.

Chapter 62

KNOCK SENSORS BASED ON LEAD-FREE PIEZOCERAMICS

Yu. I. Yurasov[1], A. V. Pavlenko[2,3], I. A. Verbenko[2], L. A. Reznichenko[2] and H. A. Sadykov[2,]*

[1]Don State Technical University, Rostov-on-Don, Russia
[2]Research Institute of Physics, Southern Federal University, Rostov-on-Don, Russia
[3]Southern Scientific Centre RAS, Rostov-on-Don, Russia

ABSTRACT

New lead-free piesoceramic materials and knock sensors on its base were produced. Knock sensor was tested in comparison with its foreign analogues having lead elements. It was proved that new knock sensors have not shortcomings of foreign analogues but even exceed them according to some criterion.

Keywords: knock sensor, piezoceramics, niobates sodium-kalium

INTRODUCTION

Sensors based on the ferroelectric ceramic materials (FCM) are widely used in the internal combustion engines (ICE). For instance, knock sensors (KS) have been widely used in the car industry for several decades. This experience allows one to burn fuel air mix more effectively and fully in the engine cylinders. KSs of resonance and wideband kinds are set in the engine body and in case of knocking oscillate a signal and its amplitude depends on the detonation level. Controller fixes the signal and controls the timing angle for knock preventing.

---

* Corresponding author: H. A. Sadykov. E-mail: yucomp@yandex.ru.

The literature review has proved that the basic components for most modern ferroelectric ceramic materials, used in sensor structures, are lead-containing solid solutions (SS) (PbTi$_{1-x}$Zr$_x$O$_3$ (PZT), $(1-x)$Pb(Nb$_{2/3}$Mg$_{1/3}$)O$_3$ – xPbTiO$_3$ (PMN – PT), etc.). [1], that contradicts the claims of ecologically-friendly industry [2]. Thus, the issue of engineering of new toxic-free FCM is burning. These materials would be able to substitute not only well-known FCM based on lead, but also to discover new characteristics.

Scientists of the Research Institute of Physics, Southern Federal University, have been engaged in producing of lead free FCM (Table 1) and its testing in KSs.

The research was done to estimate the characteristics of KS $\varepsilon/\varepsilon_0(T)$ and tg$\delta(T)$ FCM samples under the frequency, f, (103-105) Hz under the temperature $T = (20\text{-}400)°C$. The experiment results of one of the FCM are present in Figure 1.

It is obvious quite that FCM under research is ferrielectric with diffuse phase transition (PT), and the temperature (T_{C2}) of ferroelectric PT changes slightly depending on f, that proves its weak relaxor character [3, 4].

Table 1. Electrophysical characteristics of lead-free FCM

No.	$\varepsilon_{33}^T/\varepsilon_0$	tgδ	K_p	$|d_{31}|$, pQ/N	d_{33}, pQ/N	Q_m	V_1^E, km/s
1	252	0.007	0.15	12	45	1022	5.247
2	250	0.006	0.17	13	50	1122	5.347
3	304	0.008	0.25	21	80	441	5.275
4	314	0.007	0.30	27	88	611	5.236

$\varepsilon_{33}^T/\varepsilon_0$ is the relative dielectric permittivity of polarized samples, tgδ is the dielectric loss tangent, d_{ij} are the piezoelectric moduli; K_p is the electromechanical coupling coefficient of the vibration planar mode; Q_M is the mechanical Q-factor.

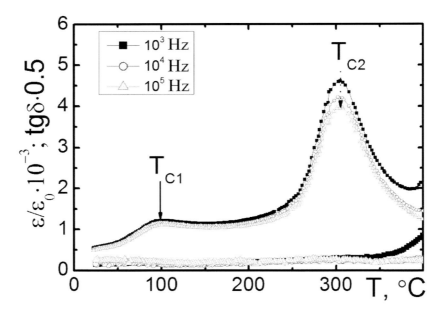

Figure 1. Characteristic curve $\varepsilon/\varepsilon_0(T)$ and tg $\delta(T)$ under $f = (10^3 - 10^5)$ Hz of one of the FCM.

The temperature of the first PT, T_{C1}, is in the operating temperature range of ICE, where the KS was fixed ($T = 90 - 120°C$). In fact, all the ferroelectric material characteristics in the field of PT are rather extreme, that is possibly one of the positive factors influencing KS efficiency while using lead-free piezoceramics [5]. Further comparative experiments of the work of KSs based on our FCM and KSs produced by "BOSH" were made [6, 7]. Sensors were set on ICE of VAZ 21214 ("NIVA"), and the signal was registered with the help of the program, indicating the work of the board computer and the recording oscillometer OWON PBS 5022S. Electrical polarity of the sensors while they were on was the same. The adjustment of the ignition angle in case of simultaneous measurements while two sensors working was switched off. The experiment results can be seen in Figures 2, 3.

Size and form comparison, registered under various engine rotational rate (Figure 2) and in the conditions of throttling (Figure 3), made it possible to conclude the efficiency of sensors. On the one hand, while comparing oscillograph traces under different engine rotational rate there was no correlation that was rather strange. On the other hand, taking into account the fact that we concentrated on the time range 25 ms it was rather difficult to state the number of knock changes.

Some changes may be knocks of crankshaft rolling. It is well-known that KS reacts during the throttling or if the fuel octan rating does not correspond to the type of engine. In this case, the board computer, reacting on the knock, changes the angle. The fact that the sensor reacted on the knock witnessed during the comparative experiment of the work of both sensors with the help of board computer.

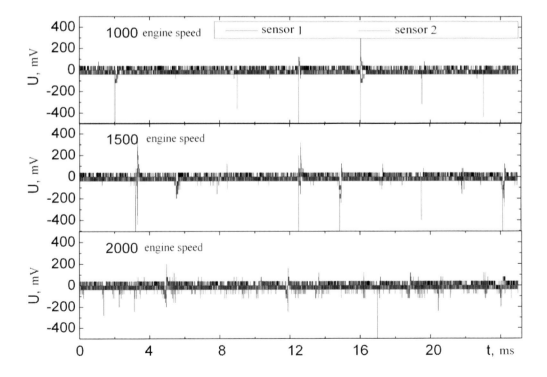

Figure 2. Oscillograph trace of sensors under fixed engine rotational rate.

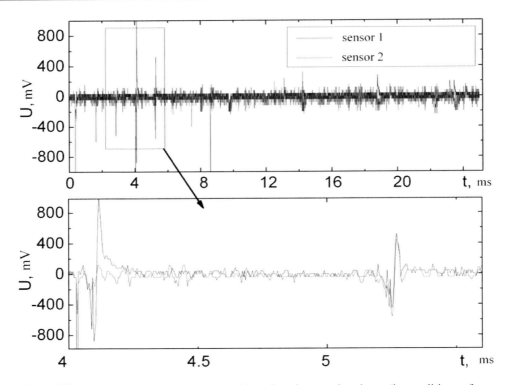

Figure 3. Oscillograph trace of sensors under accidental engine rotational rate (in conditions of throttling).

The results of the experiment showed that the knock happens, but not under every engine rotational rate, as we supposed, and signal level of the sensors coincides. While comparing KS produced by "BOSCH" and the one produced by the Research Institute of Physics, SFedU, with the help of diagnostic program, fixing the work of the automobile ignition control (Bosch M1.5.4), it was proved that with the usage of the last one the changes of the angle of ignition happen gradually and vary from 0.8 to 3.0.

To conclude, it should be noted that the research results, presented in the article, prove that our new FCM and based on it KS can be considered as rather rational substitution for Russian and foreign analogues and can be used in various fields, such as automobile construction, shipbuilding and aircraft construction. Among the advantages, we can mark its environmental friendliness, running efficiency that allows considering it as an import substitution of high quality.

Acknowledgments

This work was supported by the Ministry of Education and Science of the RF (Federal program: State contract No. 14.575.21.0007; Projects Nos. 213.01-11/2014-21, 3.1246.2014/K, 213.01-2014/012).

REFERENCES

[1] Klusacek, S., Fialka, J. Experimental methods for testing active element of piezoelectric knock sensors by using temperature dependences. *20th International Congress on Sound and Vibration (ICSV20)*, 2013, 1-7.

[2] Directive 2002/95/EC of the European Parliament and of the Council of 27 January 2003 on the restriction of the use of certain hazardous substances in electronic equipment. *Official Journal of the European Union*. 2003, No. 37, 19-23.

[3] Andryushina, I. N., Andryushin, K. P., Razumovskaya, O. N., Shilkina, L. A., Reznichenko, L. A., Yurasov, Yu. I. *Bulletin of the Russian Academy of Sciences: Physics*. 2010, vol. 74, 1127-1129.

[4] Shilkina, L. A., Kupriyanov, M. F., Fesenko, E. G., Reznichenko, L. A. *Sov. Phys. Tech. Phys.* 1977, vol. 22, 10, 1262-1265.

[5] Reznichenko, L. A., Shilkina, L. A. *Sov. Phys. Tech. Phys.* 1977, vol. 47 (2), 453–455.

[6] Bayha, K., Fries, R., Mattes, G. *Knock Sensor*. D.E. Patent 3643956 A1.

[7] Kohashi, M., Yokoi, A., Takahashi, Y., Sawada, Y. *Knock Sensor and Manufacturing Method thereof*. US Patent 7201038 B2.

In: Proceedings of the 2015 International Conference …   ISBN: 978-1-63484-577-9
Editors: Ivan A. Parinov, Shun-Hsyung Chang et al.   © 2016 Nova Science Publishers, Inc.

Chapter 63

BIOLOGICAL SOLDERS FOR LASER WELDING OF BIOLOGICAL TISSUES

Alexander Gerasimenko, Levan Ichkitidze[*], *Vitaly Podgaetsky, Evgenie Pyankov, Dmitrie Ryabkin and Sergei Selishchev*

National Research University of Electronic Technology "MIET,"
MIET, Zelenograd, Moscow, Russia

ABSTRACT

The properties of biological solders (biosolders) for laser welding of biological tissues studied. As biological tissue samples, skin of pig and bovine trachea cartilage were used. The tensile strength of laser welds made using biosolders measured. The biosolder compositions based on aqueous dispersions of a bovine serum albumin (BSA) matrix and single- and multiwall carbon nanotubes (SWCNTs and MWCNTs). The use of carbon nanotubes significantly increases the laser weld strength. The use of the biosolder with the composition 25 wt.% BSA + 0.3 wt.% MWCNT or 25 wt.% BSA + 0.3 wt.% SWCNT made it possible to increase the weld strength by ~30% relative to the case without using CNTs. The potential of the biological solders consisting of the BSA nanomaterial and SWCNTs (MWCNTs) for application in laser welding of biological tissues demonstrated.

Keywords: biological solders; laser welding; laser radiation; albumin; carbon nanotubes

1. INTRODUCTION

The use of lasers in surgical practice has a number of advantages associated with the specificity of the impact of laser radiation on biological tissues. The possibility of high light

---

[*] Corresponding author: Levan Ichkitidze. National Research University of Electronic Technology "MIET," MIET, Zelenograd, Moscow, 124498, Russia. E mail: leo852@inbox.ru.

energy concentration in small volumes allows selective affecting biological tissues and dosing this effect from tissue coagulation to vaporization and cut. In this case, the contactless removal of a tissue portion performed with minimum trauma and high accuracy. Good homeostasis in the irradiated area prevents edema near a wound, which makes the postoperative period almost painless [1]. Laser radiation penetrates deep in biological tissues and activates cells, thus accelerating laser wound healing.

Lasers find various applications in medicine, including joining the dissected biological tissues of injured organs. The integrity of tissues conventionally repaired using suture materials, sewing devices, gluing composites, etc. Meanwhile, these methods are complex, painful and not always provide sufficiently strong joints. The traditional methods of joining and repair of biological tissues can be significantly improved using laser welding (LW). The so-called laser welds do not form rough, visible scars.

Laser welding performed using special biological solders, which applied to the region of a future weld. Laser solders actively absorb laser radiation. In early studies, LW was performed using high-frequency radiation and biological solders were made of filling materials based on collagen, gelatin, or their mixture [2]. However, recent studies showed that the most suitable biosolders for LW are colloidal aqueous dispersions of albumin, i.e., transport protein. Laser welding with the use of bovine serum albumin (BSA) performed on different tissues, including brain tunic [3] and ureter [4]. In all the cases, the histological study carried out in a month after the procedure showed normal epithelization and the absence of complications. However, the tensile strength $\sigma < 0.1$ MPa of the laser welds was lower than the strength of traditional joints with the use of sutures, bridges, or medical glues (e.g., sulphacrylate with the strength $\sigma \sim 0.5$ MPa) [5].

The strength and quality of laser welds can be significantly enhanced using different additives to BSA. Indeed, some organic dyes, e.g., indocyanine green (ICG) or fatty acid [6], filling the BSA matrix of a biosolder increase the absorption of laser radiation, which results in strengthening of laser welds. However, the proposed biosolders have certain disadvantages, including low strength ($\sigma \sim 0.02\text{-}0.5$ MPa) of laser welds in the first ten days after the operation as compared with the traditional methods, concomitant thermal injuries of neighboring tissues, and poor repeatability (inconsistency) of the results [7-9]. A good absorber of laser radiation with the generation wavelength $\lambda_{gen} = 970$ nm is polymeric manganese phthalocyanine (PMP). Laser welding with the use of the biosolder BSA + PMP made it possible to joint cartilaginous tissues, but tensile strength σ of these joints was low (≤ 0.5 MPa).

The observed pronounced positive effect was the increasing strength of laser welds obtained with the use of a biosolder based on BSA and carbon nanotubes (CNTs) [10, 11]. The biosolder was an aqueous ultradispersion from the BSA matrix (25 wt.%) filled with CNTs. At the concentrations $C \sim 0.1\text{-}0.2$ wt.% of single- and multiwall carbon nanotubes (SWCNTs and MWCNTs), the tensile strength of laser welds was significantly enhanced. In particular, in the optimal LW regime, the respective maximum relative strength $\sigma/\sigma_m \sim 23\text{-}32\%$ and 10-15% for bovine trachea cartilage and skin of pig was attained, where σ_m is the strength of a continuous tissue. The use of a pure BSA solder without CNTs yields the values $\sigma/\sigma_m \sim 0.15\text{-}0.3\%$. Test samples were laser welds obtained using biosolders with the composition BSA + K-354 carbon black or activated carbon. The biosolders were prepared under the same conditions as the biosolders BSA + CNTs. The observed negative effect was the reduction of the σ/σ_m value by 30-50% relative to the case of pure BSA biosolder.

The biosolder is the main element of joining biological tissues. In this work, we continue studying the strength of laser welds obtained at different LW regimes, compositions of biological solders, carbon nanotube types, and kinds of biological tissues.

2. MATERIALS

Sample preparation and experimental conditions considered in detail in [10, 11]. Here, we provide just a brief description. As a main solder material, BSA with 98% purity (Heat Shock Isolation, Amresco Inc., US) used. A 25 wt.% BSA aqueous solution was mixed in a magnetic stirrer for 3 h and 45 min under the ultrasound bath conditions at the controlled temperature $t \leq 30°C$. Then, MWCNTs or SWCNTs added to the prepared solution. The obtained dispersion BSA + CNT mixed, filtered, and decanted similarly to the BSA aqueous solution preparation procedures. The obtained dispersion (nanomaterial) then used as a laser solder. In the experiments, we used Taunit MWCNTs of the MWCNT-MD type (Komsomolets plant, Tambov, Russia) [12] and carboxylated SWCNTs of the Paste k-SWCNT-90 type (OOO Uglerod Chg) [13]. Typical 25 wt.% BSA solutions and biosolders consisting of a 25 wt.% BSA and a 0.2 wt.% MWCNT filler are shown in Figure 1.

3. METHODS

Objects of study were biological tissues, specifically, planar samples of bovine trachea cartilage and skin of pig 25-30 mm long, 4-8 mm wide, and 0.5-1.5 mm thick. Continuous samples were dumbbell-shaped and had a narrow central part with a thickness smaller than the thickness of wide ends by a factor of 2-3. This shape ensured the sample break at the center and an acceptable accuracy of measured strength σ_m.

Figure 1. General view of the biosolder: (*a*) BSA solution and (*b*) BSA + CNT dispersion.

The values of σ and σ_m were determined by dividing the maximum break force by the sample sectional area at the broken places. Before LW, the samples attached to a special stage allowing their rotation and uniform irradiation from all sides. At the strip center, a through transverse cut made; the obtained surfaces greased with the laser solder; after that, the samples fixed in the initial positions. To strengthen the biological tissue joint, a Prolene surgical mesh applied to welds in some places.

After LW, tensile strength measurements performed. Sample ends fixed by grips with sandpaper glued to their inner surface in order to prevent sample slip. The applied force measured by an AIGUZP-500N digital dynamometer with a resolution of 0.1 N. For each biological solder type (BSA + MWCNT-MD or BSA + SWCNT), the measurements performed on no less than five identical samples of the same biological tissue. The measurements yielded the averaged σ values.

The laser welding procedure consisted in pointwise positioning of the laser spot along the laser weld line with the biosolder. Depending on the solder composition, the irradiation time varied within 10-200 s at a radiation power of 1-6 W at the fiber end.

Figure 2 shows an original automated LW setup consisting of a power supply unit and a laser head. Two heads operating at generation wavelengths of 810 and 970 nm used[1]. The head involves a pilot laser operating at a wavelength of 530 nm for focusing the laser beam at the welded tissue place. Laser radiation from the laser head transmitted to the welded place by an optical fiber ~ 600 μm in diameter. At the optical fiber end, there is a collimator for narrowing a laser beam to the diameter equal to the laser weld width.

Thermal IR radiation from the welded tissue passes to a concave prism surface coated with germanium and inclined by 45° with respect to the optical axis of the radiation. Then, the IR radiation is reflected from the germanium coating and focused on semiconductor IR sensor, which is set perpendicular to the optical axis of the incident radiation. The signal from the IR sensor perpendicular to the optical axis of the incident laser radiation with information about the laser weld temperature is supplied to a microcontroller by an I2c channel programmed to maintain a required weld temperature range, depending on the tissue type (generally, from 38 to 65°C).

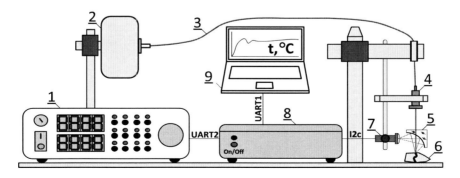

Figure 2. Automated setup for laser welding of biological tissues: (*1*) power supply unit, (*2*) laser head, (*3*) optical fiber, (*4*) collimator, (*5*) prism, (*6*) biological tissue, (*7*) IR sensor, (*8*) microcontroller, and (*9*) computer.

---

[1] Diode lasers with generation wavelengths of 810 and 970 nm (the so-called scalpel lasers) are widely used in laser surgery [14].

The microcontroller serves for controlling the power supply unit with laser head by varying a current strength in the latter. The control implemented by a UART2 communication standard at a data transfer rate of 9600 bit/s, a row length of 8 bit, and one stop bit without parity check. The temperature at the welded place controlled using a PID algorithm. A program for dynamic changing the PID algorithm coefficients (proportional P, integral I, and differential D) was developed. These coefficients depend on optical characteristics of a tissue and a laser solder. The temperature control error is no more than 0.5°C, which is acceptable for LW. The entire LW process is controlled and all information recorded, stored, and processed by a computer.

4. RESULTS AND DISCUSSION

The stages of LW of a biological tissue (skin of pig) illustrate in Figure 3. The biosolder used was the dispersion (nanomaterial) with the composition 25 wt.% BSA + 0.1 wt.% MWCNTs. The laser welding was performed at a radiation power of ~ 5 W/cm^2 for an irradiation time of 10-30 s at a laser weld temperature of 62 ± 1°C. After LW, laser weld strength σ measured and compared with continuous tissue strength σ_m, i.e., the relative strength σ/σ_m determined. The average values of σ and σ/σ_m for the laser welds made using different biosolders on different biological tissues are given in Table 1.

We can see that the laser weld strength parameter enhanced by about an order of magnitude when the CNTs introduced in a solder. The literature data on these parameters obtained with the use of biosolders consisting of 47 wt.% BSA (skin of pig *in vivo* [15]) and 50 wt.% BSA (dog bowel *in vitro* [16]) are multiply lower than the parameters obtained by us. Such a strong difference is certainly due to the use of CNT-based biosolders in LW. The strengthening of laser welds especially pronounced on the cartilaginous tissue.

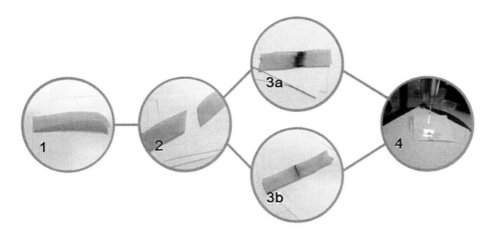

Figure 3. Laser welding of biological tissue (skin of pig): (*1*) continuous sample, (*2*) cut sample, (3a) laser weld obtained with the use of the Prolene surgical mesh, (3b) laser weld without surgical mesh, (*4*) laser welding procedure (pilot laser beam is colored in green; black color of the weld place is due to the natural solder color).

The use of biosolders consisting of BSA and CNTs yielded a relative laser weld strength of up to 30-35%, which is comparable with the relative suture strength parameters ($\sigma \sim$ 150-700 MPa, $\sigma_m \sim$ 1000-3000 MPa, and $\sigma/\sigma_m \sim$ 15-50%) characteristic of traditional methods of joining metal constructions (clips, bolts, plates, etc.) [17].

In medical practice, the joints of biological tissues are often strengthened using surgical meshes. The use of this approach in LW yielded insignificant (3-8%) additional strengthening of tissue joints. Such a minor σ increment can be related, in particular, to insufficient adhesion of the aqueous dispersion BSA + CNT to the surgical mesh, which was confirmed by microphotographs of the weld place where the mesh was applied.

Figure 4 shows a typical electron microscopy image of the biosolder with the composition BSA + MWCNT. One can see nanotubes and their agglomerates.

The analogous picture of Figure 4 is observed for a bulk nanocomposite material (BNCM) obtained by drying the laser solder BSA+CNT under laser irradiation. In this case, the mechanical parameters (a hardness of $\sim$ 300 MPa and a tensile strength of $\sim$ 30 MPa) of the BNCM are similar to the parameters of the well-known materials polymethylmethacrylate and aluminum and the density is 1.25 g/cm^3. Since the laser weld and BNCM were prepared from the same dispersion (BSA + CNT) under the same conditions, we can expect a similar physical mechanism of their strengthening under laser irradiation. For instance, under laser irradiation, a strong frame for attaching albumin molecules can be formed from CNTs or their agglomerates. Preliminary biosolder allergy tests were made on two dwarf rabbits (Figure 5). The mass of each rabbhe introduced preparations were a 25 wt.% BSA aqueous solution (I) and a biosolder dispersion with the composition 25 wt.% BSA + 0.3 wt.% MWCNTs (II).

Table 1. Tensile strength of laser welds of different biological tissues with the use of different solders

Laser biosolder compositions	Tensile strength			
	Skin of pig $\sigma_m = (12 \pm 4)$ MPa		Bovine cartilage $\sigma_m = (6 \pm 2)$ MPa	
	σ, MPa	σ/σ_m, %	σ, MPa	σ/σ_m, %
25 wt.% BSA	0.25 ± 0.1	2.1	0.3 ± 0.1	5
25 wt.% BSA + + 0.3 мас.% MWCNT	1.8 ± 0.3	15	1.9 ± 0.6	32
25 wt.% BSA + + 0.3 wt.% MWCNT + + surgical mesh	2.2 ± 0.6	18	2.1 ± 0.6	35
25 wt.% BSA+ + 0.3 wt.% SWCNT	$2,1 \pm 0.6$	17	1.8 ± 0.6	30
25 wt.% BSA + + 0.3 wt.% SWCNT + + surgical mesh	2.2 ± 0.5	18	2.3 ± 0.6	38
BSA [15]	0.025-0.05 (skin of pig)	–	–	–
BSA [16]	0.43 (dog bowel)	–	–	–

Figure 4. Electron microscopy pattern of the biosolder BSA + MWCNTs.

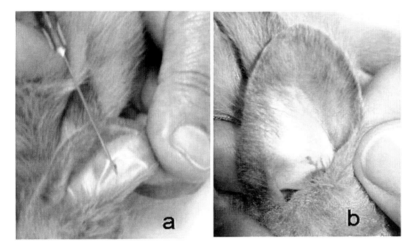

Figure 5. Cartilages of rabbit ears one month after introduced 25 wt.% BSA aqueous solution (a, right ear) and after introduced biosolder dispersion 25 wt.% BSA + 0.3 wt.% MWCNTs (b, right ear).

For comparison, preparation I was introduced in the right ear and preparation II, in the left ear. The rabbit health status controlled for a month; the healing was fast and no allergic reactions observed. General blood tests before and after the introduction of preparations showed no deviations. Thus, introduction of the albumin solution and biosolder consisting of albumin and carbon nanotubes in rabbit's ears caused no allergic reactions, which allows us to make a preliminary conclusion about biocompatibility of these preparations.

CONCLUSION

The obtained positive results, including a significant (by about an order of magnitude) increase in the laser weld tensile strength at laser welding with the use of the biosolders containing CNTs, belong to the *in vitro* regime. In view of this, we believe that the analogous *in vivo* experiments with biological tissues will also yield an increase in the laser weld strength. Similar *in vivo* experiments showed [15] that the strength of a laser weld on skin of pig significantly increased through several weeks after the operation and approached the strength of a continuous skin. The properties of the skin recovered faster than when using traditional methods.

The high tensile strength (~2 MPa) and relative strength (~30%) of the laser weld were attained by using the developed biological solder containing carbon nanotubes (≤0.3 wt.%). These nanoparticles are responsible for strengthening of the laser weld, although the physical mechanism of this strengthening is still understudied. In our case, the strengthening is apparently due to the formation of a BSA + CNT nanocomposite in the laser weld, whose mechanical strength exceeds that of the albumin matrix by a factor of 5-7 [18-21]. Indeed, the Vickers hardness H_v ~ 300-350 MPa and tensile strength σ ~ 30 MPa of this bulk composite nanomaterial approach the properties of the human spongy bone tissue (H_v ~ 500 MPa and σ ~ 15-50 MPa).

We suggest that carbon nanotubes exposed to laser radiation form a strong frame and albumin molecules attached to it. This mechanism possibly works at the formation of a laser weld with the use of the biological solders based on the aqueous albumin dispersion and carbon nanotubes.

The results of this study demonstrate the potential of laser solder application in laser welding of biological tissues and are of great importance for medical practice.

ACKNOWLEDGMENTS

We are grateful to Profs. S.A. Tereshenko for useful discussions, N.N. Iakovleva and I.B. Rimshan for help in the experiments.

This work was financially supported by the Ministry of Education and Science of the Russian Federation (agreement No. 14.575.21.0044, RFMEFI57514X0044).

REFERENCES

[1] Omi, T., Bjerring, P., Sato, S., Kawada, S., Hankins, R. W., Honda, M. *J. Cosmet. Laser Ther.* 2004, vol. 6, 156-162.

[2] Sawyer, P. N. Method for Welding Biological Tissue. *United States Patent no. 5824015*, 1998.

[3] Forer, B., Vasilyev, T., Brosh, T., Kariv, N., Gil, Z., Fliss, D. M., Katzir A. *Lasers in Surgery and Medicine.* 2005, vol. 37(4), 286-292.

[4] Forer, B., Vasilyev, T., Gil, Z., Brosh, T., Kariv, N., Katzir A., Fliss, D. M. *Skull Base: an Inerdisciplinary approach*, 2007, vol. 17(1), 17-23.

[5] Marchenko, V. T. Clinical and Morphological Aspects of the New Glue "Sulfacrylate" surgery of the chest and abdomen. *The Dissertation for the Degree of Doctor of Medical Sciences.* Novosibirsk. 2004, pp. 1-212 (In Russia).

[6]   Andreeva, I. V., Bagratashvili, V. N., Ichkitidze, L. P., Podgaetsky, V. M., Ponomareva, O. V., Savranskii, V. V., Selishchev, S. V. *Biomedical Engineering*, 2009, vol. 43(6), 241-248.

[7]   Simhon, D., Ravid, A., Halpern, M., Cilesiz, I., Brosh, T., Kariv, N., Leviav, A., Katzir, A. *Lasers in Surgery and Medicine*. 2001, vol. 29, 265-273.

[8]   Mcnally, K. M., Sorg, B. S., Chan, E. K., Welch, A. J., Dawes, J. M., Owen, E. R. *Lasers in Surgery and Medicine*, 2000, vol. 26, 346-356.

[9]   Simhon, D., Brosh, T., Halpern, M., Ravid, A., Vasilyev, T., Kariv, N., Katzir, A.; Nevo, Z. *Lasers in Surgery and Medicine*. 2004, vol. 35, 1-11.

[10] Gerasimenko, A. Yu., Gubarkov, O. V., Ichkitidze, L. P., Podgaetsky, V. M., Selishchev, S. V., Ponomareva, O. V. *Semiconductors*. 2011, vol. 45 (13), 93-98.

[11] Ichkitidze, L. P., Komlev, I. V., Podgaetsky, V. M., Ponomareva, O. V., Selishchev, S. V., The method of Laser Welding of Biological Tissue. *Russian Federation Patent No. 2425700*, 2011.

[12] www.nanotam@yandex.ru.

[13] Krestinin, A. V., Kharitonov, A. P., Shul'ga, Yu. M., Gigalina, O. M., Knerelman, E. I., Brzhezinskaya, M. M., Vinogradov, A. S., Zvereva, G. I., Kislov, M. B., Martinenko, V. M., Korobov, I. I., Davidova, G. I., Gigalina, V. G., Kiselov, N. A. *Nanotechnologies in Russia*. 2009, vol. 4, 60-78.

[14] http://www.ntoire-polus.ru.

[15] Simhon, D., Halpern, M., Brosh, T., Vasilev, T., Ravid, A., Tennenbaum, T., Nevo, Z., Katzir, A. *Annals of Surgery*. 2007, vol. 245 (2), 206-213.

[16] Stevenson, F. J. *Crops Soils*, 1979, vol. 31, 14-16.

[17] Bleustein, C. B., Felsen, D., Poppas, D. P. *Lasers in Surgery and Medicine*. 2000, vol. 27(2), 82-86.

[18] http://www.dako.com/28829_connection_14.pdf.

[19] Ageeva, S. A., Eliseenko, V. I., Gerasimenko, A. Yu., Ichkitidze, L. P., Podgaetsky, V. M., Selishchev, S. V. *Biomedical Engineering*. 2011, vol. 44(6), 233-236.

[20] Gerasimenko, A. Yu., Dedkova, A. A., Ichkitidze, L. P., Podgaetsky, V. M., Selishchev, S. V. *Optics and Spectroscopy*, 2013, vol. 115(2), 283-289.

[21] Ichkitidze, L., Podgaetsky, V., Prihodko, A., Selishchev, S., Blagov, E., Galperin, V., Shaman, Y., Tabulina, L. *Materials Sciences and Applications*. 2012, vol. 3(10), 728-732.

In: Proceedings of the 2015 International Conference …   ISBN: 978-1-63484-577-9
Editors: Ivan A. Parinov, Shun-Hsyung Chang et al.   © 2016 Nova Science Publishers, Inc.

Chapter 64

HIFU TRANSDUCERS DESIGNS AND TREATMENT METHODS FOR HEMOSTASIS OF DEEP ARTERIAL BLEEDING

A. N. Rybyanets[1,], A. E. Berkovich[2], T. V. Rybyanets[3], D. I. Makariev[1] and A. N. Reznitchenko[1]*

[1]Institute of Physics, Southern Federal University,
Rostov-on-Don, Russia
[2]Sankt-Petersburg Polytechnic University, Sankt-Petersburg, Russia
[3]Academy of Biology and Biotechnologies, Southern Federal University,
Rostov-on-Don, Russia

ABSTRACT

The purpose of this study was to evaluate the feasibility of high intensity focused ultrasound (HIFU) for hemostasis of deep arterial bleeding. In this paper, the design, development and evaluation of an ultrasound applicator capable of creating thermal lesions in the arterial vessels were presented. New effective HIFU transducers designs and ultrasonic methods for deep arterial bleeding were developed and estimated. Mathematical modeling of the HIFU transducers and acoustic fields were performed. The experiments were made on acoustic vascular phantoms as well as on lamb's femoral artery at a standard protocol. During ultrasound exposure, arterial blood flow was temporarily stopped using intravascular balloon. Postponed hemostasis was observed at lamb's femoral artery experiments for all HIFU treatments. We have demonstrated that HIFU can be used to stop active bleeding from vascular injuries including punctures and lacerations. The results of theoretical modeling, acoustic measurements, and in-vivo vascular experiments prove the efficacy, safety and selectivity of developed HIFU transducers and methods used for enhancing of tissue lysis and hemostasis.

---

[*] Corresponding author: A. N. Rybyanets. E-mail: arybyanets@gmail.com.

1. INTRODUCTION

Ultrasound has found usage in all aspects of the medical field, including diagnostic, therapeutic, and surgical applications. The use of ultrasound as a valuable diagnostic and therapeutic tool in several fields of clinical medicine is now so well established that it can be considered essential for good patient care [1]. However, remarkable advances in ultrasound imaging technology over last decade have permitted us now to envision the combined use of ultrasound both for imaging/diagnostics and for therapy. Traditional therapeutic applications of ultrasound include the treatment of soft tissue and bone injuries, wound healing, hyperthermic cancer treatment, focused ultrasound surgery of Parkinson's disease, glaucoma and retinal detachment and for sealing traumatic capsular tears, benign prostatic hyperplasia, the liver, the kidney, prostate and bladder tumours, vascular occlusion therapy, and tool surgery [1, 2]. Therapeutic transducers are usually made of low loss lead zirconate-titanate (PZT) or recently from 1-3 connectivity type piezocomposites [3]. They are mounted in a light-weight, hand-held waterproof housing, and are typically air-backed. In the past decade, with the advent of faster processing, specialized contrast agents, understanding of nonlinear wave propagation, novel real-time signal and image processing as well as new piezoelectric materials, processing technologies and ultrasound transducer designs and manufacturing, ultrasound imaging and therapy have enjoyed a multitude of new features and clinical applications [1, 4].

The paper presents the results on development and experimental study of different high intensity focused ultrasound (HIFU) transducers. Technological peculiarities of the HIFU transducer design as well as theoretical and numerical models of such transducers and the corresponding HIFU fields are discussed. Several HIFU transducers of different design have been fabricated using different advanced piezoelectric materials. Acoustic field measurements for those transducers have been performed using a calibrated fiber optic hydrophone and an ultrasonic measurement system (UMS).

The results of *ex vivo* experiments with different tissues as well as *in vivo* experiments with blood vessels are presented that prove the efficacy, safety and selectivity of the developed HIFU transducers and methods.

2. APPLICATIONS OF HIFU FOR HEMOSTASIS

Acoustic hemostasis may provide an effective method in surgery and prehospital settings for treating trauma and elective surgery patients. Application of HIFU therapy to hemostasis was primarily initiated in an attempt to control battlefield injuries on the spot. High-intensity ultrasound (ISA = 500-3000 W/cm^2) is usually adopted for hemostasis. Many studies on animal models have been successful for both solid organ and vascular injuries. The thermal effect has a major role in hemostasis. The proposed mechanisms of its action are as follows. Structural deformation of the parenchyma of a solid organ due to high temperature induces a collapse of small vessels and sinusoids or sinusoid-like structures. Heat also causes coagulation of the adventitia of vessels, and subsequently, fibrin-plug formation. The mechanical effect of acoustic cavitation also appears to play a minor role in hemostasis. Microstreaming induces very fine structural disruption of the parenchyma to form a tissue homogenate that acts as a seal and induces the release of coagulation factors. No statistically significant hemolysis or changes in the number of white blood cells and platelets have been observed when blood is exposed to HIFU with intensities up to 2000 W/cm^2 [4]. The main drawback of the hemostasis applications is low ultrasound absorption ability of blood and, as a result, low heating and coagulation rate at real blood flow. In this section, HIFU transducer design, on-linear acoustic field calculations and in-vivo experiments on blood vessels confirming enhanced hemostasis are described.

3. THEORETICAL CALCULATIONS AND NUMERICAL MODELING OF HIFU

The characterization of medical acoustic devices that operate at high output levels has been a research topic and an issue of practical concern for several decades [5]. The importance of nonlinear effects has been considered and addressed even at diagnostic levels of ultrasound [6]. Numerical modeling has been used to predict high amplitude acoustic fields from medical devices. One advantage of modeling is that it can be used to determine the acoustic field in both water and tissue. Numerical algorithms, most commonly based on the nonlinear parabolic Khokhlov-Zabolotskaya-Kuznetsov (KZK) equation, have been developed and applied to the nonlinear fields of lithotripters, unfocused ultrasonic piston sources, diagnostic ultrasonic transducers operating in tissue harmonic imaging mode, focused ultrasound sources, and HIFU sources. For strongly focused fields non-linear models such as Westervelt equation can be used, which is a generalization of the classical wave equation to the nonlinear case in the approximation of the absence of back propagating waves. Even more complex models based on the solution of the full nonlinear wave equation have been developed [7]. However, these approaches require large computing power and time-consuming calculations (up to several days) on supercomputers, i.e., practically inapplicable to practical problems. This difficulty can be significantly reduced by using the evolution equation for the quasi-plane wave. The corresponding equation in nonlinear acoustics equation is known as the KZK equation [6, 7].

The characterization of medical acoustic devices that operate at high output levels has been a research topic and an issue of practical concern for several decades [1]. The importance of nonlinear effects has been considered and addressed even at diagnostic levels of ultrasound [5]. In lithotripsy and HIFU, these effects are critical as acoustic pressures of up to 100 MPa or higher can be reached; such pressures are two or even three orders higher in magnitude than diagnostic ultrasound.

3.1. HIFU Transducers Design

HIFU transducer comprised 1.6 MHz spherical element made from porous piezoceramics [8-12] with 80 mm aperture and 40 mm centre hole having radius of curvature 54 mm. The piezoelement was sealed in custom-designed cylindrical housing filled with the mineral oil providing acoustic contact and cooling of the element.

The housing had an acoustic window made of very thin (0.15 mm) PVC membrane. Centre opening was reserved for ultrasonic imaging transducer (Figure 1).

3.2. Acoustic Field Calculations

Calculations of acoustic fields of HIFU transducers were made using the models and algorithms described above. Figure 2 shows two-dimensional distributions of heat sources power in HIFU transducer's acoustic axis plane. Power density levels are represent in absolute values (kW/cm^3).

In Figure 3 acoustic pressure signals in the focus calculated at different initial intensities for 1.6 MHz frequency are shown. It is obvious that even at initial intensity level of 5 W/cm^2 non-linear effects lead to pressure profile asymmetry that transforms to a shock front in focus at initial intensity 20 W/cm^2 that give rise to extreme heating [6].

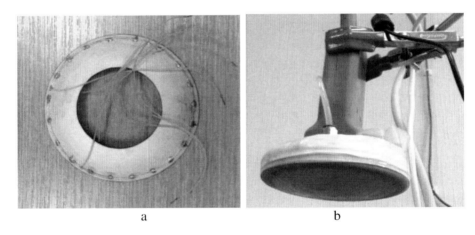

Figure 1. Focusing piezoelement (a) and assembled HIFU ultrasonic transducer (b).

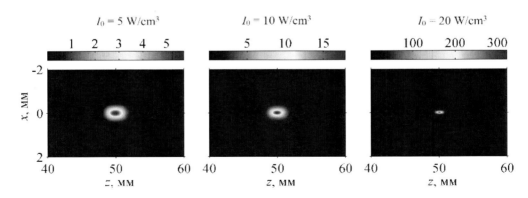

Figure 2. Two-dimensional distributions of heat sources power in acoustic axis plane of HIFU transducer for 1.6 MHz frequency.

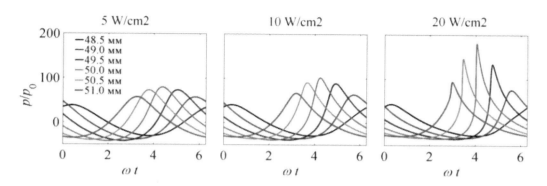

Figure 3. Acoustic pressure signals in the focus calculated at different initial intensities for 1.6 MHz frequency.

4. Ex Vivo Experiments on Tissues

Fresh porcine liver, mussel and adipose tissues were obtained from a butcher within 24 h of slaughter. A single element spherical PZT transducer (1.6 MHz, 80 mm aperture and 40 mm centre hole has been used for experiments. The acoustic intensity for porous PZT transducer was 750 W/cm^2 (ISAL). The samples were placed in an oil bath and positioned right under the transducer such that focal point was placed inside the sample. The samples were irradiated by the harmonics frequency HIFU for 3-20 s at different duty cycles (from 1/2 to 1/100) and burst lengths (from 10 to 200 cycles) of the signals. After exposure the samples were sectioned along the beam axis respectively to compare the dimensions of the lesions.

The photographs of thermal and cavitational lesions in the mussle, liver and porcine samples induced by HIFU are shown in Figure 4.

5. *In Vivo* Experiments on Blood Vessels

The experiments were made on lamb's femoral artery at a standard protocol. During ultrasound exposure arterial blood flow was temporarily stopped using intravascular balloon. Ultrasonic transducer with 1.6 MHz frequency described in previous sections was used for experiments. All acoustic measurements were performed in 3D Scanning System (UMS3) using a fiber-optic hydrophone (FOPH 2000) from Precision Acoustics Ltd. Waveforms from the hydrophones and the driving voltage were recorded using a digital oscilloscope LeCroy. The transducer was driven by a function generator Agilent 33521B and a linear rf-amplifier E&I model 2400L RF and operates in a CW mode. The acoustic intensity in the focal plane measured in water tank at 5000 W/cm^2 (I$_{SAL}$) was kept for the object treatment. After sonication procedure and angiography study, the samples of femoral artery were extracted to confirm hemostasis and disclose vessel thrombus. The X-ray image of blood vessels obtained using contrast agents and photograph of dissected femoral artery are shown in Figure 4.

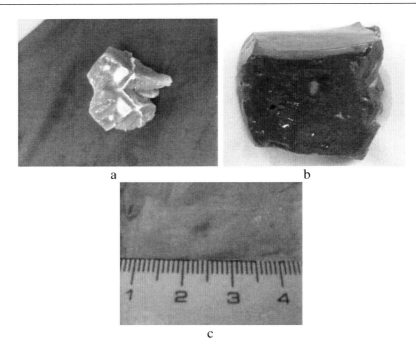

Figure 4. Thermal (a, b) and cavitational (c) lesions in the mussel, liver and porcine samples induced by HIFU transducers. Treatment parameters: (a) cw – exposure time = 3 s and 9 s, (b) cw – exposure time = 3 s, (d) burst mode – duty cycle = 1/20, burst length = 10 cycles, exposure time = 9 s.

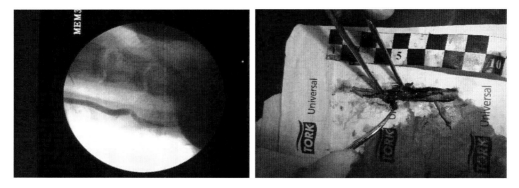

Figure 4. *Angiography image* of blood vessels showing ultrasound hemostasis and photograph of vessel thrombus in dissected femoral artery.

CONCLUSION

The results of theoretical modeling and experimental study of different HIFU transducers were presented. *Ex vivo* experiments in tissues (fresh porcine adipose tissue, bovine liver) and *in vivo* experiments in lamb's femoral artery were carried out using different protocols.

The results of theoretical modeling and tissue experiments prove the efficacy, safety, and selectivity of the developed HIFU transducers and methods enhancing the tissue lysis and hemostasis and can be used for various therapeutic, surgical and cosmetic applications. We have demonstrated that HIFU can be used to stop active bleeding from vascular injuries including punctures and lacerations. Using HIFU transducers, operated at a frequency of 1 or 2 MHz in continuous mode with intensities of 2,000-5,000 W/cm^2, we were able to stop bleeding from major blood vessels that were punctured with an 18- or a 14-gauge needle. Postponed hemostasis was observed at lamb's femoral artery experiments for all HIFU treatments. We have demonstrated that HIFU can be used to stop active bleeding from vascular injuries including punctures and lacerations. Those methods and transducers can be used also for various therapeutic, surgical and cosmetic applications.

ACKNOWLEDGMENTS

Work supported by the RSF grant No. 15-12-00023.

REFERENCES

[1] *Physical Principles of Medical Ultrasonics.* Second Edition. C. R. Hill, J. C. Bamber, G. R. ter Haar (Eds.), John Wiley and Sons Ltd., 2004.

[2] Ter Haar, G. R. *Prog. Biophys. Mol. Biol.* 2007, vol. 93, 111-129.

[3] Rybyanets, A. N. In: *Piezoelectrics and Related Materials: Investigations and Applications*, Ivan A. Parinov (Ed.). Nova Science Publishers Inc. NY. 2012, Chapter 5, 143-187.

[4] Rybyanets, A. N. *IEEE Trans. UFFC.* 2011, vol. 58, 1492-1507.

[5] Dyson, M., Brookes, M. In: *Ultrasound*, Lerski, R. A., Morley, P. (Eds.), Pergamon, Oxford, 1983, 61-66.

[6] Khokhlova, V., Ponomarev, A., Averkiou, M., Crum, L. *Acoustical Physics.* 2006, vol. 52, 481-489.

[7] Khokhlova, V. A., Souchon, R., Tavakkoli, J., Sapozhnikov, O. A., Cathignol, D. *J. Acoust. Soc. Am.* 2001, vol. 110, 95-108.

[8] Rybyanets, A. N. *Ferroelectrics.* 2011, vol. 419, 90-96.

[9] Rybyanets, A. N., Rybyanets, A. A. *IEEE Trans. UFFC.* 2011, vol. 58, 1757-1774.

[10] Rybyanets, A. N., Naumenko, A. A., Shvetsova, N. A. In.: *Nano- and Piezoelectric Technologies, Materials and Devices*, Ivan A. Parinov (Ed.). Nova Science Publishers Inc. NY. 2013, Chapter 1, 275-308.

[11] Rybyanets, A. N. *IEEE Trans. UFFC*, 2011, vol. 58, 1492-1507.

[12] Rybyanets, A. N. *Ferroelectrics*, 2011, vol. 419, 90-96.

In: Proceedings of the 2015 International Conference …   ISBN: 978-1-63484-577-9
Editors: Ivan A. Parinov, Shun-Hsyung Chang et al.   © 2016 Nova Science Publishers, Inc.

Chapter 65

THEORETICAL MODELING AND EXPERIMENTAL STUDY OF HIGH INTENSITY FOCUSED ULTRASOUND TRANSDUCERS: AN UPDATE

S. A. Shcherbinin[1,], A. A. Naumenko[1], N. A. Shvetsova[1], A. E. Berkovich[2] and A. N. Rybyanets[1]*

[1]Institute of Physics, Southern Federal University, Rostov-on-Don, Russia
[2]Sankt-Petersburg Politechnical University, Sankt-Petersburg, Russia

ABSTRACT

The recent advances in the field of physical acoustics, imaging technologies, piezoelectric materials and ultrasonic transducer designs have lead to emerging of a novel methods and apparatus for ultrasonic diagnostics, therapy and anesthetics as well as expansion of traditional applications fields. The chapter presents the results of theoretical modeling and experimental study of different HIFU transducers. Exact solutions for acoustic fields of single and multi-element spherical focusing transducers in linear approximation were calculated using Rayleigh integral. For strongly focused nonlinear field calculations, parabolic approximation of KZK equation was used. HIFU transducers comprising a spherical piezoceramic caps immersed in acoustic contact fluids were fabricated and tested. Acoustic field measurements for HIFU transducers were made using calibrated fiber optic hydrophone and ultrasonic measurement system (UMS). *In-vitro* experiments on tissues (fresh porcine adipose tissue and bovine liver) and *in-vivo* experiments on blood vessels were made using HIFU transducers at different protocols. The results of theoretical modeling and tissue experiments, which prove the efficiency, safety and selectivity of developed HIFU transducers and methods enhancing a tissue lysis and hemostasis were presented.

---

* Corresponding author: S. A. Shcherbinin. Institute of Physics, Southern Federal University, Rostov-on-Don, Russia. E-mail: step_scherbinin@list.ru.

1. INTRODUCTION

Ultrasound has found usage in all aspects of the medical field, including diagnostic, therapeutic, and surgical applications. The use of ultrasound as a valuable diagnostic and therapeutic tool in several fields of clinical medicine is now so well established that it can be considered essential for good patient care [1]. However, remarkable advances in ultrasound imaging technology over last decade have permitted us now to envision the combined use of ultrasound both for imaging/diagnostics and for therapy. Traditional therapeutic applications of ultrasound include the treatment of soft tissue and bone injuries, wound healing, hyperthermic cancer treatment, focused ultrasound surgery of Parkinson's disease, glaucoma and retinal detachment and for sealing traumatic capsular tears, benign prostatic hyperplasia, the liver, the kidney, prostate and bladder tumours, vascular occlusion therapy, and tool surgery [1, 2]. Therapeutic transducers are usually made of low loss lead zirconate-titanate (PZT) or recently from 1-3 connectivity type piezocomposites [3]. They are mounted in a light-weight, hand-held waterproof housing, and are typically air-backed. In the past decade, with the advent of faster processing, specialized contrast agents, understanding of nonlinear wave propagation, novel real-time signal and image processing as well as new piezoelectric materials, processing technologies and ultrasound transducer designs and manufacturing, ultrasound imaging and therapy have enjoyed a multitude of new features and clinical applications [1, 3].

The Chapter presents the results on development and experimental study of different high intensity focused ultrasound (HIFU) transducers. Technological peculiarities of the HIFU transducer design as well as theoretical and numerical models of such transducers and the corresponding HIFU fields are discussed. Several HIFU transducers of different design have been fabricated using different advanced piezoelectric materials. Acoustic field measurements for those transducers have been performed using a calibrated fiber optic hydrophone and an ultrasonic measurement system (UMS). The results of *ex vivo* experiments with different tissues as well as *in vivo* experiments with blood vessels are present that prove the efficiency, safety and selectivity of the developed HIFU transducers and methods.

2. THEORETICAL CALCULATIONS AND NUMERICAL MODELING OF HIFU

The characterization of medical acoustic devices that operate at high output levels has been a research topic and an issue of practical concern for several decades [1]. The importance of nonlinear effects has been considered and addressed even at diagnostic levels of ultrasound [4]. In lithotripsy and HIFU, these effects are critical as acoustic pressures of up to 100 MPa or higher can be reached; such pressures are two or even three orders higher in magnitude than diagnostic ultrasound.

Numerical modeling has been used to predict high amplitude acoustic fields from medical devices. One advantage of modeling is that it can be used to determine the acoustic field in both water and tissue. Numerical algorithms, most commonly based on the nonlinear parabolic Khokhlov-Zabolotskaya-Kuznetsov (KZK) equation, have been developed and applied to the nonlinear fields of lithotripters, unfocused ultrasonic piston sources, diagnostic ultrasonic transducers operating in tissue harmonic imaging mode [1], focused ultrasound sources [5], and HIFU sources [6]. For highly focused sources, the Rayleigh integral can be used to capture diffraction effects more accurately in the model. More comprehensive models based on full nonlinear wave equations have also been developed, but they are much more computationally intensive.

2.1. HIFU Transducers Design

HIFU transducer comprise 1.6-2 MHz spherical element made from porous piezoceramics [7-10] with 80 mm aperture and 40 mm center hole having radius of curvature 54 mm. The piezoelement was sealed in custom-designed cylindrical housing filled with the mineral oil providing acoustic contact and cooling of the element.

The housing had an acoustic window made of very thin (0.15 mm) PVC membrane. Center opening was reserved for ultrasonic imaging transducer (Figure 1).

2.2. Acoustic Field Calculations

Calculations of acoustic fields of HIFU transducers were made according models and algorithms developed in [3-5]. Figure 2 shows two-dimensional distributions of acoustic intensity in HIFU transducer's acoustic axis plane. Intensity levels represent in absolute values (kW/cm^2). Calculations were made for two frequencies 1.6 MHz (a) and 2 MHz (b), respectively.

In Figure 3 acoustic pressure signals in the focus calculated at different initial intensities for 1.6 MHz and 2 MHz are shown. It is obvious that even at initial intensity level 5 W/cm^2 non-linear effects lead to pressure profile asymmetry that transforms to a shock front in focus at initial intensity 20 W/cm^2 leading to extreme heating.

Figure 1. Focusing piezoelement (a) and assembled HIFU ultrasonic transducer (b).

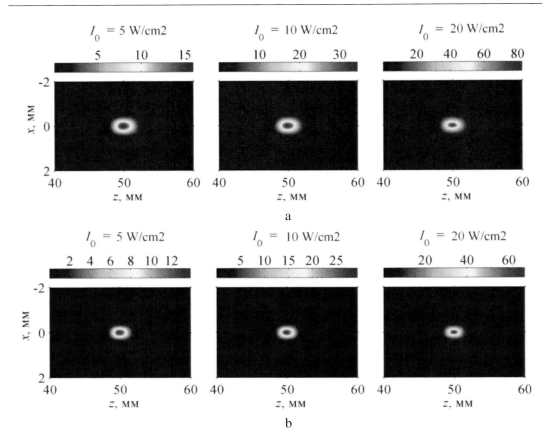

Figure 2. Two-dimensional distributions of acoustic intensity in acoustic axis plane of HIFU transducer for 1.6 MHz (a) and 2 MHz (b) frequencies.

2.3. Applications of HIFU for Hemostasis

Acoustic hemostasis may provide an effective method in surgery and prehospital settings for treating trauma and elective surgery patients. Application of HIFU therapy to hemostasis was primarily initiated in an attempt to control battlefield injuries on the spot. High-intensity ultrasound (ISA = 500-3000 W/cm^2) is usually adopted for hemostasis.

Many studies on animal models have been successful for both solid organ and vascular injuries [1].

The thermal effect has a major role in hemostasis. The proposed mechanisms of its action are as follows. Structural deformation of the parenchyma of a solid organ due to high temperature induces a collapse of small vessels and sinusoids or sinusoid-like structures. Heat also causes coagulation of the adventitia of vessels, and subsequently, fibrin-plug formation. The mechanical effect of acoustic cavitation also appears to play a minor role in hemostasis. Microstreaming induces very fine structural disruption of the parenchyma to form a tissue homogenate that acts as a seal and induces the release of coagulation factors [1]. No statistically significant hemolysis or changes in the number of white blood cells and platelets have been observed when blood is exposed to HIFU with intensities up to 2000 W/cm^2 [2].

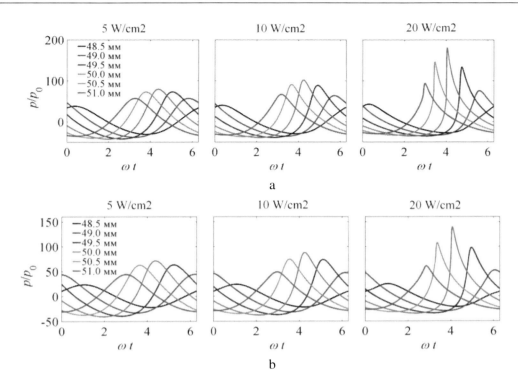

Figure 3. Acoustic pressure signals in the focus calculated at different initial intensities for 1.6 MHz (a) and 2 MHz (b) frequencies.

The main drawback of the hemostasis applications is low ultrasound absorption ability of blood and, as a result, low heating and coagulation rate at real blood flow. In this section, HIFU transducer design, nonlinear acoustic field calculations and *in vivo* experiments on blood vessels confirming enhanced hemostasis are described.

3. *In Vivo* Experiments on Blood Vessels

The experiments were made on lamb's femoral artery at a standard protocol. During ultrasound exposure arterial blood flow was temporarily stopped using intravascular balloon. Ultrasonic transducer with 2 MHz frequency described in previous sections was used for experiments. All acoustic measurements have been performed in 3D Scanning System (UMS3) using the fiber-optic hydrophone (FOPH 2000) from Precision Acoustics Ltd. Waveforms from the hydrophones and the driving voltage were recorded using a digital oscilloscope Lecroy.

The transducer was driven by a function generator Agilent 33521B and a linear RF amplifier E&I model 2400L RF and operates in a CW mode. The acoustic intensity in the focal plane measured in water tank at 5000 W/cm$_2$ (ISAL) was kept for the object treatment.

After sonification procedure and angiography study the samples of femoral artery were extracted to confirm hemostasis and disclose vessel thrombus. The X-ray image of blood vessels obtained using contrast agents and photograph of dissected femoral artery are shown in Figure 4.

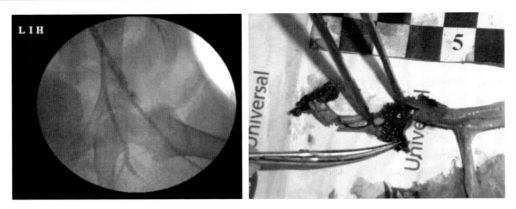

Figure 4. Angiography image of blood vessels showing ultrasound hemostasis and photograph of vessel thrombus in dissected femoral artery.

CONCLUSION

The results of theoretical modeling and experimental study of different HIFU transducers were presented. *In-vitro* experiments on tissues (fresh porcine adipose tissue, bovine liver) and *in-vivo* experiments on lamb's femoral artery were made at different protocols. The results of theoretical modeling and tissue experiments prove the efficiency, safety and selectivity of developed HIFU transducers and methods enhancing of tissue lysis and hemostasis and can be used for various therapeutic, surgical and cosmetic applications.

ACKNOWLEDGMENTS

Work supported by the RSF grant no. 15-12-00023.

REFERENCES

[1] *Physical Principles of Medical Ultrasonics*. Second Edition. C. R. Hill, J. C. Bamber and G. R. ter Haar. (Eds.), John Wiley and Sons Ltd. 2004.
[2] Ter Haar, G. R. *Prog. Biophys. Mol. Biol.*, 2007, vol. 93, 111-129.
[3] Rybyanets, A. N. In: *Piezoelectrics and Related Materials: Investigations and Applications*, Ivan A. Parinov (Ed.). Nova Science Publishers Inc. NY. 2012, Chapter 5, 143-187.
[4] Rybyanets, A. N. *IEEE Trans. UFFC*. 2011, vol. 58, 1492-1507.
[5] Khokhlova, V., Ponomarev, A., Averkiou, M., Crum, L. *Acoustical Physics*. 2006, vol. 52, 481-489.
[6] Khokhlova, V. A., Souchon, R., Tavakkoli, J., Sapozhnikov, O. A., Cathignol, D. *J. Acoust. Soc. Am.* 2001, vol. 110, 95-108.
[7] Rybyanets, A. N. *Ferroelectrics*. 2011, vol. 419, 90-96.
[8] Rybyanets, A. N., Rybyanets, A. A. *IEEE Trans. UFFC*. 2011, vol. 58, 1757-1774.

[9]   Rybyanets, A. N. In: *Piezoceramic Materials and Devices*, Ivan A. Parinov (Ed.). Nova Science Publishers Inc. NY. 2010, Chapter 3, 113-174.

[10] Rybyanets, A. N., Naumenko, A. A., Shvetsova, N. A. In: *Nano- and Piezoelectric Technologies, Materials and Devices*, Ivan A. Parinov (Ed.). Nova Science Publishers Inc. NY. 2013, Chapter 1, 275-308.

In: Proceedings of the 2015 International Conference …   ISBN: 978-1-63484-577-9
Editors: Ivan A. Parinov, Shun-Hsyung Chang et al.   © 2016 Nova Science Publishers, Inc.

Chapter 66

MODELING OF STRESS STATE OF WELDED JOINT IN PIPELINE TAKING INTO ACCOUNT UNEVEN DISTRIBUTION OF MECHANICAL PROPERTIES

V. V. Deryushev[1], E. E. Kosenko[1], A. V. Cherpakov[2], V. V. Kosenko[1], M. M. Zaitseva[1], L. E. Kondratova[1] and S. V. Teplyakova[1,]*

[1]Rostov State University of Civil Engineering, Rostov-on-Don, Russia
[2]Vorovich Mathematics, Mechanics and Computer Sciences Institute,
Southern Federal University, Rostov-on-Don, Russia

ABSTRACT

In this paper, we present calculation of stress state of the welded joint in a pipeline taking into account uneven distribution of mechanical properties along the welde length on the base of finite element modeling. An analysis of the stress state in elastic-plastic deformation allows us to refine the limit stress state in the case of critical load.

Keywords: pipe, welded joint, stress state, finite element modeling

1. INTRODUCTION

Evaluation of stress state of pipelines for various purposes, taking into account a variety of factors influence and complexity of the various areas of pipeline construction is very important.

---

[*] Corresponding author: A.V. Cherpakov, Vorovich Mathematics, Mechanics and Computer Sciences Institute, Southern Federal University Rostov-on-Don, Russia. E-mail: alex837@yandex.ru.. E-mail: alex837@yandex.ru.

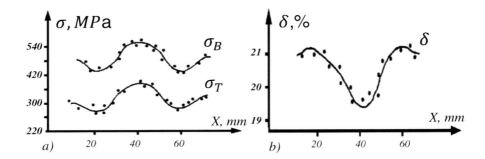

Figure 1. Test dependencies of mechanical characteristics on the sample length: (a) σ_T, σ_B are the yield strength and tensile strength, respectively, (b) δ is the elongation.

One of the effective methods to ensure performance of pipelines is to improve methods for calculating the strength on the base of most complete account of the actual operating conditions and damage of the pipe [1].

Among computational methods for determining stresses in complex shape bodies in the elastic and plastic stages of loading, the most widely uses finite element method [2]. Experience-based finite element modeling is quite successful, with the example of operational using the license, for example the ANSYS simulation package. It allows one to simplify approaches to modeling and computation, and also to perform a visualization process of results [3, 4, 5].

In this method, the body is divided into elements whose size is taken: the smaller size ensures the greater expected stress gradient in the sample area. The stress and strain fields within each element is taken uniform. The calculation reduces to solving a system of equations expressing the equilibrium lattice sites. For a given load, boundary conditions and known solution of the set of the stiffness matrix equations allows one to find nodal forces. Then, based on them, we can define moving nodes within each element to determine the strain and stress. In the case of using the criterion of ductility, we can also determine the material state (elastic or plastic).

One of the most critical parts of the pipe structures are loaded welds. Welds have geometric nonlinearity and a certain distribution of mechanical properties along their length, which can be determined experimentally. Information on the distribution of mechanical properties along the length of the weld joint [6] will be clarified allowable stress state on the considered part of pipeline. In critical development of pressures in the pipeline, as "permissible stresses," we can select stress σ_T above the yield strength, but does not exceed the stress limit of the viscous terms of destruction σ_{destr} [7].

2. METHODS

2.1. Objective

In this paper, we calculate stress state of a welded joint, taking into account the uneven distribution of mechanical properties along the weld length on the base of finite element modeling. The anisotropy of properties is determined by corresponding dependencies of mechanical properties on length of the welded joint, defined experimentally (Figure 1) [5].

2.2. Object of Study

The estimated model presents a quarter pipe cut symmetrical in respect to the longitudinal weld pipe with a length 5 mm, a diameter 1020 mm and a wall thickness 20 mm. The absenting parts are replaced by couples.

We represent a cross-section of the horizontal tube (Figure 2) filled with a fluid at rest. Denote the pressure Px in the center pipe at 0. The load is the uniformly distributed pressure on inner surface of the pipe (15 MPa).

The mechanical properties of the construction material are the following: Steel 09G2S (9MnSi5), elastic modulus $E = 2.06 \times 10^5$ MPa, shear modulus $G = 8.08 \times 10^4$ MPa; Poisson's ratio $\nu = 0.3$, density $\rho = 7.85 \times 10^{-6}$ kg/mm^3

To account for the elastic-plastic deformation, we used bilinear relationship "stress - strain," in which the yield strength of the material deforming linearly (plastic module) is equal to $H = E_T /(1 - E_T / E) = 105000.1$, where E_T is the tangent modulus of elasticity, $\delta = 0.05$).

The calculation was performed using the iterative procedure of Newton – Raphson method (in which a series of linear approximations converge to the actual nonlinear solution).

2.3. Finite Element Modeling

The calculations provide the software package MSC/NASTRAN, providing the physically and geometrically nonlinear studies, taking into account the elastic-plastic behavior of steel pipe.

As a criterion of plasticity, we used elastic-plastic Mises criterion and bilinear theory of plasticity.

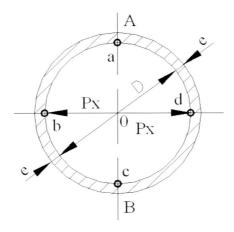

Figure 2. Scheme of the pipe.

The following boundary conditions are given:

1. introduced an additional connection nodes to compensate the reaction of the absenting parts;
2. all nodes into plane of the weld section remained along an axis normal to the section under consideration (*x*-axis);
3. all nodes into plane of the pipe section fixed along an axis normal to the section under consideration (*y*-axis);
4. all nodes into planes of the pipe segment section fixed along an axis normal to the section under consideration (*z*-axis).

Let us use the following method of calculation of welded structures at finite element modeling:

1. to develop a geometric model of the welded parts with the simulation of the welds;
2. to perform a given grid partitioning;
3. to conduct a partitioning into finite elements and parts of the welds;
4. to give boundary conditions, loads and design.

The calculations were carried out in a physically and geometrically nonlinear statement by using the elastic-plastic behavior of steel pipe. The total number of items was equal to 1613 and units – 3568 (see Figure 3).

2.4. Analysis

The stress distribution for the model pipe of 09G2S steel at a pressure 7.5 MPa is as follows: the maximum stresses occur near the edge of the weld, and the maximum value is near the lower edges of the weld.

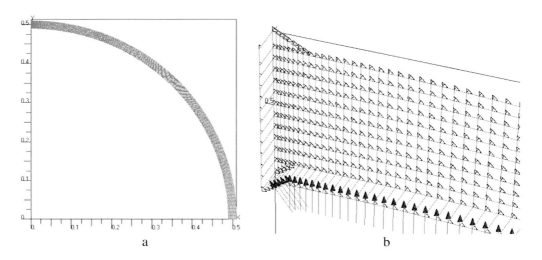

Figure 3. (a) FE-model of quarter pipe, (b) FE-model of pipe with weld located inside of the load pipe.

The minimum values of the stresses at the top and bottom edges along the axis of the weld. In the heat affected zone, the stress values are lesser than those on the base metal and in the weld middle (Figure 4). The stress distribution for the model pipe of 09G2S steel at a pressure 12.6 MPa is as follows: the maximum stresses occur in the lower part of the weld, near the top edge of the weld and on the boundary of HAZ – base metal on the inner surface of the pipe, and the maximum value is around the top edge. The minimum values of the stresses locate at the top and bottom edges along the axis of the weld. In the heat affected zone, stress values are lesser than those on the base metal and in the middle of the weld (Figure 5).

We determine the allowable stress σ_a in the pipeline on the base of the inequality:

$$\sigma_{T\min} \cdot K_{H\,coff} > \sigma_a > \sigma_{T\min} \cdot K_{disp} \tag{1}$$

where K_{disp} is the dispersion coefficient of the material yield strength for the welded joint:

$$K_{disp} = \frac{\sigma_{T\max}}{\sigma_{T\min}}, \tag{2}$$

$\sigma_{T\min}$, $\sigma_{T\max}$ are minimum and maximum characteristics of the distribution of the yield strength of the material along the length of the weld joint, respectively, $K_{H\,coff}$ is the hardening coefficient, which may depend on the elongation and toughness.

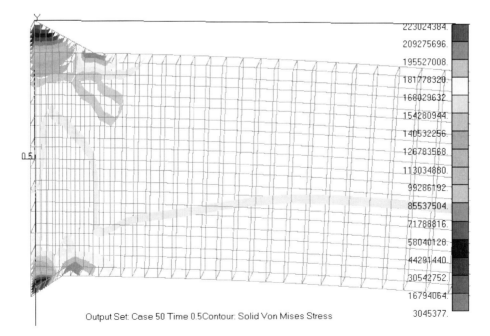

Figure 4. Stress distribution in the model of steel pipe 09G2S at operating pressure of 7.5 MPa (von Mises stress).

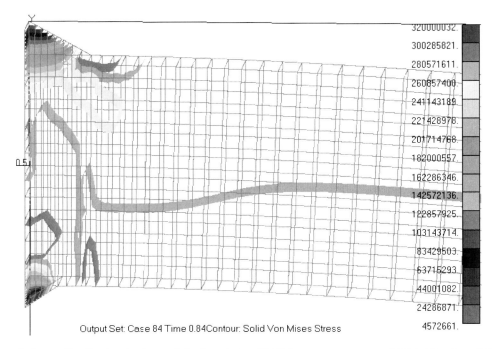

Figure 5. Stress distribution in the model of steel pipe 09G2S at a test pressure of 12.6 MPa (von Mises stress).

Table 1. Calculation results

Pressure, MPa	Steel mark	$\sigma_{T min}$, MPa	$\sigma_{T max}$, MPa	$K_{H\,coff}$	Load stress, σ_l, MPa	Stress concentration factor at the design pressure	
						value	location
7.5	09G2S	280	380	1.07	223.0	1.35	The lower edge of the weld
10.05	09G2S	280	380	1.07	298.8	1.33	The lower edge of the weld
12.6	09G2S	280	380	1.07	320.0	1.18	The upper edge of the weld

This approach will allow for a new approach to the question of verifying the strength of the structural elements of the welded joint of the pipeline.

The stress distribution in steel pipes for the model 09G2S from the condition (1) implies maximum stresses (10.5 MPa) occur around the edges of the weld, and the maximum value locates at the bottom edge of the weld. Minimum stress values cituate at the top and bottom edges of the weld axis. Stresses values in the heat affected zone are lesser than in the base metal and in the middle of weld (Figure 6).

According to the value found, the stress determines magnitude of the hydraulic pressure (load at a certain stage of loading, causing stresses satisfying the values found). The calculation results are present in Table 1.

CONCLUSION

The application of calculations, based on the finite element method, has allowed to clarify the distribution of stresses in the welded joint of the pipeline, taking into account the uneven distribution of mechanical characteristics along the length of the weld joint, and choose a satisfactory load during its operation.

ACKNOWLEDGMENTS

This work was supported in part by the Southern Federal University (No. 213.01.-2014/03VG).

REFERENCES

[1] Frolov, A. V., Shulanbaeva, L. T., Sunagatov, M. F., Gumerov, A. K. Evaluation of the stress state of underground pipelines, taking into account changes in the ground during the operation. *Problems of Gathering, Treatment and Transportation of Oil and Oil Products*. 2010, vol. 1(79), 61-66 (in Russian).

[2] Postnov, V. A., Harhurim, I. Y. *The Finite Element Method in the Calculation of Ship Structures*. Sudostroenie, Leningrad. 1974, pp. 1-342 (in Russian).

[3] Glushchenko, A. V., Morgunov, K. P. *Vestnik of State University of Marine and River Flot Named after Admiral S. O. Makarov*. 2012, No. 2, 13-18 (in Russian).

[4] Burtseva, O. A., Kosenko, E. E., Kosenko, V. V., Nefedov, V. V., Cherpakov, A. V. *On-line Journal "Engineering Journal of Don."* 2011, No. 4 (in Russian).

[5] Cherpakov, A. V., Kayumov, R. A., Kosenko, E. E., Mukhamedova, I. Z. *Vestnik Kazan Technological University*. 2014, vol. 17(10), 182-184 (in Russian).

[6] Belen'kii, D. M., Vernezi, N. L., Cherpakov, A. V. *Welding International*. 2004, vol. 18 (3), 213-215.

[7] Belen'kii, D. M., Vernezi, N. L., Cherpakov, A. V. *"Gas Industry."* 2005, No. 10. 83-92 (in Russian).

In: Proceedings of the 2015 International Conference … ISBN: 978-1-63484-577-9
Editors: Ivan A. Parinov, Shun-Hsyung Chang et al. © 2016 Nova Science Publishers, Inc.

Chapter 67

METHOD AND COMPUTER SOFTWARE FOR DEFINITION OF STRESS-STRAIN STATE IN LAYERED ANISOTROPIC CONSTRUCTIONS AT PULSE LOADING

I. P. Miroshnichenko[*]

Don State Technical University, Rostov-on-Don, Russia

ABSTRACT

The paper presents a novel theoretical method and computer software to define stress-strain state in specific regions of layered cylindrical and elliptic constructions, manufactured of isotropic and transverse-isotropic composite materials. The study takes into account features of all wave types propagating in these materials, wave processes in the layered structures and influence of a surface curvature of the considered constructions at given spatial-time distribution of pulse sources.

Keywords: stress-strain state, structural materials, composite materials, diagnostics

1. INTRODUCTION

Intense development and using new types of construction materials (composite, polymeric, etc.) lead to activation of development of novel analytical methods for analysis of their stress-strain state in different calculations at all stages of a life time cycle of these materials (manufacture, quality control, exploring, etc.).

The aim of this study is the development a novel theoretical method and computer software to define fields of displacements, stresses and strains in specific regions of layered cylindrical and elliptic constructions, manufactured of isotropic and transverse-isotropic

---

[*] E-mail: ipmir2011@yandex.ru.

composite materials. The study takes into account features of all wave types propagating in these materials, wave processes in the layered structures and influence of a surface curvature of the considered constructions at given spatial-time distribution of pulse sources of local or distributed pulse loading at internal and external surfaces of treated constructions.

2. METHODS

The paper [1] developed a generalized method of scalarization of dynamic elastic fields in transverse-isotropic media allowing one to describe dynamic elastic fields of displacements, stresses and strains in constructions manufactured of transverse-isotropic materials. This approach used three scalar functions corresponding to quasi-longitudinal, quasi-transverse and transverse waves, i.e., all types of waves propagating in transverse-isotropic materials.

Based on tensor relationships of the method [1], we developed a method for definition of stress-strain state in layered cylindrical and elliptic constructions, manufactured of isotropic and transverse-isotropic composite materials. The method takes into account all types of waves proagating in the materials, wave processes in layered structures and influence of surface curvature of the considered constructions at specific spatial-time distribution of sources of local and distributed pulse loading at external and/or internal surfaces of the treated constructions.

We have proposed definition methods of stress-strain state in layered cylindrical constructions, manufactured of isotropic and transverse-isotropic composite materials for specific two-dimensional cases in dependence on orientation of material symmetry axis of the transverse-isotropic materials of layers.

We have proposed definition methods of stress-strain state in elliptic bar, space with elliptic cavity and thick-thickness elliptic tube for partial two-dimensional cases taking into account an orientation of material symmetry axis of the transverse-isotropic materials.

We have considered in detail a common case of excitation of elastic waves in layered elliptic constructions, manufactured of transverse-isotropic construction materials.

In the problem solution, we used a method of division of the exciting sources into spatial and time harmonics. In this spectral representation, we defined amplitudes of all harmonics in each of layers by using "coupling" fields at boundaries of the layers, and then transited to spatial-time representation by using repeated inverse Fourier transform. The papers [2-8] describe in detail the developed and above-mentioned methods.

The developed computer software [9-15] realized the proposed methods [2-8] for numerical solution of corresponding problems. We have developed original program software [16-19] for calculation of high transcendent functions, namely Mathieu functions, modified Mathieu functions, modified Mathieu functions in the form of series of products of the Bessel functions and in the form of series of the Hankel functions, used in the problem solutions for elliptic constructions.

By using the pointed computer software, we performed numerical modeling of stress-strain state in two-layer constructions, manufactured of transverse-isotropic construction materials at local pulse loading on external surface of the treated construction.

Comparison of results of the numerical modeling with experimental results confirmed permissibility of using the proposed analytical methods and computer software in solution of applied problems.

We have developed original computer software [20-21] for modeling of processes of the state diagnostics for layered constructions, manufactured of transverse-isotropic materials, based on acoustic active methods of non-destructive control at single or repeated influences of probing pulses in the form of Gauss's function, locally-distributed on external and internal surfaces of the treated constructions. We used successively the computer software [20, 21] for *a priori* and *a post priori* analysis of results of state diagnostics of the two-layered cylindrical constructions at ultrasound quality control of their processing.

In a whole, we have developed and probed novel analytical method and computer software for definition of stress-strain state in layered cylindrical and elliptic constructions with anisotropic layers at repeated pulse probing influence on their external and internal surfaces. We obtained these results by using the generalized scalarization method of dynamic elastic fields of displacements, stresses and strains in transverse-isotropic composite media. We also took into account features of all wave types, propagating in transverse-isotropic composite materials, wave processes in layered constructions, and considering surface curvature of the constructions (cylindrical and elliptic) at given spatial-time distribution of sources of the local pulse probing loading.

Above results allows us to define numerically dynamic elastic fields of displacements, stresses and strains in layered cylindrical and elliptic constructions (strengthened frame constructions, fuselage, wing, blade, etc.) with anisotropic layers at repeated pulse probing loading with given spatial-time distribution of its sources on external and internal surfaces of the treated constructions. These results obtain as during loading process as after its ending, taking into account the pointed features of wave processes.

In application to the problems of state diagnostics of construction materials, the pointed results allow us to reduce significantly a time of the result processing of non-destructive control by using acoustic methods. Their information increases for account of *a priory* analysis of the calculated dependencies of displacement change on external and internal surfaces of control objects and due to comparative analysis with measurement results.

The obtained results discussed at different international scientific-technical conferences and innovative salons (see e.g., [22-30]). The proposed computer software [9-21] probed successively to solve lot of actual scientific and industrial problems.

It is expedient to use the obtained results at designing and exploring constructions from advanced anisotropic composite materials, quality control of their fabrication and state diagnostics at various stages of life cycle in machinery-building, aircraft-building, shipbuilding, fuel-energetic complex, etc.

ACKNOWLEDGMENTS

This study has been supported by the Russian Foundation for Basic Research grant No. 13-08-00754.

REFERENCES

[1] Sizov, V. P. Izvestiya of USSR Academy of Science. *Mechanics of Solids,* 1988, No 5, 55-58.

[2] Miroshnichenko, I. P.; Sizov, V. P. Izvestia of Higher School. North-Caucases Region. *Natural Sciences*, 1999, No 2, 19-22 (in Russian).

[3] Miroshnichenko, I. P.; Sizov, V. P., Izvestia of Higher School. North-Caucases Region. *Natural Sciences*, 1999, No 3, 40-42 (in Russian).

[4] Miroshnichenko, I. P.; Sizov, V. P., Izvestiya of Russian Academy of Science. *Mechanics of Solids*, 2000, No 1, 97-104.

[5] Sizov, V. P.; Miroshnichenko, I. P. Elastic Waves in Layered Anisotropic Structures, LAP LAMBERT Academic Publishing, Saarbrucken (Germany), 2012, pp. 1 – 270.

[6] Sizov, V. P.; Miroshnichenko, I. P. Use of Generalized Method of Scalarization of the Dynamic Elastic Fields in Transverse Isotropic Media. In: Physics and Mechanics of New Materials and their Applications. S. H. Chang, I. A. Parinov (Eds.), Chapter 17. New York: Nova Science Publishers, 2013. pp. 203-211.

[7] Miroshnichenko, I. P. Modeling Methods of Stress-Strain State in Layered Constructions from Anisotropic Materials at Pulsed Loading. In: Advanced Materials. Physics, Mechanics and Applications, S.-H. Chang, I. A. Parinov, V. Yu. Topolov (Eds.), Springer Proceedings in Physics, vol. 152, Chapter 14. Springer, Heidelberg, New York, Dordrecht, London, 2014, pp. 163 – 179.

[8] Miroshnichenko, I. P.; Sizov, V. P. Definition of Stress-Strain State of Layered Anisotropic Elliptic Construction under Action of Impulse Load, In: Advanced Materials - Studies and Applications, I. A. Parinov, S. H. Chang, S. Theerakulpisut (Eds.). Chapter 22. New York: Nova Science Publishers, 2015. pp. 353-390.

[9] Korobchak, I. V.; Miroshnichenko, I. P. Computer program for definition of stress-strain state in layered cylindrical construction. Certificate on Russian state registration of program for computer No. 2011610785, 2011 (in Russian).

[10] Korobchak, I. V.; Miroshnichenko, I. P. Computer program for definition of stress-strain state in layered cylindrical construction under action of local pulse load. Certificate on Russian state registration of program for computer No. 2011612071, 2011 (in Russian).

[11] Petrov A. M.; Miroshnichenko, I. P. Computer program for definition of stress-strain state in elastic half-space under action of pulse load on circular region on its surface. Certificate on Russian state registration of program for computer No. 2013615307 РФ, 2013 (in Russian).

[12] Miroshnichenko, I. P.; Petrov A. M. Computer program for definition of stress-strain state in multilayer cylindrical construction with transverse-isotropic layers under action of pulse load with given spatial-time distribution of its sources. Certificate on Russian state registration of program for computer No. 2013615308 РФ, 2013 (in Russian).

[13] Miroshnichenko, I. P. Computer program for definition of stress-strain state in multilayer cylindrical construction. Certificate on Russian state registration of program for computer No. 2014613314 РФ, 2014 (in Russian).

[14] Miroshnichenko, I. P. Computer program for definition of displacements and stresses in layered cylindrical construction under action of combined pulse load. Certificate on

Russian state registration of program for computer No. 2014613315 РФ, 2014 (in Russian).

[15] Miroshnichenko, I. P. Computer program for definition of displacements and stresses in layered cylindrical construction under action of pulse load. Certificate on Russian state registration of program for computer No. 2014613316 РФ, 2014 (in Russian).

[16] Miroshnichenko, I. P. Computer program for definition of Mathieu functions and their derivatives. Certificate on Russian state registration of program for computer No. 2008614102 РФ, 2008 (in Russian).

[17] Miroshnichenko, I. P. Computer program for definition of modified Mathieu functions in the form of series of products of the Bessel functions and their derivatives. Certificate on Russian state registration of program for computer No. 2009615589 РФ, 2009 (in Russian).

[18] Miroshnichenko, I. P. Computer program for definition of modified Mathieu functions in the form of series of the Hankel functions and their derivatives. Certificate on Russian state registration of program for computer No. 2010611891 РФ, 2010 (in Russian).

[19] Miroshnichenko, I. P. Computer program for definition of modified Mathieu functions in the form of hyperbolic functionsв and their derivatives. Certificate on Russian state registration of program for computer No. 2010611892 РФ, 2010 (in Russian).

[20] Miroshnichenko, I. P.; Parinov, I. A. Computer program for modeling of process of the material state diagnostics in monolayer construction. Certificate on Russian state registration of program for computer No. 2012661166 РФ, 2012 (in Russian).

[21] Miroshnichenko, I. P.; Parinov, I. A. Computer program for modeling of process of the material state diagnostics in layer construction. Certificate on Russian state registration of program for computer No. 2012661167 РФ, 2012 (in Russian).

[22] Miroshnichenko, I.P.; Sizov, V.P. On Use of Generalized Method of Scalarization of the Dynamic Elastic Fields in Transverse Isotropic Media. In: Russian-Taiwanese Symposium "Physics and Mechanics of New Materials and Their Applications." Abstracts and Schedule. Rostov-on-Don: Southern Federal University, 2012. p. 57.

[23] Miroshnichenko, I.P. On Modeling Processes of State Diagnostics of the Layered Anisotropic Materials. In: 2013 International Symposium on "Physics and Mechanics of New Materials and Underwater Applications" (PHENMA 2013). Abstracts and Schedule, Kaohsiung, Taiwan, June 5-8, 2013, P. 78.

[24] Miroshnichenko, I.P. Polzunov Almanac. 2014, No. 1. P. 114-116 (in Russian).

[25] Miroshnichenko, I.P. Scientific-Methodic Apparatus for Determination of the Stress-Strain State in Layered Anisotropic Constructions. In: 2014 International Symposium on "*Physics and Mechanics of New Materials and Underwater Applications*" (PHENMA 2014), Abstracts and Schedule, Khon Kaen, Thailand, March 27-29, 2014. p. 59.

[26] Miroshnichenko, I.P. Computer software for study of stress-strain state in layered anisotropic cylindrical and elliptic constructions at pulse loads. In: Information Technologies in Science and Education. Moscow: NOU IKT Press, 2014, p. 55-58 (in Russian).

[27] Miroshnichenko, I. P.; Parinov, I. A. Program complex for modeling wave processes in layered anisotropic constructions at state diagnostics by methods of non-destructive control. In: Catalog of the XVII Moscow International Salon of Inventions and Innovative Technologies "Archemede-2014," April 1-4, 2014, Moscow, 2014, p. 22-2.

[28] Miroshnichenko, I.P.; Sizov, V.P. Scientific-Methodic Apparatus for Determination of the Stress-Strain State in Layered Anisotropic Constructions. In: Innovative Technologies in Science and Education, September 4-7, 2014, Rostov-on-Don, 2014, p. 300 (in Russian).

[29] Miroshnichenko, I.P. Definition of stress-strain state in layered anisotropic constructions with account of features of wave processes. In: Wave and Vibrowave Technologies in Machinery-building, Metal-processing and other Industrial Branches, October 7-10, 2014, Rostov-on-Don, 2014, p. 113-115 (in Russian).

[30] Miroshnichenko, I.P. Method and Computer Software for Definition of Stress-Strain State in Layered Anisotropic Constructions at Pulse Loading. In: Abstracts and Schedule of the 2015 International Conference on Physics and Mechanics of New Materials and Their Applications (PHENMA 2015). I.A. Parinov, S.-H. Chang (Eds.). Southern Federal University Press: Rostov-on-Don, 2015. pp. 161-162.

In: Proceedings of the 2015 International Conference … ISBN: 978-1-63484-577-9
Editors: Ivan A. Parinov, Shun-Hsyung Chang et al. © 2016 Nova Science Publishers, Inc.

Chapter 68

BASED ON ANALYTICAL MODELING IDENTIFYING THE LOCATION OF MULTIPLE CRACKS IN THE ROD CONSTRUCTION

A. N. Soloviev[1,2,3], A. V. Cherpakov[3,] and I. A. Parinov[3]*

[1]Don State Technical University, Rostov-on-Don, Russia
[2]Southern Scientific Center, Russian Academy of Sciences,
Rostov-on-Don, Russia
[3] Vorovich Mathematics, Mechanics and Computer Sciences Institute,
Southern Federal University, Rostov-on-Don, Russia

ABSTRACT

The paper describes an approach to the problem of crack identification in a cantilever rod construction with multiple defects. This approach is based on the multi-parametric identification analyzing frequencies and parameters of natural oscillations of the construction. As identifiers of the defect location, we propose some parameters based on the crack curvature and bend angles of waveforms.

Keywords: identification of cracks, rod construction, numerical modeling, natural oscillations, model parameters, curvature, bend angles, identifier

1. INTRODUCTION

The problem of development of simple and reliable methods for technical control of the construction state and elements is the key problem in reducing the risk and consequences of natural and man-made disasters.

Solutions of the identification problems of damages in rod constructions are present in numerous studies, reviewed not only for a long time, but in recent years [1–6]. Among key

---

[*] alex837@yandex.ru.

issues arising in studies of damaged constructions, an important problem is the develoment of mathematical algorithms for solving the identification problems of damages.

Some modern approaches to solve this problem are present in [7]. The task solution of damage control is based on the use of correlation couples between frequency spectrum parameters of oscillations and a degree of damage of the construction elements. These links are detected by using resonance methods, developed for natural and forced oscillations, in particular acoustic control methods. In majority of the studies, authors investigated the simple construction elements (rods, beams, tubes) [8, 9].

The purpose of the study consisted in identification of the crack location in a rod by analyzing parameters of oscillation modes in the rod construction by using an analytical modeling.

2. METHODS

An object of the simulation was the solid rod with a square cross-section (length $L = 1000$ mm, height of the cross-section $h = 20$ mm, and width $a = 20$ mm) with cracks. The paper studies transverse oscillations of the rod construction with two defects. The cracks are modeled by elastic elements with certain flexural stiffnesses. The cracks located at certain points on a distance $\bar{L}_{di}$ from the rod clamping place, where $\bar{L}_{di} = L_{di}/L$, L_d is the damage location. We considered the case of the damaged rod with location of the rod cracks corresponding to values of $\bar{L}_{d1} = 0.2$ and $\bar{L}_{d2} = 0.7$. The values of stiffness of the elastic elements, modeling the cracks, we accepted as $K_{t1} = 100$, $K_{t2} = 100$. Then, we introduced the dimensionless coordinate $\bar{x} = x/L$.

Analytical modeling of the oscillations was performed by using computer software MAPLE. Figure 1 shows the model investigated.

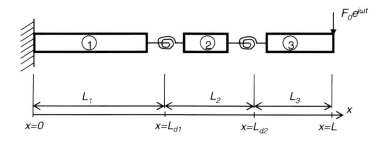

Figure 1. Model rod with defects in the form of elastic elements.

Equivalent analytical model is considered into framework of the Euler-Bernoulli beam model:

$$\frac{\partial^2}{\partial x^2}\left[EJ(x)\frac{\partial^2 u_i}{\partial x}\right] - m(x)\frac{\partial^2 u_i}{\partial t^2} + F(t)\delta(x) = 0, \qquad (1)$$

where $u_i = u_i(x,t), i = 1,2,3$ are the transverse displacements of points of the beam axis, where a subscript indicates number of the beam part, E is the elasticity modulus, $J(x)$ is the moment of inertia, $m(x)$ is the linear density, $F(t)\delta(x)$ is the force applied at the point L.

The boundary conditions for the composite structure have the form:

$$u_1(0) = 0, \; u_1'(0) = 0, \text{ at } x = 0;$$
$$u_1(L_d) = u_2(L_d), \; u_1''(L_d) = u_2''(L_d), \; u_1'''(L_d) = u_2'''(L_d),$$
$$-EJu_1''(L_d) = K_t[u_1'(L_d) - u_2'(L_d)], \text{ at } x = L_{d1};$$
$$u_2(L_d) = u_3(L_d), u_2''(L_d) = u_3''(L_d), \; u_2'''(L_d) = u_3'''(L_d),$$
$$-EJu_2''(L_d) = K_t[u_2'(L_d) - u_3'(L_d)], \text{ at } x = L_{d2};$$
$$u_3''(L) = 0; \; u_3'''(L) = F_0 / EJ, \text{ at } x = L, \tag{2}$$

where K_t is the stiffness of elastic element.

We solved the problem on natural oscillations of the cantilever. Table 1 shows the eigen-frequencies for the rod construction without cracks, $\omega_i(0)$, and with cracks, $\omega_i(d)$, as well as their relative values.

Table 1. Eigen-frequencies for the rod construction without and with cracks and their relative values

i-th mode	$\omega_i(0)$, Hz	$\omega_i(d)$, Hz	$\Delta\overline{\omega}_i = \omega_i(d)/\omega_i(0)$
1	16.798	2.140	0.127
2	105.299	16.192	0.153
3	295.523	179.270	0.607
4	579.840	379.810	0.655

The analysis of natural frequencies shows that for this rod construction with cracks, the 4-th and 3-rd natural frequencies changed most strongly; their relative values were $\Delta\overline{\omega}_4 = 0.655$ and $\Delta\overline{\omega}_3 = 0.607$, respectively.

We defined also eigen-forms (oscillation modes) of the cantilever with cracks. The oscillation amplitudes were normalized by the maximum amplitude of oscillations of the rod points along its length. Figure 2 shows the normalized eigen-forms for first four oscillation modes of the rod oscillations for given case of defects. We calculated also values of curvature $\overline{U}''$ of the oscillation modes and angle α of bend, formed by the tangents at the points of the waveform.

The values of curvature $\overline{U}''$ and angle α between tangents at the points of the waveform were calculated by using a discrete approach based on the measurement of the oscillation amplitudes at a finite number of points. Since for the defect identification, we used value of second derivative, then this method is sensitive to the measurement discreteness of the oscillation eigen-waveform. Based on the numerical experiments for the identification problem, a step of discrete segment was selected as $\Delta x = 0.01\,L$.

The analysis of the oscillation waveforms (see, Figure 2) shows that location of the crack is defined by bend of the waveform. The angles of the waveform bend are different for various positions of cracks at the same stiffness values and the same oscillation mode. These angles differ for various oscillation modes. The plots of angles of the oscillation waveform bend show that crack location is defined by sharp "peak." It forms in this case by three discrete points in a vicinity of the crack. Table 2 shows the numerical values of the curve bend of oscillation waveforms α_1 and α_2 at the points of crack dispositions $\overline{L}_{d1}$ and $\overline{L}_{d2}$. For comparison, we presented the angles of bend α_{01} and α_{02} of the waveforms at the points, disposed at a distance Δx from the points of the crack locations. Their ratios α_1/α_{01} and α_2/α_{02} are calculated as

$$\alpha_{iind} = \alpha_i / \alpha_{0i}. \qquad (3)$$

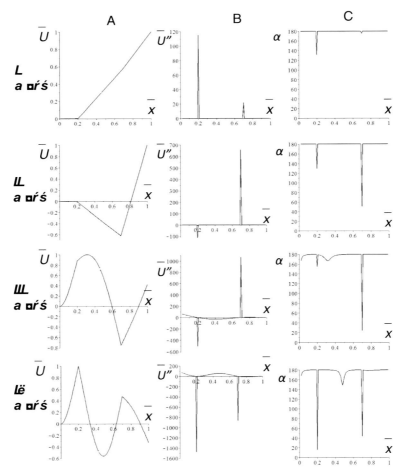

Figure 2. Normalized values of the transverse displacement $\overline{U}$ of cantilever with cracks for four oscillation modes and plots of curvature $\overline{U}''$ of the oscillation modes and angle of bend α of the oscillation waveforms.

Table 2. Parameters of the bend angles for two cracks (elastic elements with the same stiffness) and their various locations for four oscillation modes

i mode	α_{01}, grad	α_1, grad	α_1 / α_{01}	α_{02}, grad	α_2, grad	α_2 / α_{02}
1	179.9	131.34	0.73	179.99	175.29	0.97
2	179.9	130.0	0.72	179.94	50.71	0.28
3	179.96	155.31	0.86	179.99	23.91	0.13
4	179.99	-15.49	0.08	179.97	43.35	0.24

Analysis of ratios of the angles for two neighboring points shows that the parameter α_{iind} can characterize the crack location. The value of the identifier α_{iind} equal to ratio of the angles at adjacent points is less than 1.

Analysis of the plots of curvature of the waveforms for four oscillation modes, similarly to the plots of the bend angles, shows sharp "peak," formed by three discrete points in a vicinity of the crack. The values of curvature both as at the crack location points ($\overline{U}''_1$ and $\overline{U}''_2$) as at the points, located on the distance Δx from the cracks ($\overline{U}''_{01}$ and $\overline{U}''_{02}$) are present in Table 3. The ratios of curvature values at the crack locations to the curvatures at the adjacent points to the cracks ($\overline{U}''_1 / \overline{U}''_{01}$ and $\overline{U}''_2 / \overline{U}''_{02}$) may also be identifiers $\overline{U}''_{iind}$ of the damage places:

$$\overline{U}''_{iind} = \overline{U}''_i / \overline{U}''_{0i} . \tag{4}$$

In this case, change of the identifiers in absolute value took place from 190.1 to 2749.8 for four cases of the crack dispositions and various oscillation modes.

Table 3. Parameters of curvature $\overline{U}$ at the points of crack locations and in their vicinity for four oscillation modes

i-th mode	$\overline{U}''_{01}$	$\overline{U}''_1$	$\overline{U}''_1 / \overline{U}''_{01}$	$\overline{U}''_{02}$	$\overline{U}''_2$	$\overline{U}''_2 / \overline{U}''_{02}$
1	0.041	114.9	2749.8	0.008	21.615	2637
2	-0.005	-123.7	2215.2	0.241	657.5	2731
3	2.640	-502.4	-190.1	-0.470	1068.9	-2269
4	0.590	-1476.1	-2500	2.4	-860.0	-357

CONCLUSION

1. By considering the identification process of crack locations in rod and analyzing various parameters of waveforms, we can use as identifier of the crack locations a sign of the "sharp" bend of the oscillation waveform.
2. As an additional identifier of the crack dispositions, we can select the ratio of bend angles α_{iind} or the ratio of curvatures $\overline{U}''_{iind}$ at two adjacent points of corresponding waveform.

3. The value of the identifier based on analysis of the oscillation waveform curvature $\overline{U}''_{iind}$ is more sensitive compared with the identification parameter based on the bend angles α_{iind}, formed at the oscillation waveform points.

ACKNOWLEDGMENTS

This work was supported in part by the Southern Federal University (grant No. 213.01.-2014/03VG), the Russian Foundation for Basic Research (grants Nos. 14-38-50915, 14–08–00546-a). I. A. Parinov acknowledges financial support from the Ministry of Education and Sciences of Russia in the framework of "Organization of Scientific Research" Government Assignment (grant No. 654).

REFERENCES

[1] Del Grosso, A.; Lanato, F.; A critical review of recent advances in monitoring data analyses and interpretation for civil structures. *Proc. Eur. Conf. on Structural Control (4ECSC). St.-Peterburg.* 2008, *vol. 1*, 320-327.

[2] Akopyan, V.; Soloviev, A.; Cherpakov, A.; Chapter 4. Parameter Estimation of Pre-Destruction State of the Steel Frame Construction Using Vibrodiagnostic Methods. In: *Mechanical Vibrations: Types, Testing and Analysis.* A. L. Galloway (Ed.). Nova Science Publishers, New York. 2010, 147-161.

[3] Dimarogonas, A. D.; *Eng. Fract. Mech.* 1996, *vol. 55*, 831–857.

[4] Doebling, S. W.; Farmer, C. R.; Prime, H. B.; *The Shock and Vibration Digest.* 1998, *vol. 30(2)*, 91-100.

[5] Giraldo, D.F.; A *Structural Health Monitoring Framework for Civil Structures.* PhD Thesis in the Washington University, Department of Civil Engineering. Saint Jouris, Missouri, 2006.

[6] Krasnoshchekov, A. A.; Sobol, B.V.; Soloviev, A. N.; Cherpakov, A.V.; *Russian Journal of Non-Destructive Testing.* 2011, *vol. 47(6)*, 412–419.

[7] Matveev, V. V.; Bovsunovsky, A. P.; *J. Sound Vibration.* 2002, *vol. 249(1)*, 23-40.

[8] Akopyan, V.A.; Soloviev, A.N.; Cherpakov, A.V.; Shevtsov, S.N.; *Russian Journal of Non-Destructive Testing.* 2013, *vol. 49(10)*, 34 – 39.

[9] Akopyan, V. A.; Rozhkov, E. V.; Soloviev, A. N.; Shevtsov, S. N.; Cherpakov, A. V.; *Identification of Damage in Elastic Structures: Approaches, Methods, and Analysis.* Southern Federal University Press, Rostov-on-Don. 2015, pp. 1 – 74 (In Russian).

In: Proceedings of the 2015 International Conference … ISBN: 978-1-63484-577-9
Editors: Ivan A. Parinov, Shun-Hsyung Chang et al. © 2016 Nova Science Publishers, Inc.

Chapter 69

VIBRODIAGNOSTICS OF TRUSS MODEL WITH DAMAGES: EXPERIMENT

A. N. Soloviev [1,2,3], A. V. Cherpakov [3,], E. V. Rozkov [3] and I. A. Parinov [3]*

[1]Don State Technical University, Rostov-on-Don, Russia
[2]South Scientific Center, Russian Academy of Sciences,
Rostov-on-Don, Russia
[3]Vorovich Mathematics, Mechanics and Computer Sciences Institute,
Southern Federal University, Rostov-on-Don, Russia

ABSTRACT

The described approach of vibrodiagnostics is used to study a model of bridge truss with defects. The technique of experimental research and laboratory setup are developed. The study of vibrations is performed at various points of the model. Based on data, obtained for frequency response, the truss model with damages is investigated. This approach can be used to develop methods for vibrodiagnostics of damage conditions, for example, of bridges and trusses.

Keywords: vibrodiagnostics, identification, multiple defects, rod truss model, laboratory setup, information complex, damped oscillations, AFC

1. INTRODUCTION

Various technological and climatic factors can lead to damage of construction elements and a whole structure. The problem of control, especially hazardous in developing countries, often connected with impossibility to detect defects by science-based methods and absence of

---

* E-mail: alex837@yandex.ru.

appropriate diagnostic equipment, can lead to the fracture of constructions. It requires to develop and improve scientific approaches for control methods of construction damage.

Investigations in the field of vibrodiagnostics, development of new approaches and solution of the problems of damage identification in various constructions are in the center of many papers (see, for instance [1-4]). Theoretical and experimental research approaches and techniques, developed for vibrodiagnostics of rod models, are present in Refs. [5-8]. At the same time, the variety of technical systems has led to creation of diffent models. Their theoretical description is not always adequate to experimental data. Some modern approaches to solve the problem of identification of defects, based on the vibrodiagnostics methods, are present in Refs. [7-9]. The problems of control are based on the use of correlation parameters for oscillation frequency spectra at given damage degree of construction elements. These damages are identified by using available resonance methods and forced oscillations, in particular acoustic methods of non-destructive control, and in the most studies investigated and probed on standard construction elements.

By solving the problem of damage control in these models and statement of the criteria for finding defects, an additional information for the identification method is the spectrum of natural frequencies in a certain area. This spectrum is derived from analytical and experimental data for a model construction with given defects, which locate in certain places [10, 11]. To estimate the degree of construction damage, we can use the method, described in Ref. [12] and based on analysis of changes in spectral characteristics of the construction. In this paper, deviations of natural frequencies for different oscillation modes are stated in dependence on damage of the construction.

2. OBJECTS, METHODS AND RESULTS

The aim of the paper is to study the model of rod truss with defects by using the vibrodiagnostic method.

2.1. Object of Study

As the object of study, we chose a model section of truss bridge with rod elements. The construction elements used are the solid steel bars (from Steel 20) with rectangular cross-section (a length of the rod element $L = 250$ mm, a height of the cross-section $h = 8$ mm and a width $a = 4$ mm). Junction at nodes of the truss was performed by welding. The right point of the model was fixed on rigid support; the left point had a possibility of free horizontal displacement (Figure 1). The defect was in the form of rectangular notch with a width of 1 mm and a height $t = 5$ or 7 mm; it located at the junction of rods in the truss lower rod (crk, see Figure 1). Depth of the defect is calculated in relation to the height of the cross-section of the rod element. For symmetrical model, we had two surface defects on both lower rods of the truss (see Figure 2,a). In this paper, we studied transverse vibrations of the rod truss with two defects. At the point dat, shown in Figure 1, we collected information on the model vibrations.

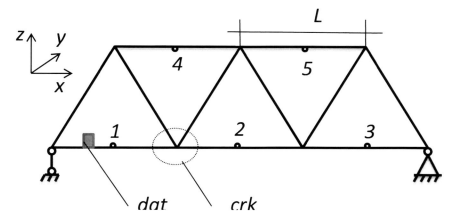

Figure 1. Scheme of truss model; *dat* is the place of sensor-accelerometer location, *crk* is the place of defect location at the node.

The numbers 1-5 in Figure 1 indicate the points of applied pulse or impact load, fixed by sensor. The common shape of the tested model and real node with defect of size $t = 7$ mm are shown in Figure 2.

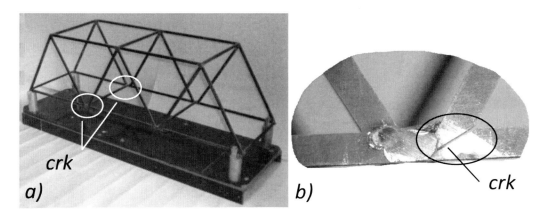

Figure 2. Common shape of the full-scale model of the bridge section (a), fragment of defect construction at the node (b).

The studies were conducted using laboratory measurements and information setup MIC-3, modeling truss elements with defects and sizes of structural damage. Block-scheme of the measurement system is shown in Figure 3.

The setup includes the following equipment: optical laser displacement meter RF-603; vibration sensor ADXL-103 for measuring acceleration of studied model elements; frequency CH3-64; electromagnetic exciter; LFO AFG3022B; power amplifier LV-102; computer; matching device of the optical sensor; measurement device ADC/DAC L-Card E14-440; PowerGraph software for the measurement device – analog-to-digital converter (ADC); original computer software developed by us for studying the truss damage.

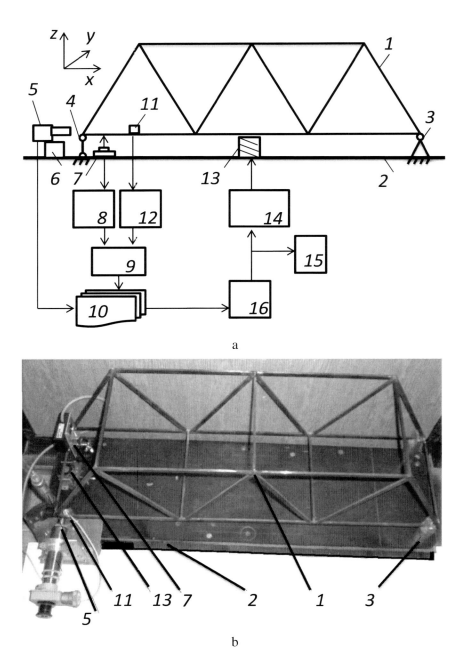

Figure 3. Block-scheme of measurement system (a) and full-scale model of truss with test elements of setup (b); these elements are: 1 – test model; 2 – base; 3, 4 – truss support; 5 – digital microscope; 6 – arm of the microscope; 7 – optical sensor; 8 – matching device of the optical sensor; 9 – analog-converter; 10 – PC; 11 – vibration sensor; 12 – matching device of vibration sensor; 13 – electromagnetic vibration exciter; 14 – power amplifier; 15 – frequency counter; 16 – low-frequency oscillator.

2.2. Methods of Research

Dynamic loading of the truss structure was carried out with an electromagnetic biasing device 13, mounted on base 2 of the setup in a vicinity of free edge of model at lower rod of the truss. Excitation of electromagnetic oscillations of the loading device 13 was conducted by low-frequency oscillator 16 and power amplifier 14. Powerful vibration exciter provided dynamics of the natural oscillations of the object with defects in the range $10 - 1000$ Hz.

The measurement part of dynamic load included: frequency counter 15, microscope 5, clampening bracket 6 for calibration of sensor and measurement of the test sample length of test model 1, sensor of impact loading 11, optical sensor 7 and matching devices 8, 12. Loads at various points of the construction were defined by using the vibration device 11 and optical sensor 7. The oscillation frequency with significant resonances was recorded by the frequency counter 15. Moreover, we performed the Fourier analysis of oscillations and calculated a frequency response by using computer software. The point of disposition of the sensor is shown in Figure 1. The signals from the primary transducers 7 and 11 after matching devices 8 and 12 transferred to the ADC E14-440, where they digitized and stored as computer data. PowerGraph software was applied to record, process and store the signals recorded by the analog-to-digital converter. Computer software, developed by us (VibroGraph), calculated frequency response and quantitative characteristics of the truss damage.

2.3. Results of Research

Diagnostics of the model construction conducted on the base of the research results of oscillations of the test section. These results were obtained during the construction vibroexcitation by using test measurements and data of MIC-3 setup, as well as the construction damage was estimated.

At first stage, we solved the problem of estimating the sensitivity of amplitude-frequency characteristic (AFC) of the test model (Figure 1) in order to define a place of application of the vibration load and location of the sensors. Figure 4,a shows registered waveforms at various points and an example of the frequency spectrum (Figure 4,b) of oscillations, registered at location of vibration sensor at the point 1 (see Figure 1). Based on analysis of numerical data, we obtained amplitude-frequency characteristics at different location points and amplitudes of vibroexcitation and refined resonance frequencies.

The second stage of the research was performed for nodes of the test model (see Figure 2) (by comparing with a case of symmetric location of the damages). We successively introduced notches on rods and studied two cases with the damage depth $t = 5$ mm and $t = 7$ mm.

In various points of the model, we obtained amplitude-time characteristics of the oscillations and calculated the amplitude-frequency characteristics by using the fast Fourier transform. At analysis of the model with notches, based on frequency response data, we identified three lower resonance frequencies. The results are present in Table 1. The deviations of resonance frequencies in dependence on relative depth of the notch, are shown in Figure 5.

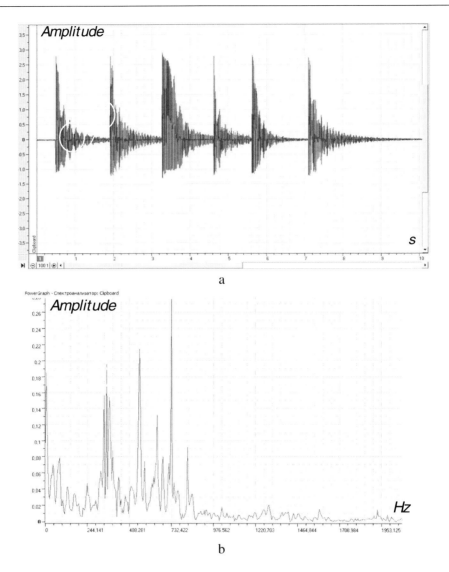

Figure 4. (a) Forms of damped oscillations, derived from the pulse excitation at different points of truss; in recording, it was used vibration sensor, mounted at the point 3 in Figure 1; (b) frequency spectrum of vibroexcitation at the pont 1; vibration sensor located also at the point 1.

Table 1. Natural resonance frequency and relative deviations of three lower oscillation modes for cases of defect presence and absence
(Δt is the relative depth of notch)

i-th mode	$t = 0$ ($\Delta t = 0\%$)		$t = 5$ mm ($\Delta t = 62.5\%$)		$t = 7$ mm ($\Delta t = 87.5\%$)	
	ω, Hz	$\Delta\omega$, %	ω, Hz	$\Delta\omega$, %	ω, Hz	$\Delta\omega$, %
Mode 1	30.7	100	29.3	95.4	27.1	88.3
Mode 2	51.0	100	46.3	90.8	44.2	86.7
Mode 3	88.1	100	78.0	88.5	69.9	79.3

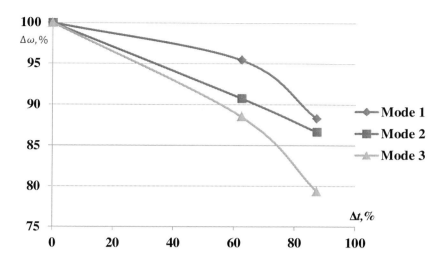

Figure 5. Dependence of deviations of three lower resonance frequencies for model truss section on relative depth of defect in the case of impact loading.

Analysis of influence of the damage degree (depth of notch) on the resonance frequency for three lower oscillation modes showed that the most sensitive to the damage degree was third oscillation mode of the studied model. When the depth of notch was 7 mm, we observed a deviation of the resonance frequency more than 20%.

The relative damage ξ was calculated by using the formula [12]:

$$\xi = 100 \frac{1}{n}\sum_{1}^{n} \xi_i; \quad \xi_i = 1 - (\omega_i^t / \omega_i^{t=0})^2, \qquad (1)$$

where n is the number of the considered oscillation modes ($n = 3$), ω_i^t and $\omega_i^{t=0}$ are the oscillation frequency of the construction with and without defect, respectively.

Table 2. The relative damage at various construction defects

i-th mode	$\xi_i(\Delta t = 62.5\%)$, %	$\xi_i(\Delta t = 87.5\%)$, %
Mode 1	9.0	22.0
Mode 2	18.0	25.0
Mode 3	21.0	37.0
average relative damage, $\bar{\xi}$	15.8	28.0

Table 2 shows data for each mode separately and average relative damage, $\bar{\xi}$, of a whole truss. For the size of defects $t = 5$ mm ($\Delta t = 62.5\%$), the average relative damage reached 15.8%, and for the size of defects $t = 7$ mm ($\Delta t = 87.5\%$), this value was 28.0%.

Conclusion

1. This study of damaged rod truss by vibrodiagnostic method is based on analysis of frequency response of the construction.
2. Methods and examples of using the information and measurement setup to identify the damage degree of the truss are considered.
3. This approach of damage diagnostics can be used, for example, to develop methods of non-destructive control and to state damage conditions for bridge trusses.

Acknowledgments

This work was supported in part by the Southern Federal University (grant No. 213.01.-2014/03VG), the Russian Foundation for Basic Research (grants Nos. 14-38-50915, 14–08–00546-a). I. A. Parinov acknowledges financial support from the Ministry of Education and Sciences of Russia in the framework of "Organization of Scientific Research" Government Assignment (grant No. 654).

References

[1] Del Grosso, A.; Lanato, F.; A critical review of recent advances in monitoring data analyses and interpretation for civil structures. *Proc. Eur. Conf. on Structural Control (4ECSC). St.-Peterburg.* 2008, *vol. 1*, 320-327.

[2] Akopyan, V.; Soloviev, A.; Cherpakov, A.; Chapter 4. Parameter Estimation of Pre-Destruction State of the Steel Frame Construction Using Vibrodiagnostic Methods. In: *Mechanical Vibrations: Types, Testing and Analysis*. A. L. Galloway (Ed.). Nova Science Publishers, New York. 2010, 147-161.

[3] Friswell, M. I.; *Phil. Trans. R. Soc. A.* 2007, *vol. 365*, 393–410.

[4] Akopyan, V.A.; Cherpakov, A.V.; Soloviev, A.N.; Kabelkov, A.N.; Shevtsov, S.N.; *Izvestiya VUZov. North-Caucasus Region. Technical Sciences.* 2010, *No 5*, 21 – 28 (In Russian).

[5] Cherpakov, A.V.; Akopyan, V.A.; Soloviev, A.N.; Rozhkov, E.V.; Shevtsov, S.N.; *Vestnik of DSTU.* 2011, *vol. 11(3)*, 312-318 (In Russian).

[6] Bocharova, O.V.; Lyzhov, V.A.; Andzhikovich, I.E.; *Vestnik SSC RAS.* 2013, *vol. 9(2)*, 11-15 (In Russian).

[7] Yesipov, Yu. V.; *Journal of Applied Mechanics and Technical Physics.* 2013, *vol. 54(2)*, 190-195.

[8] Postnov, V.A.; *Mechanics of Solids.* 2000, *No.6*, 155-160.

[9] Il'gamov, M.A.; Khakimov, A.G.; *Russian Journal of Nondestructive Testing.* 2009, *vol. 45(6)*, 430-435.

[10] Cherpakov, A.V.; Soloviev, A.N.; Gricenko, V.V.; Mohanty, S.C.; Parshin, D.Y.; Butenko, U.I.; Bocharova, O.V.; *On-line Journal "Engineering Journal of Don."* 2014, *No. 4*, 5 p. (In Russian).

[11] Burtseva, O.A.; Chipco, S.A.; Kaznacheeva, O.K.; Cherpakov, A.V.; *European Journal of Natural History*. 2012, *No. 4*, 39-44.

[12] Kotlyarevskiy, V.A.; *Journal of Oil and Gas Constructions*. 2014, *No. 2*, 46-52 (In Russian).

In: Proceedings of the 2015 International Conference … ISBN: 978-1-63484-577-9
Editors: Ivan A. Parinov, Shun-Hsyung Chang et al. © 2016 Nova Science Publishers, Inc.

Chapter 70

ANALYSIS OF FUNCTIONAL AND TEST DEFORMATION RESPONSES OF FRAME-ROD MODELS FOR EARLY DIAGNOSTICS OF INTEGRAL METALLIC CONSTRUCTIONS

E. V. Saulina, Yu. V. Esipov and A. I. Cheremisin

South Scientific Center of the Russian Academy of Sciences,
Rostov-on-Don, Russia

ABSTRACT

By introducing and registering the constructive form of metal (steel 3) seven and eleven span frame and solid rod made layouts systematize the experimental informative parameters. Which used for subsequent quick discernment of the specific technical condition of the layout of the studied set of conditions (source created defects in the form of bending elements, torsion layout). On the basis of the method of serial impact testing according to established and proposed and tested new informative parameters (criteria) to assess the degree of tension and (or) damage solid structures (without cutting). Which presented in the form of frequency shifts of the peaks areas of the Fourier image of the deformation of responses and measures of dissipation (blur phase diagrams) deformation vibrations. These criteria used for functional test and diagnosis of solid structures without cutting.

Keywords: integral designs, tensely deformed states, the control over deformation responses, informative parameters, the Fourier the analysis, thin ferroelectric films

1. INTRODUCTION

Owing to technology and asymmetry of connection of elements and action of external factors of environment, in difficult, and especially in welded, designs arise the internal tensely deformed states (TDS). As a rule, these statuses can be premises and (or) the reasons of defects and damages. In the field of the linear deformation of materials of elements, oscillations of such constructions considered as the linear. However, in some perturbation of the tensely deformed states is the source of as linear oscillations and unstable effects. These oscillations detected through a nonlinear deformation processes, as well as elastic, elastic-plastic and plastic hysteresis effects [1, 2, 3]. One of the reasons of development of tensely deformed status in constructions and accumulation of damages to their nodes of conjugation are elasto-plastic and plastic hysteresis effects. The study of these effects can give rich information on a construction status in general.

2. CLASSIFICATION OF INTRODUCED AND STUDIED STATES AND REGIMES OFSUSPENSION OF THE 7- AND 11-SPAN MODELS

The studied types tensely deformed states and consistently incorporated in models "accumulated" defects were described combination of the following technical conditions: C = {C0, C1, C2, C3}. C0 condition - a condition originally created (by electric welded) frame-core design, square frame and the core elements, which were prepared: the electric-cutting and electric welding. Condition C1 - defect 1: elastoplastic compression 7.5 mm opposite upper rods of the second span. Condition C2 - defect 2: (axially symmetric elastic-plastic torsion) spans 4 - 6 layout at an angle of 10 degrees. Condition C3 - defect 3: (axially symmetric elastic-plastic torsion) spans 4 - 6 layout at an angle of 25 degrees. The loss of form of the layout simulated and recorded during the experiment by introducing the following damage models (condition C1, C2, C3):

1) Thecentral deflection of mullions (from one to several on different spans) to a maximum of 15 mm is consistent with the relative importance of elastic-plastic deformation to a value of 0.06;
2) Axially symmetrical (rotary) twists at an angle of 25 degrees. This led to transverse deviation of the longitudinal axis of the mock-up to 110 mm with a length of eleven spans of the layout 2380 mm.

3. MODIFICATION AND PROBING OF THE METHOD OF SERIAL IMPACT TESTING AND FOURIER ANALYSIS OF DEFORMATION RESPONSES OF THE CONSTRUCTION

Modification of the method of testing [4, 5] was administered, and further the following operations and requirements.

1) A distinction (in shock) normalizing the Fourier image of the deformation response;
2) There was a quasi-linear increase in pulse u impact force, whose magnitude expressed in units of 1 to 5;
3) The conditions were set identical punches implementation: a) the duration (in the range of 0.2 to 0.4 ms); b) the centering of the point of impact; c) on the single condition "bounce" of the ball.

4) In cases of blunders or emissions during registration bending vibrations (due to electromagnetic interference, electrical interference, spurious vibration and mechanical stress) performed rejection results.

Consequently, the relative error of each series of varying pulse force of impact does not exceed 2%. These measures have improved the accuracy of the comparative analysis of linear and non-linear region of the Fourier image of the responses in terms of impact of acquisition and dissipation of deformation energy in the layout design [6, 7]. This is important in an environment where the estimated value of the energy of linear deformation vibrations exceeds the energy of nonlinear oscillations of two or more orders of magnitude.

4. ANALYSIS OF FOURIER IMAGES, CONSTRUCTION OF PHASE DIAGRAMS AND INTRODUCTION OF NEW INFORMATIVE PARAMETERS

In the construction and analysis of the phase diagrams of the deformation response elements and layouts that were originally created in the first (cut electric-welded by electric means or) state we established a sign in the form of "splitting" of the chart. and the "splitting" of increased with increasing the force of impact. Phase diagrams for describing the basic conditions and modes suspension span eleven layout shown in the Figure 1.

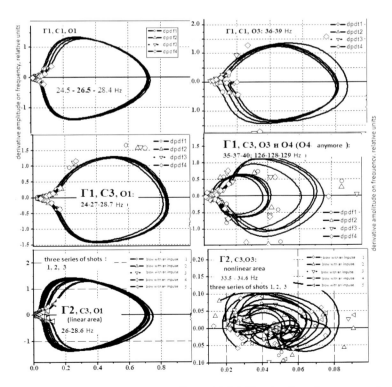

Figure 1. Phase diagrams of linear (left) and nonlinear (right) Fourier domain response in the initial layout (C1) and "the injury" (C3) state.

From the standpoint of "sensitivity" to changes in tension and loss of form solid layout chosen and the following areas:

1) an area of O3 deformation response, which in the free suspension (G2) emerged as a nonlinear variation with frequency shift df = 3 Hz and dissipation measure Q j (u) ≤ 50%;
2) the linear region with a monotonic decrease O1: Q j (u) (G1, G2): 98 - 4 (%)

CONCLUSION

Based on the development and approbation of a method of shock testing of intense constructions on non-linear deformation images the following results received.

1) Information parameters for distinguishing of intense statuses of a prototype and (or) elements of rod construction, which treat, are set:
   A. The relation of the area of non-linear area of a phase portrait in damaged (strained) and to the area of non-linear area of a phase portrait in the initial status of construction;
   B. The relation of the area of non-linear area to the area of the linear area of a phase portrait of deformation responses in damaged (strained) constructions.
   C. The frequency shift of peaks (maxima) of the linear areas of Fourier of an image of deformation responses of construction (element).
2) The following criteria for distinguishing of the intense (damaged) statuses of construction constructed:
   a. Existence of non-linear area of Fourier of an image in deformation responses of construction (element) to serial shock testing as qualitative criterion;
   b. A condition of finding of an index η in the field of values: $\eta \geq 0.1$;
   c. A condition of finding of shift Δf on frequency from 0.5 to 2 and more (Hz) of a maximum of the linear areas of Fourier of images of deformation responses of construction (element) where criteria and (in) are the quantitative criteria of distinguishing.

This work was supported by grant RFFI № 14-08-00546.

REFERENCES

[1] Ekaterina V. Saulina, Yury V. Esipov, Experimental Substantiation of Early Diagnostic' Criteria Of Complex Intense Structures. *International Symposium on "Physics and Mechanics of New Materials and Underwater Applications" (PHENMA2013, 5-8 June, 2013, Kaohsiung, Taiwan).*
[2] Esipov Yu, Mukhortov you. M., go I. Evaluation type status bar elements of structures on the basis of registration in the deformation response of serial impact excitation. *Defectoscopy.* 2010, vol. 46. 7, 76-81.

[3] Esipov Yu An experimental approach to the construction of the spectral criterion for the diagnosis of periodic core design based on ferroelectric strain gauges. *Applied Mechanics and Technical Physics*. 2013, vol. 54, 156-160.

[4] Saulina E.V., Esipov Y.V. The method of distinguishing between states of stress models of structural integrity of the deformation response. *Control. Diagnostics*. 2014, vol. 1, 34-39.

[5] Adams, D., M. Nataraju. A nonlinear dynamical systems framework for structural diagnosis and prognosis. *International Journal of Engineering Science.*2002, vol. 40, 1919–1941.

[6] Douglas E. Adams, and Charles R. Farrar. Classifying Linear and Nonlinear Structural Damage Using Frequency Domain ARX Models. *Structural Health Monitoring*. 2002, vol.1, 185–201.

[7] Sedmak S., Sedmak A., Arsi M., Tuma J. An Experimental Verification Of Numerical Models For The Fracture And Fatigue Of Welded Structures. *Materials and Technology*. 2007, vol.41, 173–178.

In: Proceedings of the 2015 International Conference … ISBN: 978-1-63484-577-9
Editors: Ivan A. Parinov, Shun-Hsyung Chang et al. © 2016 Nova Science Publishers, Inc.

Chapter 71

MATHEMATICAL MODELING OF PLANE-PARALLEL ELECTROCHEMICAL ELECTROLYTIC CELL WITH PERFORATED CATHODE

A. N. Gerasimenko[1], Yu. Ya. Gerasimenko[1,], E. Yu. Gerasimenko[1] and A. I. Emelyanov[1]*

[1]Don State Technical University,
Rostov-on-Don, Russia

ABSTRACT

This chapter concernsmathematical modeling of plane-parallel electrochemical electrolyzes with a perforated cathode. We have performed mathematical modeling of plane-parallel electric and concentration fields of electrolyte in electrolyzes with planar electrodes, coinciding with boundary surfaces of opposite faces.

Keywords: mathematical modeling, electrolyte, concentration, cathode, plate, electrical current density, mass-transfer

1. MATHEMATICAL MODEL

We consider the plan-parallel of electrolyte in an electrolytic cell with planar electrodes, coinciding with the boundary surfaces of opposite faces. A cathode is made in the form of flat strips and it is situated at the same plane. By modeling the electrolytic cell with depth a, we take the following physical assumptions [1-4]:

(i) electrolytic cell walls are made of dielectric (Figure 1),
(ii) the electrical conductivity of the electrolyte is a constant,
(iii) the diffusion coefficient of the electrolyte is a constant,

---

* E-mail: knagna_olga@inbox.ru.

(iv) the kinetics of electrode processes of electrolytic cell is controlled at diffusion stage in the electrolyte;
(v) mass-transfer in the electrolyte is described by hyperbolic diffusion equation with relaxation constant τ_r,
(vi) electric field in electrolyte is potential,
(vii) electric current density on the surfaces of all electrodes is distributed uniformly.

The mathematical statement of the initial-boundary problem is present as follows:

$$\frac{\partial c}{\partial t} + \tau_r \frac{\partial^2 c}{\partial^2 t} = D \frac{\partial^2 c}{\partial^2 x}; \tag{1}$$

$$c(x;y;0) = c_0 \ ; \ \frac{\partial c}{\partial t}(x;y;0) = 0 \ ; \ \frac{\partial c}{\partial y}(x;0;t) = 0 \ ; \ \frac{\partial c}{\partial x}(0;y;t) = 0 \ ; \ y \in (0;d_1) \ ;$$

$$\frac{\partial c}{\partial x}(0;y;t) = \alpha N \frac{I(t)}{\Theta_1 a} \ ; \ y \in (d_1;d_1+\Theta_1) \ ; \ \frac{\partial c}{\partial x}(0;y;t) = 0 \ ; \ y \in (d_1+\Theta_1;d_2) \ ;$$

$$\frac{\partial c}{\partial x}(0;y;t) = (1-\alpha) N \frac{I(t)}{\Theta_2 a} \ ; \ y \in (d_2;d_2+\Theta_2) \ ; \ \frac{\partial c}{\partial x}(0;y;t) = 0 \ , \ y \in (d_2+\Theta_2;h) \ ;$$

$$\frac{\partial c}{\partial x}(l;y;t) = 0 \ , \ y \in (0;b) \ ; \ \frac{\partial c}{\partial x}(l;y;t) = N \frac{I(t)}{\Theta a} \ ; \ y \in (b;b+\Theta) \ ; \ \frac{\partial c}{\partial x}(l;y;t) = 0 \ ;$$

$$y \in (b+\Theta;h) \ ,$$

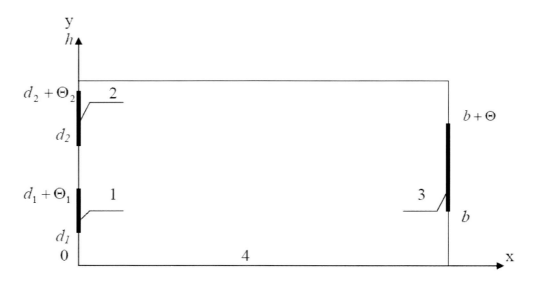

Figure 1. Construction of electrolytic cell: 1, 2 – cathodes; 3 – plate; 4 – electrolyte.

where τ_r is the relaxation constant, D is the diffusion coefficient, $N > 0$ is the electrode kinetic constant, α is the part of external current for first cathode, c_0 is the initial concentration.

We shall solve this problem by the Laplace operator method [5, 6]. We introduce the next compliances:

$$c(x;y;t) \stackrel{\circ\circ}{=} C(x;y;p) \stackrel{\circ}{;} \quad I(t) \stackrel{\circ\circ}{=} I(p)$$

The original initial-boundary problem reduces to the boundary problem for concentration reflection $\overset{\circ}{C}(x;y;p)$:

$$\frac{\partial^2 \overset{\circ}{C}(x;y;p)}{\partial x^2} + \frac{\partial^2 \overset{\circ}{C}(x;y;p)}{\partial y^2} - \frac{p(1+\tau_r p)}{D}\overset{\circ}{C}(x;y;p) = -\frac{(1+\tau_r p)c_0}{D} \tag{2}$$

$$\frac{\partial \overset{\circ}{C}(x;0;p)}{\partial y} = 0 \tag{3}$$

$$\frac{\partial \overset{\circ}{C}(x;h;p)}{\partial y} = 0 \tag{4}$$

$$\frac{\partial \overset{\circ}{C}(0;y;p)}{\partial x} = 0$$
$$y \in (0;d_1) \tag{5}$$

$$\frac{\partial \overset{\circ}{C}(0;y;t)}{\partial x} = \alpha N \frac{\overset{\circ}{I}(t)}{\Theta_1 a}, \ y \in (d_1;d_1 + \Theta_1); \tag{6}$$

$$\frac{\partial \overset{\circ}{C}(0;y;p)}{\partial x} = 0, \ y \in (d_1 + \Theta_1;d_2); \tag{7}$$

$$\frac{\partial \overset{\circ}{C}(0;y;t)}{\partial x} = (1-\alpha) N \frac{\overset{\circ}{I}(t)}{\Theta_2 a}, \ y \in (d_2;d_2 + \Theta_2); \tag{8}$$

$$\frac{\partial \overset{\circ}{C}(0;y;p)}{\partial x} = 0, \ y \in (d_2 + \Theta_2;h); \tag{9}$$

$$\frac{\partial \overset{\circ}{C}(l;y;p)}{\partial x} = 0, \ y \in (0;b);$$

(10)

$$\frac{\partial \overset{\circ}{C}(l;y;t)}{\partial x} = N\frac{I(t)}{\Theta a}, \ y \in (b;b+\Theta);$$

(11)

$$\frac{\partial \overset{\circ}{C}(l;y;p)}{\partial x} = 0, \ y \in (b+\Theta;h)$$

(12)

We introduce new unknown function as follows:

$$\overset{\circ}{U}(x;y;p) = \overset{\circ}{C}(x;y;p) - \frac{c_0}{p}$$

(13)

Relationship (13) allows one to reduce problem (2) – (12) to the following:

$$\frac{\partial^2 \overset{\circ}{U}(x;y;p)}{\partial x^2} + \frac{\partial^2 \overset{\circ}{U}(x;y;p)}{\partial y^2} - \frac{p(1+\tau_r p)}{D}\overset{\circ}{U}(x;y;p) = 0$$

(14)

$$\frac{\partial \overset{\circ}{U}(x;0;p)}{\partial y} = 0$$

(15)

$$\frac{\partial \overset{\circ}{U}(x;h;p)}{\partial y} = 0$$

(16)

$$\frac{\partial \overset{\circ}{U}(0;y;p)}{\partial x} = 0; \ y \in (0;d_1)$$

(17)

$$\frac{\partial \overset{\circ}{U}(0;y;t)}{\partial x} = \alpha N\frac{I(t)}{\Theta_1 a}$$
$$y \in (d_1;d_1+\Theta_1)$$

(18)

$$\frac{\partial \overset{\circ}{U}(0;y;p)}{\partial x} = 0, \ y \in (d_1+\Theta_1;d_2)$$

(19)

$$\frac{\partial \overset{\circ}{U}(0;y;t)}{\partial x} = (1-\alpha)N\frac{I(t)}{\Theta_2 a}$$
$$y \in (d_2; d_2 + \Theta_2) \tag{20}$$

$$\frac{\partial \overset{\circ}{U}(0;y;p)}{\partial x} = 0, \ y \in (d_2 + \Theta_2; h) \tag{21}$$

$$\frac{\partial \overset{\circ}{U}(l;y;p)}{\partial x} = 0, \ y \in (0;b) \tag{22}$$

$$\frac{\partial \overset{\circ}{U}(l;y;t)}{\partial x} = N\frac{I(t)}{\Theta a}, \ y \in (b;b+\Theta) \tag{23}$$

$$\frac{\partial \overset{\circ}{C}(l;y;p)}{\partial x} = 0, \ y \in (b+\Theta; h) \tag{24}$$

Differential equation (14) with similar boundary conditions (15), (16) allows one to define eigen-functions of problem (14) – (24) in the form:

$$Y_k(y) = \begin{cases} \dfrac{1}{\sqrt{h}}, k = 0 \\ \sqrt{\dfrac{2}{h}}\cos\dfrac{k\pi}{h}y, k = 1,2,... \end{cases} \tag{25}$$

We present [7 – 9] solution $\overset{\circ}{U}(x;y;p)$ of the problem (14)-(24) in form of Fourie series on eigen-functions (25):

$$\overset{\circ}{U}(x;y;p) = \beta_0(x;p)Y_0(y) + \sum_{k=1}^{\infty}\beta_k(x;p)Y_k(y) \tag{26}$$

Then we substitute expression (26) into differential equation (14) and get auxiliary equations for coefficients $\beta_0(x;p)$ and $\beta_k(x;p)$:

$$\beta_0''(x;p) - \frac{p(1+\tau_r p)}{D}\beta_0(x;p) = 0 \tag{27}$$

$$\beta_k{''}(x;p) - \left(\frac{k^2\pi^2}{h^2} + \frac{p(1+\tau_r p)}{D}\right)\beta_k(x;p) = 0 \tag{28}$$

The common solutions of Equations (27), (28) are the functions:

$$\beta_0(x;p) = M_0(p)sh\sqrt{\mu_0(p)}x + N_0(p)ch\sqrt{\mu_0(p)}x \tag{29}$$

$$\beta_k(x;p) = M_k(p)sh\sqrt{\mu_k(p)}x + N_k(p)ch\sqrt{\mu_k(p)}x \tag{30}$$

where

$$\mu_0(p) = \frac{p(1+\tau_r p)}{D} \quad \mu_k(p) = \frac{k^2\pi^2}{h^2} + \frac{p(1+\tau_r p)}{D}$$

The substitution of (29), (30) into (26) leads to the following representation:

$$\overset{\circ}{U}(x;y;p) = \left(M_0(p)sh\sqrt{\mu_0(p)}x + N_0(p)ch\sqrt{\mu_0(p)}x\right)Y_0(y) +$$

$$+ \sum_{k=1}^{\infty}\left(M_k(p)sh\sqrt{\mu_k(p)}x + N_k(p)ch\sqrt{\mu_k(p)}x\right)Y_k(y). \tag{31}$$

Then we find unknown coefficients $M_0(p)$, $N_0(p)$, $M_k(p)$ and $N_k(p)$ by using representation (31) and non-uniform boundary conditions (17) – (24) in the form:

$$M_0(p) = \frac{N\overset{\circ}{I}(p)}{a\sqrt{h}\cdot\sqrt{\mu_0(p)}} \tag{32}$$

$$M_k(p) = \sqrt{\frac{2}{h}}\cdot\frac{2N\overset{\circ}{I}(p)}{a}\cdot\frac{h}{k\pi}\cdot\frac{\dfrac{\alpha}{\Theta_1}\sin\dfrac{k\pi(2d_1+\Theta_1)}{h\cdot2}\sin\dfrac{k\pi\Theta_1}{h\cdot2} + \dfrac{(1-\alpha)}{\Theta_2}\sin\dfrac{k\pi(2d_2+\Theta_2)}{h\cdot2}\sin\dfrac{k\pi\Theta_2}{h\cdot2}}{\sqrt{\mu_k(p)}}, \tag{33}$$

$$N_0(p) = -\frac{N\overset{\circ}{I}(p)sh\dfrac{\sqrt{\mu_0(p)}l}{2}}{a\sqrt{h}\cdot\sqrt{\mu_0(p)}ch\dfrac{\sqrt{\mu_0(p)}l}{2}} \tag{34}$$

$$N_k(p) = \frac{\sqrt{\frac{2}{h}} \cdot \frac{2NI(p)}{\Theta a} \cdot \frac{h}{k\pi} \sin\frac{k\pi(2b+\Theta)}{h\cdot 2} \sin\frac{k\pi\Theta}{h\cdot 2}}{\sqrt{\mu_k(p)}sh\sqrt{\mu_k(p)}l} -$$

$$-\frac{\sqrt{\frac{2}{h}} \cdot \frac{2NI(p)h}{ak\pi}\left(\frac{\alpha}{\Theta_1}\sin\frac{k\pi(2d_1+\Theta_1)}{h\cdot 2}\sin\frac{k\pi\Theta_1}{h\cdot 2} + \frac{(1-\alpha)}{\Theta_2}\sin\frac{k\pi(2d_2+\Theta_2)}{h\cdot 2}\sin\frac{k\pi\Theta_2}{h\cdot 2}\right)}{\sqrt{\mu_k(p)}sh\sqrt{\mu_k(p)}l}ch\sqrt{\mu_k(p)}l.$$

(35)

After substitution of coefficients (32)-(35) into common solution (31), we obtain:

$$U(x;y;p) = \frac{NI(p)}{a\sqrt{h}} \cdot \frac{sh\sqrt{\mu_0(p)}\left(x-\frac{l}{2}\right)}{\sqrt{\mu_0(p)}ch\sqrt{\mu_0(p)}\frac{l}{2}}Y_0(y)+$$

$$+\sum_{k=1}^{\infty}\sqrt{\frac{2}{h}} \cdot \frac{2NI(p)h}{ak\pi} \cdot \left[\frac{\frac{1}{\Theta}\sin\frac{k\pi(2b+\Theta)}{h\cdot 2}\sin\frac{k\pi\Theta}{h\cdot 2}ch\sqrt{\mu_k(p)}x}{\sqrt{\mu_k(p)}ch\sqrt{\mu_k(p)}l} - \frac{\left(\frac{\alpha}{\Theta_1}\sin\frac{k\pi(2d_1+\Theta_1)}{h\cdot 2}\sin\frac{k\pi\Theta_1}{h\cdot 2} + \frac{(1-\alpha)}{\Theta_2}\sin\frac{k\pi(2d_2+\Theta_2)}{h\cdot 2}\sin\frac{k\pi\Theta_2}{h\cdot 2}\right)ch\sqrt{\mu_k(p)}(l-x)}{\sqrt{\mu_k(p)}ch\sqrt{\mu_k(p)}l}\right]Y_k(y).$$

(36)

Now we can substitute the function $U(x;y;p)$ into formula (13) and get concentration reflection in the form:

$$C(x;y;p) = U(x;y;p) + \frac{c_0}{p}$$

(37)

The formulae (36) and (37) allow one to calculate the concentration field of the electrolyte $C(x;y;p)$. For fulfillment of the calculation procedure, it is necessary to know a law of current $I(t)$ change. The simple example corresponds to the constant current I_0. It allows one to define exact meaning of the concentration $c(x;y;t)$ with help from the approach presented in [10].

REFERENCES

[1]   Gerasimenko, Yu. Ya. Mathematical Modeling of Electrochemical Systems. SRSTU Press, Novocherkassk. 2009, pp. 1 – 314.

[2]   Gerasimenko, Yu. Ya.; Gerasimenko, E. Yu. Mathematical Modeling of Electrochemical System with Diffusive Hyperbolic Control of Electrode Kinetics. IWK Information Technology and Electrical Engineering – Devices and Syistems, Materials and Technology for the Future, Germany, *Ilmenau*, 7 – 10 September 2009, p. 391-392.

[3]   Gerasimenko, Yu. Ya.; Tsygulev, N. I.; Gerasimenko, A. N.; Gerasimenko, E. Yu.; Fugarov, D. D.; Purchina, O. A. *Life Science Journal*, 2014, vol. 11(12), 265-269.

[4] Gerasimenko, Yu. Ya.; Gerasimenko, E. Yu. IWK Information Technology and Electrical Engineering – Devices and Syistems, Materials and Technology for the Future, Germany, *Ilmenau,* 7 – 10 September 2009. p. 389-390.

[5] Beyer, W. H. CRS Standard Mathematical Tables and Formulae. CRS Press, Boca Raton, 1991, pp. 1 – 609.

[6] Davis, B. Integral Transforms and Their Applications. Springer-Verlag, New York, 1978.

[7] Butkcov, E. Mathematical Physics. Addison-Wesley, Reading, Mass., 1968.

[8] Zwillinger, D. Handbook of Differential Equations. Academic Press, Boston, 1989, pp. 1 – 673.

[9] Farlow, S. J. Partial Differential Equations for Scientists and Engineers. John Wiley and Sons, New York, 1982.

[10] Lavrentyev, M. A.; Shabat, B. V. Methods of Theory of Functions of Complex Variable. *Nauka,* Moscow, 1974, pp. 1 – 320.

In: Proceedings of the 2015 International Conference ...   ISBN: 978-1-63484-577-9
Editors: Ivan A. Parinov, Shun-Hsyung Chang et al.   © 2016 Nova Science Publishers, Inc.

Chapter 72

ALGORITHM OF REDISTRIBUTION OF CONNECTIONS BETWEEN OUTPUTS ON THE BASE OF SWARM INTELLECT

O. A. Purchina[*], *A. Y. Poluyan, D. D. Fugarov and Y. N. Bugaeva*

Don State Technical University, Rostov-on-Don, Russia

ABSTRACT

From a mathematical point of view routing is the most complicated problem of selection from a vast number of optimal solution choices. Development of methods and algorithms for solving the routing problem is carried out for many years though the issue is still relevant. This is due to the fact that, first of all, this is a nondeterministic polynomial time complete problem (so called NP-complete problem), and thus to develop a universal algorithm for finding the exact optimal solution within a reasonable time is quite a challenging task. The emergence of new, more sophisticated computer equipment, giving powerful computing resources, as well as exclusive standards of the projected devices is the driving force behind the development of new algorithms for solving the routing problem. There are several approaches to solve NP-complete problems.

Keywords: swarm intellect, ant colony, adaptive behavior, collective alternative adaptation, connections permutation, optimization

1. INTRODUCTION

Due to the high complexity and dimensionality of the routing problem, design engineering usually bases on employment of a hierarchical approach. There are two basic

---

[*] knagna_olga@inbox.ru.

routing levels: global and detailed routings. The goal of the global routing is to distribute connections over sub-areas. Detailed routing consists in implementing the connections in each sub-area. Usually detailed routing is divided into the channel routing and the routing of the switch-boxes. It is noted that a final result of the routing depends on a large extent of the initial distribution of connections for pins. Consequently, the routing problem is proposed as a task of pins permutation [5, 8, 16].

The connection permutation between pins is possible in case where the pins are functionally equivalent. Two pins (or terminal row) call functionally equivalent, if the switching of corresponding connected circuits does not change the logical function of the circuit [13]. The goal of permutation is to reduce the density of the routing areas, to decrease the length of the connections, to reduce the number of crossings, to improve the integration density, etc.

The connections permutation phase usually performs after the global routing phase. The main purpose of connections permutation between the pins is to increase the routability index for the next step of a detailed routing. The criteria based on the length of the conductors (total length of communications, length of the longest communication) have acquired widespread acceptance [7]. The length l_t of half the perimeter of a rectangle, contouring the pins of the circuit, uses to estimate the total length of the circuit. Although these criteria are easily computable and can be integrated into an optimization tool, they roughly simulate the routability index of the circuit. In the work [16] instead of using the criterion of the total length of conductors, based on the simulation of routing that does not depend on the topology, i.e., is TP-free, a target function, which allows one more accurately assess the conductors' concentration, is used. Switchboard is split into an ordered set of areas $\Theta = < \Theta_k | \Theta_{k+1} \subset \Theta_k, k =1, 2,..., N >$. Each area is included into the next area with the proper order of magnitude. The formal criterion writes as

$$Q^1 = \sum_{k=0}^{N} Q(\Theta_k)$$

where $Q(\Theta_k)$ is the number of circuits crossing the area border.

In the works [11, 17, 18], usually the reference grid $X \times Y$ is superimposed on the circuit area at routing simulation. Routing biddings simulated for each edge of the grid. To account for the close connection between the allocation and routing tasks the special local criteria proposed. These criteria are based on the topology simulation, i.e., TP-based, providing building of the routing trees on the routing grid.

Recently, the methods based on the application of artificial intelligence techniques are used increasingly for solving various "complicated" problems, which include also the problems of distribution (or binding) of the circuits with pins. Especially, the rapid growth of interest is observed in the development of algorithms inspired by natural systems [13 − 15, 22, 23]. One of the latest trends of such approaches are connected with multi-agent methods of intelligent optimization, based on the simulation of collective intelligence [3, 4].

The architecture and operation framework of biological control systems that ensure the ability of animals to specialize and adapt to the constantly changing conditions of the external environment are the subject of active research in the leading scientific centers. Among them, the swarm intellect methods are developing most actively [6, 21, 28], in which a set of relatively simple agents develop a strategy of their behavior without the global control. The idea of applying the ant colony (AC) algorithm bases on simulating the behavior of ants, associated with their ability to quickly find the shortest path from the anthill to the food source [6]. By moving, the ant marks the path by pheromone, and this information uses by other ants to choose the path. The ant colony algorithm bases on a simulation of ants' movement on the problem-solving graph. Distance, covered by the ant, is displayed as soon as the ant visits all the nodes of the graph. The process of finding solutions by ant-based algorithm is iterative. As for bees, the basic mechanisms of their behavior are as follows. First, some "scout" bees fly out of the hive in a random direction to seek out sources of nectar. After that, other bees fly to found nectar sources. At that, the number of bees flying to the source depends on the amount of available nectar. It is supposed that greater quantity of nectar in the source leads the more bees fly to this source, while the "scout" bees fly again to look for other sources, and the process repeats.

This article modifies methodology for the representation of combinatorial problems in the form of swarm algorithms in accordance with the mathematical models of bees and ants behavior [12].

The work [19] outlines a method for solving the problem of the connections distribution between the pins based on the simulation of adaptive behavior of ant colony.

The advantage of the ant colony algorithm guarantees convergence to the optimal solution, as well as a higher rate of finding optimal solution as compared to traditional methods. The shortcomings include the fact that the proposed algorithm is quite dependent on the initial search parameters, which choose experimentally, as well as the lack of a detailed study of the search space. The work [10] outlines a method for solving the problem of connections distribution between the pins, based on the simulation of collective alternative adaptation. The experimental studies have shown that the use of this algorithm gives a significant reduction in the initial density of the channel, as well as the total length of the horizontal fragments and the total number of non-removable connection intersections with each other [26].

To enhance the strengths and mitigate the weaknesses of the considered methods, we propose a paradigm of swarm algorithm based on integration of the two models: (i) an adaptive behavior of ant colony and (ii) collective alternative adaptation.

2. METHODOLOGY

The problem of connections permutation between the pins of a special class is of principal interest. This concerns the problem on connections permutation in the channel or switching unit. Connections permutation produces before the detailed routing phase and it aims at reducing the length of the connections inside the channel, as well as the density of the channel that facilitates the routing process [19].

2.1. Problem Statement

Let the set of pins $V = < v_i |\ i = 1, 2,..., n >$ are connects to the set of terminal server couples $T^0 = < t_i |\ i=1,2,..., n >$, respectively. A terminal is an endpoint of a connection linking it to the pin. Preliminarily, based on the analysis of the circuit, a set A of the equivalent pins group called terminal row forms: $A = \{A_e |\ e = 1, 2,..., n_e\}$. Here $A_e = \{a_{ej} |j = 1, 2,..., n_{ej}\}$ is the terminal row, T^*_e is the group of terminals connected to a terminal row A_e. One must find a valid permutation of terminals T^* at which the criterion has a better value. In turn, a valid permutation of terminals T^* is the set of admissible permutations of terminals T^*_e, either of which defines the connection of terminals to the appropriate terminal row A_e. $T^* = \cup\ T^*_e$, $T^*_i \cap T^*_j = \varnothing$.

2.2. Organization of Search Procedures Based on Simulation of Adaptive Behavior of Ant Colony

The ant colony method can be applied to any combinatorial problem, which is consistent with the following requirements [8].
Problem representation:

(i) the solution space must be represented as a graph with a set of nodes and edges between the nodes;
(ii) the correspondence should be set up between the solution of the combinatorial problem and the route in the graph.

It is necessary to develop the rules (methods) for

(i) initial distribution of ants in the nodes of the graph;
(ii) building a valid alternative solutions (route in a graph);
(iii) determining the probability of movement of an ant from one node to another;
(iv) updating pheromones on the edges (nodes) of the graph;
(v) pheromone evaporation.

The search for solutions is carried out on the family of complete sub-graphs $G = \{G_e | e = 1, 2,..., n_e\}$, $G_e = (X_e, U_e)$. The nodes of the X_e set correspond to the terminals of a T_e^* set, connected to the terminal row A_e.

In general, the search for a valid permutation of the terminals T^* is carried out by the community of ants cluster $Z = \{z_e | e = 1, 2,..., n_e\}$. The number of ants is equal to the number of terminal rows A_e. At each iteration, each ant z_e of Z-cluster builds on a corresponding graph $G_e = (X_e, U_e)$ of its particular solution, i.e., valid permutation of terminals T_e^*.

The general solution, i.e., a valid permutation of terminals T^*, is determined by a set of admissible permutations of terminals T_e^*, built by ants of a single cluster. In other words, the number of solutions, generated by the ants at each iteration, is equal to the number of ant clusters.

Simulation of the ants behavior in a problem of finding the permutation of terminals T^* associates with the distribution of pheromone on the edges of the G graphs family. Initially, the same (small) amount of pheromone Φ/v lays at all edges of the G graphs family, where $v = |\cup U_e|$. The parameter Φ specifies, *a priori*. The problem-solving process is iterative. Each iteration l consists of three phases. The number of solutions n_k, which form by the ants at each iteration, is given. This serves basis to form n_k clusters of Z_k. At the first stage, each ant z_{ek} of each cluster Z_k builds a concrete solution, i.e., M_{ek}-route on the corresponding graph $G_e = (X_e, U_e)$. In other words, the ants of Z_k-cluster build cluster of routes $M_k = \{M_{ek}|e =1, 2,..., n_e\}$. The sequence of X_e nodes set in the route M_{ek} corresponds to the permutation of terminals T_{ek}^*. For a set (cluster) of routes, built by the ants of a single cluster Z_k and corresponding to the k-th solution, we determine the value of the optimization criterion F_k.

At the second stage, each ant $z_{ek} \in Z_k$ lays pheromone on the edges of the built route in the graph G_e in amount proportional to the criterion F_k. We use ant-cycle method of ant systems. In this case, all ants of all clusters lay pheromone, simultaneously. The third step refers to the total evaporation of pheromone on the edges of the family of solution-searching graphs G. After performing all of the steps at the iteration, we find the cluster with the best solution, which saves. Then we proceed to the next iteration.

2.3. The Ant Colony Algorithm

(1) A set of terminal rows $A = \{A_e| e = 1, 2,..., n_e\}$ forms in accordance with the initial data, based on the circuit analysis. Then the value of n_e is determined.

$A_e = \{a_{ej}|j = 1, 2,...,n_{ej}\}$ is the terminal row.

(2) The solution-searching graphs $G = \{G_e|e = 1, 2,..., n_e\}$, $G_e = (X_e, U_e)$ are generated. By this, the nodes of set X_e correspond to the terminals of the set T_e^*, connected to a terminal row A_e. $|A_e|=|X_e|$

(3) The initial amount of pheromone Φ/v, lays on all edges (or nodes) of the graph G_e, where $v = |\cup U_e|$. Next the values of the parameters Q_i, α, and ρ are defined.

(4) The number of clusters n_k allows one to form n_k clusters of ants.

The number of iterations n_l is defined.

We give $l = 0$ (where l is the number of iterations).

(5) $l = l + 1$ (the next iteration is selected).

We give $k = 0$ (where k is the cluster number).

(6) $k = k + 1$ (the next cluster is selected).

We give $e = 1$ (e is the index of a subset of equivalent pins A_e).

(7) The rout M_{ek} is constructed employing the ant colony algorithm.

(8) If $e < n_e$, than $e = e + 1$. Another ant in the cluster is selected and we go to the step (7) to build the next route, otherwise due to the fact, that all routes of Z_k-cluster are built, we go to step (9).

(9) The objective function $F_k(l)$ is calculated for a set of routes, built by the ants of a single cluster Z_k on l-th iteration. If $k < n_k$, we go to step (6), otherwise, due to the fact that all the routes for all clusters Z_k are built, we go to step (10).

(10) The solution with the best value of the optimization criterion F_k* and the set of routes, built by the ants of the single Z_k-cluster, are determined. *We give $k = 0$.*

(11) $k = k + 1$. The same amount of pheromone lays at all edges of all the routes, built by ants of a single cluster Z_k at the l-th iteration, $h_k(l) = \lambda/F_k(l)$.

(12) If $k < n_k$, we go to step (11), otherwise we go to step (13).

(13) After each agent has formed a solution and laid the pheromone, the total evaporation of pheromone on the edges of the G graphs takes place at the 3-rd stage in accordance with formula: $h_{ij}(t+1) = h_{ij}(t)(1-\rho) + \varphi_{ij}(l)$, where $h_{ij}(t)$ is the pheromone level on the edge (i, j), ρ is the coefficient of renewal.

(14) If $l < n_l$, we go to step (5), otherwise we go to step (15).

(15) The end of the algorithm.

2.4. The Ant Colony Algorithm

(1) Choose (randomly) a node x_0 at the graph $G_e = (X_e, U_e)$ for the initial location of an ant $z_{ek} \in Z_{k..}$

(2) $t = 1$, t is the $M_{ek}(t)$ routing step at the G_e-graph.

The node x_0 is included in the route $M_{ek}(t)$.
We give $x_p(t) = x_0$, where $x_p(t)$ is the last node of the $M_{ek}(t)$-route.

(3) A set of nodes $X_{ek}(t) \in X_e$ neighboring to $x_p(t)$ forms. By this, each of the nodes $x_i \in X_{ek}(t)$ can be added to the $M_{ek}(t)$ route under formation.

(4) For each node $x_i \in X_{ek}(t)$, the parameter h_{iek} is calculated, indicating the total level of pheromone laid on the edge of the G_e graph, which connects x_i with the node $x_p(t)$.

(5) The probability $P_{iek}(t)$ of the inclusion of the node $x_i \in X_{ek}(t)$ in the route $M_{ek}(t)$ under formation is defined as the ratio $P_{iek}(t) = h_{iek} / \sum_i h_{iek}$

(6) The agent with the probability $P_{iek}(t)$ selects one of the nodes $x_i \in X_{ek}(t)$, which is included in the $M_{ek}(t)$-route.

(7) We give $x_p(t) = x_i$.

(8) If the route $M_{ek}(t)$ is fully formed ($X_{ek}(t) = \varnothing$), we go to step (10), otherwise we go to step (8).

(9) We give $t = t + 1$ and go to step (3).

(10) The end of the algorithm.

Time complexity of this algorithm depends on the lifetime of the colony l (number of iterations), number of graph nodes n, and the number of ants m. It is defined as $O(l \cdot n^2 \cdot m)$.

2.4. Organization of Search Procedures Based on Simulation of Collective Alternative Adaptation

Adaptation is the ability of a living organism or a technical system to change its state and behavior (parameters, structure, algorithm and performance) depending on changes in external environment conditions through the collection and use of information about the environment [27].

As a rule, the adaptive system is understood as a system, which operates under expected uncertainty and changing external conditions, while the information about these conditions, obtained in the course of operation, is used to improve the efficiency of the system's performance.

Adaptation is a special case of control with stable targets. The main goals of adaptation are associated with extreme requirements for an adaptation object in the form of maximizing the efficiency of its performance.

The adapting exposure may be of diverse nature. One can change the parameters of adaptation object, as well as its structure. In the first case, we have parametric adaptation, while in the second case arises structural adaptation. Depending on the type of variable parameters, parametric adaptation can be continuous, discrete and binary. It is obvious that structural adaptation is a deeper and more radical, because the relevant changes affect the most hidden part, namely the structure of the adaptable object. Structural adaptation is usually accompanied by parametric adaptation, since each structure has its own parameters, which also require adaptation.

Adaptation of a structure is possible, first, by its slight changes having the evolutionary nature (evolutionary adaptation), and secondly, by choosing one of the alternative structures of an object (alternative adaptation).

The process of exploratory adaptation has sequential multistep nature, at which adapting impact on the object that increases its efficiency and optimizes quality criteria, is determined at each stage. In our case the initial commutation of the connections to the pins is given. At each iteration, under adapting impact, group switching of connections is fulfilled without changing the logical function of the circuit. As a model of a trainable system, M. Tsetlin has proposed probabilistic trainable automatic machine, called the adaptation automatic machine. The state of the adaptation automatic machine corresponds to a specific alternative of adapting impact on the object. In the adaptation process, based on feedback from the external environment, the automatic machine turns to a state that corresponds to the best alternative of the adapting impact on the object [9].

For each circuit, at the points of location of the connected pins we introduce the attractive interaction forces, acting on the connected terminals. The magnitude and direction of these forces characterize the expectation of the connection to the switch. Then the force $F_{ij} = \alpha r_{ij}$ is applied from the terminal connection, attached to the pin v_j, towards the terminal connection, attached to the pin v_i, whereas the force $F_{ji} = -\alpha r_{ij}$ is applied from the terminal connection of the same circuit, attached to the pin v_j, where r_{ij} is the distance between the terminals, α is a coefficient. Note that F_{ij} is a vector. Since the circuit is connected to several pins, the terminal connection attached to the pin v_i experiences action of the total force defined as $F_i = \sum_j F_{ij}$.

The terminals t_k connected to the pins v_i are the adaptation objects. In our case, we used three types of adaptation impacts: (i) switching to the left, (ii) switching to the right, and (iii) no switch. The attractive interaction forces, acting at the terminal, change due to switching. The local goal of a specific object t_k is to reach the state at which the attraction force, acting in the v_i, would be equal to 0. The global goal of the set of objects is to reach the state of the environment, providing favorable conditions for the subsequent routing (minimizing the density of the switchboard and the number of intersections). At the each iteration in some selected pairs of equivalent pins a reciprocal switching of terminals carries out based on the analysis of the current situation.

When switching terminals, the following combinations of the states of corresponding pair AA are evident: $\rightarrow \leftarrow$; $\rightarrow 0$; $0 \leftarrow$. Here $\rightarrow$ and $\leftarrow$ means tendency of AA to switch to the pin, located in the route from the right, or from the left, respectively, while 0 means neutral state.

2.5. Integrating the Models of Adaptive Behavior of Ant Colony and Collective Alternative Adaptation

The main goal when developing a new paradigm of swarm algorithm is to integrate the metaheuristic algorithm, embedded in the collective alternative adaptation and the ant colony algorithm.

The searching process begins with the construction of solutions searching graphs $G = \{G_e | e = 1, 2, \ldots, n_e\}$, $G_e = (X_e, U_e)$ in accordance with the paradigm of the ant colony algorithm. At the initial stage, on all edges of each graph G_e, the same (small) amount of pheromone Φ/v, is laid, where $v = |U|$.

The solutions searching process is iterative. Each of iterations consists of three steps. At the first stage of each iteration the set of $M^*(l) = \cup M_{ek}^*(l)$ routes forms on the G_e graphs. The construction of each route $M_{ek}^*(l) \in M^*(l)$ is carried out in two steps. The first step deals with the procedures of ant colony algorithm. Each ant $z_{ek} \in Z_k$ forms at the edges of the G_e-graph its own route $M_{ek}(l)$. Then the solution $R_k(l)$, corresponding to the cluster of routes $M_k(l) = \{M_{ek}(l) | e = 1, 2, \ldots, n_e\}$, constructed by the ants of Z_k-cluster, determines, and assessment of the solution $F_k(l)$ is carried out. At the second stage, the solution $R_k(l)$ is treated as an initial solution, and the solution $R_k^*(l)$ is formed with an assessment of $F_k^*(l)$ using collective alternative adaptation algorithm. After constructing the aggregates of routes $M_k(l)$ by all the ants and building solutions $R_k^*(l)$ on the base of each aggregate using the algorithm of collective alternative adaptation of solutions $R_k^*(l)$, each solution $R_k^*(l)$ represents into an aggregate of routes $M_k^*(l)$ on the solutions searching graphs G_e. At this stage, pheromone lays on the edges of each set of routes $M_k^*(l) \in M^*(l)$, built on the solutions searching graphs G_e in accordance with assessments of the routes $F_k^*(l)$. The third step corresponds to the evaporation of pheromone on the edges of the set of G graphs. The best solution, found after performing l iterations, saves.

3. RESULTS AND DISCUSSION

The analysis of existing approaches for solving the routing problem revealed the following. The mathematical models of the problem stated and studied by the standard methods of optimization, such as linear programming techniques, dynamic programming method, etc. Within this formulation, it is possible, in theory, to obtain a global result. However, since no one standard method does not exclude the possibility of cycling in locally optimal regions, these methods turn out unacceptable for solving real-dimensionality problems [2, 24].

In this treatment, the developers of the algorithms had devised algorithms based on intelligent optimization methods.

To enhance the convergence and provide the ability to exit from the local optima, we propose to base the search procedures on simulation of collective adaptation. The priority of criteria may change by simple modifications of the environment response. The developed search engine, which integrates alternative adaptation heuristics and ant colony algorithm, allows the use of a hierarchical control strategy. The article presents the technology of converting the population during the transition from one generation to another that allows us to secure the best solutions.

Comparative analysis with other algorithms for solving problems of this class carried out using standard test examples and schemes (the benchmarks). The majority of global routing programs use various versions of greedy heuristics, or time-consuming methods of linear integer programming. To solve this problem, authors of the work [1, 20] proposed an algorithm that first generated the circuits routing order, based on the result of a single-layer routing, and then solved the problem of distribution across layers (one circuit was selected at each step) using dynamic programming. Weak point of these approaches is the sequencing problem of traceable circuits.

CONCLUSION

Based on the self-organization and self-learning ideas, the search adaptation methods were considered using trainable automatic machines, simulating the behavior of design objects in a random environment.

The article considered adaptation structures, their architecture and transition mechanisms.

The article highlighted key issues and tasks solved in the course of original formulation of design engineering problems in the form of adaptive search processes that allowed formalizing and streamlining the research and development of adaptive search mechanisms in relation to the current task.

The numerical solutions, obtained using hybrid algorithm, were better by 3% than those known previously. Under the new approach, the probability of obtaining the optimal solution was 0.9. The total estimation of time complexity lies within the limits of $O(n^2) - O(n^3)$. In the following studies, it is assumed to speed up the process of complex systems synthesis by $10 - 15\%$, as well as to increase the routability index for the next step of a detailed routing by 10% as compared to the existing analogues, through the use of an integrated approach to solving the problem.

REFERENCES

[1] Cho, M.; Pan, D. Box outer: A new global router based on box expansion and progressive ILP. *Proceedings of the Design Automation Conference*, 2006.

[2] Cho, M.; Lu, K.; Yuan, K.; Pan, D. Box router 2: Architecture and implementation of a hybrid and robust global router. *Proceedings of the International Conference on Computer-Aided Design,* 2007.

[3] Cong, J., et al., *Microarchitecture Evolution with Physical Planning*, DAC, 2003.

[4] Dorigo, M.; Stützle, T. *Ant Colony Optimization.* Cambridge, MA, MIT Press, 2004.

[5] Emelyanov, V.; Kureichik, V.M.; Kureichik V.V. Theory and Practice of Evolutionary Modeling. Fizmatlit, Moscow, 2003 (in Russian).

[6] Engelbrecht, A. *Fundamentals of Computational Swarm Intelligence.* Chichester, John Wiley and Sons, 2005.

[7] Gerasimenko, Y.; Fugarov, D.; Purchina, O. *Life Science Journal*, 2014, *vol. 11*, 12 p.

[8] Kureichik, V.; Lebedev, B.; Lebedev, O. *Search Adaptation: Theory and Practice.* Fizmatlit, Moscow, 2006 (in Russian).

[9] Kureichik, V.; Lebedev, B.; Lebedev, V. *Adaptation in the Topology Design Problems. on Development of Perspective Micro- and Nanoelectronic Systems.* IPPM RAS, Moscow, 2010 (in Russian).

[10] Kureichik, V.; Lebedev, B.; Lebedev, V. *Bulletin of the Russian Academy of Sciences. Theory and Controlling Systems.* 2013, *vol. 1*, 84-101 (in Russian).

[11] Lebedev, B. *Distribution of the Switchboard Resources. Automation of Electronic Equipment Design.* Taganrog, Taganrog Radio Engineering Institute, 1988, *vol. 1*, 89-92 (in Russian).

[12] Lebedev, B. *Intellectual Procedures of VLSI Topology Synthesis.* Taganrog, TRTU Publishing house, 2003 (in Russian).

[13] Lebedev, O. *Bulletin of the Southern Federal University, Technical Sciences*, 2009, *vol. 4(93)*, 46-52 (in Russian).

[14] Lebedev, V. Method of bee colony for combinatorial problems on graphs. *Proceedings of the 13th National Conference on Artificial Intelligence.* Fizmatlit, Moscow, 2012 (in Russian).

[15] N/A.

[16] Lebedev, O. *Models of Adaptive Behavior of Ant Colony in the Design Tasks.* Taganrog, SFU Publishing House. 2013 (in Russian).

[17] Lebedev, B.; Lebedev, V. *Bulletin of the Southern Federal University*, 2010, *vol. 7*, 32-39 (in Russian).

[18] Lebedev, B.; Lebedev, O. Multilayer global routing by collective adaptation method. *Proceedings of the 5th All-Russian Scientific-Technical Conference on the Problems of Development of Perspective Micro- and Nanoelectronic Systems*, 2012 (in Russian).

[19] Lebedev, B.; Lebedev, O. *Bulletin of the Southern Federal University*, 2012, *vol. 7*, 27-35 (in Russian).

[20] Lebedev, B.; Lebedev, V. *Bulletin of the Southern Federal University*, 2013, *vol. 7*, 11-17 (in Russian).

[21] Lee, T.; Wang, T. *IEEE Transactions on Computer-Aided Design of Integrated Circuits and Systems*, 2008, *vol. 27(9)*, 1643-1656.

[22] Lučić, P.; Teodorović, D. *International Journal on Artificial Intelligence Tools*, 2003, *vol. 12*, 375-394.

[23] Mazumder, P.; Rudnick, E. *Genetic Algorithm for VLSI Design, Layout and Test Automation*. Pearson Education, India, 2003.

[24] McConnell, J. *Foundations of Modern Algorithms*. Technosphere, Moscow, 2004 (in Russian).

[25] Pan, M.; Chu, C. Fast route 2: A high-quality and efficient global router. *Proceedings of the Asia South Pacific Design Automation Conference*, 2007, *vol. 27(9)*, 1643–1656.

[26] Purchina, O.; Lebedev, O. Connections permutation between the pins in the channel by swarm intelligence methods. *Proceedings of the Congress on Intelligent Systems and Information Technologies*. Fizmatlit, Moscow, 2012 (in Russian).

[27] Purchina, O. Adaptive algorithm for permutation of the connections between the pins. *Proceedings of the Congress on Intelligent Systems and Information Technologies*. Fizmatlit, Moscow, 2013 (in Russian).

[28] Red'ko, V. *Evolutionary Cybernetics*. Nauka, Moscow, 2001 (in Russian).

[29] Tarasov, V. *From Multi-Agent Systems to Intelligent Organizations: Philosophy, Psychology, and Computer Science*. Editorial URSS, Moscow, 2002 (in Russian).

In: Proceedings of the 2015 International Conference …   ISBN: 978-1-63484-577-9
Editors: Ivan A. Parinov, Shun-Hsyung Chang et al.   © 2016 Nova Science Publishers, Inc.

Chapter 73

SUSTAINABLE LIVELIHOOD FRAMEWORK ANALYSIS FOR IMPROVING LOCAL SECTOR INDUSTRIES, CASE STUDY: LOCAL SECTOR INDUSTRIES AT LEDOK KULON, EAST JAVA-INDONESIA

R. A. Retno Hastijanti[*]

Department of Architecture, University of 17 Agustus 1945 Surabaya,
East Java, Indonesia

ABSTRACT

Ledok Kulon is a peri-urban section of Bojonegoro city. The meaning of "Ledok" is the lowest area of Bojonegoro. By studying its geographic condition, the Government understood that the Ledok Kulon is a gateway for floods to enter Bojonegoro city, so they decided that the Ledok Kulon is an unsafe environment for its inhabitants. To reduce the flooding, the solution is to build an embankment in the Ledok Kulon area. Unfortunately, this requires the relocation of Ledok Kulon inhabitants to other safe areas, but the inhabitants have refused this plan. Their reason is that their livelihood has been settled there for over 4 generations, which allows them to be brick makers. So, from the inhabitants' opinion, the flooding is an advantage that gives them new sediment for making the brick. Then the conflict arises. But finally, the government can reduce the conflict with a win-win solution. The Government of Bojonegoro Regency decided to build an embankment, not just for preventing the flow of the flood into the housing area of Bojonegoro but also for catching more sediment, so the brick makers have the advantage to process it. Based on field observation and Sustainable Livelihood Framework analyzing, we can conclude that the strategy and the policy of the Government to developing the unsafe living environment has some advantages for Ledok Kulon's brick makers. The primary advantage is to improve the local informal sector of material industries, which can support the economic conditions of Ledok Kulon.

---

[*] Email: retnohasti@untag-sby.ac.id.

Keywords: sustainable livelihood framework analysis, local sector industries, unsafe livelihood environment

INTRODUCTION

Ledok Kulon is a peri-urban interface of Bojonegoro city. The meaning of "Ledok" itself is the lowest area, while the meaning of "Kulon" is the West. So the name of this area provides the explanation of its geographic condition, the lowest area at the West of Bojonegoro city. It is located at the riverbank of Bengawan Solo Great River. It is shaped like a small river cape. Because the soil contains a lot of clay and lime; the main occupation in this area is brickmaking. The brick makers community, here, has stood for 4 generations. Because of its geographic condition, every rainy season, the big flood comes from the Bengawan Solo Great River, and enters Bojonegoro City from Ledok Kulon. Retno [1] observed that such conditions are not positive for other occupations. The impact of this condition is decreasing economic growth. The UNCHS's report [2] stated the usual conditions that happen in this pheri-urban area. But the inhabitants that have lived here for 4 generations still survive and choose to live at this location, because floods have given them new sediment for brick making. Retno [2] analyzed that if the flood didn't come,, there wouldn't be materials for the brick making process. When the flood comes, they evacuate to the nearest place that is higher than Ledok Kulon. For a while, they become refugees in their own environment. And then, the problem is expanding a multi-dimensional view from a usual environmental problem. Retno [4] elaborated that the Government of Bojonegoro city did a study about Ledok Kulon's livelihood environment, and concluded that the area is an unsafe area to live. They also realized that the Ledok Kulon is the gateway for floods to enter surrounding areas. Bojonegoro itself is the lowest area of any surrounding cities. It is reasonable to think about an embankment in order to make the city safe for all the people, but because the gate is at the Ledok Kulon area, the solution has involved the decision of its inhabitants. Another problem has occurred. There is a conflict of interest between the government and the Ledok Kulon inhabitants.

From the background above, we can conclude that for the people who live in Ledok Kulon, their location is not problematic and harmless. From The government's view, there was a big problem occurring here, by managing the environment of the peri-urban interface. In this case, it involves the environment of Ledok Kulon. Another big problem has occurred, connected with the development of Bojonegoro city itself, since the flood in every rainy season isn't just a usual flood, but it can become a disaster. The height of the water level can reach almost 3-4 m from the lowest land and reach 15m in central Bojonegoro city. It is certainly necessary to build the embankment for this condition. But again, they have to face the Ledok Kulon inhabitants. The conflict can embrace one importance aspect, which is affordable land for embankment building since the embankment building will need a wide area, which is the housing area of the Ledok Kulon inhabitants and their livelihood environment. So the solution is the relocation of all the Ledok Kulon inhabitants. Then the relocation proposal was made.

The community rejected the proposal so the conflict arises. The Government has to find another way to manage and solve the problem. They have to reach their goal by exploring alternative solutions and bargaining. After a decade of years, the embankment was built (2000-2005). It is located in the middle of the village, and became the new road too. The community becomes split in two, but their sense of togetherness has not diminished one bit. They stick together in one whole community. Since then, the population Ledok Kulon can live more comfortably and is able to build a better life. They finally have a safe living environment. The entire population of the Bojonegoro City also feels this condition. Bojonegoro city now can have better planning and is free of disaster risk. From Biel [5] viewed, it's shown that the Government of the city has succeeded in managing its pheri-urban area. And from Davila [6] viewed, what the Government of this city has done is apply Sustainable Livelihood Framework analysis.

METHODS: DEVELOPING UNSAFE LIVELIHOOD ENVIRONMENT

Human Assets

When the relocation proposal was announced, the numerous brick makers numbered only 100 people in 1990. They are full-skilled brick makers. They learn how to process the bricks, traditionally inherited from their ancestors. They also learned that the current changes very rapidly, both in terms of technology and in terms of demand. So, they also learn to develop new technologies of the brick manufacturing process, so that production can be of better quality.

The brick makers from Ledok Kulon are known as the best brick makers, not only in their surrounding area, but also in their region. They make husk-bricks and fire it by husk or wood as fire material. The other inhabitants, who don't work as brick makers, chose other occupations, which are: fried tofu makers, cattle breeders, and laborers in the city or other.

Social Assets

The unity of the Ledok Kulon community is key for living in specific areas like this. The power of the community reflects their social networks and relationships. There are 21 RT's (*Rukun Tetangga* mean is the peer group of households that are excluded by physical aspects, like roads or building blocks. The total average is more than 20 households per *Rukun Tetangga* or RT), which are divided into 2 RW's (RW is *Rukun Warga* with contents of 20 0r 21 RTs) for their formal institution. Other support groups for this formal institution are *karang taruna* (the collaboration of boys and girls in social activities), and *dasa wisma* (grouping for women of ten households). The Community Group Representatives is a higher formal institution. Its members are a blend of the people that represent the community and the people that represent the Government. Besides that, there are nucleus families and extended families for their traditional institution based on family relationship. The extended family relationship made the network expand, not only inside the homeland, but also outside the region that establishes trade networks for bricks. Both of the two social institutions strengthened the community and made them solid in every condition.

In bridging the conflict on the construction of the embankment, the Community Group Representatives played a very big role. Through various processes of discussion and assistance to various alternative solutions, they could eventually convince the population of Ledok Kulon that the construction of the embankment would not damage or eliminate livelihood for brick makers. With the process of social discussion, they then found a variety of solutions that would benefit not only the community, but also the government. And more importantly, the government was finally able to eliminate conflicts that could arise, and be able to meet their responsibilities as the holder of the autonomy of urban development.

Natural Assets

The most important asset is the geographic conditions, which are lower land, the riverbank, clay and lime, and the cape shape. This asset is a potential asset for the development of the brick maker. Another asset that has two opposite sides is flooding. On the one hand it is very useful to bring material for making bricks while on the other hand, it is seen as disastrous. Natural other assets: banana plants are very easy to grow along the riverbanks. People use the bananas to make traditional cakes and can increase their income, although slightly.

Physical Assets

The only 'big' road here is the main rural road that divided the village into 2 parts of linear housing. The housing itself is a Javanese traditional settlement that had a linear housing pattern. Only a few households have an electrical generator to access the electricity, while the others just have gas lamps. There are a lot of workshops, which produce the brick named 'linggan', located along the riverbank. The total of the 'linggan' at that time were almost 80. They are free to access the lot for brick ingredients as well as a blending and processing area.

Financial Assets

In this case, it is necessary to distinguish between the financial assets of the general public and the assets of local government. Due to their poor economic situation, the public does not have sufficient financial assets to be able to save for their future.

Vulnerability Context

Every rainy season, the floods will come and force the community of Ledok Kulon, to evacuate. Due to the financial condition of the community is very limited, so the government must take care of them during the evacuation period. This makes the government bear the loss, because it must spend extra finances to take care of the refugees, and they are not able to plan better development of the region. The only solution is to build embankments in the area of Ledok Kulon.

Transforming Structures and Processes

The important structures that are possible to handle the problem are the Government and the Community Group Representatives. They influenced the inhabitants by bargaining using some alternative solutions that could advantage all. The agreement resulted, and the solutions are the relocation has been canceled if the inhabitants allowed the Government to build the embankment; the embankment will be built at the middle of the village, because if it is built along the riverbank, it will destroy the brick makers livelihood environment; the allocation site that will be the location for the embankment is both the community site (the rural main road) and the private site (more than half of the inhabitant front land). So, the Government needs to buy the private land at a cheap price because of the Government financial limitations. And the inhabitants have agreed to that as their participation for this project. Besides that, they will help as the free of charge workers for this project. The consequence of this building is that the village will be separated into 2 parts. One of the parts, which is located nearer the riverbank, will still be flooded in every rainy season, and the other wouldn't. For fairness, the Government would raise their taxes for the site located inside the embankment. They are not only agreeing but also promising they will help their neighbors whose site is still flooded (it can be done because of their stronger family relationship/social aspects).

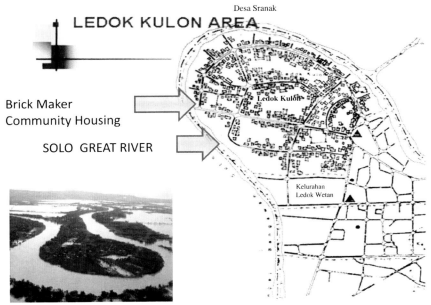

Figure 1. Ledok Kulon Area.

The project of embankment building is done not only in an effective way, but also in a good relationship situation between the Government and the community. The other advantages will appear when it is done completely. The Ledok Kulon becomes a safe and healthy place for living. So, the migration into this land could not be prohibited. In a little time, the area became a growth peri-urban interface. The land price becomes higher and the other occupation that could not be done before, now occur in a lot of variation. The other occupations that occurred are the fried tofu makers, the cow breeders, and the chicken breeders.

Table 1. Module Standard of Indonesian Brick Dimension (SII-0021-78)

Module	Thickness (mm)	Width (mm)	Length (mm)
M-5a	65	90	190
M-5b	65	140	220
M-6	55	110	220

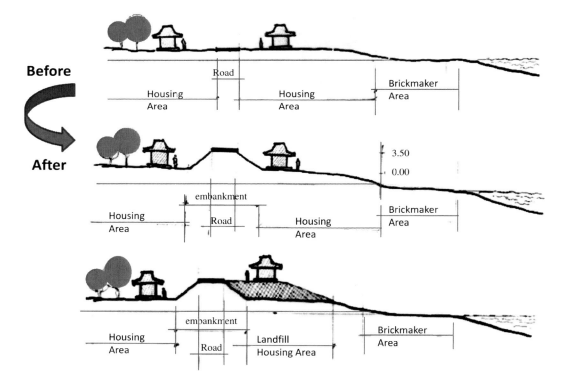

Figure 2. The build of embankment.

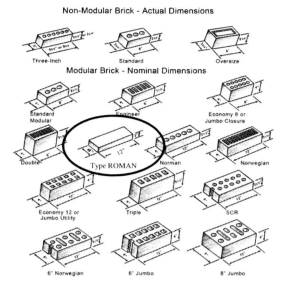

Figure 3. Brick type, produced by Ledok Kulon Village.

CONCLUSION

The increase of economical growth of Ledok Kulon has been influencing the economic welfare of Bojonegoro city. Bojonegoro city has been the primary potential market for brick, and other product from Ledok Kulon, because it is not only the nearest big city from Ledok Kulon but also the Capital city for Bojonegoro Regency. From here, the expanding market will be easier. The market networks for the brick are Lamongan Regency and Tuban Regency. So, the product has been known in remote areas. Because of the changed Ledok Kulon that has been predicted by the Government after the agreement of the embankment building, the Government has set the role of the Ledok Kulon in their Master Plan (1990-2010). They set Ledok Kulon as part of city development area that functioned as the center of trading activity and part of the medium density housing area. The Road embankment has set to connect to other new road, which had higher hierarchy, to the central city area. This road became the gate to new Ledok Kulon. The other facility and public utilities that are needed by the inhabitants of Ledok Kulon have been entered, which were electricity, telecommunication, other public building, etc.

The key lessons from the case study are: (a) the sameness of the vision between the Government and the community in terms of the primary asset of the livelihood environment as the key for solving the problem; (b) the good will of the Government for not insisting on their plan for the community has effected the best agreement /best bargain for all (win-win solution); (c) by focusing on the social assets, which is the strength of family relationship of the community, the Government has established the best process to reach the best agreement; (d) by entering the Ledok Kulon area into the Master Plan, the Government has made the rural-urban linkages more visible, the role of the peri-urban more clearly, and the development of the area under development more predictable; and (e) by developing an unsafe livelihood environment, the Government has been improving the local informal sector of material industries, and making the Ledok Kulon a brick industrial area.

REFERENCES

Hastijanti, Retno, 1995. *The Ledok Kulon Village, The Influenced of The Housing Aspect to The Village Housing Pattern*, paper presented at internal seminar on Housing System and Housing Affordability, Architectural Department of ITS, Surabaya, November 1995.

UNCHS (Habitat), 2001. *The State of The World's Cities 2001*, UNCHS (Habitat) Publications Unit, Nairobi, Kenya.

Hastijanti, Retno, 1996. *Developing The Local Material Industries by Improving Inter-Regency Network Planning, Case Study at The Ledok Kulon*, paper presented at internal seminar on Construction Sector, The Challenge and The Opportunity, Architectural Department of ITS, Surabaya, March 1996.

Hastijanti, Retno, 2007. *The Housing Tipology of The Ledok Kulon Village, Bojonegoro, East Java-Indonesia*, paper presented at internal seminar, Architectural Department of Untag Surabaya, September 2007.

Biel R, 2010. *Toward Transition* in Levene, Mark, Rob Johnson and Penny Roberts, eds, History at The End of The World History, Climate Change and The Possibility of Closure, Penrith, 234 – 250.

Davila, Julio D., 2009. *Urban Poverty and Rural-Urban Linkages*, chapter 5 in Uri Raich (Editor), Municipal Development in Mozambique. Lessons from The First Decade, vol. II, The World Bank, Washington DC.

In: Proceedings of the 2015 International Conference ... ISBN: 978-1-63484-577-9
Editors: Ivan A. Parinov, Shun-Hsyung Chang et al. © 2016 Nova Science Publishers, Inc.

Chapter 74

CLUSTER DEVELOPMENT OF SMALL AND MEDIUM MANUFACTURING INDUSTRY IN SURABAYA CITY, EAST JAVA, INDONESIA

*Erni Puspanantasari Putri**

Department of Industrial Engineering, Khon Kaen University (KKU) Thailand;
Department of Industrial Engineering, University of 17 Agustus 1945 (UNTAG)
Surabaya, Indonesia

ABSTRACT

The purpose of this research is to study the cluster development of small and medium manufacturing industries in Surabaya City, East Java, Indonesia. Surabaya is the capital city of East Java province, Indonesia as well as the largest metropolitan city in the province. Surabaya is the second largest city in Indonesia after Jakarta. As a metropolitan city, Surabaya became the center of business, commerce, industry, and education in eastern Indonesia. Surabaya and the surrounding region is a region with the most rapid economic development in the east of Indonesia and one of the most advanced with an average economic growth of 7.5% per year. The method used in this study is the method of Efficiency Theory, that is how effective is the development of the cluster based SMMI. Geographically it is highly populated which makes it easy to find buyers looking for suppliers of products. The labor pool allows employers to hire them as needed. Entrepreneurs are flocking to Surabaya, it is easier to relate directly with the market, easier, indeed to access market information, technology and training. This is a crucial element in the cluster, the entrepreneur groups are more likely to cooperate for the purchase of raw materials, marketing, manufacturing facilities, training and research. The results of this research show that Cluster Development of SMMI in Surabaya City, East Java, Indonesia should become the Creative Village. The Creative Village is one of the breakthroughs of community empowerment. Residents are invited to get involved in the development in accordance with the potential of their respective regions. The Creative Village will allow economic growth for the villagers, and by this I mean a group of

---

* KKU Thailand, UNTAG Surabaya Indonesia; Email:erniputri@untag-sby.ac.id.

people gathered together to create, produce and potentially generate an economy through a commodity, in the Creative Village.

Keywords: cluster development, SMMI, creative village

INTRODUCTION

The economic crisis that hit Indonesia in mid-1997 caused the national economy to decrease. Economic growth before the crisis was high, it was not supported by solid industrial structure. Inequality of economic structure occurs because such a rapid growth in the Manufacturing Industry Large Scale (MIL) is not supported by the Small and Medium Manufacturing Industry (SMMI), so that the MIL received a very severe blow due to the crisis, in fact this crisis had a huge impact on the Indonesian economy as a whole. However, the economic crisis did not make the SMMI crash alongside the large companies. Whilst the big companies found it difficult to operate, the SMMI continued as usual, some corporate sectors of the SMMI actually benefited due to the depreciation of the rupiah against the dollar.

The Small and Medium Manufacturing Industry (SMMI) has a strategic role in the national economic development of Indonesia. Besides a role in economic growth and employment, SMMI also plays a role in the distribution of developmental outcomes. The SMMI should be increased (up graded) and active, in order to go forward and compete with other economic players. If not, SMMI in Indonesia, which is the heart of the Indonesian economy, will not be able to progress and develop [1].

METHODS

Definition of the SMMI Cluster

The SMMI cluster is the concentration of a number of similar SMMI (sectoral) and adjacent regions (geographically). Besides sectoral and geographical concentration, suppliers, and the relationship between cluster members, with other stakeholders, there are also links to related industry and supported industry.

Cluster Benefits for the SMMI

The SMMI members of a cluster will benefit from the so-called collective efficiency which consists of passive benefits (external economics) and active benefits (joint action). Passive benefit means that without doing anything the SMMI will benefit from the existence of clusters. Active benefits are the benefits of cooperation (joint action) between cluster members. Clusters will perform of production chains and enter global value chains. With this partnership members of the cluster will be able to increase their competitiveness [2].

SMMI Cluster Development an Effective Way

According to the theory of efficiency the development of the SMMI based clusters is effective. Geographically a greater population makes it easier for buyers looking for suppliers of products. The labor pool allows employers to hire them as needed. Entrepreneurs are flocking to all SMMIs because it is easier for them to relate directly to the market and easier for them to actively access the market for information technology and training. This is a crucial element in the cluster, entrepreneur groups are more likely to cooperate for the purchase of raw materials, marketing, manufacturing facilities, training and research.

Cluster Development of the SMMI Center

The development of industrial clusters should be encouraged. The presence of the synergy of clusters from upstream to downstream activities, or between the main activities and auxiliary activities, raw materials supply and marketing outlets will accelerate the business dynamics within the cluster. The essence of the strategy is the creation of integrated and solid clusters to build a synergy to achieve economic growth with a wide base.

Regional autonomy at this point proves that the SMMI sector plays an important role in national economic development. It opens opportunities for local government facilities to members for regional excellence development (potential territory), including the development of the SMMI.

The SMMI development focuses on driving regional economic groups, among which are the SMMI supporters, the SMMI export-oriented group, and the SMMI's new initiative. In order to achieve the expected goals, there needs to be adjustments to the type of industrial developmental purpose, the level of industrial and technological development, as well as considering the condition and potential of the region.

Key Factors and Determinants in Cluster Development

The key factors and determinants, which create a competitive advantage for the cluster's development of the SMMI centers making them effective and dynamic, are as follows: (a) Factor conditions, which includes: natural carrying capacity, human resources, infrastructure, capital resources, and information technology (b) The SMMI product demand (c) Related and Supporting Industries Related Industries, which includes: Capital Institution, Research and Development Institution, Institute of Education/Training, Cooperative of the SMMI craftsmen.

Supporting Industries, which include: Feeder Industries of raw material, machinery and equipment productions (d) Firm Strategy, Structure and Rivalry (e) Change, that includes: domestic and foreign policy and (f) The role of the Government [3].

RESULT AND DISCUSSION

Surabaya City, East Java, Indonesia

Surabaya is the capital city of East Java province, Indonesia, as well as the largest metropolitan city in the province. Surabaya is the second largest city in Indonesia after Jakarta. As a metropolitan city, Surabaya became the center of business, commerce, industry, and education in eastern Indonesia [9]. Surabaya and the surrounding region is a region with the most rapid economic development in the East and one of the most advanced in Indonesia with an average economic growth of 7.5% per year. Surabaya city consists of 31 districts, which are divided into five regions. In each district, there are urban villages which now look lush and green. The villages are part of the city (usually, populated by low-income people) [4].

Creative Villages in Surabaya

Understanding Surabaya better with all the dynamics behind the bustle of metropolitan activity. A group of people has gathered to create, produce and potentially generate economic commodity in the creative village. The Department of Trade and Industry in the city of Surabaya, has noticed some potentially growing villages by the peculiarities of their work, accompanied by the huge success in the welfare of the community itself. In particular Bag Village at Morokrembangan District, Paving Village at Pakal District, Cakes Village at Rungkut District, Slippers-Shoes Village at Osowilangun Region and Ribbon Patchwork Village at Tambak Asri.

Bag Village at Morokrembangan District
The biggest bag center in East Java, is in Tanggulangin, which is in the southern part of Sidoarjo. However, in New Gadukan Road, Morokrembangan, Surabaya there is a village that has turned everyday citizens into producing various kinds of bags. This bag village was founded around 1978. Starting with 6-10 people who were experts in making bags of various materials, their creativity eventually influenced other citizens. Now there are 68 craftsmen who produce bags in the area.

Paving Village at Pakal District
Paving stones were chosen as a material for yards and streets, for its variety, because it gives space for drainage, and because it is easy to repair. There was such an increased demand for paving stones that many big companies producing these materials. In the midst of these busy factories making stone paving you will also find it produced in Pakal Village.

Cakes Village at Rungkut District
Starting from a hobby of making cakes, housewives at the Flat Penjaringan Sari, Rungkut District, developed their hobby into a business. They produce a variety of recipes for delicious cakes which have economic value. Cake Village began with three home industries which have produced a variety of cakes since 2001. Over time, other people around the flat have since also set up in the cake making industry.

Slippers-Shoes Village at Osowilangun Region

When you are shopping for shoes or slippers at the mall or grocery center at the Surabaya, do not be surprised if the product has been made by the shoe-slipper craftsmen of the Osowilangun Region. In this village thousands or even millions of products are made by the local people, it is an inherited skill. Shoes and slippers were made in home industries, without a brand sold to shoe and slipper traders then labeled, branded and marketed.

Ribbon Patchwork Village at Tambak Asri

Villagers at Tambak Asri, a suburb of Surabaya have succeeded in producing creative crafts and have managed to penetrate the foreign market. In the 1980s, the land was still covered with lush trees. The atmosphere was dark, there was no light except obliquely. Gradually this region was disturbed by human life. It started with just one or two people they opened the land and then settled down to live there. This village was then called the Village of 1001 Nights. Now the citizens are increasingly creative with ribbon patchwork craft [5].

The Role of Surabaya City Government in Developing SMMI

The Role of Surabaya City Government in developing Creative Villages is to develop a business climate, the provision of technical assistance, financial aid, facilities and infrastructure, empowering human resources through education and training, strengthening and institutionalizing development, and encouraging national and international cooperation.

SMMI development is reached by selecting and prioritizing industrial sectors to become the focus of development, that is:

a) Regional Economic Mobilization of the SMMI group, that is the industry developed in an area, which will allow the local economy to grow quickly, and involves the participation of the community at large.
b) Supporting the SMMI groups, that is industries that produce intermediate products.
c) Export Orientation SMMI groups, that is an industry which has had the opportunity to supply the world market with the products produced.
d) New Initiatives within the SMMI groups, that is those SMMI based on science and technology. [6].

CONCLUSION

Cluster Development of Small and Medium Manufacturing Industries in Surabaya city, East Java, Indonesia is to develop the Creative Village. The Creative Village is one of the breakthroughs of community empowerment. Residents are invited to get involved in the development in accordance with the potential of their respective regions. The Creative Village will improve the economic growth of the villagers. Surabaya City Government encourages villages in Surabaya in the creative industry where the products will improve the local economy. If the creative industries can be encouraged to grow it would be a revival of the creative industries in Surabaya which would impact on the economic growth of the local community and the country as a whole.

ACKNOWLEDGMENTS

The author would like to thank:

1) Assoc. Prof. Dr. Danaipong Chetchotsak, as my advisor of PhD Program in Industrial Engineering Department, Khon Kaen University Thailand.
2) Assoc. Prof. Dr. Kanchana Sethanan who provided me with the scholarship of Research unit on System Modeling for Industry (SMI), Faculty of Engineering, Khon Kaen University Thailand.

REFERENCES

[1] Erni Puspanantasari Putri, Supardi and Indahati, 2013. Product Quality Improvement of Small and Medium Manufacturing Industry Craftsmen to Increase Competitiveness Business, *Proceedings of 2nd SciTech International Seminar: The Role of Innovative Technology in Global Competitive Era, Welcoming ASEAN Community 2015, Faculty of Engineering, University of 17 Agustus 1945 Surabaya* pp 106-113.

[2] Erni Puspanantasari Putri and Indahati, 2007. Guide Business Assessment of Industry Centers in Sidoarjo with Clustering Approach As Base For Developing Regional Excellence Through Small and Medium Industry Development with Effective and Competitive, *Research Report of Young Lecturer, Directorate General of Higher Education, Ministry of National Education.*

[3] Erni Puspanantasari Putri and Supardi, 2014. Establish the Regional Excellence through the Cluster Development of Small and Medium Manufacturing Industry Center with Effective and Dynamic, *Journal of Advanced Materials Research, KKU International Engineering, Trans Tech Publication Inc., Material Science and Engineering*, 931-932 (2014) pp 1701-1704.

[4] Surabaya City, Wikipedia bahasa Indonesia, http://id.wikipedia.org/wiki/Kota_Surabaya.

[5] Kampung Kreatif Surabaya, http://wisata.suarasurabaya.net/news/2013 /124995-Kampung-Kreatif-Surabaya.

[6] Direktur Jenderal Industri dan Dagang Kecil Menengah, 2003, *Kebijakan dan Strategi Pengembangan Industri*, SNPI-ITB, Bandung.

EDITORS' CONTACT INFORMATION

Dr. Ivan A. Parinov, DrSc

Vorovich Mathematics, Mechanics and Computer Sciences Institute,
Southern Federal University, Chief Research Fellow,
Rostov-on-Don, Russia
Tel.: 7-863-2975224
Email: ppr@math.rsu.ru, parinov_ia@mail.ru

Dr. Shun-Hsyung Chang, PhD

Department of Microelectronics Engineering, Professor,
National Kaohsiung Marine University,
Kaohsiung City, Taiwan, R.O.C.
Tel.: +886-7-3617141, ext. 3363
Email: stephenshchang@me.com

Dr. Vitaly Yu. Topolov, DrSc

Southern Federal University,
Physics Department Professor,
Rostov-on-Don, Russia
Tel.: 7-863-2975127
Email: vutopolov.sfedu.ru, vitaliyyju@rambler.ru

INDEX

#

^{57}Fe Mössbauer spectra, 31, 32

A

absorption spectra, xiv, 4, 5, 207, 208, 209, 210, 212, 213
absorption spectroscopy, 4
acoustic emission, 324, 326, 377, 379, 382
acoustic medium, 439, 440, 442, 443, 445
acoustics, xiv, 53, 487, 493
actuator(s), 38, 123, 132, 424
AFM, 93, 226, 272
Al2O3 particles, 49, 53, 188
albumin, xiv, 305, 306, 307, 310, 312, 475, 476, 480, 481, 482
alkaline-earth elements, 151, 157, 160
amplitude, 21, 53, 217, 241, 242, 254, 326, 328, 354, 355, 356, 380, 416, 420, 435, 439, 440, 443, 444, 445, 447, 451, 469, 487, 494, 517, 525
amplitude-frequency characteristic (AFC), 525
anisotropy, 9, 11, 14, 48, 123, 126, 129, 131, 134, 138, 151, 161, 162, 166, 408, 426, 440, 503
antiferromagnetic phase transition, 31, 32
auxetics, 447

B

beams, 21, 516
bending, 4, 337, 338, 347, 453, 531, 533
biocompatibility, 481
biocompatible materials, 312
biological control, 546
biological solders, xiv, 312, 475, 476, 477, 482
biomedical applications, 306
biotechnology, 332

bismuth ferrite, xiv, 259, 263, 265, 267, 272, 275, 283, 284, 286
bismuth–lanthanum manganite, 289
blowing agent, 170, 171
Boltzmann constant, 236, 257, 268
bonding, 70
bonds, 97, 153, 326, 367, 368, 369, 370, 371, 372, 373, 374
Boundary integral equations, 359, 361, 364, 365
boundary value problem, 294, 315
Boundary-Element Method, 359
BST, 61, 62
bulk composite nanomaterial, xiv, 305, 306, 309, 312, 482

C

cadmium, 99
calibration, 327, 525
carbon nanotubes, xiv, 305, 306, 310, 312, 475, 476, 481, 482
cartilaginous, 476, 479
cathode, 293, 537, 539
CEC, 105
ceramic(s), xiv, 9, 10, 11, 12, 13, 17, 18, 19, 23, 24, 26, 27, 28, 33, 34, 37, 39, 46, 47, 48, 51, 53, 54, 55, 56, 61, 65, 67, 69, 70, 72, 100, 115, 128, 132, 145, 151, 152, 154, 157, 161, 162, 163, 164, 169, 170, 171, 172, 173, 175, 183, 184, 185, 186, 187, 188, 193, 195, 204, 207, 208, 209, 216, 221, 228, 233, 235, 237, 240, 241, 243, 245, 254, 259, 260, 267, 268, 276, 277, 278, 280, 284, 285, 286, 287, 291, 292, 332, 384, 385, 386, 387, 388, 389, 397, 407, 408, 409, 410, 411, 412, 418, 469, 470
ceramic grain size, 72, 169, 172
ceramic materials, 9, 11, 46, 55, 169, 292, 384, 469, 470

574   Index

ceramic technology, 10, 24, 28, 55, 56, 152, 157, 162, 207, 216, 221, 260, 268, 276, 284
characterization techniques, 141, 142
chemical, xiii, 37, 38, 48, 58, 70, 75, 82, 96, 103, 105, 152, 192, 207, 208, 223, 236, 245, 257, 259, 270, 294, 296, 297, 298, 306, 326, 370
chemical characteristics, 58, 105
chemical interaction, 48
chemical properties, 105
cluster development, 565, 566
clustering, 32, 35, 192, 226
clusters, 240, 548, 549, 550, 566, 567
composite(s), xiv, 47, 48, 49, 51, 53, 69, 70, 71, 72, 73, 115, 116, 118, 119, 120, 121, 123, 124, 125, 126, 128, 129, 131, 132, 133, 134, 135, 136, 137, 138, 164, 173, 185, 305, 306, 309, 311, 312, 323, 326, 383, 384, 385, 392, 408, 409, 411, 412, 439, 440, 441, 442, 443, 445, 476, 482, 509, 510, 511, 517
composite materials, 48, 306, 383, 384, 392, 440, 509, 510, 511
composition, 4, 10, 11, 24, 26, 28, 38, 39, 56, 57, 62, 70, 90, 91, 96, 97, 144, 152, 154, 159, 163, 164, 172, 173, 175, 176, 178, 191, 192, 196, 200, 208, 222, 223, 224, 225, 228, 229, 233, 234, 235, 236, 240, 241, 242, 254, 257, 259, 260, 262, 263, 264, 270, 276, 283, 284, 290, 292, 307, 380, 383, 475, 476, 478, 479, 480
compounds, 24, 37, 38, 59, 98, 159, 172, 200, 203, 221, 233, 286
compressibility, 317
compressible liquid, 453, 454, 455, 456
compression, 332, 532
concentration, 11, 12, 65, 67, 105, 151, 152, 153, 172, 188, 192, 193, 194, 195, 196, 199, 200, 203, 208, 211, 216, 217, 219, 222, 223, 224, 234, 236, 245, 246, 247, 249, 260, 262, 268, 269, 270, 271, 272, 276, 277, 279, 284, 285, 287, 293, 294, 295, 296, 297, 299, 305, 306, 307, 311, 312, 377, 381, 382, 409, 410, 411, 417, 419, 420, 462, 463, 466, 476, 506, 537, 539, 543, 546, 566
conductance, 293, 298, 301
conduction, 75, 76, 78, 82, 84, 98, 144, 178
conductivity, 76, 77, 82, 84, 90, 105, 152, 228, 241, 249, 276, 290, 297, 301, 309, 311
conductors, 210, 546
connectivity, 124, 126, 132, 392, 393, 397, 486, 494
connectivity patterns, 132
constant load, 308
constant rate, 241
construction, 200, 332, 435, 437, 440, 443, 457, 462, 472, 501, 503, 509, 510, 511, 512, 513, 515, 516,

517, 521, 522, 523, 525, 527, 528, 532, 533, 534, 535, 552, 560
cooling, 32, 33, 42, 87, 116, 145, 178, 208, 209, 216, 217, 218, 226, 228, 241, 268, 279, 291, 324, 385, 487, 495
cooling process, 279
creative village, 566, 568
crystal growth, 4
crystal structure, 23, 24, 169, 170, 192, 222, 234, 240, 275, 280, 284, 290
crystalline, 24, 75, 162, 164, 195, 259, 378
crystalline solids, 378
crystallite grains, 207, 208, 211
crystallites, 170, 171, 172, 207, 210, 286
crystallization, 89, 90, 91
crystals, xiv, 34, 48, 75, 90, 95, 96, 183, 185, 235, 416
curvature, 332, 355, 448, 487, 495, 509, 510, 511, 515, 517, 518, 519, 520
cycles, 240, 241, 242, 243, 418, 489, 490
cycling, 239, 240, 241, 242, 243, 553
cytotoxicity, 312

D

damped oscillations, 521, 526
damping, 53, 212, 409, 447, 451
decomposition, 5, 7, 39, 43, 104, 144, 162, 178, 263, 385, 449, 450, 457
decomposition temperature, 39, 144, 162, 178, 385
defect, 14, 26, 27, 199, 249, 262, 270, 355, 356, 377, 379, 381, 382, 419, 433, 435, 438, 515, 517, 522, 523, 526, 527, 532
defective state, 191
deficiency, 58
deflections, xiv, 337, 338, 344, 345, 350
deformation, 87, 157, 191, 192, 193, 195, 196, 234, 241, 246, 292, 317, 326, 368, 371, 372, 374, 377, 378, 379, 380, 382, 434, 439, 440, 441, 486, 496, 531, 532, 533, 534, 535
density functional theory, 58
destruction, 21, 48, 154, 162, 164, 193, 284, 323, 324, 326, 328, 332, 377, 379, 381, 382, 502
diagnostics, xiv, 47, 151, 326, 328, 377, 382, 407, 433, 435, 486, 493, 494, 509, 511, 513, 528
dielectric characteristics, 234, 237, 245, 253, 268, 270
dielectric constant, 57, 193, 195, 268, 401, 409, 410, 424, 462
dielectric permeability, 201, 202
dielectric permittivity, 33, 38, 39, 42, 46, 57, 67, 116, 119, 120, 121, 124, 153, 159, 207, 208, 212,

Index 575

216, 217, 222, 228, 242, 243, 276, 284, 410, 424, 470

dielectric spectra, 13, 151, 152, 154, 236, 270, 287

dielectrics, 91, 93, 419

differential equations, 75, 360, 371, 424

differential scanning, 276, 284

differential scanning calorimeter, 276, 284

diffraction, 19, 24, 25, 57, 172, 193, 218, 246, 260, 276, 284, 353, 355, 357, 495

diffusion, 10, 34, 82, 91, 92, 203, 217, 246, 257, 294, 295, 324, 419, 462, 466, 537, 538, 539

diffusivity, 276, 284, 287

discharge with runaway electrons, 17

discrete circular ribs, 439, 440, 445

discreteness, 517

discretization, 359, 362, 463

discs, 71, 96, 152, 200, 234

dislocation, 24, 25, 26, 378

dispersion, 13, 22, 33, 34, 151, 152, 160, 183, 184, 185, 186, 188, 208, 210, 212, 213, 217, 218, 248, 249, 256, 268, 270, 277, 287, 290, 305, 307, 308, 309, 312, 320, 325, 383, 385, 386, 388, 389, 402, 424, 430, 477, 479, 480, 481, 482, 505

displacement, 133, 142, 144, 176, 178, 316, 317, 355, 356, 360, 362, 364, 399, 400, 401, 405, 419, 426, 429, 430, 435, 439, 440, 442, 443, 444, 447, 451, 464, 511, 518, 522, 523

domain structure, 28, 157, 210, 416

domains, 157, 207, 210, 212, 239, 240, 243, 316, 343, 344, 416, 417

E

elastic coefficients, 142, 384, 400

elastic deformation, 325, 374

elastic limit, 377, 378

elasticity modulus, 517

electric circuits, 302

electric conductivity, 306, 311

electric current, 293, 297, 298, 538

electric current density, 293, 538

electric dynamic fatigue, 239

electric dynamic fatigue (EDF), 239

electric field, 13, 42, 58, 71, 116, 124, 125, 132, 133, 142, 145, 152, 176, 178, 212, 222, 234, 236, 239, 240, 241, 242, 243, 248, 254, 276, 294, 296, 297, 309, 385, 399, 400, 401, 402, 405, 415, 416, 417, 418, 419, 420, 421, 424, 426, 538

electrical characteristics, 11, 12, 69

electrical conductivity, 9, 14, 24, 63, 256, 537

electrical properties, 200, 277, 280, 286

electrical resistance, 97

electrodes, 33, 37, 38, 42, 46, 56, 70, 71, 72, 73, 96, 97, 116, 119, 145, 152, 163, 178, 208, 210, 216, 228, 240, 241, 243, 299, 385, 416, 417, 423, 425, 427, 537, 538

electrolyte, 293, 294, 295, 296, 297, 298, 299, 537, 538, 543

electromagnetic, 183, 208, 210, 311, 424, 434, 523, 524, 525, 533

electromagnetic waves, 210

electromechanical coupling factor (ECFs), 42, 44, 56, 57, 73, 124, 131, 134, 136, 138, 151, 159, 166, 222, 409, 410, 411

electron, 10, 11, 34, 49, 58, 163, 185, 260, 311, 381, 385, 407, 409, 463, 464, 480

electron microscopy, 381, 480

electron paramagnetic resonance, 58

electrons, 10, 11, 17, 262, 462, 464

electrophysical properties, 59, 191, 195, 196, 223, 407, 412

energy, xiv, 4, 32, 38, 46, 76, 82, 97, 123, 129, 138, 184, 193, 207, 208, 210, 212, 213, 320, 367, 372, 440, 453, 454, 457, 463, 476, 533

engineering, 299, 306, 315, 332, 359, 433, 470, 545, 553

environment, 9, 55, 77, 83, 97, 161, 221, 298, 324, 332, 454, 531, 533, 547, 551, 552, 553, 557, 558, 561, 563

environmental problems, 104, 221

epitaxial films, 424

EXAFS, 3, 4, 5, 6, 7

exposure, 39, 71, 77, 93, 193, 307, 309, 485, 489, 490, 497, 551

external environment, 551

external influences, xiii, 151, 253, 415

F

F-2ME, polymer, 115, 116, 118, 120, 121

fabrication, 37, 47, 48, 53, 88, 95, 148, 149, 162, 166, 169, 181, 222, 226, 265, 306, 309, 384, 392, 407, 408, 511

Fermi level, 463, 464

ferrite, 259, 263, 265, 267, 272, 275, 276, 283, 284, 286

ferroelectic hardness, 157

ferroelectric ceramic, 9, 48, 58, 60, 115, 116, 128, 151, 155, 161, 169, 172, 173, 238, 242, 408, 469, 470

ferroelectric epitaxial film, 423

ferroelectrics, 9, 61, 101, 132, 169, 170, 171, 172, 173, 199, 208, 215, 236, 239, 240, 248, 257, 415, 416

ferromagnetic, 259, 276, 284, 289, 292

ferromagnetism, 292
field of sound pressure, 439
Figures of merit (FOMs), vi, 116, 123, 124, 126, 127, 128, 129
fillers, 306, 307
filling materials, 476
film, 17, 18, 19, 20, 21, 22, 28, 75, 76, 77, 78, 79, 89, 91, 92, 93, 95, 96, 97, 98, 99, 100, 240, 323, 324, 326, 328, 423, 424, 425, 426, 427, 430, 434
film formation, 78, 91
film thickness, 17, 21, 91, 100
films, xiv, 17, 18, 19, 20, 22, 28, 75, 76, 77, 78, 79, 89, 91, 93, 96, 97, 98, 99, 324, 332, 424, 431, 531
filter, 95, 99, 100, 101, 200, 477
filters, 96, 99, 100, 423
finite element method (FEM), 81, 82, 142, 324, 338, 345, 346, 384, 391, 400, 424, 434, 502, 507
finite element modeling, xiv, 391, 392, 396, 501, 502, 503, 504
force oscillations, 447
Fourier analysis, 525
fragility, 222
fragments, 217, 235, 278, 286, 294, 324, 547
frequency analysis, 324, 326, 428
friction, xiv, 320, 323, 324, 325, 326, 327, 328
friction coefficient, 324, 327, 328
friction transfer film, 324
fundamental system of equations of a semiconductor, 461, 462

G

glass ceramics, 221, 223
glassy dielectric junction, 89, 90, 91, 92, 93
glassy films, 89, 91, 92
glow discharge, 18
grain boundaries, 10, 11, 28, 164, 240, 292
grain size, 4, 10, 11, 28, 33, 39, 43, 45, 46, 59, 72, 73, 169, 172, 228, 235, 241, 246, 248
grain structure, 9, 14, 61, 96, 172, 192, 253, 254, 259, 275, 280, 283
Green's function method, 423, 424, 427, 428, 430
Green's matrix, 315, 316, 319, 320

H

hardness, xiv, 49, 76, 90, 157, 160, 305, 306, 308, 309, 310, 312, 331, 332, 333, 335, 385, 415, 416, 420, 421, 480, 482
heat conduction, 75, 76, 78, 82, 98
heat conductivity, 76
heat release, 78

heat removal, 325
heat transfer, 76, 77, 83, 84, 325
heating, 4, 5, 6, 7, 33, 39, 77, 84, 98, 203, 208, 209, 216, 217, 218, 220, 228, 234, 241, 254, 259, 277, 278, 279, 307, 324, 331, 332, 333, 335, 336, 487, 488, 495, 497
heating rate, 39
helium, 3, 4, 5, 6, 7, 33
hemostasis, xiv, 485, 486, 489, 490, 493, 496, 497, 498
heterogeneity, 10, 11, 154, 292, 316, 379, 440
heterogeneous medium, 315, 316, 317
high-energy milling, 38, 46
high-frequency diffraction, 353, 357
high-speed integrated photodetector, 461
horizontal directed crystallization (HDC), 89, 90
hysteresis, 237, 241, 242, 246, 248, 249, 254, 279, 399, 402, 403, 404, 405, 532
hysteresis loop, 241, 242, 246, 249, 404, 405

I

ICE, 469, 471
IDT, 95, 96, 97, 99, 100, 416, 423, 424, 427
imperfection, 99, 377
implants, 306, 312
import substitution, 472
impurity(ies), 24, 25, 28, 58, 59, 76, 90, 222, 247, 248, 260, 261, 264, 265, 277, 280, 284, 286, 306, 419, 462
indentation, xiv, 331, 332, 333, 334, 335, 336
indentation speed, 331, 332, 334
informal sector, 557, 564
information complex, 521
information technology, 567
informative parameters, 531
infrastructure, 567, 569
initial stage of destruction, 324, 328
injuries, 476, 485, 486, 490, 494, 496
Instron, 380
instrumented indentation, 332, 334, 335, 336
integral designs, 531
integral represenation, 315
integrated circuits, 95, 461
integrated optical commutation system, 461, 462, 465, 466
integration, 100, 338, 456, 546, 547
interdigital transducer, 95, 96, 97, 100, 423, 424, 431
intragranular porosity, 169, 170
intraphase transitions, 191
inversion, 359, 360, 361, 362
ion bombardment, 97

Index

ions, 19, 24, 32, 34, 35, 39, 58, 97, 192, 226, 243, 247, 248, 249, 259, 262, 270
iron, 34, 35, 36, 226, 233, 243, 245, 267, 272, 310
irradiation, 307, 308, 478, 479, 480
isotropic media, 510

K

Kirchhoff plate theory, 337, 338
knock sensor, xiv, 469, 473
Kramers-Kronig relations, 183, 184, 185, 188

L

laboratory measurement, 523
Laplace transform, 359, 360, 361, 362
laser annealing, xiv, 75, 76, 79
laser processing, 81, 82, 83, 84, 87
laser radiation, 75, 76, 77, 78, 84, 85, 305, 306, 307, 309, 311, 312, 462, 463, 475, 476, 478, 482
laser scanning velocity, 81, 82, 84, 85, 86, 87
laser welding, xiv, 100, 312, 314, 475, 476, 478, 479, 481, 482
lasers, 462, 466, 475, 478
lattice misfit strain, 423, 424, 430
lattice parameters, 219
lead iron niobate, 36, 226, 233, 245, 267, 272
lead titanate, 48, 151, 161, 162, 166, 172, 241, 408, 415, 416, 420
lead zirconate titanate (PZT), 170, 416
lead-free ceramics, 55
lead-free ferroelectrics, 9, 61, 132
leonardite, vi, 103, 104, 105, 106, 107, 108, 109, 110
lesions, 485, 489, 490
LFA, 276, 284
liquid phase, 4, 11, 14, 38, 39, 59, 235, 246, 316, 317
lithium, 96, 97, 100, 416
local sector industries, 558
localization, 204, 234, 236, 247, 249, 253, 379, 419
lossy piezoelectric materials, 183, 184

M

macrostructural inhomogeneities, 169, 172
magnesium, 48, 408, 423, 424, 425, 430, 431
magnetic characteristics, 267, 289, 290
magnetic field, 4, 227, 290, 292
magnetic properties, 32, 35, 226
magnetic structure, 284
magnetization, 34, 225, 226, 227, 228, 229, 230, 290
magnetodielectric effect, 289
magnetoelectric materials, 265, 276

magnetoresistance, 289, 292
magnetostriction, 289, 290, 291
manganese, 24, 234, 238, 476
man-made disasters, 515
manufacturing, 73, 81, 82, 89, 93, 95, 96, 99, 259, 486, 494, 559, 565, 567
mass-transfer, 293, 537, 538
mathematical methods, 336
mathematical modeling, xiii, 82, 293, 440, 537
matrix, xiv, 10, 47, 48, 49, 52, 53, 62, 116, 120, 123, 124, 125, 126, 133, 134, 143, 145, 148, 177, 180, 183, 184, 185, 188, 192, 268, 306, 307, 309, 311, 312, 315, 316, 319, 320, 321, 356, 385, 389, 401, 402, 408, 409, 412, 441, 451, 475, 476, 482, 502
mechanical model, 367, 375
mechanical performances, 309
mechanical properties, xiv, 48, 56, 58, 59, 199, 305, 306, 331, 332, 336, 374, 377, 408, 501, 502, 503
mechanical quality factor, 42, 47, 51, 52, 55, 57, 72, 162, 163, 176, 193, 222, 400
mechanical stress, 124, 125, 132, 133, 166, 193, 212, 240, 263, 417, 533
mechanically activated powders, 31, 32
MEMS, v, xiv, 37, 100, 423
metal hydroxides, 207, 208
metals, 38, 56, 306, 324, 379
meter, 39, 57, 152, 193, 201, 222, 234, 246, 254, 268, 276, 284, 523
method of local approximations, 337
methodology, 81, 464, 547
microdefects, 377, 382
microelectronics, 76, 79, 89, 93, 332
microgeometry, xiv, 115, 116, 121
microhardness, 308
micrometer, 210
micronutrients, 104
microphotographs, 308, 309, 480
microscope, 10, 11, 253, 260, 276, 284, 309, 311, 524, 525
microscopy, 93, 381, 382, 481
microstructural inhomogeneities, 169, 172, 424
microstructure(s), 4, 9, 10, 11, 14, 23, 24, 27, 37, 38, 49, 63, 161, 162, 164, 169, 170, 171, 172, 185, 240, 259, 263, 264, 275, 276, 285, 286, 385, 408
microwave energy absorption, 207
migration, 28, 152, 294, 419, 562
models, xiv, 324, 338, 373, 377, 391, 392, 396, 397, 402, 405, 425, 439, 440, 443, 445, 462, 466, 486, 487, 488, 494, 495, 496, 522, 532, 535, 547, 552
modifiers, 62, 96, 151, 153, 279, 284, 287
modules, 100, 399, 405

modulus, 84, 157, 186, 193, 201, 202, 217, 222, 306, 325, 331, 332, 333, 334, 335, 340, 372, 378, 379, 388, 409, 427, 434, 448, 503
molecular weight, 25
molecules, 104, 480, 482
moments, 337, 338, 348, 441, 465
monolayer, 513
morphology, 79, 91, 93, 107
morphotropic transition region (MTR), 170
Mössbauer data, 229
Mössbauer spectra, 31, 32, 34, 228, 229, 268, 272
multi-core ultra-large-scale integrated circuit, 461
multiferroic(s), xiv, 24, 31, 32, 34, 36, 226, 228, 229, 233, 245, 259, 267, 275, 280, 283, 284
multifractal analysis, 259, 260
multilayered ceramics, 38
multilayered structure, 440
multiple defects, 515, 521

N

nanocomposites, 3
nanoelectronics, 332
nanomaterials, xiii, xiv, 306, 309, 312
nanometers, 4, 379
nanoparticles, 3, 4, 5, 6, 7, 169, 170, 173, 482
nanopores, 169
nanostructures, xiii, 97, 101
nanotechnology, 76, 81
nanotube, 305, 311, 312, 477
natural oscillations, 515, 517, 525
neuroblastoma, 312
neutral, 320, 552
niobates, xiv, 9, 48, 61, 160, 247, 253, 408, 469
niobates of barium-strontium, 253
niobium, 14, 18, 24, 56
nitrogen, 104, 105, 106, 108
nodes, 342, 502, 504, 522, 525, 532, 547, 548, 549, 550
non-polar, 14, 124, 254
numerical analysis, 82, 87
numerical modeling, 424, 510, 511, 515
numerical simulation, 75, 79, 81, 87
numerical-analytical solution, 337, 338, 350

O

optical microscopy, 49, 385
optical properties, 17, 18
optimization, 25, 391, 392, 393, 394, 396, 436, 545, 546, 549, 550, 552, 553
optimization method, 553

oscillations, xiv, 124, 151, 193, 199, 207, 210, 315, 316, 318, 323, 327, 342, 345, 380, 426, 434, 435, 440, 441, 442, 444, 447, 448, 449, 451, 453, 454, 455, 457, 515, 516, 517, 521, 522, 525, 526, 527, 532, 533
oscillograph, 471
Osmocote, 105, 108, 109, 110, 111
oxidation, 24, 28, 234
oxide films, 75, 76, 79
oxygen, 14, 17, 18, 19, 23, 28, 45, 58, 63, 104, 153, 243, 286

P

$Pb(Fe_{1/2}Nb_{1/2})O_3$, viii, 225, 239
$PbFe_0Ta_{0.5}O_3$, 31, 32
$PbFe_0Ta_{0.5}O_3$ (PFT), 31, 32
$PbMn_{1/2}Nb_{1/2}O_3$, v, 23, 28
$PbNb_2O_6$, 161, 162, 164, 165, 166
$PbTiO_3$, 9, 39, 132, 144, 151, 154, 155, 159, 161, 162, 164, 165, 166, 172, 228, 229, 230, 240, 241
PCR, 128, 144, 145, 166, 185, 186, 187, 188, 207, 208, 242, 384, 385, 386, 387, 388
PCR-1, 185, 186, 187, 188, 207, 208, 384, 385, 386, 387, 388
permittivity, 33, 38, 39, 42, 46, 57, 67, 72, 73, 116, 119, 120, 121, 124, 143, 152, 153, 159, 177, 207, 208, 210, 212, 213, 216, 217, 222, 228, 229, 234, 242, 243, 246, 248, 253, 268, 276, 283, 284, 290, 400, 401, 410, 424, 426, 470
perovskite oxide, 32
PFT, 31, 32, 33, 34
pH, 4, 105
phase boundaries, 55, 56, 57, 58, 59, 260
phase composition, 38, 39, 57, 191, 196, 260, 276, 283, 284
phase diagram, xiv, 55, 56, 57, 58, 59, 61, 62, 153, 159, 192, 199, 215, 218, 220, 225, 226, 267, 268, 271, 272, 531, 533
phase transformation, 215, 220
phase transition, 11, 13, 31, 32, 34, 35, 57, 199, 200, 203, 204, 208, 215, 216, 220, 225, 226, 227, 228, 229, 230, 236, 237, 248, 249, 254, 255, 257, 262, 268, 269, 270, 272, 283, 289, 379, 430, 470
phosphate, 104, 105
phosphorus, 105, 106, 108
photodetector, xiv, 461, 462, 463, 464, 465, 466
photolithography, 95, 96, 97, 100
physical and mechanical properties, 332, 440
physical and topological model, 461, 462, 465, 466
physical mechanisms, 184, 185, 377, 383, 389, 419
physical properties, 61, 62, 75, 169, 184, 260, 325
physicochemical properties, 90, 91

piezo-active composite, 121, 123, 129, 132

piezoceramics, xiv, 49, 69, 70, 71, 72, 73, 96, 141, 142, 144, 145, 148, 159, 161, 162, 163, 164, 165, 166, 175, 176, 178, 179, 180, 183, 185, 186, 207, 208, 209, 210, 212, 213, 383, 384, 385, 391, 392, 393, 394, 395, 396, 397, 399, 400, 408, 410, 415, 416, 417, 418, 419, 420, 421, 469, 471, 487, 495

piezoelectric ceramics, 38, 39, 43, 45, 70, 72, 96, 158, 159, 221, 222, 242, 243, 384, 392, 408

piezoelectric coefficient(s), 9, 11, 14, 42, 56, 57, 115, 116, 119, 120, 121, 124, 125, 127, 131, 132, 134, 135, 136, 138, 159, 175, 176, 185, 385

piezoelectric crystal, 97, 325

piezoelectric properties, 9, 48, 55, 61, 64, 72, 115, 116, 143, 151, 161, 162, 164, 166, 177, 199, 203, 223, 407, 408

piezoelectricity, 141, 400, 401, 424

piezoelectrics, xiii, xiv, 72, 132, 143, 148, 177, 180, 188, 415, 420

p-i-n photodiode, 461, 462, 464, 465, 466

pipe, 501, 502, 503, 504, 505, 506

pipeline, xiv, 501, 502, 505, 506, 507

plane problem, 337

plastic deformation, 379, 381, 501, 503, 532

plate, xiv, 100, 175, 176, 293, 308, 337, 338, 340, 345, 346, 347, 350, 453, 454, 455, 456, 457, 537, 538

platelets, 487, 496

platinum-based catalysts, 3

point defects, 220, 240

Poisson ratio, 427

polarization, 18, 72, 119, 132, 133, 193, 195, 207, 217, 220, 236, 239, 240, 241, 242, 243, 248, 249, 254, 292, 399, 400, 401, 405, 409, 416, 417, 418, 419, 420

polycrystalline materials, 377, 380

polyether, 71

polymer(s), xiv, 47, 48, 69, 70, 71, 72, 73, 115, 116, 119, 120, 121, 123, 124, 125, 126, 127, 128, 129, 131, 132, 133, 306, 324, 325, 326, 400, 439, 440, 443

polymer composites, 116, 124, 440

polymer materials, 324

polymer matrix, 120, 124, 125, 127, 129

polymeric composites, xiii

polymeric materials, 373

polymerization, 70

polymethylmethacrylate, 480

polystyrene, 104

polyurethane, 126, 128, 129, 132, 135, 136, 137

polyvinyl alcohol, 241

polyvinyl chloride, 104

pore, 10, 120, 123, 125, 126, 144, 145, 162, 169, 170, 171, 172, 178, 185, 360, 361, 384, 385, 392, 393

pore size, 144, 169, 170, 171, 172, 178, 185, 384, 393

pores, 10, 28, 53, 103, 108, 110, 116, 119, 120, 124, 126, 162, 170, 171, 172, 173, 186, 286, 316, 317, 386, 388, 392

poroelasticity, 360, 362

porosity, 48, 51, 71, 73, 96, 115, 119, 121, 129, 144, 161, 162, 163, 164, 165, 166, 169, 170, 175, 176, 178, 185, 208, 209, 306, 317, 360, 383, 384, 385, 386, 391, 392, 393, 397, 408, 409, 410, 411

porous ferroelectric ceramics, 161, 169, 172, 173

porous material, 172, 315, 364

porous piezoceramics, xiv, 69, 71, 73, 144, 145, 148, 162, 163, 164, 165, 166, 178, 179, 180, 183, 383, 385, 392, 393, 394, 395, 396, 397, 487, 495

poroviscoelasticity, xiv, 359, 362

power generation, 103, 105

praseodymium, 260, 262

precipitation, 207, 208

predicate, 73

propagation, 96, 184, 188, 325, 327, 359, 415, 416, 419, 420, 423, 424, 425, 434, 440

pyroelectric and dielectric properties, 215, 239, 242

PZT, vii, x, xiv, 9, 28, 37, 38, 47, 48, 49, 52, 53, 55, 56, 71, 73, 116, 128, 132, 142, 144, 145, 157, 160, 161, 163, 166, 170, 171, 172, 173, 175, 176, 178, 185, 187, 188, 191, 192, 195, 199, 200, 202, 203, 204, 215, 217, 219, 242, 243, 383, 384, 391, 392, 393, 400, 401, 407, 408, 409, 410, 411, 412, 415, 416, 417, 418, 419, 420, 470, 486, 489, 494

R

radiation, 4, 7, 17, 18, 24, 33, 39, 57, 76, 80, 82, 84, 88, 99, 100, 170, 193, 200, 216, 222, 228, 234, 241, 246, 258, 260, 276, 284, 305, 307, 309, 311, 318, 323, 326, 329, 379, 440, 442, 462, 463, 464, 476, 478, 479

rare earth elements, 284

rare-earth ions, 259, 262

RAS, 31, 469, 528, 554

raw materials, 222, 565, 567

reactions, 234, 246, 290, 368, 370, 371, 372, 374, 481

reagents, 32, 38, 56

recrystallization, 58, 247

redistribution, 270

relaxation process, 97, 415, 416, 420, 421, 461, 466

relaxation rate, 229

relaxation times, 373, 415, 421

580 Index

relaxor-ferroelectric, 123, 125, 129
relaxor-ferroelectric single crystal, 123
resin, 103, 104, 106, 107, 108, 109, 110
resonance modes, 141, 142, 145, 148, 176, 178, 180, 386, 392, 393, 400
resonator, 99, 142, 177, 425, 464
resources, 424, 545, 567
retinal detachment, 486, 494
rhenium, 98
ROC, 122, 129, 138, 239
rods, 124, 128, 145, 367, 391, 392, 393, 410, 516, 522, 525, 532
room temperature, 4, 5, 17, 18, 24, 34, 39, 62, 127, 132, 172, 226, 234, 241, 243, 246, 254, 270, 272, 275, 276, 277, 280, 283, 284, 290, 307, 308, 333, 423, 424, 430
rotations, 138, 417, 419

S

sapphire, xiv, 75, 76, 77, 78, 79, 89, 90, 91, 92, 93
sapphire substrate surface, 89
sapphire surface, 75, 76, 78, 79
Scanning Electron Microscopy, 24, 39, 103, 107
scattering, 19, 24, 25, 26, 32, 34, 35, 47, 48, 49, 51, 52, 53, 54, 183, 184, 186, 187, 188, 278, 353, 385, 386, 388, 408, 409, 417
scattering losses, 183, 184, 188
scattering of ultrasonic waves, 353
SEM micrographs, 110, 144, 186, 384, 409
semiconductor, 62, 95, 100, 461, 462, 463, 464, 466, 478
sensitivity, 18, 34, 47, 48, 53, 134, 138, 142, 143, 162, 163, 166, 178, 200, 243, 384, 391, 399, 407, 525, 534
sensors, xiv, 97, 132, 138, 162, 166, 332, 353, 392, 424, 433, 434, 469, 471, 472, 473, 525
serum albumin, 305, 306, 310, 312, 475, 476
shape, 5, 10, 28, 38, 48, 72, 119, 120, 144, 171, 178, 207, 208, 209, 212, 241, 261, 279, 286, 309, 325, 353, 354, 368, 383, 391, 392, 402, 453, 454, 477, 502, 523, 560
shear, 28, 145, 148, 163, 180, 220, 338, 339, 350, 372, 378, 385, 392, 393, 401, 409, 440, 443, 451, 503
shearing force, 337, 349, 350
shell, 4, 439, 440, 442, 443, 444, 445, 448, 449, 450, 451, 452
single crystals, 32, 33, 34, 95, 96, 97, 116, 123, 132, 200, 240, 243, 424
single crystals (SC), 124, 125, 126, 127, 128, 129, 132, 133, 134, 135, 136, 137, 138, 258
SiO2, 82, 84, 222, 223

slow-release fertilizer, 103, 104, 105, 106, 107, 108, 109, 110
SMMI, 565, 566, 567, 569
solid phase, 62, 98, 151, 152, 162, 234, 246, 268, 290, 316
solid phase synthesis, 62, 151, 152, 162, 268
solid solution(s), xiv, 9, 10, 11, 12, 14, 27, 34, 37, 39, 44, 55, 56, 60, 61, 62, 63, 64, 65, 67, 124, 151, 152, 153, 154, 157, 159, 160, 161, 162, 170, 172, 173, 191, 192, 194, 195, 196, 199, 200, 201, 202, 203, 204, 215, 221, 222, 223, 224, 225, 226, 228, 229, 230, 234, 240, 241, 247, 253, 257, 259, 260, 261, 263, 267, 268, 270, 271, 272, 276, 277, 283, 284, 285, 289, 290, 470
solid state, 31, 32, 38, 55, 56
solid state sintering, 31, 32
spatial defects, xiv, 353
specific heat, 77, 287
specific surface, 92
spectroscopy, 47, 49, 141, 142, 143, 162, 166, 175, 176, 177, 183, 184, 185, 243, 286, 384, 385, 389, 407, 408
spherulite, 28
spongy bone, 482
stability, 14, 23, 28, 91, 154, 279, 378, 428
stabilization, 72, 151, 152, 200, 216, 223, 245
stages of deformation, 377, 378, 380, 382
stimulation, 311, 312
stoichiometry, 56, 57, 245
strain, 83, 133, 142, 176, 240, 367, 373, 377, 378, 379, 380, 381, 382, 399, 400, 401, 402, 403, 405, 423, 424, 426, 428, 429, 430, 442, 502, 503, 535
strain characteristics, 377
strength, 4, 11, 71, 90, 115, 116, 121, 212, 229, 235, 305, 306, 307, 308, 309, 310, 311, 312, 332, 377, 381, 382, 433, 475, 476, 477, 478, 479, 480, 481, 482, 502, 503, 505, 506, 563
stress, 82, 116, 124, 125, 132, 133, 142, 166, 176, 316, 317, 338, 360, 367, 368, 377, 378, 399, 400, 401, 402, 405, 426, 442, 449, 501, 502, 503, 504, 505, 506, 507, 509, 510, 511, 512, 513, 514, 533, 535
stress state, 338, 501, 502, 507
stress-strain state, 449, 509, 510, 511, 512, 513, 514
strontium, 17, 18, 39, 253, 431
strontium barium niobate, 17, 18
structural changes, 6, 374
structural characteristics, 285
structural materials, 382, 509
structural transitions, 58
structure, xiv, 5, 9, 10, 11, 14, 17, 18, 23, 24, 26, 27, 28, 53, 55, 58, 61, 62, 63, 75, 76, 77, 78, 79, 82, 96, 132, 151, 152, 154, 157, 169, 170, 171, 172,

191, 192, 195, 199, 207, 209, 210, 212, 215, 220, 222, 228, 233, 234, 235, 237, 240, 241, 245, 246, 247, 249, 253, 254, 257, 259, 262, 263, 264, 265, 268, 270, 275, 276, 277, 278, 280, 284, 286, 290, 311, 312, 323, 326, 328, 367, 374, 379, 380, 392, 410, 411, 416, 423, 425, 427, 430, 440, 461, 462, 463, 464, 465, 466, 517, 521, 525, 551, 566

structure formation, 79

substrate(s), 17, 18, 19, 20, 22, 75, 82, 83, 84, 87, 89, 91, 92, 93, 95, 97, 98, 99, 100, 294, 295, 297, 298, 301, 308, 332, 415, 416, 423, 424, 425, 426, 427, 430, 431

surface acoustic wave (SAW), 95

surface acoustic waves, xiv, 415, 423, 431

surface area, 72, 142, 176

surface energy, 325

surface layer, 242, 323, 324, 325, 328

surface of tribocontact, 323

surface tension, 453, 454, 457

surface wave field, x, 315, 433, 434

sustainable livelihood framework analysis, 558

symmetry, 19, 57, 124, 143, 177, 192, 215, 218, 219, 225, 226, 241, 263, 276, 277, 284, 401, 510

synthesis, xiv, 3, 4, 10, 18, 24, 25, 28, 36, 37, 38, 39, 46, 55, 56, 60, 62, 151, 152, 162, 193, 200, 207, 208, 241, 245, 247, 253, 259, 260, 268, 276, 284, 553

T

technology, xiv, 10, 23, 24, 28, 37, 38, 55, 56, 69, 70, 71, 73, 75, 81, 91, 95, 97, 99, 100, 116, 144, 151, 152, 157, 160, 161, 162, 166, 178, 207, 213, 216, 221, 245, 260, 268, 276, 284, 385, 433, 486, 494, 531, 553, 559, 565, 567, 569

temperature annealing, 92

temperature dependence, 34, 79, 82, 152, 153, 200, 201, 203, 216, 228, 229, 248, 268, 416, 473

temperature field, 76, 323, 324, 325, 328

tensile strength, 305, 308, 309, 310, 312, 475, 476, 478, 480, 481, 482, 502

tension, 332, 378, 379, 380, 531, 534

ternary oxides, 24

tetrahydrofuran, 4

theory of elasticity, 83, 337, 367

thermal analysis, 84

thermal decomposition, 3, 4

thermal evaporation, 97

thermal expansion, 48, 83, 90, 91, 92, 93, 200, 408

thermal properties, 84, 275, 280, 283

thermal stability, 95, 96

thermal treatment, 37, 380

thermalization, 76

thermodynamic equilibrium, 463

thin ferroelectric films, 424, 531

thin films, 17, 18, 22, 97, 99, 100, 240, 424, 430

titanate, 18, 38, 48, 151, 161, 162, 163, 166, 170, 172, 176, 241, 400, 408, 415, 416, 420, 423, 424, 430, 431, 486, 494

transducer, 37, 47, 48, 53, 95, 96, 124, 142, 325, 326, 327, 384, 399, 407, 408, 412, 423, 425, 486, 487, 488, 489, 493, 494, 495, 496, 497

transition points, 377, 379, 382

transition temperature, 32, 35, 151, 204, 225, 228, 229

trinary systems, 157

U

ultrasonic attenuation, 183, 184

ultrasound, xiv, 55, 59, 477, 485, 486, 487, 489, 490, 494, 495, 496, 497, 498, 511

uniaxial tension, 377, 378, 380

unsafe livelihood environment, 558, 564

V

vertical rectangular plate, 453

vibration, 51, 52, 73, 92, 222, 409, 439, 440, 445, 447, 451, 470, 523, 524, 525, 526, 533

vibration damping, 447, 451

vibroacoustics, xiv, 439, 445

Vickers hardness, 308, 482

viscoelastic properties, 359

volume fraction, 48, 49, 51, 52, 53, 115, 116, 120, 125, 126, 128, 132, 133, 134, 185, 187, 188, 407, 409

W

waste, 103, 104

wave, x, 95, 134, 142, 144, 177, 178, 184, 185, 208, 307, 309, 311, 315, 316, 317, 318, 320, 324, 325, 353, 354, 355, 356, 357, 359, 385, 423, 425, 428, 430, 433, 434, 435, 442, 451, 453, 454, 455, 456, 457, 486, 487, 494, 495, 509, 510, 511, 513, 514

wave process, 315, 320, 509, 510, 511, 513, 514

wave tank, 453

welded joint, xiv, 501, 502, 505, 506, 507

X

x, T-phase diagram, 225, 226

XAFS, 3, 4

X-ray analysis, 24, 260
X-ray diffraction (XRD), 20, 24, 25, 33, 57, 58, 62, 170, 172, 200, 204, 222, 228, 234, 235, 241, 246, 259, 276, 278, 381, 382

Y

yield point, 377, 378, 381

Z

ZTP-19, 38, 39, 40, 41, 42, 43, 44, 45, 46, 207, 208, 210, 212, 213
ZTPSt-2, 38, 39, 40, 41, 42, 43, 44, 45, 46
ZTS-19, vi, 115, 116, 118, 119, 120, 121